Classical and Quantum Dynamics

Walter Dittrich Martin Reuter

Classical and Quantum Dynamics

From Classical Paths to Path Integrals

Third Edition

Springer

Prof. Dr. W. Dittrich
University of Tübingen
Institute of Theoretical Physics
Auf der Morgenstelle 14
72076 Tübingen
Germany

Prof. Dr. Martin Reuter
University of Mainz
Institute of Physics
Staudinger Weg 7
55099 Mainz
Germany
reuter@thep.physik.uni-mainz-de

Library of Congress Cataloging-in-Publication Data

Dittrich, Walter. Classical and quantum dynamics: from classical paths to path integrals/ Walter Dittrich, Martin Reuter.—3. Ed. p.cm.—(Advanced texts in physics, ISSN 1439-2674)Originally published: Classical and quantum dynamics. 2nd corr. And enl. Ed. Berlin : New York Springer-Verlag, c1994. Includes bibliographical references and index ISBN 3540420665

ISSN 1439-2674
ISBN 3-540-420665-5 Springer-Verlag Berlin Heidelberg New York
ISBN 3-540-56245-1 2nd edition Springer-Verlag Berlin Heidelberg New York

Springer-Verlag Berlin Heidelberg New York
Springer is a part of Springer Science+Business Media

Cover-design: *design & production* GmbH, Heidelberg

Dittrich & Reuter: Classical and Quantum Dynamics, Third edition

First Indian Reprint 2011

ISBN 978-81-322-0253-0

This edition is published by Springer (India) Private Limited, (a part of Springer Science+Business Media), Registered Office: 3rd Floor, Gandharva Mahavidyalaya, 212, Deen Dayal Upadhyaya Marg, New Delhi – 110 002, India

Printed in India by Rashtriya Printers, Delhi.

Preface to the Third Edition

In this third edition the major purposes and emphasis are still the same, but there are extensive additions. These consist mainly in the chapter on the action principle in classical electrodynamics and the functional derivative approach, which is set side by side to the path integral formulation. A further major augmentation is a chapter on computing traces in the context of the WKB-propagator. Finally, we have corrected some (not only typographical) errors of the previous editions.

Tübingen and Mainz
Februar 2001

Walter Dittrich
Martin Reuter

Preface to the First Edition

This volume is the result of the authors' lectures and seminars given at Tübingen University and elsewhere. It represents a summary of our learning process in nonlinear Hamiltonian dynamics and path integral methods in nonrelativistic quantum mechanics. While large parts of the book are based on standard material, readers will find numerous worked examples which can rarely be found in the published literature. In fact, toward the end they will find themselves in the midst of modern topological methods which so far have not made their way into the textbook literature.

One of the authors' (W.D.) interest in the subject was inspired by Prof. D. Judd (UC Berkeley), whose lectures on nonlinear dynamics familiarized him with Lichtenberg and Lieberman's monograph, *Regular and Stochastic Motion* (Springer, 1983). For people working in plasma or accelerator physics, the chapter on nonlinear physics should contain some familiar material. Another influential author has been Prof. J. Schwinger (UCLA); the knowledgeable reader will not be surprised to discover our appreciation of Schwinger's Action Principle in the introductory chapters. However, the major portion of the book is based on Feynman's path integral approach, which seems to be the proper language for handling topological aspects in quantum physics.

Our thanks go to Ginny Dittrich for masterly transforming a long and complex manuscript into a readable monograph.

Tübingen and Hannover
January 1992

Walter Dittrich
Martin Reuter

Contents

Introduction

The subject of this monograph is classical and quantum dynamics. We are fully aware that this combination is somewhat unusual, for history has taught us convincingly that these two subjects are founded on totally different concepts; a smooth transition between them has so far never been made and probably never will.

An approach to quantum mechanics in purely classical terms is doomed to failure; this fact was well known to the founders of quantum mechanics. Nevertheless, to this very day people are still trying to rescue as much as possible of the description of classical systems when depicting the atomic world. However, the currently accepted viewpoint is that in describing fundamental properties in quantum mechanics, we are merely borrowing names from classical physics. In writing this book we have made no attempt to contradict this point of view. But in the light of modern topological methods we have tried to bring a little twist to the standard approach that treats classical and quantum physics as disjoint subjects.

The formulation of both classical and quantum mechanics can be based on the principle of stationary action. Schwinger has advanced this principle into a powerful working scheme which encompasses almost every situation in the classical and quantum worlds. Our treatment will give a modest impression of the wide range of applicability of Schwinger's action principle.

We then proceed to rediscover the importance of such familiar subjects as Jacobi fields, action angle variables, adiabatic invariants, etc. in the light of current research on classical Hamiltonian dynamics. It is here that we recognize the important role that canonical perturbation theory played before the advent of modern quantum mechanics.

Meanwhile, classical mechanics has been given fresh impetus through new developments in perturbation theory, offering a new look at old problems in nonlinear mechanics like, e.g., the stability of the solar system. Here the KAM theorem proved that weakly disturbed integrable systems will remain on invariant surfaces (tori) for most initial conditions and do not leave the tori to end up in chaotic motion.

At this stage we point to the fundamental role that adiabatic invariants played prior to canonical quantization of complementary dynamical variables. We are reminded of torus quantization, which assigns each adiabatic invariant an integer multiple of Planck's constant. All these semiclassical quantization procedures have much in common with Feynman's path integral or, rather, approximations thereof. Indeed, Feynman's path integral methods are ideally suited to follow a quantum mechanical system – if certain restrictions are enforced – into its classical realm. Consequently it is one of our main goals to apply Feynman's path integral and other geometrical methods to uncover the mystery of the zero point energy (Maslov anomaly) of the quantum harmonic oscillator.

That quantum and classical mechanics are, in fact, disjoint physical worlds was clear from the very beginning. Present-day experience is no exception; it is rather embarrassing to find out that an important geometric phase in a cyclic adiabatic quantal process has been overlooked since the dawn of quantum mechanics. This so-called Berry phase signals that in nonrelativistic as well as relativistic quantum theory, geometrical methods play an eminent role.

The appearance of topology in quantum mechanics is probably the most important new development to occur in recent years. A large portion of this text is therefore devoted to the geometric structure of topologically nontrivial physical systems. Berry phases, Maslov indices, Chern-Simons terms and various other topological quantities have clearly demonstrated that quantum mechanics is not, as of yet, a closed book.

1. The Action Principles in Mechanics

We begin this chapter with the definition of the action functional as time integral over the Lagrangian $L(q_i(t), \dot{q}_i(t); t)$ of a dynamical system:

$$S\{[q_i(t)]; t_1, t_2\} = \int_{t_1}^{t_2} dt\, L(q_i(t), \dot{q}_i(t); t)\ . \tag{1.1}$$

Here, q_i, $i = 1, 2, \ldots, N$, are points in N-dimensional configuration space. Thus $q_i(t)$ describes the motion of the system, and $\dot{q}_i(t) = dq_i/dt$ determines its velocity along the path in configuration space. The endpoints of the trajectory are given by $q_i(t_1) = q_{i1}$, and $q_i(t_2) = q_{i2}$.

Next we want to find out what the actual dynamical path of the system is. The answer is contained in the principle of stationary action: in response to infinitesimal variation of the integration path, the action S is stationary, $\delta S = 0$, for variations about the correct path, provided the initial and final configurations are held fixed. On the other hand, if we permit infinitesimal changes of $q_i(t)$ at the initial and final times, including alterations of those times, the only contribution to δS comes from the endpoint variations, or

$$\delta S = G(t_2) - G(t_1)\ . \tag{1.2}$$

Equation (1.2) is the most general formulation of the action principle in mechanics. The fixed values G_1 and G_2 depend only on the endpoint path variables at the respective terminal times.

Again, given a system with the action functional S, the actual time evolution in configuration space follows that path about which general variations produce only endpoint contributions. The explicit form of G is dependent upon the special representation of the action principle. In the following we begin with the one that is best known, i.e.,

1) Lagrange: The Lagrangian for a point particle with mass m, moving in a potential $V(x_i, t)$, is

$$L(x_i, \dot{x}_i; t) = \frac{m}{2}\dot{x}_i^2 - V(x_i, t)\ . \tag{1.3}$$

Here and in the following we restrict ourselves to the case $N = 3$; i.e., we describe the motion of a single mass point by $x_i(t)$ in real space. The dynamical variable $x_i(t)$

denotes the actual classical trajectory of the particle which is parametrized by t with $t_1 \leq t \leq t_2$.

Now we consider the response of the action functional (1.1) with respect to changes in the coordinates and in the time, $\delta x_i(t)$ and $\delta t(t)$, respectively. It is important to recognize that, while the original trajectory is being shifted in real space according to

$$x_i(t) \to x_i'(t') = x_i(t) + \delta x_i(t) \tag{1.4}$$

the time-readings along the path become altered locally, i.e., different at each individual point on the varied curve – including the endpoints. This means that our time change is *not* a global ($\delta t(t) = \text{const.}$) rigid time displacement, equally valid for all points on the trajectory, but that the time becomes changed locally, or, shall we say, gauged, for the transported trajectory. All this indicates that we have to supplement (1.4) by

$$t \to t'(t) = t + \delta t(t)\ , \tag{1.5}$$

where the terminal time changes are given by $\delta t(t_2) = \delta t_2$, and $\delta t(t_1) = \delta t_1$.

To the time change (1.5) is associated the change in the integration measure in (1.1) given by the Jacobi formula

$$d(t+\delta t) = \frac{d(t+\delta t)}{dt}\, dt = \left(1 + \frac{d}{dt}\,\delta t(t)\right) dt \tag{1.6}$$

or

$$\delta(dt) := d(t+\delta t) - dt = dt \frac{d}{dt}\,\delta t(t)\ . \tag{1.7}$$

If the time is not varied, we write δ_0 instead of δ; i.e., $\delta_0 t = 0$ or $[\delta_0, d/dt] = 0$. The variation of $x_i(t)$ is then given by

$$\delta x_i(t) = \delta_0 x_i(t) + \delta t \frac{d}{dt}(x_i(t)) \tag{1.8}$$

since up to higher order terms we have

$$\begin{aligned}\delta x_i(t) &= x_i'(t') - x_i(t) = x_i'(t+\delta t) - x_i(t) = x_i'(t) + \delta t \frac{dx_i'(t)}{dt} - x_i(t)\\ &= (x_i'(t) - x_i(t)) + \delta t \frac{dx_i}{dt} =: \delta_0 x_i(t) + \delta t \frac{dx_i}{dt}\ .\end{aligned}$$

Similarly,

$$\delta \dot{x}_i(t) = \delta_0 \dot{x}_i(t) + \delta t \frac{d}{dt}\,\dot{x}_i \tag{1.9}$$

$$\begin{aligned}&= \delta_0 \dot{x}_i + \frac{d}{dt}(\delta t \dot{x}_i) - \dot{x}_i \frac{d}{dt}\,(\delta t)\\ &= \frac{d}{dt}\left(\delta_0 + \delta t \frac{d}{dt}\right) x_i - \dot{x}_i \frac{d}{dt}\delta t = \frac{d}{dt}(\delta x_i) - \dot{x}_i \frac{d}{dt}\,\delta t\ .\end{aligned} \tag{1.10}$$

The difference between δ and δ_0 acting on t, $x_i(t)$ and $\dot{x}_i(t)$ is expressed by the identity

$$\delta = \delta_0 + \delta t \frac{d}{dt} . \tag{1.11}$$

So far we have obtained

$$\delta S = \int_{t_1}^{t_2} [\delta(dt)L + dt\delta L] = \int_{t_1}^{t_2} dt \left[\overbrace{L\frac{d}{dt}(\delta t)}^{\frac{d}{dt}(L\delta t)-\delta t\frac{dL}{dt}} + \delta L \right]$$

$$= \int_{t_1}^{t_2} dt \left[\frac{d}{dt}(L\delta t) + \left(\delta L - \delta t \frac{dL}{dt} \right) \right] = \int_{t_1}^{t_2} dt \left[\frac{d}{dt}(L\delta t) + \delta_0 L \right] , \tag{1.12}$$

since, according to (1.11) we have

$$\delta L = \delta_0 L + \delta t \frac{d}{dt} L . \tag{1.13}$$

The total variation of the Lagrangian is then given by

$$\begin{aligned}
\delta L &= \delta_0 L + \delta t \frac{d}{dt} L = \frac{\partial L}{\partial x_i} \delta_0 x_i + \frac{\partial L}{\partial \dot{x}_i} \delta_0 \dot{x}_i + \delta t \frac{dL}{dt} \\
&= \frac{\partial L}{\partial x_i} \delta_0 x_i + \frac{\partial L}{\partial \dot{x}_i} \delta_0 \dot{x}_i + \delta t \left(\frac{\partial L}{\partial x_i} \dot{x}_i + \frac{\partial L}{\partial \dot{x}_i} \ddot{x}_i + \frac{\partial L}{\partial t} \right) \\
&= \frac{\partial L}{\partial x_i} \left(\delta_0 + \delta t \frac{d}{dt} \right) x_i + \frac{\partial L}{\partial \dot{x}_i} \left(\delta_0 + \delta t \frac{d}{dt} \right) \dot{x}_i + \delta t \frac{\partial L}{\partial t} \\
&= \frac{\partial L}{\partial x_i} \delta x_i + \frac{\partial L}{\partial \dot{x}_i} \delta \dot{x}_i + \frac{\partial L}{\partial t} \delta t .
\end{aligned}$$

Now we go back to (1.3) and substitute

$$\frac{\partial L}{\partial x_i} = -\frac{\partial V(x_i, t)}{\partial x_i} , \quad \frac{\partial L}{\partial \dot{x}_i} = m\dot{x}_i , \quad \frac{\partial L}{\partial t} = -\frac{\partial V}{\partial t} , \tag{1.14}$$

so that we obtain, with the aid of (1.10):

$$\delta L = -\frac{\partial V}{\partial t} \delta t - \frac{\partial V}{\partial x_i} \delta x_i + m\dot{x}_i \frac{d}{dt} \delta x_i - m\dot{x}_i^2 \frac{d}{dt} \delta t . \tag{1.15}$$

Our expression for δS then becomes

$$\delta S = \int_{t_1}^{t_2} dt \left[m\dot{x}_i \frac{d}{dt} \delta x_i - \frac{\partial V}{\partial t} \delta t - \frac{\partial V}{\partial x_i} \delta x_i + (L(t) - m\dot{x}_i^2) \frac{d}{dt} \delta t \right] . \tag{1.16}$$

We can also write the last expression for δS a bit differently, thereby presenting explicitly the coefficients of δx_i and δt:

$$\delta S = \int_{t_1}^{t_2} dt \left\{ \frac{d}{dt}\left[m\frac{dx_i}{dt}\delta x_i - \left(\frac{m}{2}\left(\frac{dx_i}{dt}\right)^2 + V\right)\delta t\right] \right.$$
$$\left. -m\frac{d^2x_i}{dt^2}\delta x_i - \frac{\partial V}{\partial x_i}\delta x_i - \frac{\partial V}{\partial t}\delta t + \delta t\frac{d}{dt}\left[\frac{m}{2}\left(\frac{dx_i}{dt}\right)^2 + V\right]\right\}, \tag{1.17}$$

or with the definition

$$E = \frac{\partial L}{\partial \dot{x}_i}\dot{x}_i - L = \frac{m}{2}\left(\frac{dx_i}{dt}\right)^2 + V(x_i, t)\ , \tag{1.18}$$

$$\delta S = \int_{t_1}^{t_2} dt \frac{d}{dt}\left[m\frac{dx_i}{dt}\delta x_i - E\delta t\right]$$
$$+ \int_{t_1}^{t_2} dt\left[-\delta x_i\left(m\frac{d^2x_i}{dt^2} + \frac{\partial V}{\partial x_i}\right) + \delta t\left(\frac{dE}{dt} - \frac{\partial V}{\partial t}\right)\right]. \tag{1.19}$$

Since δx_i and δt are independent variations, the action principle $\delta S = G_2 - G_1$ implies the following laws:

$$\delta x_i : \ m\frac{d^2x_i}{dt^2} = -\frac{\partial V(x_i, t)}{\partial x_i}\ , \qquad \text{(Newton)}\ , \tag{1.20}$$

i.e., *one* second-order differential equation.

$$\delta t : \ \frac{dE}{dt} = \frac{\partial V}{\partial t}\ , \tag{1.21}$$

so that for a static potential, $\partial V/\partial t = 0$, the law of the conservation of energy follows: $dE/dt = 0$.

$$\text{Surface term:} \quad G = m\frac{dx_i}{dt}\delta x_i - E\delta t\ . \tag{1.22}$$

2) Hamiltonian: As a function of the Hamiltonian,

$$H(x_i, p_i; t) = \frac{p_i^2}{2m} + V(x_i, t)\ , \tag{1.23}$$

the Lagrangian (1.3) can also be written as ($p_i := \partial L/\partial \dot{x}_i$):

$$L = p_i\frac{dx_i}{dt} - H(x_i, p_i; t)\ . \tag{1.24}$$

Here, the independent dynamical variables are x_i and p_i; t is the independent time-parameter variable. Hence the change of the action is

$$\delta S = \delta\int_{t_1}^{t_2} dt\left[p_i\frac{dx_i}{dt} - H(x_i, p_i; t)\right]$$
$$= \int_{t_1}^{t_2} dt\left[p_i\frac{d}{dt}\delta x_i + \frac{dx_i}{dt}\delta p_i - \delta H - H\frac{d}{dt}\delta t\right]. \tag{1.25}$$

Upon using

$$\delta H = \left(\frac{\partial H}{\partial x_i} \delta x_i + \frac{\partial H}{\partial p_i} \delta p_i \right) + \frac{\partial H}{\partial t} \delta t \ , \tag{1.26}$$

where, according to (1.23): $\partial H/\partial x_i = \partial V/\partial x_i$ and $\partial H/\partial p_i = p_i/m$, we obtain

$$\begin{aligned} \delta S &= \int_{t_1}^{t_2} dt \frac{d}{dt}[p_i \delta x_i - H\delta t] \\ &+ \int_{t_1}^{t_2} dt \left[-\delta x_i \left(\frac{dp_i}{dt} + \frac{\partial V}{\partial x_i} \right) + \delta p_i \left(\frac{dx_i}{dt} - \frac{p_i}{m} \right) + \delta t \left(\frac{dH}{dt} - \frac{\partial H}{\partial t} \right) \right] . \end{aligned} \tag{1.27}$$

The action principle $\delta S = G_2 - G_1$ then tells us here that

$$\delta p_i : \quad \frac{dx_i}{dt} = \frac{\partial H}{\partial p_i} = \frac{p_i}{m} \ , \tag{1.28}$$

$$\delta x_i : \quad \frac{dp_i}{dt} = -\frac{\partial H}{\partial x_i} = -\frac{\partial V}{\partial x_i} \ . \tag{1.29}$$

Here we recognize the two first-order Hamiltonian differential equations.

$$\delta t : \quad \frac{dH}{dt} = \frac{\partial H}{\partial t} \ . \tag{1.30}$$

$$\text{Surface term:} \quad G = p_i \delta x_i - H\delta t \ . \tag{1.31}$$

Let us note for later use:

$$\delta S = G_2 - G_1 = [p_i \delta x_i - H\delta t]_{t_2} - [p_i \delta x_i - H\delta t]_{t_1} \ . \tag{1.32}$$

Compared with ($x_1 := \{x_i(t_1)\}, x_2 := \{x_i(t_2)\};\ i = 1, 2, 3$)

$$\delta S = \frac{\partial S}{\partial x_1} \delta x_1 + \frac{\partial S}{\partial x_2} \delta x_2 + \frac{\partial S}{\partial t_1} \delta t_1 + \frac{\partial S}{\partial t_2} \delta t_2 \tag{1.33}$$

(1.32) yields

$$p_1 = -\frac{\partial S}{\partial x_1} \ , \qquad H(x_1, p_1; t_1) = \frac{\partial S}{\partial t_1} \tag{1.34}$$

or

$$H\left(x_1, -\frac{\partial S}{\partial x_1}, t_1 \right) - \frac{\partial S}{\partial t_1} = 0 \ . \tag{1.35}$$

In the same manner, it follows that:

$$p_2 = \frac{\partial S}{\partial x_2} \ , \qquad H\left(x_2, \frac{\partial S}{\partial x_2}, t_2 \right) + \frac{\partial S}{\partial t_2} = 0 \ . \tag{1.36}$$

Obviously, (1.35) and (1.36) are the Hamilton–Jacobi equations for finding the action S. In this way we have demonstrated that the action (1.1) satisfies the Hamilton–Jacobi equation. (Later on we shall encounter S again as the generating function of a canonical transformation $(q_i, p_i) \to (Q_i, P_i)$ of the $F_1(q_i, Q_i, t)$-type.

3) Euler–Maupertuis (Principle of Least Action): This principle follows from the Lagrangian representation of the action principle:

$$\delta S = \delta \int_{t_1}^{t_2} dt\, L = \left[m \frac{dx_i}{dt} \delta x_i - E\delta t \right]_1^2 , \tag{1.37}$$

if we introduce the following restrictions:

a) L should not be explicitly time dependent; then the energy E is a conserved quantity both on the actual and the varied paths; b) for the varied paths, $\delta x_i(t)$ should vanish at the terminal points: $\delta x_i(t_{1,2}) = 0$. What remains is

$$\delta \int_{t_1}^{t_2} dt\, L = -E(\delta t_2 - \delta t_1) . \tag{1.38}$$

But under the same restrictions we have, using (1.18),

$$\int_{t_1}^{t_2} dt\, L = \int_{t_1}^{t_2} dt \frac{\partial L}{\partial \dot{x}_i} \dot{x}_i - E(t_2 - t_1) , \tag{1.39}$$

the variation of which is given by

$$\delta \int_{t_1}^{t_2} dt\, L = \delta \int_{t_1}^{t_2} dt \frac{\partial L}{\partial \dot{x}_i} \dot{x}_i - E(\delta t_2 - \delta t_1) . \tag{1.40}$$

Comparing (1.40) with (1.38), we get, taking into consideration $p_i := \partial L/\partial \dot{x}_i$:

$$\delta \int_{t_1}^{t_2} dt\, p_i \frac{dx_i}{dt} = 0 . \tag{1.41}$$

If, in addition, we assume the potential to be independent of the velocity, i.e., that

$$\frac{\partial T}{\partial \dot{x}_i} \dot{x}_i = 2T , \tag{1.42}$$

then (1.41) takes on the form

$$\delta \int_{t_1}^{t_2} dt\, T = 0 , \tag{1.43}$$

or

$$\int_{t_1}^{t_2} dt\, T = \text{Extremum}. \tag{1.44}$$

Thus the Euler–Maupertuis Principle of Least Action states: The time integral of the kinetic energy of the particle is an extreme value for the path actually selected compared to the neighboring paths with the same total energy which the particle will travel between the initial and final position at any time $-t$ is varied! This variation in time can also be expressed by writing (1.43) in the form [see also (1.7)]:

$$\delta \int_{t_1}^{t_2} dt\, T = \int_{t_1}^{t_2} dt \left(T \frac{d}{dt} \delta t + \delta T \right) . \tag{1.45}$$

In N-dimensional configuration space, (1.41) is written as

$$\delta \int_{t_1}^{t_2} \sum_{i=1}^{N} \frac{\partial L}{\partial \dot{q}_i} \dot{q}_i \, dt = 0 \, , \tag{1.46}$$

or

$$\delta \int_1^2 \sum_{i=1}^{N} p_i \, dq_i = 0 \, . \tag{1.47}$$

If we parametrize the path in configuration space between 1 and 2 using the parameter ϑ, then (1.47) is written

$$\delta \int_{\vartheta_1}^{\vartheta_2} \sum_{i=1}^{N} p_i \frac{dq_i}{d\vartheta} \, d\vartheta = 0 \, . \tag{1.48}$$

On the other hand, it follows from the Hamiltonian version of the action principle in its usual form with vanishing endpoint contributions $\delta q_i(t_{1,2}) = 0$, $\delta t(t_{1,2}) = 0$ in $2N$-dimensional phase space:

$$\tilde{\delta} \int_{t_1}^{t_2} dt \left[\sum_{i=1}^{N} p_i \frac{dq_i}{dt} - H \right] = 0 \tag{1.49}$$

One should note the different role of δ in (1.46) – the time is also varied and $\tilde{\delta}$, which stands for the conventional virtual (timeless) displacement.

With the parametrization ϑ in (1.49), the expression

$$\tilde{\delta} \int_{\vartheta_1}^{\vartheta_2} d\vartheta \left[\sum_{i=1}^{N} p_i \frac{dq_i}{d\vartheta} - H \frac{dt}{d\vartheta} \right] = 0 \tag{1.50}$$

can, by introducing conjugate quantities,

$$q_{N+1} = t \, , \quad p_{N+1} = -H \, , \tag{1.51}$$

be reduced formally to a form similar to (1.48):

$$\tilde{\delta} \int_{\vartheta_1}^{\vartheta_2} \sum_{i=1}^{N+1} p_i \frac{dq_i}{d\vartheta} \, d\vartheta = 0 \,. \tag{1.52}$$

Besides the fact that in (1.52) we have another pair of canonical variables, the different roles of the two variation symbols δ and $\tilde{\delta}$ should be stressed. δ refers to the paths with constant $H = E$, whereas in the $\tilde{\delta}$ variation, H can, in principle, be any function of time. $\tilde{\delta}$ in (1.52) applies to $2N + 2$-dimensional phase space, while δ in (1.48) applies to configuration space.

If, in the case of the principle of least action, no external forces are involved, i.e., we set without loss of generality $V = 0$, then E as well as T are constants. Consequently, the Euler–Maupertuis principle takes the form

$$\delta \int_{t_1}^{t_2} dt = 0 = \delta t_2 - \delta t_1 \,, \tag{1.53}$$

i.e., the time along the actual dynamical path is an extremum.

At this point we are reminded of Fermat's principle of geometrical optics: A light ray selects that path between two points which takes the shortest time to travel.

Jacobi proposed another version of the principle of least action. It is always useful when one wishes to construct path equations in which time does not appear. We derive this principle by beginning with the expression for the kinetic energy of a free particle in space:

$$T = \frac{1}{2} \sum_{i,k=1}^{3} m_{ik} \frac{dx_i}{dt} \frac{dx_k}{dt} \,, \tag{1.54}$$

where m_{ik} are the elements of the mass tensor, e.g. $m_{ik} = m\delta_{ik}$.

In generalized coordinates in N-dimensional configuration space, we then have

$$T = \frac{1}{2} \frac{(ds)^2}{(dt)^2} \,, \tag{1.55}$$

with the line element

$$(ds)^2 = \sum_{i,k=1}^{N} m_{ik}(q_1, q_2, \ldots, q_N) dq_i dq_k \tag{1.56}$$

and position-dependent elements m_{ik}; for example, from

$$T = \frac{m}{2} \frac{(dr)^2 + r^2 (d\vartheta)^2 + (dz)^2}{(dt)^2} \tag{1.57}$$

we can immediately see that

$$\overleftrightarrow{m} = \begin{pmatrix} m & 0 & 0 \\ 0 & mr^2 & 0 \\ 0 & 0 & m \end{pmatrix} .$$

The m_{ik} take over the role of the metric tensor in configuration space. At this point mechanics becomes geometry.

Writing (1.55) in the form $dt = ds/\sqrt{2T}$ we can restate (1.43) as

$$\delta \int_{t_1}^{t_2} dt\, T = 0 = \delta \int_1^2 ds \sqrt{T} \,. \tag{1.58}$$

Here, we substitute $T = H - V(q_i)$ to obtain Jacobi's principle:

$$\delta \int_1^2 \sqrt{H - V(q_i)}\, ds = 0 \,, \tag{1.59}$$

or, with (1.56):

$$\delta \int_1^2 \sqrt{H - V(q_i)} \sqrt{\sum_{i,k=1}^{N} m_{ik}(q_j) dq_i dq_k} = 0 \,. \tag{1.60}$$

In the integrand, only the generalized coordinates appear. If we parametrize them with a parameter ϑ, we get

$$\int_{\vartheta_1}^{\vartheta_2} \sqrt{H - V} \sqrt{m_{ik} \frac{dq_i}{d\vartheta} \frac{dq_k}{d\vartheta}}\, d\vartheta = \text{Extremum} \,. \tag{1.61}$$

Since ϑ is not constrained in any way, we can construct the Euler equations for the integrand using the conventional variation procedure. The solutions to these equations yield the trajectories in parameter representation.

A comparison of Fermat's and Jacobi's principles is appropriate here. If we apply the principle of least time (1.53) to a light ray in a medium with index of refraction $n(x_i)$ and, due to

$$\frac{v}{c} = \frac{1}{n(x_i)} \,, \qquad v dt = ds \,, \qquad dt = \frac{n(x_i)}{c} ds \tag{1.62}$$

get the expression

$$\delta \int_1^2 ds\, n(x_i) = 0 \,, \tag{1.63}$$

then it is obvious from a comparison with Jacobi's principle (1.59) that the quantity $\sqrt{(E - V)}$ can be looked at as "index of refraction" for a massive particle.

4) Schwinger: Here we use x_i, p_i, t and v_i as the variables to be varied. We shall immediately see, however, that v_i does not satisfy an equation of motion, i.e., $dv_i/dt = \ldots$ does not appear; therefore v_i is not a dynamical variable (just like ϕ and $\boldsymbol{B}$ in the canonical version of electrodynamics). Schwinger writes

$$L = p_i\left(\frac{dx_i}{dt} - v_i\right) + \frac{1}{2}mv_i^2 - V(x_i, t) \tag{1.64}$$

$$= p_i\frac{dx_i}{dt} - H(x_i, p_i, t)\ , \tag{1.65}$$

with H given by

$$H = p_i v_i - \tfrac{1}{2}mv_i^2 + V(x_i, t)\ . \tag{1.66}$$

The variation of the action now gives

$$\delta S = \int_{t_1}^{t_2} dt\left[p_i\frac{d}{dt}\delta x_i - \frac{\partial H}{\partial t}\delta t - \frac{\partial V}{\partial x_i}\delta x_i + \left(\frac{dx_i}{dt} - v_i\right)\delta p_i\right.$$
$$\left.+(-p_i + mv_i)\delta v_i - \left(p_i v_i - \frac{1}{2}mv_i^2 + V\right)\frac{d}{dt}\delta t\right],$$

or

$$\delta S = \int_{t_1}^{t_2} dt\frac{d}{dt}[p_i\delta x_i - H\delta t] + \int_{t_1}^{t_2} dt\left[-\delta x_i\left(\frac{dp_i}{dt} + \frac{\partial V}{\partial x_i}\right)\right.$$
$$\left.+\delta p_i\left(\frac{dx_i}{dt} - v_i\right) + \delta v_i(-p_i + mv_i) + \delta t\left(\frac{dH}{dt} - \frac{\partial H}{\partial t}\right)\right]. \tag{1.67}$$

With the definition of H in (1.66), the action principle yields

$$\delta x_i:\ \frac{dp_i}{dt} = -\frac{\partial H}{\partial x_i} = -\frac{\partial V}{\partial x_i}\ , \tag{1.68}$$

$$\delta p_i:\ \frac{dx_i}{dt} = \frac{\partial H}{\partial p_i} = v_i\ . \tag{1.69}$$

There is no equation of motion for v_i: no dv_i/dt.

$$\delta v_i:\ -p_i + mv_i = -\frac{\partial H}{\partial v_i} = 0\ , \tag{1.70}$$

$$\delta t:\ \frac{dH}{dt} = \frac{\partial H}{\partial t}\ . \tag{1.71}$$

$$\text{Surface term:}\quad G = p_i\delta x_i - H\delta t\ . \tag{1.72}$$

Schwinger's action principle contains the Lagrangian and Hamiltonian versions as special cases. So when we write

$$H(x_i, p_i, v_i, t) = p_i v_i - \frac{m}{2}v_i^2 + V(x_i, t)$$
$$\equiv \frac{p_i^2}{2m} + V(x_i, t) - \frac{1}{2m}(p_i - mv_i)^2 \tag{1.73}$$

and introduce $v_i = p_i/m$ as *definition* of v_i, we return to the Hamiltonian description. On the other side we can also write L in (1.65) as

$$L = p_i \frac{dx_i}{dt} - p_i v_i + \frac{m}{2} v_i^2 - V(x_i, t) = \frac{m}{2} \left(\frac{dx_i}{dt} \right)^2 - V(x_i, t) + (p_i - m v_i) \left(\frac{dx_i}{dt} - v_i \right) - \frac{m}{2} \left(\frac{dx_i}{dt} - v_i \right)^2 , \tag{1.74}$$

and if we now *define:* $v_i = dx_i/dt$, then the Lagrangian description follows.

Once again: Schwinger's realization of the action principle is distinguished by the introduction of additional variables for which no equations of motion exist.

Finally, we should like to briefly discuss the usefulness of the surface terms $G_{1,2}$. These offer a connection between the conservation laws and the invariants of a mechanical system (Noether).

Let us assume that our variation of the action vanishes under certain circumstances: $\delta S = 0$. We then say that the action, which remains unchanged, is invariant under that particular variation of the path. The principle of stationary action then states:

$$\delta S = 0 = G_2 - G_1 , \tag{1.75}$$

i.e., G has the same value, independent of the initial and final configurations.

In particular, let us assume that the action (Hamiltonian version) is invariant for a variation around the actual path for which it holds that

$$\delta x_i(t_{1,2}) = 0 , \quad \frac{d}{dt}(\delta t) = 0 : \ \delta t = \text{const.} = \varepsilon . \tag{1.76}$$

Then it follows from the invariance of S under infinitesimal constant time translation:

$$\delta S = 0 = G_2 - G_1 = -H(t_2)\delta t_2 + H(t_1)\delta t_1 = -(H_2 - H_1)\, \varepsilon , \tag{1.77}$$

the conservation of energy:

$$H(t_2) = H(t_1) , \quad \text{meaning} \quad \frac{dH}{dt} = 0 . \tag{1.78}$$

Similarly, the conservation law for linear momentum follows if we assume that the action of the system is invariant under constant space translation and the change of the terminal times vanishes:

$$\delta x_i = \delta \varepsilon_i = \text{const.} , \quad \delta t(t_{1,2}) = 0 . \tag{1.79}$$

$$\delta S = 0 = G_2 - G_1 = (p_i \delta x_i)_2 - (p_i \delta x_i)_1 = (p_{i2} - p_{i1}) \delta \varepsilon_i \tag{1.80}$$

or

$$p_i(t_2) = p_i(t_1) , \quad \text{meaning} \quad \frac{dp_i}{dt} = 0 . \tag{1.81}$$

Now let

$$H = \frac{p_i^2}{2m} + V(r) , \tag{1.82}$$

i.e., the potential may only depend on the distance $r = \sqrt{x_i^2}$. Then no space direction is distinguished, and with respect to rigid rotations $\delta\omega_i =$ const. and

$$\delta t(t_{1,2}) = 0 , \quad \delta x_i = \varepsilon_{ijk}\delta\omega_j x_k , \tag{1.83}$$

we obtain

$$\delta S = \delta \int_{t_1}^{t_2} dt \left[p_i \frac{dx_i}{dt} - \frac{p_i^2}{2m} - V(\sqrt{x_i^2}) \right] = 0 . \tag{1.84}$$

Let us prove explicitly that $\delta S = 0$.

$$\delta\left(p_i \frac{dx_i}{dt}\right) - \delta\left(\frac{p_i^2}{2m}\right) = \delta p_i \frac{dx_i}{dt} + p_i \frac{d}{dt}\delta x_i - \frac{p_i}{m}\delta p_i = p_i \frac{d}{dt}\delta x_i ,$$

where we used $dx_i/dt = p_i/m$, since our particle travels on thc correct classical path; thus we are left with

$$p_i \frac{d}{dt}\delta x_i = p_i \frac{d}{dt}\varepsilon_{ijk}\delta\omega_j x_k = \frac{1}{m}\varepsilon_{ijk}\delta\omega_j p_i p_k = 0 , \tag{1.85}$$

where again, $\dot{x}_k = p_k/m$ has been applied together with the total antisymmetry of ε_{ijk}.

The remaining variation is

$$\begin{aligned} \delta V &= \frac{\partial V}{\partial x_i}\delta x_i = \frac{\partial V}{\partial x_i}\varepsilon_{ijk}\delta\omega_j x_k = \frac{x_i}{r}\frac{\partial V}{\partial r}\varepsilon_{ijk}\delta\omega_j x_k \\ &= \frac{1}{r}\frac{\partial V}{\partial r}\varepsilon_{ijk}\delta\omega_j x_i x_k = 0 . \end{aligned} \tag{1.86}$$

Because

$$\begin{aligned} \delta S = 0 = G_2 - G_1 &= (p_i\delta x_i)_2 - (p_i\delta x_i)_1 = (p_i\varepsilon_{ijk}\delta\omega_j x_k)_2 - (p_i\varepsilon_{ijk}\delta\omega_j x_k)_1 \\ &= \delta\omega_i \left\{[(\boldsymbol{r}\times\boldsymbol{p})_i]_2 - [(\boldsymbol{r}\times\boldsymbol{p})_i]_1\right\} \end{aligned} \tag{1.87}$$

this implies the conservation of angular momentum:

$$\boldsymbol{L}(t_2) = \boldsymbol{L}(t_1) , \quad \text{meaning} \quad \frac{d\boldsymbol{L}}{dt} = 0 . \tag{1.88}$$

Conversely, the conservation of angular momentum corresponds to the invariance, $\delta S = 0$, under rigid rotation in space. The generalization of this statement is: if a conservation law exists, then the action S is stationary with respect to the infinitesimal transformation of a corresponding variable. The converse of this statement is also true: if S is invariant with respect to an infinitesimal transformation, $\delta S = 0$, then a corresponding conservation law exists.

2. The Action Principle in Classical Electrodynamics

The main purpose of this chapter is to consider the formulation of a relativistic point particle in classical electrodynamics from the viewpoint of Lagrangian mechanics. Here, the utility of Schwinger's action principle is illustrated by employing three different kinds of action to derive the equations of motion and the associated surface terms.

The first choice is rather standard and is given in covariant language through the invariant action

$$S = \int_{\tau_1}^{\tau_2} d\tau \left[-m_0 c^2 + \frac{\mathrm{e}}{c} \frac{dx^\mu}{d\tau} A_\mu(x) \right]. \tag{2.1}$$

Here, τ denotes the proper time, and $d\tau = \frac{1}{c}\sqrt{-dx^\mu dx_\mu} = \sqrt{1 - \frac{v^2}{c^2}}\, dt$ is the invariant proper time interval. The variation of S is given by

$$\delta S = \int_1^2 \left[-m_0 c^2 \delta(d\tau) + \frac{\mathrm{e}}{c} \delta(dx^\mu) A_\mu(x) + \frac{\mathrm{e}}{c} dx^\mu \delta A_\mu(x) \right] \tag{2.2}$$

with

$$\begin{aligned}
\delta(d\tau) &= \frac{1}{c} \frac{1}{\sqrt{-dx^\mu dx_\mu}} \left(-dx^\mu d(\delta x_\mu) \right) = -\frac{1}{c^2} \frac{dx^\mu}{d\tau} d(\delta x_\mu) \\
&= -\frac{1}{c^2} d\left(\frac{dx^\mu}{d\tau} \delta x_\mu \right) + \frac{1}{c^2} \frac{d^2 x^\mu}{d\tau} \delta x_\mu,
\end{aligned}$$

$$\begin{aligned}
A_\mu \delta(dx^\mu) &= d(A_\mu \delta x^\mu) - dA_\mu \delta x^\mu, \\
dx^\mu \delta A_\mu &= dx^\mu \frac{\partial A_\mu}{\partial x^\nu} \delta x^\nu = \partial_\mu A_\nu dx^\nu \delta x^\mu
\end{aligned}$$

so that δS takes the form

$$\begin{aligned}
G_2 - G_1 = \delta S = \int_1^2 \Bigg[& d\left(m_0 \frac{dx^\mu}{d\tau} \delta x_\mu + \frac{\mathrm{e}}{c} A^\mu \delta x_\mu \right) \\
& + \left(-m_0 \frac{d^2 x^\mu}{d\tau} + \frac{\mathrm{e}}{c} \partial^\mu A_\nu dx^\nu - \frac{\mathrm{e}}{c} dA^\mu \right) \delta x_\mu \Bigg].
\end{aligned} \tag{2.3}$$

From here we obtain the equation of motion,

$$m_0\frac{d^2x^\mu}{d\tau^2} = \frac{\mathrm{e}}{c}\partial^\mu A_\nu\frac{dx^\nu}{d\tau} - \frac{\mathrm{e}}{c}\frac{dA^\mu}{d\tau} = \frac{\mathrm{e}}{c}(\partial^\mu A_\nu - \partial_\nu A^\mu)\frac{dx^\nu}{d\tau}$$
$$= \frac{\mathrm{e}}{c}F^\mu{}_\nu\frac{dx^\nu}{d\tau}, \tag{2.4}$$

which we recognize as the Lorentz equation. The surface term is identified as

$$G = \left(m_0\frac{dx^\mu}{d\tau} + \frac{\mathrm{e}}{c}A^\mu\right)\delta x^\mu \quad (\equiv p^\mu\delta x_\mu). \tag{2.5}$$

With the aid of (2.4) we can easily verify that $\frac{dx^\mu}{d\tau}\frac{dx_\mu}{d\tau}$ remains constant ($= -c^2$):

$$\frac{dx_\mu}{d\tau}\frac{d^2x^\mu}{d\tau^2} = \frac{1}{2}\frac{d}{d\tau}\left(\frac{dx^\mu}{d\tau}\frac{dx_\mu}{d\tau}\right) = \frac{\mathrm{e}}{m_0c}F^\mu{}_\nu\frac{dx^\nu}{d\tau}\frac{dx_\mu}{d\tau} = 0$$

because of the antisymmetry of $F_{\mu\nu}$.

Now, τ is not a measure of proper time unless $\frac{dx^\mu}{d\tau}\frac{dx_\mu}{d\tau} = -c^2$, as assumed so far. If, more generally, τ serves as an independent parameter other than proper time, we can write as an equivalent action:

$$S = \int_{\tau_1}^{\tau_2} d\tau\left[\frac{m_0}{2}\left(\frac{dx^\mu}{d\tau}\frac{dx_\mu}{d\tau} - c^2\right) + \frac{\mathrm{e}}{c}\frac{dx^\mu}{d\tau}A_\mu\right]. \tag{2.6}$$

With the additional requirement that G be independent of $\delta\tau$, we can carry out our variational procedure and end up with

$$\begin{aligned}\delta S = \int_1^2 \Bigg[& d\left(m_0\frac{dx^\mu}{d\tau}\delta x_\mu + \frac{\mathrm{e}}{c}A^\mu\delta x_\mu - \frac{1}{2}m_0\left(\frac{dx^\mu}{d\tau}\frac{dx_\mu}{d\tau} - c^2\right)\delta\tau\right)\\ & + \left(-m_0\frac{d^2x^\mu}{d\tau} + \frac{\mathrm{e}}{c}F^\mu{}_\nu dx^\nu\right)\delta x_\mu + \left(\frac{m_0}{2}d\left(\frac{dx^\mu}{d\tau}\frac{dx_\mu}{d\tau}\right)\right)\delta\tau\Bigg]\\ = G_2 - G_1 &\end{aligned} \tag{2.7}$$

with $G = \left(m_0\frac{dx^\mu}{d\tau} + \frac{\mathrm{e}}{c}A^\mu\right)\delta x_\mu \quad (\equiv p^\mu\delta x_\mu)$,

as before. Since G was required to be independent of $\delta\tau$, we obtain

$$\delta\tau: \quad \frac{dx^\mu}{d\tau}\frac{dx_\mu}{d\tau} - c^2 = 0. \tag{2.8}$$

As a consequence, the last term in (2.7) multiplying $\delta\tau$ vanished automatically:

$$\frac{dx^\mu}{d\tau}\frac{dx_\mu}{d\tau} = \text{const.}$$

Again we find for the equation of motion:

$$\delta x_\mu : \qquad m_0 \frac{d^2 x^\mu}{d\tau^2} = \frac{\mathrm{e}}{c} F^\mu_{\ \nu} \frac{dx^\nu}{d\tau}. \tag{2.9}$$

Incidentally, the structure of the integrand of equation (2.6) can easily be derived by use of the method of Lagrange multipliers. For this reason, we introduce a new Lagrangian by

$$\overline{L} = L + \frac{\lambda(\tau)}{2}\left(\dot{x}^2 + c^2\right) \tag{2.10}$$

where $L = -m_0 c^2 + \frac{\mathrm{e}}{c}\dot{x}_\mu A^\mu$.

Variation of the associated action leads us directly to the Euler–Langrange equations:

$$\frac{\partial L}{\partial x^\mu} - \frac{d}{d\tau}\left(\frac{\partial L}{\partial \dot{x}^\mu}\right) - \frac{d}{d\tau}(\lambda \dot{x}_\mu) = 0 \tag{2.11}$$

and the variation with respect to the multiplier λ gives

$$\dot{x}^2 + c^2 = 0. \tag{2.12}$$

Altogether we have five equations for the five unknowns $x^\mu(\tau)$ and $\lambda(\tau)$. Multiplying equation (2.11) by $\dot{x}^\mu$ and using $\dot{x}_\mu \ddot{x}^\mu = 0$, which follows from $\dot{x}^2 = -c^2$, we find

$$\dot{x}^\mu \frac{\partial L}{\partial x^\mu} - \dot{x}^\mu\left(\frac{\partial L}{\partial \dot{x}^\mu}\right) - \underbrace{\dot{x}^\mu \frac{d}{d\tau}(\lambda \dot{x}^\mu)}_{=\frac{d}{d\tau}(\lambda \dot{x}_\mu \dot{x}^\mu) = -c^2 \dot{\lambda}} = 0$$

or

$$\frac{d\lambda}{d\tau} = -\frac{1}{c^2}\left[\dot{x}^\mu \frac{\partial L}{\partial x^\mu} - \dot{x}^\mu\left(\frac{\partial L}{\partial \dot{x}^\mu}\right)\right]$$

which can be integrated to yield

$$\lambda = \int d\tau \dot{\lambda} = \frac{-1}{c^2}\left[\int d\tau \frac{\partial L}{\partial x^\mu}\frac{dx^\mu}{d\tau} - \int d\tau \dot{x}^\mu \frac{d}{d\tau}\left(\frac{\partial L}{\partial \dot{x}^\mu}\right)\right]. \tag{2.13}$$

Since our original Lagrangian has the form $L = L(x^\mu, \dot{x}^\mu)$, we have

$$\frac{dL}{d\tau} = \frac{\partial L}{\partial x^\mu}\dot{x}^\mu + \frac{\partial L}{\partial \dot{x}^\mu}\ddot{x}^\mu \tag{2.14}$$

so that we can continue to write for $\lambda(\tau)$

$$\begin{aligned}
\lambda(\tau) &= -\frac{1}{c^2}\left[L - \int d\tau \frac{\partial L}{\partial \dot{x}^\mu}\ddot{x}^\mu - \int d\tau \dot{x}^\mu \frac{d}{d\tau}\left(\frac{\partial L}{\partial \dot{x}^\mu}\right)\right] \\
&= -\frac{1}{c^2}\left[L - \int d\tau \frac{d}{d\tau}\left(\dot{x}^\mu \frac{\partial L}{\partial \dot{x}^\mu}\right)\right] \\
&= -\frac{1}{c^2}\left[L - \dot{x}^\mu \frac{\partial L}{\partial \dot{x}^\mu}\right].
\end{aligned} \tag{2.15}$$

In our particular case with $\frac{\partial L}{\partial \dot{x}^\mu} = \frac{\mathrm{e}}{c} A_\mu$, we obtain

$$\lambda = -\frac{1}{c^2}\left[-m_0 c^2 + \frac{\mathrm{e}}{c}\dot{x}^\mu A_\mu - \frac{\mathrm{e}}{c}\dot{x}^\mu A_\mu\right] \\ = m_0.$$

Remarkably, λ has become independent of τ by now. Therefore, according to (2.10), the modified Lagrangian L is given by

$$\overline{L} = -m_0 c^2 + \frac{\mathrm{e}}{c}\dot{x}A + \frac{m_0}{2}(\dot{x}^2 + c^2) \\ = \frac{m_0}{2}(\dot{x}^2 - c^2) + \frac{\mathrm{e}}{c}\dot{x}A$$

which brings us back to the integrand of equation (2.6).

Finally, as a third version of the action we consider

$$S = \int_{\tau_1}^{\tau_2} d\tau \left[p_\mu\left(\frac{dx^\mu}{d\tau} - v^\mu\right) + \frac{m_0}{2}\left(v^\mu v_\mu - c^2\right) + \frac{\mathrm{e}}{c}v^\mu A_\mu\right] \tag{2.16}$$

which, after variation of the independent variables, x_μ, p_μ, v_μ and τ leads to

$$\delta S = \delta\int_1^2 \left[p_\mu dx^\mu - p_\mu v^\mu d\tau + \frac{m_0}{2}\left(v^\mu v_\mu - c^2\right)d\tau + \frac{\mathrm{e}}{c}v^\mu A_\mu d\tau\right] \tag{2.17}$$

$$= \int_1^2 \Big[d(p_\mu \delta x^\mu) + d\left(\delta\tau\left(-p_\mu v^\mu + \frac{m_0}{2}(v_\mu v^\mu - c^2) + \frac{\mathrm{e}}{c}v^\mu A_\mu\right)\right) \\ + \delta x_\mu\left(-dp^\mu + \frac{\mathrm{e}}{c}v_\nu \partial^\mu A_\nu d\tau\right) + \delta p_\mu\left(dx^\mu - v^\mu d\tau\right) \\ + \delta v_\mu\left(-p^\mu d\tau + m_0 v^\mu d\tau + \frac{\mathrm{e}}{c}A^\mu d\tau\right) \\ + \delta\tau d\left(p^\mu v_\mu - \frac{m_0}{2}\left(v^\mu v_\mu - c^2\right) - \frac{\mathrm{e}}{c}v^\mu A_\mu\right)\Big] \tag{2.18}$$

$$= G_2 - G_1.$$

The surface term yields

$$G = p_\mu \delta x^\mu \tag{2.19}$$

provided that G is again independent of $\delta\tau$, which implies

$$p_\mu v^\mu = \frac{m_0}{2}(v^\mu v_\mu - c^2) + \frac{\mathrm{e}}{c}v^\mu A_\mu \tag{2.20}$$

so that the term in (2.18) that multiplies $\delta\tau$ is automatically zero.
The equations of motion follow from

$$\delta x_\mu : \quad \frac{dp^\mu}{d\tau} = \frac{\mathrm{e}}{c}\partial^\mu A_\nu v^\nu \tag{2.21}$$

$$\delta p_\mu : \quad \frac{dx^\mu}{d\tau} = v^\mu. \tag{2.22}$$

There is no dynamical equation of motion for v^μ since there is no term proportional to $\frac{dv^\mu}{d\tau}$ in the action. Thus

$$\delta v_\mu : \qquad p^\mu = m_0 v^\mu + \frac{\mathrm{e}}{c} A^\mu, \tag{2.23}$$

which, together with (2.21) and (2.22) can be combined into

$$m_0 \frac{d^2 x^\mu}{d\tau^2} = \frac{\mathrm{e}}{c} F^\mu{}_\nu \frac{dx^\nu}{d\tau}. \tag{2.24}$$

Furthermore, substituting equations (2.22) and (2.23) into (2.20) gives us back the constraint

$$\frac{dx^\mu}{d\tau}\frac{dx_\mu}{d\tau} = v^\mu v_\mu = -c^2. \tag{2.25}$$

In summary, we were led to three equivalent actions which express the dynamics of a relativistic point particle in presence of an external electromagnetic field.

3. Application of the Action Principles

We begin this chapter by deriving a few laws of nonconservation in mechanics. To this end we first consider the change of the action under rigid space translation $\delta x_i = \delta\varepsilon_i$ and $\delta t(t_{1,2}) = 0$. Then the noninvariant part of the action,

$$S = \int_{t_1}^{t_2} dt \left[p_i \frac{dx_i}{dt} - \frac{p_i^2}{2m} - V(x_i, t) \right] , \tag{3.1}$$

is given by

$$\delta V(x_i, t) = \frac{\partial V}{\partial x_i} \delta x_i \ ,$$

and thus it immediately follows for the variation of S that

$$\delta S = \int_{t_1}^{t_2} dt \left[-\frac{\partial V(x_i, t)}{\partial x_i} \delta x_i \right] = G_2 - G_1 = \int_{t_1}^{t_2} dt \frac{d}{dt} (p_i \delta x_i) \ ,$$

or

$$\int_{t_1}^{t_2} dt \left[\frac{d}{dt} p_i + \frac{\partial V}{\partial x_i} \right] \delta\varepsilon_i = 0 \ .$$

Here we recognize Newton's law as nonconservation of the linear momentum:

$$\frac{dp_i}{dt} = -\frac{\partial V(x_i, t)}{\partial x_i} \ . \tag{3.2}$$

Now it is straightforward to derive a corresponding law of nonconservation of the angular momentum. To do so, we need the variation of (3.1) under $\delta x_i = \varepsilon_{ijk} \delta\omega_j x_k$ with constant $\delta\omega_j$ and again, $\delta t(t_{1,2}) = 0$. As before, only $V(x_i, t)$ contributes to the variation :

$$\delta V(x_i, t) = \frac{\partial V}{\partial x_i} \delta x_i = \frac{\partial V}{\partial x_i} \varepsilon_{ijk} \delta\omega_j x_k = \delta\omega_i \varepsilon_{ijk} x_j \frac{\partial V}{\partial x_k} = \delta\omega_i (\boldsymbol{r} \times \nabla V)_i \ .$$

The variation $\delta[p_i(dx_i/dt) - (p_i^2/2m)]$ makes no contribution.

Then we obtain

$$\delta S = -\int_{t_1}^{t_2} dt(\boldsymbol{r} \times \nabla V)_i \delta\omega_i = G_2 - G_1 = \int_{t_1}^{t_2} dt \frac{d}{dt} p_i(\delta\boldsymbol{\omega} \times \boldsymbol{r})_i$$
$$= \int_{t_1}^{t_2} dt \frac{d}{dt} (\boldsymbol{r} \times \boldsymbol{p})_i \delta\omega_i$$

or

$$\int_{t_1}^{t_2} dt \left[(\boldsymbol{r} \times \nabla V)_i + \frac{d}{dt} (\boldsymbol{r} \times \boldsymbol{p})_i \right] \delta\omega_i = 0 \, .$$

Upon using the definition $\boldsymbol{L} = \boldsymbol{r} \times \boldsymbol{p}$ and $\boldsymbol{F} = -\nabla V$ we have immediately

$$\frac{d\boldsymbol{L}}{dt} = \boldsymbol{N} = \boldsymbol{r} \times \boldsymbol{F} \, . \tag{3.3}$$

As a further example we consider a particle in an $1/r$-potential with $r = \sqrt{x_i^2}$ and $k = \text{const.}$:

$$S = \int_{t_1}^{t_2} dt \left[p_i \frac{dx_i}{dt} - \frac{p_i^2}{2m} + \frac{k}{r} \right] \, . \tag{3.4}$$

The special form of the variations of δx_i and δp_i is now given by the rigid displacements ($\delta\varepsilon_i = \text{const.}$, $\delta t(t_{1,2}) = 0$):

$$\delta x_j = \delta\varepsilon_i \left[\frac{1}{mk} \delta_{ij} x_k p_k - x_i p_j - \varepsilon_{ijk} L_k \right] \, , \qquad L_k = \varepsilon_{klm} x_l p_m \tag{3.5}$$

$$\delta p_j = \delta\varepsilon_i \left[\frac{x_i x_j}{r^3} - \delta_{ij} \frac{1}{r} - \frac{1}{mk} (p_i p_j - \delta_{ij} p^2) \right] \, . \tag{3.6}$$

Here, in contrast to our former examples, δp_i is not arbitrary anymore. The calculation of δS with the help of (3.5) and (3.6) is performed in the usual way and yields, after a few steps (here is an exercise):

$$\delta S = 2\delta\varepsilon_i \int_{t_1}^{t_2} dt \left(-\frac{d}{dt} \right) \left(\frac{x_i}{r} \right) . \tag{3.7}$$

So the action principle then reads:

$$\delta S = -2\delta\varepsilon_i \int_{t_1}^{t_2} dt \frac{d}{dt} \left(\frac{x_i}{r} \right) = \int_{t_1}^{t_2} dt \frac{d}{dt} p_i \delta x_i \, . \tag{3.8}$$

For the integrand on the right-hand side we get

$$p_i \delta x_i = \frac{2}{mk} \delta\varepsilon_i \left[p_k x_k p_i - p^2 x_i \right] = \frac{2}{mk} \delta\varepsilon_i (\boldsymbol{L} \times \boldsymbol{p})_i \, . \tag{3.9}$$

Our final result is, therefore:

$$2\delta\boldsymbol{\varepsilon}\cdot\int_{t_1}^{t_2}dt\left[-\frac{d}{dt}\left(\frac{\boldsymbol{r}}{r}\right)-\frac{d}{dt}\frac{1}{mk}(\boldsymbol{L}\times\boldsymbol{p})\right]=0\,. \tag{3.10}$$

So we have proved that the Runge–Lenz vector A is a conserved quantity in the Coulomb problem:

$$\boldsymbol{A}:=\frac{1}{mk}\boldsymbol{L}\times\boldsymbol{p}+\frac{\boldsymbol{r}}{r}:\quad \frac{d\boldsymbol{A}}{dt}=0\,. \tag{3.11}$$

In our series of standard examples, the harmonic oscillator is still missing. The paths of a particle in the three-dimensional oscillator potential,

$$V(r)=\frac{1}{2}kr^2=\frac{m}{2}\omega^2r^2 \tag{3.12}$$

with $k=m\omega^2$ and $r^2=x_i^2$ are, as in the Kepler (Coulomb) problem, closed. In the case of the $1/r$-potential, the presence of closed paths is attributed to the existence of the conserved Runge–Lenz vector. This suggests searching for additional conserved quantities in the harmonic oscillator. The well-known constants of motion are the energy and the angular momentum:

$$E=\frac{1}{2m}(p_i^2+m^2\omega^2x_i^2)\,,\quad \frac{dE}{dt}=0\,, \tag{3.13}$$

$$L_i=\varepsilon_{ijk}x_jp_k\,,\quad \frac{dL_i}{dt}=0\,. \tag{3.14}$$

We now wish to prove that the following tensor (nine elements) of the Runge–Lenz type is also a constant of motion:

$$A_{ij}:=\frac{1}{2m}(p_ip_j+m^2\omega^2x_ix_j)\,. \tag{3.15}$$

Here we need not limit ourselves to three space dimensions. In the following we thus consider the isotropic N-dimensional harmonic oscillator:

$$H=\frac{1}{2m}\sum_{i=1}^{N}p_i^2+\frac{m}{2}\omega^2\sum_{i=1}^{N}x_i^2\,. \tag{3.16}$$

The variations δx_i and δp_i are now given by ($\delta\eta_{ik}=\text{const.}$)

$$\delta x_i=\frac{1}{2m}\delta\eta_{jk}(\delta_{ij}p_k+p_j\delta_{ik})\,, \tag{3.17}$$

$$\delta p_i=-\frac{m\omega^2}{2}\delta\eta_{jk}(\delta_{ij}x_k+x_j\delta_{ik})\,. \tag{3.18}$$

The variation δS is then obtained in the form

$$\delta S = \delta\eta_{jk} m\omega^2 \int_{t_1}^{t_2} dt \left[-\frac{d}{dt}(x_j x_k)\right] = G_2 - G_1 = \int_{t_1}^{t_2} dt \frac{d}{dt}(p_i \delta x_i) . \qquad (3.19)$$

In (3.19) we need $p_i \delta x_i = (1/m)\delta\eta_{jk} p_j p_k$, so that our variation (3.19) reads

$$\delta\eta_{jk} \int_{t_1}^{t_2} dt \frac{d}{dt}\left[m\omega^2 x_j x_k + \frac{1}{m} p_j p_k\right] = 0 ,$$

or, using (3.15):

$$\overset{\leftrightarrow}{A} := \frac{1}{2m}(\boldsymbol{pp} + m^2\omega^2 \boldsymbol{rr}) : \quad \frac{d\overset{\leftrightarrow}{A}}{dt} = 0 . \qquad (3.20)$$

The virial theorem in mechanics also provides a good example of an application. Here we begin with the variation

$$\delta x_i = \delta\varepsilon x_i , \quad \delta p_i = -\delta\varepsilon p_i , \qquad (3.21)$$

in

$$\delta S = \delta \int_{t_1}^{t_2} \left[p_i \frac{dx_i}{dt} - T(p) - V(x_i)\right] , \qquad (3.22)$$

where $T(p)$ denotes the kinetic energy $T(p) = p_i^2/2m$. The term $p_i(dx_i/dt)$ in (3.22) remains unchanged under (3.21):

$$\delta\left(p_i \frac{dx_i}{dt}\right) = \delta p_i \frac{dx_i}{dt} + p_i \frac{d}{dt}\delta x_i = -\delta\varepsilon p_i \frac{dx_i}{dt} + p_i \frac{d}{dt}(\delta\varepsilon x_i) = 0 . \qquad (3.23)$$

But $H = T(p) + V(x_i)$ changes according to

$$\delta H = \frac{1}{m} p_i \delta p_i + \frac{\partial V}{\partial x_i}\delta x_i = -\frac{\delta\varepsilon}{m} p_i^2 + \delta\varepsilon \frac{\partial V}{\partial x_i} x_i = \delta\varepsilon\left(-2T + x_i \frac{\partial V}{\partial x_i}\right) . \qquad (3.24)$$

Applying the action principle yields

$$\begin{aligned}\delta S &= \delta \int_{t_1}^{t_2} dt \left[p_i \frac{dx_i}{dt} - H\right] = \int_{t_1}^{t_2} dt \left[\delta\varepsilon\left(2T - x_i \frac{\partial V}{\partial x_i}\right)\right] \\ &= G_2 - G_1 = \int_{t_1}^{t_2} dt \frac{d}{dt}(p_i \delta x_i) = \int_{t_1}^{t_2} dt \frac{d}{dt}(\delta\varepsilon x_i p_i) ,\end{aligned}$$

so that the theorem we seek follows:

$$\frac{d}{dt}(x_i p_i) = 2T - x_i \frac{\partial V}{\partial x_i} . \qquad (3.25)$$

In particular for the Kepler problem with $V(r) = -k/r$ we find, with the aid of $x_i(\partial V/\partial x_i) = k/r = -V(r)$:

$$\frac{d}{dt}(x_i p_i) = 2T + V . \tag{3.26}$$

We now come to the calculation of the action functional for a few simple cases, e.g., for a free particle in one dimension or a particle under the influence of a constant force. Here we want to apply the action principle exclusively: $\delta S = G_2 - G_1$.

Let $\delta t_1 = 0$. If we then use $H = p^2/2m$ in $G = p\delta x - H\delta t$ we have

$$G_2 = p(t_2)\delta x_2 - \frac{p^2(t_2)}{2m}\delta t_2 \tag{3.27}$$

$$G_1 = p(0)\delta x(0) , \quad x(0) = x_1 . \tag{3.28}$$

At this point we need the solutions to Hamilton's equations

$$\dot{x}(t) = \frac{\partial H}{\partial p} = \frac{p(t)}{m} , \tag{3.29}$$

$$\dot{p}(t) = -\frac{\partial H}{\partial x} = 0 . \tag{3.30}$$

Clearly we obtain $p(t) = p(t_{1,2}) = \text{const.}$ and

$$x(t) = x(0) + \frac{p(0)}{m} t \equiv x_1 + \frac{p}{m} t .$$

When we solve this for p we get $p = m[x(t_2) - x_1]/t_2$ or

$$\frac{p^2}{2m} = \frac{m}{2t_2^2}(x_2 - x_1)^2 .$$

Finally we end up with a total differential for δS:

$$\delta S = G_2 - G_1 = p\delta x_2 - p\delta x_1 - \frac{p^2}{2m}\delta t_2 = m\frac{x(t_2) - x_1}{t_2}\delta(x_2 - x_1) \tag{3.31}$$

$$- \frac{m}{2}\frac{(x_2 - x_1)^2}{t_2^2}\delta t_2 = \delta\left[\frac{m}{2}\frac{(x_2 - x_1)^2}{t_2}\right] ,$$

or

$$S = \frac{m}{2}\frac{(x_2 - x_1)^2}{t_2} + c .$$

The constant c is determined from the condition $\lim_{t_2 \to t_1 = 0} S\{[x_i]; t_1, t_2\} = 0$. This yields $c = 0$. If we then refrain from setting $t_1 = 0$, the action for a free particle of mass m is given by

$$S = \frac{m}{2}\frac{(x_2 - x_1)^2}{t_2 - t_1} . \tag{3.32}$$

The second example for calculating S from the action principle directly concerns a particle in presence of a constant force F:

$$H = \frac{p^2}{2m} - Fx \,. \tag{3.33}$$

The corresponding equations of motion are

$$\dot{x}(t) = \frac{\partial H}{\partial p} = \frac{p(t)}{m} \,, \qquad \dot{p}(t) = -\frac{\partial H}{\partial x} = F \,,$$

with the initial conditions given at $t_1 = 0 : x(0) = x_1$, $p(0) = p_1$. The solutions are obviously expressed in

$$p(t) = Ft + p_1$$
$$x(t) = x_1 + \frac{p_1}{m}t + \frac{1}{2}\frac{F}{m}t^2 \,.$$

Again we need the following:

$$p_2 = p_1 + Ft_2 = \frac{m}{t_2}\left(x_2 - x_1 - \frac{Ft_2^2}{2m}\right) + Ft_2 = \frac{m}{t_2}\left(x_2 - x_1 + \frac{Ft_2^2}{2m}\right)$$
$$\frac{p_2^2}{2m} = \frac{m}{2t_2^2}\left(x_2^2 - 2x_1x_2 + x_1^2 - \frac{Ft_2^2}{m}(x_2 - x_1) + \frac{F^2t_2^4}{4m^2}\right) .$$

If we now continue our calculation as for a free particle, we get

$$\delta S = G_2 - G_1 = p_2\delta x_2 - \left(\frac{p_2^2}{2m} - Fx_2\right)\delta t_2 - p_1\delta x_1$$
$$= \delta\left\{\frac{m}{2}\frac{(x_2 - x_1)^2}{t_2} + \frac{1}{2}Ft_2(x_1 + x_2) - \frac{F^2t_2^3}{24m}\right\}$$

or

$$S = \frac{m}{2}\frac{(x_2 - x_1)^2}{t_2 - t_1} + \frac{1}{2}F(t_2 - t_1)(x_1 + x_2) - \frac{F^2}{24m}(t_2 - t_1)^3 \,. \tag{3.34}$$

We still want to prove that the actions (3.32) and (3.34) do indeed satisfy the Hamilton–Jacobi equations (1.34) and (1.36). To show this, we build the following partial derivatives:

$$p_2 = \frac{\partial S}{\partial x_2} = m\frac{(x_2 - x_1)}{t_2 - t_1} \,, \qquad p_1 = -\frac{\partial S}{\partial x_1} = m\frac{(x_2 - x_1)}{t_2 - t_1} \,,$$

from which follows: $p_1 = p_2$, $x_2 = (p_1/m)(t_2 - t_1) + x_1$. Later we will show that S is a generating function for the canonical transformation $(x_2, p_2) \to (x_1, p_1)$:

$$\begin{pmatrix} x_1 \\ p_1 \end{pmatrix} = \begin{pmatrix} 1 & -\dfrac{t_2 - t_1}{m} \\ 0 & 1 \end{pmatrix} \begin{pmatrix} x_2 \\ p_2 \end{pmatrix} . \tag{3.35}$$

Furthermore, we have to demonstrate that $H(x_2, \partial S/\partial x_2) + \partial S/\partial t_2 = 0$.

$$H\left(x_2, \frac{\partial S}{\partial x_2}\right) = \frac{1}{2m}\left(\frac{\partial S}{\partial x_2}\right)^2 = \frac{m}{2}\frac{(x_2 - x_1)^2}{(t_2 - t_1)^2},$$
$$\frac{\partial S}{\partial t_2} = -\frac{m}{2}\frac{(x_2 - x_1)^2}{(t_2 - t_1)^2}.$$

Addition of these two expressions does, indeed, give zero. The same can be shown for

$$H\left(x_1, -\frac{\partial S}{\partial x_1}\right) - \frac{\partial S}{\partial t_1} = 0.$$

Similar steps can be performed with the action (3.34):

$$p_2 = \frac{\partial S}{\partial x_2} = m\frac{(x_2 - x_1)}{t_2 - t_1} + \frac{F}{2}(t_2 - t_1),$$
$$p_1 = -\frac{\partial S}{\partial x_1} = m\frac{(x_2 - x_1)}{t_2 - t_1} - \frac{F}{2}(t_2 - t_1).$$

These equations can be rewritten as

$$p_2 = p_1 + F(t_2 - t_1),$$
$$x_2 = x_1 + \frac{p_1}{m}(t_2 - t_1) + \frac{F}{2m}(t_2 - t_1)^2.$$

The action S in (3.34) is, correspondingly, the generating function of the canonical transformation

$$\begin{pmatrix} x_1 \\ p_1 \end{pmatrix} = \begin{pmatrix} 1 & -\dfrac{t_2 - t_1}{m} \\ 0 & 1 \end{pmatrix} \begin{pmatrix} x_2 - F\dfrac{(t_2 - t_1)^2}{2m} \\ p_2 - F(t_2 - t_1) \end{pmatrix}. \tag{3.36}$$

It can be seen that the Hamilton–Jacobi equations are also satisfied.

We are now going to complicate the previous example by allowing the external force to become time dependent so that the Lagrangian reads

$$L = \frac{m}{2}\dot{x}^2 + F(t)x$$

with the equation of motion:

$$\ddot{x} = \frac{1}{m}F(t) \equiv \ddot{G}.$$

Of course, we could proceed as before, using the action principle. However, to bring a little variety into our calculation, we decide to compute the action directly from its very definition as the time integral of the Lagrangian. We will see that in this kind of calculation we have to solve the equations of motion before we can do the integration. In the sequel we need

$$\dot{x} = \int_{t_1}^{t} dt' \frac{F(t')}{m} + a \equiv \dot{G}(t) + a$$

$$x(t) = \int_{t_1}^{t} dt' \int_{t_1}^{t'} dt'' \frac{F(t'')}{m} + a(t - t_1) + b \equiv G(t) + a(t - t_1) + b \,.$$

The constants a and b follow from

$$x(t_1) = x_1 = \overbrace{G(t_1)}^{=0} + b: \; b = x_1$$

$$x(t_2) = x_2 = G(t_2) + a(t_2 - t_1) + x_1: \; a = \frac{1}{t_2 - t_1}[(x_2 - x_1) - G(t_2)] \,.$$

Furthermore: $\dot{x}(t_1) = a$, $\dot{x}(t_2) = \dot{G}(t_2) + a$.

These results will be used in the action when we write:

$$S = \int_{t_1}^{t_2} dt\, L = \int_{t_1}^{t_2} dt \left[\frac{m}{2}\dot{x}^2 + F(t)x\right]$$

$$= \left[\frac{m}{2}x\dot{x}\right]_{t_1}^{t_2} - \frac{1}{2}\int_{t_1}^{t_2} dt\, x \underbrace{m\frac{d^2x}{dt^2}}_{=F} + \int_{t_1}^{t_2} dt\, F(t)x = \frac{m}{2}[x_2\dot{x}(t_2) - x_1\dot{x}(t_1)]$$

$$+ \frac{1}{2}\int_{t_1}^{t_2} dt\, F(t)x = \frac{m}{2}[a(x_2 - x_1) + x_2\dot{G}(t_2)] + \frac{1}{2}\int_{t_1}^{t_2} dt\, F(t)x \,.$$

Next, the time integral can be rewritten as

$$\frac{1}{2}\int_{t_1}^{t_2} dt\, F(t)x(t) = \frac{1}{2}\int_{t_1}^{t_2} dt\, F(t)G(t) + a\frac{m}{2}[t_2(\dot{G}(t_2) + a) - t_1 a - (x_2 - x_1)]$$
$$- \frac{m}{2}at_1\dot{G}(t_2) + \frac{m}{2}x_1\dot{G}(t_2) \,,$$

so that

$$S = \frac{1}{2}\int_{t_1}^{t_2} dt\, F(t)G(t) + m\dot{G}(t_2)x_2$$
$$+ \frac{m}{2}[(x_2 - x_1) - G(t_2)]^2 \frac{1}{(t_2 - t_1)} - \frac{m}{2}G(t_2)\dot{G}(t_2)$$

We note that the remaining time integral can be expressed as

$$\frac{1}{2}\int_{t_1}^{t_2} dt\, F(t)G(t) = \frac{m}{2}\int_{t_1}^{t_2} dt \frac{d}{dt}\dot{G}\,G = \frac{m}{2}\left[G(t_2)\dot{G}(t_2) - \underbrace{\dot{G}(t_1)}_{=0}G(t_1)\right]$$
$$- \frac{m}{2}\int_{t_1}^{t_2} dt\, \dot{G}^2(t) \,.$$

Finally we arrive at

$$
\begin{aligned}
S &= \int_{t_1}^{t_2} dt \left[\frac{m}{2}\dot{x}^2 + F(t)x\right] = \frac{m}{2}[(x_2 - x_1) - G(t_2)]^2 \frac{1}{(t_2 - t_1)} \\
&- \frac{m}{2}\int_{t_1}^{t_2} dt\, \dot{G}^2(t) + m\dot{G}(t_2)x_2 = \frac{m}{2(t_2 - t_1)}\left[(x_2 - x_1) - \int_{t_1}^{t_2} dt \int_{t_1}^{t} dt' \frac{F(t')}{m}\right]^2 \\
&- \frac{1}{2m}\int_{t_1}^{t_2} dt \left(\int_{t_1}^{t} dt' F(t')\right)^2 + x_2 \int_{t_1}^{t_2} dt F(t) .
\end{aligned}
$$

Next, we present the results for the one-dimensional harmonic oscillator and for a particle with charge e and mass m in a constant magnetic field in z-direction.

$$
S = \frac{m\omega}{2}\left[(x_2^2 + x_1^2)\cot[\omega(t_2 - t_1)] - \frac{2x_1x_2}{\sin[\omega(t_2 - t_1)]}\right] , \qquad \omega = \sqrt{\frac{k}{m}} ; \tag{3.37}
$$

$$
\begin{aligned}
S &= \frac{m}{2}\left\{\frac{(z_2 - z_1)^2}{t_2 - t_1} + \omega 2 \cot\left(\frac{\omega(t_2 - t_1)}{2}\right)[(x_2 - x_1)^2 + (y_2 - y_1)^2]\right. \\
&\left. + \omega(x_1y_2 - x_2y_1)\right\}, \qquad \omega = \frac{eB}{mc} .
\end{aligned} \tag{3.38}
$$

We start out with the Lagrangian:

$$
L(x, \dot{x}) = \frac{m}{2}\,\dot{x}^2 - \frac{m}{2}\,\omega^2 x^2 .
$$

The equation of motion follows from

$$
\frac{d}{dt}\left(\frac{\partial L}{\partial \dot{x}}\right) = \frac{\partial L}{\partial x} : \; \ddot{x} + \omega^2 x = 0
$$

and has the solution

$$
x(t) = A\sin(\omega t + \alpha) . \tag{3.39}
$$

A bit later we need

$$
x(t_{1,2}) = A\sin(\omega t_{1,2} + \alpha) . \tag{3.40}
$$

Since $p = \partial L/\partial \dot{x} = m\dot{x}$, the Hamiltonian reads

$$
H = p\dot{x} - L = \frac{p^2}{2m} + \frac{m}{2}\omega^2 x^2 .
$$

Now the action can be simplified by using the equation of motion $\ddot{x} + \omega^2 x = 0$ in

$$
\begin{aligned}
S &= \int_{t_1}^{t_2} dt\, L(x, \dot{x}) = \frac{m}{2}\int_{t_1}^{t_2} dt \left[\left(\frac{dx}{dt}\right)^2 - \omega^2 x^2\right] \\
&= \left[\frac{m}{2}x\dot{x}\right]_{t_1}^{t_2} - \frac{m}{2}\int_{t_1}^{t_2} dt\, \underbrace{x(t)\left(\frac{d^2}{dt^2} + \omega^2\right)x(t)}_{=0} \\
&= \frac{m}{2}[x(t_2)\dot{x}(t_2) - x(t_1)\dot{x}(t_1)] .
\end{aligned} \tag{3.41}
$$

In (3.41) we need to eliminate $\dot{x}(t_2)$, $\dot{x}(t_1)$ in terms of $x(t_2)$, $x(t_1)$. To achieve this, let us rewrite (3.39) in the following form:

$$\begin{aligned} x(t) &= A\sin(\omega t + \alpha) = A\sin[\omega(t-t_1) + (\omega t_1 + \alpha)] \\ &= A\sin[\omega(t-t_1)]\cos(\omega t_1 + \alpha) + A\sin(\omega t_1 + \alpha)\cos[\omega(t-t_1)] \,. \end{aligned}$$

Using (3.40) again we can continue to write

$$x(t) = \frac{1}{\omega}\dot{x}(t_1)\sin[\omega(t-t_1)] + x_1\cos[\omega(t-t_1)] \,.$$

For the particular value $t = t_2$ we then find

$$\dot{x}(t_1) = \frac{\omega}{\sin[\omega(t_2-t_1)]}[x_2 - x_1\cos[\omega(t_2-t_1)]] \,. \tag{3.42}$$

Similarly,

$$\dot{x}(t_2) = \frac{\omega}{\sin[\omega(t_2-t_1)]}[-x_1 + x_2\cos[\omega(t_2-t_1)]] \,. \tag{3.43}$$

In (3.41) we need

$$\begin{aligned} x(t_2)\dot{x}(t_2) &= \frac{\omega}{\sin[\omega(t_2-t_1)]}[x_2^2\cos[\omega(t_2-t_1)] - x_2x_1] \,, \\ x(t_1)\dot{x}(t_1) &= \frac{\omega}{\sin[\omega(t_2-t_1)]}[-x_1^2\cos[\omega(t_2-t_1)] + x_2x_1] \,. \end{aligned}$$

Taking the difference of these expressions yields the predicted

$$S = \frac{m\omega}{2\sin[\omega(t_2-t_1)]}[(x_2^2 + x_1^2)\cos[\omega(t_2-t_1)] - 2x_2x_1] \tag{3.44}$$

or, with $T = t_2 - t_1$:

$$S = \frac{m\omega}{2\sin(\omega T)}[(x_2^2 + x_1^2)\cos(\omega T) - 2x_2x_1] \,, \quad \omega T \neq n\pi \,. \tag{3.45}$$

Next in the list of standard problems, we compute the classical action for a charged particle in a uniform magnetic field in z-direction. The Lagrangian has the form

$$\begin{aligned} L &= \frac{m}{2}(\dot{x}^2 + \dot{y}^2 + \dot{z}^2) + \frac{eB}{2c}(x\dot{y} - y\dot{x}) \,, \quad \boldsymbol{A} = \frac{1}{2}\boldsymbol{B}\times\boldsymbol{r} \\ &= \frac{m}{2}[(\dot{x}^2 + \dot{y}^2 + \dot{z}^2) + \omega(x\dot{y} - y\dot{x})] \,, \quad \omega = \frac{eB}{mc} \,. \end{aligned} \tag{3.46}$$

The z-coordinate satifies the equation of motion of a free particle. The associated classical action is therefore given by (3.32):

$$S[z] = \int_{t_1}^{t_2} dt \frac{m}{2}\dot{z}^2 = \frac{m}{2}\frac{(z_2 - z_1)^2}{t_2 - t_1} \,. \tag{3.47}$$

The motion perpendicular to the z-axis follows from

$$\frac{d}{dt}\frac{\partial L}{\partial \dot{x}} - \frac{\partial L}{\partial x} = m\ddot{x} - \frac{m}{2}\omega\dot{y} - \frac{m}{2}\omega\dot{y} = 0 \quad : \quad \ddot{x} = \omega\dot{y} \tag{3.48}$$

$$\frac{d}{dt}\frac{\partial L}{\partial \dot{y}} - \frac{\partial L}{\partial y} = m\ddot{y} + \frac{m}{2}\,\omega\dot{x} + \frac{m}{2}\,\omega\dot{x} = 0 \quad : \quad \ddot{y} = -\omega\dot{x}\ . \tag{3.49}$$

Equation (3.49) is solved by $\dot{y} = -\omega x + \omega C$ which, when substituted in (3.48), yields

$$\ddot{x} = -\omega^2 x + \omega^2 C\ . \tag{3.50}$$

Here we make the usual ansatz,

$$x(t) = A'\sin(\omega t) + B'\cos(\omega t) + C \tag{3.51}$$

which produces

$$\dot{y}(t) = -\omega A'\sin(\omega t) - \omega B'\cos(\omega t) - \omega C + \omega C$$

and therefore

$$y(t) = A'\cos(\omega t) - B'\sin(\omega t) + D\ . \tag{3.52}$$

Using the initial conditions $x(t_1) = x_1$, $y(t_1) = y_1$, we get

$$x(t) = A\sin[\omega(t-t_1)] + B\cos[\omega(t-t_1)] + x_1 - B\ ,$$
$$y(t) = A\cos[\omega(t-t_1)] - B\sin[\omega(t-t_1)] + y_1 - A\ .$$

Taking the time derivative of these equations yields

$$\dot{x}(t) = A\omega\cos[\omega(t-t_1)] - B\omega\sin[\omega(t-t_1)] \tag{3.53}$$

$$\dot{y}(t) = -A\omega\sin[\omega(t-t_1)] - B\omega\cos[\omega(t-t_1)]\ . \tag{3.54}$$

The fixed end points at t_2 give us in addition

$$x(t_2) = x_2 = A\sin[\omega(t_2-t_1)] + B\cos[\omega(t_2-t_1)] + x_1 - B\ ,$$
$$y(t_2) = y_2 = A\cos[\omega(t_2-t_1)] - B\sin[\omega(t_2-t_1)] + y_1 - A\ .$$

Writing $t_2 - t_1 = T$, $\sin\varphi = 2\sin(\varphi/2)\cos(\varphi/2)$, $\cos\varphi - 1 = 2\sin^2(\varphi/2)$ we get

$$x_2 = 2A\sin\frac{\omega T}{2}\cos\frac{\omega T}{2} - 2B\sin^2\frac{\omega T}{2} + x_1\ ,$$
$$y_2 = -2A\sin^2\frac{\omega T}{2} - 2B\sin\frac{\omega T}{2}\cos\frac{\omega T}{2} + y_1\ ,$$

or

$$(x_2 - x_1)\sin\frac{\omega T}{2} + (y_2 - y_1)\cos\frac{\omega T}{2} = -2B\sin\frac{\omega T}{2}$$

from which follows

$$B = -\frac{1}{2\sin(\omega T/2)}\left[(x_2 - x_1)\sin\frac{\omega T}{2} + (y_2 - y_1)\cos\frac{\omega T}{2}\right] . \tag{3.55}$$

Likewise,

$$(x_2 - x_1)\cos\frac{\omega T}{2} - (y_2 - y_1)\sin\frac{\omega T}{2} = 2A\sin\frac{\omega T}{2}$$

or

$$A = \frac{1}{2\sin(\omega T/2)}\left[(x_2 - x_1)\cos\frac{\omega T}{2} - (y_2 - y_1)\sin\frac{\omega T}{2}\right] . \tag{3.56}$$

Finally we have to compute the action;

$$\begin{aligned} S &= \int_{t_1}^{t_2} dt\, L = \frac{m}{2}\int_{t_1}^{t_2} dt\left[\left(\frac{dx}{dt}\right)^2 + \left(\frac{dy}{dt}\right)^2 + \omega(x\dot{y} - y\dot{x})\right] \\ &= \frac{m}{2}[x\dot{x} + y\dot{y}]_1^2 - \frac{m}{2}\int_{t_1}^{t_2} dt\left[x\underbrace{\left(\frac{d^2x}{dt^2} - \omega\dot{y}\right)}_{=0} + y\underbrace{\left(\frac{d^2y}{dt^2} - \omega\dot{x}\right)}_{=0}\right] \\ &= \frac{m}{2}[(x_2\dot{x}_2 - x_1\dot{x}_1) + (y_2\dot{y}_2 - y_1\dot{y}_1)] . \end{aligned} \tag{3.57}$$

Again, we just need to express $\dot{x}_1, \dot{x}_2, \dot{y}_1, \dot{y}_2$ in terms of x_1, x_2, y_1, y_2. This can easily be achieved with the aid of (3.53, 54) and (3.55, 56). We obtain

$$\begin{aligned} x_2\dot{x}_2 &= \frac{\omega}{2\sin(\omega T/2)}[x_2(x_2 - x_1)\cos(\omega T/2) + x_2(y_2 - y_1)\sin(\omega T/2)] , \\ x_1\dot{x}_1 &= \frac{\omega}{2\sin(\omega T/2)}[x_1(x_2 - x_1)\cos(\omega T/2) - x_1(y_2 - y_1)\sin(\omega T/2)] , \\ y_2\dot{y}_2 &= \frac{-\omega}{2\sin(\omega T/2)}[y_2(x_2 - x_1)\sin(\omega T/2) - y_2(y_2 - y_1)\cos(\omega T/2)] , \\ y_1\dot{y}_1 &= \frac{\omega}{2\sin(\omega T/2)}[y_1(x_2 - x_1)\sin(\omega T/2) + y_1(y_2 - y_1)\cos(\omega T/2)] . \end{aligned}$$

With these expressions, (3.57) turns into

$$\begin{aligned} \frac{m}{2}[(x_2\dot{x}_2 - x_1\dot{x}_1) + (y_2\dot{y}_2 - y_1\dot{y}_1)] &= \frac{m\omega}{4}[(x_2 - x_1)^2 + (y_2 - y_1)^2] \\ \times \cot\frac{\omega T}{2} + \frac{m}{2}\omega(x_1y_2 - y_1x_2) , \quad \omega &= \frac{eB}{mc} . \end{aligned}$$

Altogether then,

$$\begin{aligned} S_{\mathrm{cl}} = \frac{m}{2}\Bigg\{\frac{(z_2 - z_1)^2}{t_2 - t_1} + \frac{\omega}{2}\cot\frac{(t_2 - t_1)}{2}\left[(x_2 - x_1)^2 + (y_2 - y_1)^2\right] \\ + \omega(x_1y_2 - y_1x_2)\Bigg\} . \end{aligned} \tag{3.58}$$

Our final example is concerned with the linear harmonic oscillator that is driven by an external force $F(t)$. The calculation of the associated classical action is a bit more elaborate than anything we have encountered before. But besides being of great value, it leads us to the best of company: Feynman, too, treated the problem in his Princeton Ph.D. thesis.

So let us begin with the Lagrangian

$$L(x, \dot{x}) = \frac{m}{2}\dot{x}^2 - \frac{m}{2}\omega^2 x^2 + F(t)x \ . \tag{3.59}$$

The equation of motion follows from

$$\frac{d}{dt}\left(\frac{\partial L}{\partial \dot{x}}\right) - \frac{\partial L}{\partial x} = 0 : \ m\ddot{x} + m\omega^2 x = F(t) \ . \tag{3.60}$$

Introducing the Green's function equation

$$\left[m\frac{d^2}{dt^2} + m\omega^2\right] G(t, t') = \delta(t - t') \tag{3.61}$$

with

$$G(t, t') = \frac{1}{m\omega}\begin{cases} \sin[\omega(t - t')] \ , & t > t' \ , \\ 0 \ , & t < t' \ , \end{cases} \tag{3.62}$$

we can solve (3.60) by superimposing the homogeneous with a particular solution:

$$\begin{aligned} x(t) &= x_h(t) + x_p(t) = a\cos(\omega t) + b\sin(\omega t) \\ &\quad + \frac{1}{m\omega}\int_0^t dt' \sin[\omega(t - t')]F(t') \ . \end{aligned} \tag{3.63}$$

Let us choose $x(t_1) = x_1$ and $x(t_2) = x_2$ as initial conditions. Then we obtain

$$\begin{aligned} x(t) &= x_1\cos[\omega(t - t_1)] + A\sin[\omega(t - t_1)] \\ &\quad + \frac{1}{m\omega}\int_{t_1}^t d\tau \ \sin[\omega(t - \tau)]F(\tau) \ . \end{aligned} \tag{3.64}$$

At time t_2 (3.64) takes the value ($T := t_2 - t_1$)

$$x(t_2) = x_2 = x_1\cos(\omega T) + A\sin(\omega T) + \frac{1}{m\omega}\int_{t_1}^{t_2} d\tau \ \sin[\omega(t_2 - \tau)]F(\tau)$$

which identifies the constant A as

$$\begin{aligned} A &= (x_2 - x_1\cos(\omega T))\frac{1}{\sin(\omega T)} - \frac{1}{m\omega\sin(\omega T)} \\ &\quad \times \int_{t_1}^{t_2} d\tau \ \sin[\omega(t_2 - \tau)]F(\tau). \end{aligned} \tag{3.65}$$

$x(t)$ given in (3.64) indeed solves the differential equation (3.60).

Let us quickly check this. First we need

$$\frac{d}{dt}\int_{t_1}^{t} d\tau \, \sin[\omega(t-\tau)]F(\tau) = \omega \int_{t_1}^{t} d\tau \, \cos[\omega(t-\tau)]F(\tau) \, .$$

A second time derivative produces

$$\frac{d^2}{dt^2}\int_{t_1}^{t} d\tau \, \sin[\omega(t-\tau)]F(\tau) = -\omega^2 \int_{t_1}^{t} d\tau \, \sin[\omega(t-\tau)]F(\tau) + \omega F(t) \, .$$

Thus we obtain

$$\begin{aligned} m\ddot{x} + m\omega^2 x = m \Big\{ & -\omega^2 x_1 \cos[\omega(t-t_1)] - \omega^2 A \sin[\omega(t-t_1)] \\ & - \frac{1}{m\omega}\omega^2 \int_{t_1}^{t} d\tau \, \sin[\omega(t-\tau)]F(\tau) + \frac{\omega F(t)}{m\omega} \Big\} \\ & + m\omega^2 \Big\{ x_1 \cos[\omega(t-t_1)] + A \sin[\omega(t-t_1)] \\ & + \frac{1}{m\omega}\int_{t_1}^{t} d\tau \, \sin[\omega(t-\tau)]F(\tau) \Big\} = F(t) \, . \end{aligned}$$

Now let us define the following quantities:

$$H(t_1, t_2) := \frac{1}{m\omega \sin(\omega T)} \int_{t_1}^{t_2} d\tau \, \sin[\omega(t_2-\tau)]F(\tau) =: \frac{1}{\sin(\omega T)} S(t_2) \, , \tag{3.66}$$

$$S(t) := \frac{1}{m\omega}\int_{t_1}^{t} d\tau \, \sin[\omega(t-\tau)]F(\tau) \, , \tag{3.67}$$

$$C(t) := \omega \frac{1}{m\omega}\int_{t_1}^{t} d\tau \, \cos[\omega(t-\tau)]F(\tau) = \frac{d}{dt} S(t) \, . \tag{3.68}$$

With the abbreviations, (3.64) can be written as

$$x(t) = x_1 \cos[\omega(t-t_1)] + A \sin[\omega(t-t_1)] + S(t) \, . \tag{3.69}$$

Here, we substitute the expression for A given in (3.65) and obtain, after a few rearrangements:

$$x(t) = x_1 \frac{\sin[\omega(t_2-t)]}{\sin(\omega T)} + x_2 \frac{\sin[\omega(t-t_1)]}{\sin(\omega T)} - H(t_1, t_2)\sin[\omega(t-t_1)] + S(t) \, . \tag{3.70}$$

From here we get

$$\begin{aligned} \dot{x}(t) = & -\omega x_1 \frac{\cos[\omega(t_2-t)]}{\sin(\omega T)} + \omega x_2 \frac{\cos[\omega(t-t_1)]}{\sin(\omega T)} \\ & - H(t_1, t_2)\omega \cos[\omega(t-t_1)] + C(t) \, . \end{aligned} \tag{3.71}$$

Hence, for the action we obtain

$$S = \int_{t_1}^{t_2} dt\, L = \int_{t_1}^{t_2} dt \left[\underbrace{\frac{m}{2}\dot{x}^2}_{\text{int. b. parts}} - \frac{m}{2}\omega^2 x^2 + F(t)x \right]$$

$$= \left[\frac{m}{2}x\dot{x}\right]_{t_1}^{t_2} - \frac{1}{2}\int_{t_1}^{t_2} dt\, x \underbrace{\left(m\frac{d^2}{dt^2} + m\omega^2\right)x}_{=F} + \int_{t_1}^{t_2} dt\, F(t)x \tag{3.72}$$

$$S = \frac{m}{2}[x(t_2)\dot{x}(t_2) - x(t_1)\dot{x}(t_1)] + \frac{1}{2}\int_{t_1}^{t_2} dt\, F(t)x \,.$$

Here we need the expressions $\dot{x}(t_{1,2})$, which we obtain from (3.71):

$$\dot{x}(t_1) = -\,\omega x_1 \frac{\cos(\omega T)}{\sin(\omega T)} + \omega x_2 \frac{1}{\sin(\omega T)} - \frac{1}{m\sin(\omega T)}\int_{t_1}^{t_2} d\tau\, \sin[\omega(t_2-\tau)]F(\tau)$$

$$\dot{x}(t_2) = -\,\omega x_1 \frac{1}{\sin(\omega T)} + \omega x_2 \frac{\cos(\omega T)}{\sin(\omega T)} - \frac{\cos(\omega T)}{m\sin(\omega T)}\int_{t_1}^{t_2} d\tau \times \sin[\omega(t_2-\tau)]F(\tau) + \frac{1}{m}\int_{t_1}^{t_2} d\tau\, \cos[\omega(t_2-\tau)]F(\tau) \,.$$

The first contribution in (3.72) is then easily calculated and yields

$$\frac{m}{2}(x_2\dot{x}_2 - x_1\dot{x}_1) = \frac{m}{2}\frac{\omega}{\sin(\omega T)}\left\{(x_2^2+x_1^2)\cos(\omega T) - 2x_1x_2 + \frac{x_2}{m\omega} \times \int_{t_1}^{t_2} d\tau\, \sin[\omega(\tau-t_1)]F(\tau) + \frac{x_1}{m\omega}\int_{t_1}^{t_2} d\tau\, \sin[\omega(t_2-\tau)]F(\tau)\right\}. \tag{3.73}$$

The second half in (3.72) is also readily evaluated:

$$\frac{1}{2}\int_{t_1}^{t_2} dt\, F(t)x(t) = \frac{1}{2}\int_{t_1}^{t_2} dt\, F(t)\left[x_1\frac{\sin[\omega(t_2-t)]}{\sin(\omega T)} + x_2\frac{\sin[\omega(t-t_1)]}{\sin(\omega T)} \right.$$
$$\left. \frac{\sin[\omega(t-t_1)]}{m\omega\sin(\omega T)}\int_{t_1}^{t_2} d\tau\, \sin[\omega(t_2-\tau)]F(\tau) + \frac{1}{m\omega}\int_{t_1}^{t} d\tau\, \sin[\omega(t-\tau)]F(\tau)\right]$$
$$= \frac{m\omega}{2\sin(\omega T)}\left\{\int_{t_1}^{t_2} dt\, F(t)\left[\frac{x_1}{m\omega}\sin[\omega(t_2-t)] + \frac{x_2}{m\omega}\sin[\omega(t-t_1)] - \frac{\sin[\omega(t-t_1)]}{(m\omega)^2}\int_{t_1}^{t_2} d\tau\, \sin[\omega(t_2-\tau)]F(\tau) + \frac{\sin(\omega T)}{(m\omega)^2}\int_{t_1}^{t} d\tau\, \sin[\omega(t-\tau)]F(\tau)\right]\right\}. \tag{3.74}$$

Using the following identity,

$$\int_{t_1}^{t_2} ds F(s) \sin[\omega(s-t_1)] \int_{t_1}^{t_2} dt F(t) \sin[\omega(t_2-t)]$$
$$- \sin\omega T \int_{t_1}^{t_2} dt F(t) \int_{t_1}^{t} d\tau \sin[\omega(t-\tau)] F(\tau)$$
$$= 2\int_{t_1}^{t_2} dt \int_{t_1}^{t} ds F(t) F(s) \sin[\omega(t_2-t)] \sin[\omega(s-t_1)] \tag{3.75}$$

we finally end up with the classical action for the driven harmonic oscillator:

$$S = \frac{m\omega}{2\sin\omega T}\left\{(x_2^2+x_1^2)\cos(\omega T) - 2x_2x_1 + \frac{2x_2}{m\omega}\int_{t_1}^{t_2} dt\, F(t)\sin[\omega(t-t_1)]\right.$$
$$+ \frac{2x_1}{m\omega}\int_{t_1}^{t_2} dt\, F(t)\sin[\omega(t_2-t)] - \frac{2}{(m\omega)^2}\int_{t_1}^{t_2} dt \int_{t_1}^{t} ds\, F(t)F(s)$$
$$\left.\times \sin[\omega(t_2-t)]\sin[\omega(s-t_1)]\right\}. \tag{3.76}$$

For the rest of this chapter we want to stay with the one-dimensional harmonic oscillator but intend to give it a little twist. To motivate our procedure, let us write again

$$H(p,x) = \frac{p^2}{2m} + \frac{m}{2}\omega^2 x^2 , \tag{3.77}$$

with

$$\dot{x} = \frac{\partial H}{\partial p} = \frac{p}{m} , \quad \dot{p} = -\frac{\partial H}{\partial x} = -m\omega^2 x . \tag{3.78}$$

The action is

$$S = \int_{t_1}^{t_2} dt \left[p\dot{x} - \frac{p^2}{2m} - \frac{m}{2}\omega^2 x^2\right] . \tag{3.79}$$

Now let us study the response of S with respect to the changes ($\varepsilon =$ const., $\delta t(t_{1,2}) = 0$):

$$\delta x = \varepsilon\frac{\partial H}{\partial p} = \varepsilon\frac{p}{m} , \quad \delta p = -\varepsilon\frac{\partial H}{\partial x} = -\varepsilon m\omega^2 x . \tag{3.80}$$

Then we can readily prove that

$$\delta S = \int_{t_1}^{t_2} dt[p\delta\dot{x} + \delta p\dot{x} - \delta H] = 0 . \tag{3.81}$$

To see this, let us first write

$$p\delta\dot{x} = \frac{d}{dt}(p\delta x) - \dot{p}\delta x$$

and

$$\delta H = \frac{\partial H}{\partial p}\delta p + \frac{\partial H}{\partial x}\delta x = \frac{p}{m}(-\varepsilon m\omega^2 x) + m\omega^2 x\varepsilon\frac{p}{m} = 0 \,.$$

Therefore δS is reduced to

$$\delta S = \int_{t_1}^{t_2} dt\frac{d}{dt}(p\delta x) + \int_{t_1}^{t_2} dt[\delta p\dot{x} - \dot{p}\delta x] \,.$$

But

$$\delta p\dot{x} - \dot{p}\delta x = -\varepsilon m\omega^2 x\frac{p}{m} + m\omega^2 x\varepsilon\frac{p}{m} = 0 \,.$$

We get

$$\delta S = [p\delta x]_1^2 = G_2 - G_1 \,, \tag{3.82}$$

i.e., the usual form of the action principle.

Here it is appropriate to stress again that so far, all variations were performed around the actual classical path, i.e., for which the equations of motion are satisfied (Hamilton's equations "on-shell"). On the way to (3.82) we repeatedly used them at various places. Now we want to relax this on-shell requirement; i.e., we are still dealing with a Hamiltonian system (in our case, the one-dimensional linear harmonic oscillator), but we do not want the equations of motion to be satisfied as expressed by the right-hand sides of (3.78), $\dot{x} \neq \partial H/\partial p$, etc.

So let us consider the following general transformation of S with respect to

$$\delta p = -\varepsilon\frac{\partial H}{\partial x} \,, \qquad \delta x = \varepsilon\frac{\partial H}{\partial p} \,. \tag{3.83}$$

The parameter ε is, at this stage, independent of time. Again, we are not assuming that (3.78) is satisfied; i.e., we are talking about "off-shell" mechanics of the linear harmonic oscillator. The response of S in (3.79) under (3.83) is then given by

$$\delta S = \int_{t_1}^{t_2} dt\frac{d}{dt}(p\delta x) + \int_{t_1}^{t_2} dt[\underbrace{\delta p}_{-\varepsilon\frac{\partial H}{\partial x}}\dot{x} - \underbrace{\delta x}_{\varepsilon\frac{\partial H}{\partial p}}\dot{p} - \delta H(p,x)] \,.$$

Using

$$\delta H(p,x) = \delta p\frac{\partial H}{\partial p} + \delta x\frac{\partial H}{\partial x} = -\varepsilon\frac{\partial H}{\partial x}\frac{\partial H}{\partial p} + \varepsilon\frac{\partial H}{\partial p}\frac{\partial H}{\partial x} = 0$$

we get

$$\delta S = \varepsilon \int_{t_1}^{t_2} dt \left[\frac{d}{dt}\left(p \frac{\partial H}{\partial p} \right) - \left(\frac{\partial H}{\partial p} \dot{p} + \frac{\partial H}{\partial x} \dot{x} \right) \right] . \tag{3.84}$$

Notice that

$$\frac{\partial H}{\partial p} \dot{p} + \frac{\partial H}{\partial x} \dot{x} = \frac{dH}{dt} \neq 0 . \tag{3.85}$$

The variation of S under (3.83) is therefore given by

$$\delta S = \varepsilon \int_{t_1}^{t_2} dt \frac{d}{dt} \left[p \frac{\partial H}{\partial p} - H(p, x) \right] \tag{3.86}$$

$$= \varepsilon \left[p \frac{\partial H}{\partial p} - H \right]_1^2 . \tag{3.87}$$

For the harmonic oscillator δS is given by $\delta S = \varepsilon [p^2/m - H]_1^2$. Hence δS is a pure surface term which will be absent for closed trajectories (period T) – a case to be considered later on. Since ε is supposed to be independent of time, we may say that S is invariant – up to surface terms – under the global transformation (3.83).

Things really change substantially if we permit ε to depend on time, i.e., we elevate our "global" symmetry transformation to a local "gauge" symmetry. This requires the introduction of a "gauge potential" $A(t)$ which couples to the "matter" field (p, x) via

$$L = p\dot{x} - H(p, x) - A(t) H(p(t), x(t)) , \tag{3.88}$$

$$S_0[p, x, A] = \int_{t_1}^{t_2} dt\, L = \int_{t_1}^{t_2} dt [p\dot{x} - H(p, x) - A(t) H(p, x)] . \tag{3.89}$$

Let us prove that the action S_0 is – up to surface terms – invariant under

$$\delta p(t) = -\varepsilon(t) \frac{\partial H}{\partial x(t)} , \quad \delta x(t) = \varepsilon(t) \frac{\partial H}{\partial p(t)} , \tag{3.90}$$

$$\delta A(t) = \dot{\varepsilon} . \tag{3.91}$$

$$\delta S_0 = \int_{t_1}^{t_2} dt \left\{ \frac{d}{dt} \left[\varepsilon(t) p \frac{\partial H}{\partial p} \right] \underbrace{- \varepsilon(t) \frac{dH}{dt} - \underbrace{\delta A(t)}_{=\dot{\varepsilon}} H}_{d/dt(\varepsilon H) - \varepsilon (dH/dt)} - A \underbrace{\delta H}_{=0} \right\}$$

$$= \int_{t_1}^{t_2} dt \frac{d}{dt} \left\{ \varepsilon(t) \left[p \frac{\partial H}{\partial p} - H \right] \right\} = \left[\varepsilon(t) \left(p \frac{\partial H}{\partial p} - H \right) \right]_{t_1}^{t_2} . \tag{3.92}$$

For closed trajectories (period T) and "small" gauge transformations $\varepsilon(0) = \varepsilon(T)$, the surface term vanishes:

$$\varepsilon(0)\left[p\frac{\partial H}{\partial p} - H\right]_0^T = 0\,. \tag{3.93}$$

At this stage we add to L given in (3.88) a pure "gauge field" term and thereby introduce the so-called Chern–Simons action:

$$S_{\mathrm{CS}}[A] = k\int_{t_1}^{t_2} dt\, A(t)\,. \tag{3.94}$$

Here, k denotes an arbitrary real constant. Variation of S_{CS} simply gives

$$\delta S_{\mathrm{CS}} = k\int_{t_1}^{t_2} dt\,\delta A(t) = k\int_{t_1}^{t_2} dt\,\dot{\varepsilon}(t) = k(\varepsilon(t_2) - \varepsilon(t_1))\,. \tag{3.95}$$

Hence S_{CS} is invariant under "small" gauge transformations with $\varepsilon(t_2) - \varepsilon(t_1) = 0$. However, S_{CS} is not invariant under "large" gauge transformations with $\varepsilon(t_2) - \varepsilon(t_1) \neq 0$.

The complete action under discussion is

$$\begin{aligned} S[p,x,A] &= S_0[p,x,A] + S_{\mathrm{CS}}[A] \\ &= \int_{t_1}^{t_2} dt[p\dot{x} - H(p,x) - A(t)(H(p,x) - k)]\,. \end{aligned} \tag{3.96}$$

By the way, the equations of motion following from (3.96) are obtained from the independent variations δp, δx and δA:

$$\delta p:\ \dot{x} = (1 + A)\frac{\partial H}{\partial p}\,, \tag{3.97}$$

$$\delta x:\ \dot{p} = -(1 + A)\frac{\partial H}{\partial x}\,, \tag{3.98}$$

$$\delta A:\ H(p,x) = k\,. \tag{3.99}$$

Later we will show that under certain conditions it is possible to gauge $A(t)$ to zero, which would leave us with the usual equations of motion (3.78). However, there is still the constraint (3.99). Hence, only those trajectories (in phase space) are allowed for which H takes that constant value k which appears in the Chern–Simons action (3.94). The surfaces (3.99) (ellipses with fixed energies $k, k' \ldots$) foliate the entire phase space and, since for a certain 1-torus ($\equiv$ ellipse) with prescribed k the energy ($\equiv$ action J) is constant, a trajectory which begins on a certain torus will always remain on that torus.

Evidently $A(t)$ is not a dynamical field but is to be thought of as a Lagrangian multiplier for the constraint (3.99), $H(p,x) = k$. This is similar to the role of $A_0 \equiv \phi$ in electrodynamics, which does not satisfy an equation of motion either, but acts as a Lagrangian multiplier for Gauss' law:

$$\delta\phi:\ \delta L_{\mathrm{E.M.}} = \frac{1}{4\pi}\int d^3\boldsymbol{r}\,\delta\phi[\nabla\cdot\boldsymbol{E} - 4\pi\varrho]\,. \tag{3.100}$$

$$\to \nabla\cdot\boldsymbol{E} = 4\pi\varrho\,.$$

The analogue is $H = k$, where H is the generator of the gauge transformation for A, while k corresponds to the current of the "matter field".

In a later chapter we will pick up this topic again when discussing topological Chern–Simons quantum mechanics and the Maslov index in the context of semi-classical quantization à la Einstein–Brillouin–Keller (EBK).

4. Jacobi Fields, Conjugate Points

Let us go back to the action principle as realized by Jacobi, i.e., time is eliminated, so we are dealing with the space trajectory of a particle. In particular, we want to investigate the conditions under which a path is a minimum of the action and those under which it is merely an extremum. For illustrative purposes we consider a particle in two-dimensional real space. If we parametrize the path between points P and Q by ϑ, then Jacobi's principle states:

$$\delta \int_{\vartheta_1 \triangleq P}^{\vartheta_2 \triangleq Q} d\vartheta \sqrt{H - V(q_1, q_2)} \sqrt{\sum_{i,j=1}^{2} m_{ij}(q_k) \frac{dq_i}{d\vartheta} \frac{dq_j}{d\vartheta}} = 0 \,. \tag{4.1}$$

To save space let us simply write $g(q_1, q_2, \frac{dq_1}{d\vartheta}, \frac{dq_2}{d\vartheta})$ for the integrand. Hence the action reads

$$S\{[q_1, q_2]; \vartheta_1, \vartheta_2\} = \int_{\vartheta_1}^{\vartheta_2} d\vartheta \; g\left(q_1(\vartheta), q_2(\vartheta); \frac{dq_1}{d\vartheta}, \frac{dq_2}{d\vartheta}\right) . \tag{4.2}$$

For our further discussion it would be very convenient to choose one coordinate, e.g., q_1 instead of ϑ, to parametrize the path: $q_2(q_1)$ with $q_1^{(1)} \leq q_1 \leq q_1^{(2)}$. Thus in the following, we will be talking about the action

$$S\{[q_2]; q_1^{(1)}, q_1^{(2)}\} = \int_{q_1^{(1)}}^{q_1^{(2)}} dq_1 \; f\left(q_2(q_1), \frac{dq_2}{dq_1}; \not{q_1}\right) , \tag{4.3}$$

where we have dropped the external q_1-dependence. In this action we perform a variation around the actual classical path $\bar{q}_2(q_1)$. Let a varied path be given by

$$q_2(q_1) = \bar{q}_2(q_1) + \varepsilon\varphi(q_1) \,, \quad \varphi(q_1^{(1)}) = 0 = \varphi(q_1^{(2)}) \,. \tag{4.4}$$

Next we need

$$S[\bar{q}_2 + \varepsilon\varphi(q_1)] = = S[\bar{q}_2] + \varepsilon\delta S[\bar{q}_2] + \frac{\varepsilon^2}{2}\delta^2 S[\bar{q}_2] + \ldots \tag{4.5}$$

$$= \int_{q_1^{(1)}}^{q_1^{(2)}} dq_1 \; f\left(\bar{q}_2 + \varepsilon\varphi(q_1), \bar{q}_2' + \varepsilon\varphi'(q_1)\right)$$

$$= \int_1^2 dq_1 \left[f(\bar{q}_2, \bar{q}_2') + \frac{\partial f}{\partial q_2}\varepsilon\varphi + \frac{\partial f}{\partial q_2'}\varepsilon\varphi' \right. \tag{4.6}$$

$$\left. + \frac{1}{2}\left\{ \frac{\partial^2 f}{\partial q_2^2}\varepsilon^2\varphi^2 + 2\frac{\partial^2 f}{\partial q_2 \partial q_2'}\varepsilon^2\varphi\varphi' + \frac{\partial^2 f}{\partial q_2'^2}\varepsilon^2\varphi'^2 \right\} + \ldots \right],$$

where q_2' means $dq_2(q_1)/dq_1$ and the partial derivatives have to be evaluated along the actual path $\bar{q}_2(q_1)$. Now it is standard practice to perform an integration by parts on the third term in (4.6). The surface term drops out and the remainder together with the second term in the integrand yields Euler's equation

$$\left.\frac{\partial f}{\partial q_2}\right|_{\bar{q}_2} - \frac{d}{dq_1}\left(\left.\frac{\partial f}{\partial q_2'}\right|_{\bar{q}_2} \right) = 0\,. \tag{4.7}$$

So we are left with

$$\delta^2 S = \int_1^2 dq_1 \left[\left.\frac{\partial^2 f}{\partial q_2^2}\right|_{\bar{q}_2} \varphi^2 + 2 \left.\frac{\partial^2 f}{\partial q_2 \partial q_2'}\right|_{\bar{q}_2} \varphi\varphi' + \left.\frac{\partial^2 f}{\partial q_2'^2}\right|_{\bar{q}_2} \varphi'^2 \right]. \tag{4.8}$$

Now in order to find out whether we have a minimum of the action or just an extremum we have to know more about the sign of [...] in (4.8). If $q_2(q_1)$ is to be a minimum action trajectory, $\delta^2 S[\varphi]$ must be positive. For this reason we are looking for a function $\psi(q_1)$ which makes $\delta^2 S$ a minimum and if for this function $\delta^2 S$ is positive, we can be sure that $\delta^2 S[\varphi]$ is positive for all $\varphi(q_1)$. At this stage the question of positiveness of $\delta^2 S$ has been formulated in terms of a variational problem for $\delta^2 S[\varphi]$ itself.

We can normalize $\psi(q_1)$ so that

$$\int_1^2 dq_1 \; (\psi(q_1))^2 = 1\,, \quad \psi(q_1^{(1)}) = 0 = \psi(q_1^{(2)})\,. \tag{4.9}$$

Hence we are looking for a function $\psi(q_1)$ for which

$$\delta^2 S[\psi] = \text{ Minimum} \tag{4.10}$$

with the constraint

$$\delta \int_1^2 dq_1 (\psi(q_1))^2 = 0\,. \tag{4.11}$$

To proceed, we employ the method of Lagrangian undetermined multipliers:

$$\delta \int_1^2 dq_1 \left[F(\psi(q_1), \psi'(q_1)) - \lambda\psi^2 \right] = 0 \tag{4.12}$$

with

$$F = \left.\frac{\partial^2 f}{\partial q_2^2}\right|_{\bar{q}_2} \psi^2 + 2 \left.\frac{\partial^2 f}{\partial q_2 \partial q_2'}\right|_{\bar{q}_2} \psi\psi' + \left.\frac{\partial^2 f}{\partial q_2'^2}\right|_{\bar{q}_2} \psi'^2 . \tag{4.13}$$

The explicit variation of (4.12) yields

$$\delta \int_1^2 dq_1 (F - \lambda\psi^2) = \int_1^2 dq_1 \left(\frac{\partial F}{\partial \psi}\delta\psi + \underbrace{\frac{\partial F}{\partial \psi'}\delta\psi'}_{\underbrace{\frac{d}{dq_1}\left(\frac{\partial F}{\partial \psi'}\delta\psi\right)}_{\to 0} - \frac{d}{dq_1}\left(\frac{\partial F}{\partial \psi'}\right)\delta\psi} - \lambda 2\psi\delta\psi \right)$$

$$= \int_1^2 dq_1 \left[\left(\frac{\partial F}{\partial \psi} - \frac{d}{dq_1}\left(\frac{\partial F}{\partial \psi'}\right)\right) - \lambda 2\psi \right] \delta\psi .$$

So we obtain

$$\frac{d}{dq_1}\left(\frac{\partial F}{\partial \psi'}\right) - \frac{\partial F}{\partial \psi} = -\lambda 2\psi . \tag{4.14}$$

Here we need

$$\frac{\partial F}{\partial \psi} = 2 \left.\frac{\partial^2 f}{\partial q_2^2}\right|_{\bar{q}_2} \psi + 2 \left.\frac{\partial^2 f}{\partial q_2 \partial q_2'}\right|_{\bar{q}_2} \psi'$$

$$\frac{\partial F}{\partial \psi'} = 2 \left.\frac{\partial^2 f}{\partial q_2 \partial q_2'}\right|_{\bar{q}_2} \psi + 2 \left.\frac{\partial^2 f}{\partial q_2'^2}\right|_{\bar{q}_2} \psi'$$

$$\Rightarrow \frac{d}{dq_1}\frac{\partial F}{\partial \psi'} = 2\frac{d}{dq_1}\left(\frac{\partial^2 f}{\partial q_2 \partial q_2'}\right)\psi + 2\frac{\partial^2 f}{\partial q_2 \partial q_2'}\psi' + 2\frac{d}{dq_1}\left(\frac{\partial^2 f}{\partial q_2'^2}\right)\psi' + 2\frac{\partial^2 f}{\partial q_2'^2}\psi''.$$

When substituted in (4.14), we get

$$2\frac{d}{dq_1}\left(\frac{\partial^2 f}{\partial q_2 \partial q_2'}\right)\psi + 2\frac{d}{dq_1}\left(\frac{\partial^2 f}{\partial q_2'^2}\right)\psi' + 2\frac{\partial^2 f}{\partial q_2'^2}\psi'' - 2\frac{\partial^2 f}{\partial q_2^2}\psi = 2\lambda\psi . \tag{4.15}$$

This result can also be written in the form

$$\frac{d}{dq_1}\left(\left.\frac{\partial^2 f}{\partial q_2'^2}\right|_{\bar{q}_2} \psi' \right) + \left[\frac{d}{dq_1}\left(\left.\frac{\partial^2 f}{\partial q_2 \partial q_2'}\right|_{\bar{q}_2} \right) - \left.\frac{\partial^2 f}{\partial q_2^2}\right|_{\bar{q}_2} \right] \psi(q_1) = -\lambda\psi(q_1). \tag{4.16}$$

If we multiply both sides with $\psi(q_1)$ and integrate over q_1, we obtain

$$\int_1^2 dq_1 \Bigg[\underbrace{\psi \frac{d}{dq_1}\left(\frac{\partial^2 f}{\partial q_2'^2}\psi'\right)}_{\to -\psi'^2 \frac{\partial^2 f}{\partial q_2'^2}} + \underbrace{\psi \frac{d}{dq_1}\left(\frac{\partial^2 f}{\partial q_2 \partial q_2'}\right)\psi}_{\to -2\psi\psi' \frac{\partial^2 f}{\partial q_2 \partial q_2'}} - \psi \frac{\partial^2 f}{\partial q_2^2}\psi \Bigg] = -\lambda \underbrace{\int_1^2 dq_1 \psi^2}_{=1}$$

or

$$\delta^2 S[\psi] = \lambda \,. \tag{4.17}$$

So we know what λ is: it is the value $\delta^2 S[\psi]$ we are interested in; namely, the Lagrangian multiplier λ is the smallest value of $\delta^2 S$. Equation (4.16) together with (4.9) defines a Sturm–Liouville problem whose eigenfunctions and eigenvalues are those of $\delta^2 S$. Here $\delta^2 S$ is treated as a quadratic form ($\partial^2 f/\partial q_2'^2$ has to be positve, however). Eigenfunctions with eigenvalues higher than the lowest one do not minimize $\delta^2 S$, but $\delta^2 S$ is still stationary and satisfies (4.17). The eigenvalue problem (4.16) with (4.9) has an infinity of eigenvalues and eigenvectors λ_n and ψ_n with $n = 1, 2, \ldots$ ($\lambda_1 < \lambda_2 < \ldots$). The ψ_n form a complete orthonormal set of functions. Hence any function φ which vanishes at $q_1^{(1)}$, $q_1^{(2)}$ can be expanded in terms of the ψ_n's:

$$\varphi(q_1) = \sum_{n=1}^{\infty} a_n \psi_n(q_1) \,. \tag{4.18}$$

If we substitute this expression in (4.8), we obtain

$$\delta^2 S[\varphi] = \int_{q_1^{(1)}}^{q_1^{(2)}} dq_1 \left[\left.\frac{\partial^2 f(q_2, q_2')}{\partial q_2^2}\right|_{\bar{q}_2} \sum_{n,m} a_n a_m \psi_n \psi_m + \ldots \right]$$

and after use of the orthonormality condition $\int_1^2 dq_1\ \psi_n \psi_m = \delta_{nm}$, we arrive at

$$\delta^2 S[\varphi] = \sum_{n=1}^{\infty} \lambda_n a_n^2 \,. \tag{4.19}$$

Hence if all eigenvalues λ_n of $\delta^2 S$ are positive, then $\bar{q}_2(q_1)$ is a minimum-action trajectory. Conversely, $\bar{q}_2(q_1)$ is not a minimum-action trajectory if, for some n, $\lambda_n < 0$. This can occur for sufficiently small ε:

$$S[\bar{q}_2 + \varepsilon\psi_n] = S[\bar{q}_2] + \frac{\varepsilon^2}{2}\lambda_n + O(\varepsilon^3) < S[\bar{q}_2] \quad \text{for} \quad \lambda_n < 0 \,. \tag{4.20}$$

Let us apply our knowledge and work out an example; namely, the behavior of a particle with charge $(-e)$ in presence of a constant magnetic field which is directed in the positive z direction. Then we have as starting Lagrangian:

$$L(x_i, \dot{x}_i) = \frac{m}{2}\dot{x}_i^2 - \frac{e}{c}A_i \dot{x}_i \,. \tag{4.21}$$

Using

$$p_i = \frac{\partial L}{\partial \dot{x}_i} = m\dot{x}_i - \frac{e}{c}A_i : \dot{x}_i = \frac{1}{m}(p_i + \frac{e}{c}A_i)$$

we obtain the associated Hamiltonian:

$$H(x_i, p_i) = p_i\dot{x}_i - L = \frac{1}{2m}(p_i + \frac{e}{c}A_i)^2 = \frac{m}{2}\dot{x}_i^2 = \frac{m}{2}v^2 . \tag{4.22}$$

Since $\partial H/\partial t = 0$, energy is conserved: $H = \text{const}$. Then, according to (4.22), mv is likewise conserved. In order to study the space trajectory, we have to first rewrite our action principle:

$$\delta \int_{t_1}^{t_2} dt\ p_i \frac{dx_i}{dt} = \delta \int_1^2 dx_i \left[m\frac{dx_i}{dt} - \frac{e}{c}A_i \right] = 0 . \tag{4.23}$$

We choose $\boldsymbol{B} = B\hat{\boldsymbol{e}}_3$ and therefore the particle travels in the x-y plane on a circle with radius $\varrho = mvc/eB$ counterclockwise around the z direction. The constant $\boldsymbol{B}$-field in z direction can be obtained with the aid of the vector potential

$$\boldsymbol{A} = \tfrac{1}{2}B(-y\hat{\boldsymbol{e}}_1 + x\hat{\boldsymbol{e}}_2) = \tfrac{1}{2}Br\hat{\boldsymbol{\vartheta}} ,$$

since in a (r, ϑ)-coordinate system we have

$$\hat{\boldsymbol{\vartheta}} = (\hat{\boldsymbol{\vartheta}}, \hat{\boldsymbol{e}}_1)\hat{\boldsymbol{e}}_1 + (\hat{\boldsymbol{\vartheta}}, \hat{\boldsymbol{e}}_2)\hat{\boldsymbol{e}}_2 = -\sin\vartheta\hat{\boldsymbol{e}}_1 + \cos\vartheta\hat{\boldsymbol{e}}_2 = -\frac{y}{r}\hat{\boldsymbol{e}}_1 + \frac{x}{r}\hat{\boldsymbol{e}}_2$$

or

$$r\hat{\boldsymbol{\vartheta}} = -y\hat{\boldsymbol{e}}_1 + x\hat{\boldsymbol{e}}_2 .$$

Furthermore,

$$ds = \sqrt{(dx)^2 + (dy)^2} = \sqrt{(dr^2) + r^2(d\vartheta)^2} = \sqrt{r^2 + (dr/d\vartheta)^2}\ d\vartheta .$$

We use this information in (4.23) to write

$$\begin{aligned}
&\left[m\frac{dx_i}{dt} - \frac{e}{c}A_i \right] dx_i \\
&= \left[m\frac{dx_i}{ds}\frac{ds}{dt} - \frac{e}{c}A_i \right] \frac{dx_i}{ds}\, ds = \left[mv \overbrace{\left(\frac{dx_i}{ds}\right)^2}^{=1} - \frac{e}{c}A_i\frac{dx_i}{ds} \right] ds \\
&= \left[mv - \frac{e}{c}\frac{B}{2}\left(-y\frac{dx}{ds} + x\frac{dy}{ds}\right)\right] ds = mv\left[ds - \frac{eB}{2mvc}(-\,ydx + xdy)\right] .
\end{aligned}$$

Using $ydx = r\sin\vartheta\ rd(\cos\vartheta) = -r^2\sin^2\vartheta\ d\vartheta$ and $xdy = r^2\cos^2\vartheta\ d\vartheta$, our variational problem becomes formulated in

$$\delta \int_{\vartheta_1}^{\vartheta_2} d\vartheta\, mv \left[\sqrt{r^2(\vartheta) + \left(\frac{dr}{d\vartheta}\right)^2} - \frac{1}{2\varrho} r^2 \right] = 0\,. \tag{4.24}$$

The actual classical path is given by $\varrho = r_0 = \text{const.}$, so that $\varrho' = 0$. The integrand in (4.24) is now in the desired coordinate form (Jacobi principle), and from here on, we can follow our program and study the change in the action with respect to a small deviation from the actual trajectory $r_0 = \varrho$.

$$S = \{[r]; \vartheta_1, \vartheta_2\} = mv \int_{\vartheta_1}^{\vartheta_2} d\vartheta \left[\sqrt{r^2 + \left(\frac{dr}{d\vartheta}\right)^2} - \frac{r^2}{2\varrho} \right]. \tag{4.25}$$

For the slightly varied path we write

$$r(\vartheta) = r_0[1 + \varphi(\vartheta)]\,, \quad \varphi(\vartheta_1) = 0 = \varphi(\vartheta_2)\,. \tag{4.26}$$

In our former notation we have ($r' = dr/d\vartheta$)

$$f(r, r') = \sqrt{r^2 + r'^2} - r^2/2\varrho\,. \tag{4.27}$$

Following our earlier procedure we find

$$\begin{aligned} S &= mv \int_{\vartheta_1}^{\vartheta_2} d\vartheta\, f(r, r') = mv \int_{\vartheta_1}^{\vartheta_2} d\vartheta [f(\varrho, \overbrace{\varrho'}^{=0}) + 0(\text{Euler}) \\ &\quad + \frac{1}{2}\varrho^2 \left(\frac{\partial^2 f}{\partial r^2}\bigg|_\varrho \varphi^2 + 2 \frac{\partial^2 f}{\partial r \partial r'}\bigg|_\varrho \varphi\varphi' + \frac{\partial^2 f}{\partial r'^2}\bigg|_\varrho \varphi'^2 \right)] \\ &\overset{\substack{\varrho = r_0 \\ \varrho' = 0}}{=} mv \int_{\vartheta_1}^{\vartheta_2} d\vartheta\, \frac{1}{2} r_0 + mv \frac{\varrho^2}{2} \delta^2 S = \underbrace{\frac{mv}{2} r_0 (\vartheta_2 - \vartheta_1)}_{S_0} + mv \frac{r_0^2}{2} \delta^2 S \end{aligned}$$

or

$$S - S_0 = mv \frac{r_0^2}{2} \delta^2 S\,, \tag{4.28}$$

with

$$\delta^2 S = \int_{\vartheta_1}^{\vartheta_2} d\vartheta \left(\frac{\partial^2 f}{\partial r^2}\bigg|_\varrho \varphi^2 + 2 \frac{\partial^2 f}{\partial r \partial r'}\bigg|_\varrho \varphi\varphi' + \frac{\partial^2 f}{\partial r'^2}\bigg|_\varrho \varphi'^2 \right). \tag{4.29}$$

With the aid of (4.27) we can easily compute the partial derivatives in (4.29) for $r = \varrho = r_0$ and find

$$\delta^2 S[\varphi] = \frac{1}{r_0} \int_{\vartheta_1}^{\vartheta_2} d\vartheta\, [\varphi'^2 - \varphi^2]\,. \tag{4.30}$$

Now we expand in the complete orthonormal set of functions

$$\psi_n(\vartheta) = \sqrt{\frac{2}{(\vartheta_2 - \vartheta_1)}} \sin\left[n\pi \frac{(\vartheta - \vartheta_1)}{\vartheta_2 - \vartheta_1}\right], \quad \psi_n(\vartheta_1) = 0 = \psi_n(\vartheta_2) \tag{4.31}$$

$$\varphi = \sum_{n=1}^{\infty} a_n \psi_n . \tag{4.32}$$

Then we obtain

$$\delta^2 S[\varphi] = \sum_{n=1}^{\infty} \lambda_n a_n^2 \tag{4.33}$$

with

$$\lambda_n = \delta^2 S[\psi_n] = \int_{\vartheta_1}^{\vartheta_2} d\vartheta \left[\left.\frac{\partial^2 f}{\partial r^2}\right|_{\varrho} \psi_n^2 + \ldots \right] = \frac{1}{r_0} \int_{\vartheta_1}^{\vartheta_2} d\vartheta \left[\psi_n'^2 - \psi_n^2\right] \tag{4.34}$$

or

$$\lambda_n = \frac{1}{r_0}\left[\frac{n^2\pi^2}{(\vartheta_2 - \vartheta_1)^2} - 1\right] \equiv \lambda_n(\vartheta_2 - \vartheta_1) . \tag{4.35}$$

Finally we arrive at

$$\begin{aligned} S - S_0 &= mv\frac{r_0^2}{2}\frac{1}{r_0}\sum_{n=1}^{\infty} a_n^2 \left(\frac{n^2\pi^2}{(\vartheta_2 - \vartheta_1)^2} - 1\right) \\ &= mv\frac{r_0}{2}\sum_{n=1}^{\infty} a_n^2 \left(\frac{n^2\pi^2}{(\vartheta_2 - \vartheta_1)^2} - 1\right) . \end{aligned} \tag{4.36}$$

From (4.33) we see that for sufficiently small $(\vartheta_2 - \vartheta_1)$, i.e., $(\vartheta_2 - \vartheta_1) < \pi$, all λ_n are postive, and thus $r(\vartheta) = \varrho = r_0$ is a minimum-action trajectory: $S_0 < S$.

For $(\vartheta_2 - \vartheta_1) = \pi$ and only the lowest mode contribution in (33.32, 33), i.e., $a_n = 0$ for $n \neq 1$,

$$\psi_{n=1}^{(0)} = \sqrt{\frac{2}{\pi}} \sin(\vartheta - \vartheta_1) , \quad \lambda_{n=1} = 0 , \tag{4.37}$$

we have $\delta^2 S[\psi_1] = 0$ and consequently we obtain for this particular case, $S = S_0$. The zero-mode $\psi_1^{(0)}$ satisfies our equation (4.16):

$$\frac{d}{d\vartheta}\Bigg(\underbrace{\left.\frac{\partial^2 f}{\partial r'^2}\right|_{\varrho}}_{=1/\varrho} \psi_1^{(0)\prime}\Bigg) + \Bigg[\frac{d}{d\vartheta}\Bigg(\underbrace{\left.\frac{\partial^2 f}{\partial r \partial r'}\right|_{\varrho}}_{=0}\Bigg) - \underbrace{\left.\frac{\partial^2 f}{\partial r^2}\right|_{\varrho}}_{-1/\varrho}\Bigg]\psi_1^{(0)} = 0 \tag{4.38}$$

$$\left(\frac{d^2}{d\vartheta^2}+1\right)\psi_1^{(0)}=0\,,\quad \psi_1^{(0)}(\vartheta_1)=0=\psi_1^{(0)}(\vartheta_2)\,. \tag{4.39}$$

This equation is called the Jacobi equation and the function $\psi_1^{(0)}$ is called the Jacobi field for the problem under discussion.

Let us set $\vartheta_1 = 0$ and, therefore, $\vartheta_2 = \pi$. Starting from from $P(\triangleq \vartheta_1 = 0)$, we reach, after half a rotation around the circular orbit, the point $Q(\triangleq \vartheta_2 = \pi)$. The point Q is called a focal or conjugate point in relation to P along the circular trajectory. Once the trajectory has passed the conjugate point at $\vartheta_2 = \pi$ in relation to $\vartheta_1 = 0$, S_0 is no longer a minimum action.

So let us assume $(\vartheta_2 - \vartheta_1) > \pi$. Then, if we look at the contribution $a_n = 0$ for $n \neq 1$ in (3.33, 35), we find

$$(\vartheta_2-\vartheta_1)>\pi\ :\quad \delta^2 S[\psi_1]=\lambda_1=\frac{1}{r_0}\left[\frac{\pi^2}{(\vartheta_2-\vartheta_1)^2}-1\right]<0\,. \tag{4.40}$$

Therefore this particular example yields $S < S_0$, and thus, although S_0 is still an extremum, it is not a minimum. We also could drop lower lying modes, $a_{n_l} = 0$, and keep some of the higher lying ones, $a_{n_h} \neq 0$. In this case, $S > S_0$, so that S_0 is not a maximum either.

Since most of the time our goal is to study the time-development of a system let us repeat some of the former steps and investigate the dynamical t-dependent path $x(t)$ of the one-dimensional harmonic oscillator. Here the Lagrangian reads

$$L(x,\dot{x})=\frac{m}{2}\dot{x}^2-\frac{m}{2}\omega^2x^2\,, \tag{4.41}$$

and the action is given by

$$S\{[x];t_1,t_2\}=\int_{t_1}^{t_2}dt\ L=\int_{t_1}^{t_2}dt\left[\frac{m}{2}\dot{x}^2-\frac{m}{2}\omega^2x^2\right]. \tag{4.42}$$

As is by now routine, we look at the response of this action with respect to a displacement around the classical trajectory $\bar{x}(t)$:

$$x(t)=\bar{x}(t)+\varepsilon\eta(t)\,. \tag{4.43}$$

Again we expand the action according to

$$S[\bar{x}+\varepsilon\eta]=S[\bar{x}]+\varepsilon\delta S[\bar{x}]+\frac{\varepsilon^2}{2}\delta^2S[\bar{x}]+\dots \tag{4.44}$$

where

$$\delta^2S=\int_{t_1}^{t_2}dt\left[\left.\frac{\partial^2L}{\partial x^2}\right|_{\bar{x}}\eta^2+2\left.\frac{\partial^2L}{\partial x\partial\dot{x}}\right|_{\bar{x}}\eta\dot{\eta}+\left.\frac{\partial^2L}{\partial\dot{x}^2}\right|_{\bar{x}}\dot{\eta}^2\right]. \tag{4.45}$$

The classical action is given by (2.37). The partial derivatives can readily be obtained from (4.41) so that we have to deal with

$$\delta^2 S = \int_{t_1}^{t_2} dtm \left[\dot{\eta}^2 - \omega^2 \eta^2\right] . \tag{4.46}$$

As before, we are looking for a function that minimizes $\delta^2 S$. Let this function be $\psi(t)$, which should be normalized according to

$$\int_{t_1}^{t_2} dt(\psi(t))^2 = 1 , \quad \psi(t_1) = 0 = \psi(t_2) . \tag{4.47}$$

With the introduction of the Lagrangian multiplier λ we meet the variational problem

$$\delta \int_{t_1}^{t_2} dt \left[m\dot{\psi}^2 - m\omega^2\psi^2 - \lambda\psi^2\right] = 0 . \tag{4.48}$$

After an integration by parts we obtain the (Sturm–Liouville) eigenvalue equation

$$\frac{d}{dt}\dot{\psi} + \left(\omega^2 + \frac{\lambda}{m}\right)\psi = 0 , \quad \psi(t_1) = 0 = \psi(t_2) . \tag{4.49}$$

If we multiply this equation by ψ and integrate between t_1 and t_2, we find

$$\int_{t_1}^{t_2} dt \left(\psi\frac{d}{dt}\dot{\psi} + \omega^2\psi^2\right) = -\frac{\lambda}{m}\int_{t_1}^{t_2} dt\ \psi^2 = -\frac{\lambda}{m} , \tag{4.50}$$

or, after an integration by parts:

$$\int_{t_1}^{t_2} dt \left(\dot{\psi}^2 - \omega^2\psi^2\right) = \frac{\lambda}{m} = \frac{1}{m}\delta^2 S[\psi] . \tag{4.51}$$

The eigenfunctions and eigenvalues of the oscillator equation (4.49) are given by

$$\psi_n(t) = \sqrt{\frac{2}{(t_2 - t_1)}}\ \sin\left[n\pi\frac{(t - t_1)}{t_2 - t_1}\right] \tag{4.52}$$

$$\lambda_n = \int_{t_1}^{t_2} dt(\dot{\psi}_n^2 - \omega^2\psi_n^2) . \tag{4.53}$$

Since the ψ_n form a complete set of orthonormal functions, we expand $\eta(t)$ according to

$$\eta(t) = \sum_{n-1}^{\infty} a_n\psi_n . \tag{4.54}$$

When substituted in (4.46) this gives

$$\delta^2 S[\eta] = m\sum_{n=1}^{\infty}\lambda_n a_n^2 . \tag{4.55}$$

Substituting (4.52) in (4.53) or (4.49) we get

$$\left(-\frac{(n\pi)^2}{(t_2 - t_1)^2} + \omega^2 \right) \psi_n = -\frac{\lambda_n}{m} \psi_n ,$$

and therefore the spectrum is given by

$$\lambda_n = m \left(\frac{n^2 \pi^2}{(t_2 - t_1)^2} - \omega^2 \right) \equiv \lambda_n(t_2 - t_1) , \quad n = 1, 2, \ldots . \tag{4.56}$$

For sufficiently small $(t_2 - t_1)$ i.e., $(t_2 - t_1) < T/2 = \pi/\omega$ we have $\delta^2 S[\psi_n] = \lambda_n > 0$. In this case, $\bar{x}(t)$ is a minimum-action path.

For $n = 1$ and $(t_2 - t_1) = T/2$, we obtain, setting $t_1 = 0$, the Jacobi field

$$\delta^2 S[\psi_1^{(0)}] = 0 , \; \psi_1^{(0)}(t) = \sqrt{\frac{4}{T}} \sin\left(\frac{2\pi}{T} t \right) = \sqrt{\frac{2\omega}{\pi}} \sin(\omega t) , \; \lambda_1 \left(\frac{T}{2} \right) = 0 . \tag{4.57}$$

The conjugate points (caustics) follow from

$$\frac{n^2 \pi^2}{T^2} = \omega^2 : \; T_n = \frac{n\pi}{\omega} = n\frac{T}{2} , \quad n = 1, 2, \ldots , \tag{4.58}$$

i.e., at each half period we run through a focal point, i.e., as soon as $(t_2 - t_1) > T/2$ we do not have a minimum-action trajectory anymore.

We will close this chapter with another more intuitive derivation and interpretation of the Jacobi equation and the associated fields. For this reason, let us go back to our examples and think of the actual and the varied paths all leaving one and the same point, i.e., at $t_1 = 0$, if we consider the time development of the system. We label the emerging paths by their momenta $x(p, t)$ with $x(p, 0) = x_1$ for all p.

A measure for establishing how two neighboring paths deviate from one another as time goes on is given by the following derivative:

$$J(p, t) := \frac{\partial x(p, t)}{\partial p} . \tag{4.59}$$

Therefore, at time t, two neighboring paths are separated by the distance

$$x(p + \varepsilon, t) - x(p, t) = \varepsilon J(p, t) + O(\varepsilon^2) . \tag{4.60}$$

By definition, all of the trajectories $x(p, t)$ are extremum-action paths, which means that they satisfy the Euler–Lagrange equation

$$\frac{d}{dt}\left(\frac{\partial L}{\partial \dot{x}} \right) - \frac{\partial L}{\partial x} = 0 . \tag{4.61}$$

Let us differentiate this equation with respect to p. Then we need the following partial derivatives:

$$\frac{\partial}{\partial p} L(x(p,t), \dot{x}(p,t)) = \frac{\partial L}{\partial x} \frac{\partial x(p,t)}{\partial p} + \frac{\partial L}{\partial \dot{x}} \frac{\partial}{\partial p} \frac{d}{dt} x = \frac{\partial L}{\partial x} J + \frac{\partial L}{\partial \dot{x}} \dot{J} ,$$

$$\Rightarrow \frac{\partial}{\partial \dot{x}} \frac{\partial L}{\partial p} = \frac{\partial^2 L}{\partial x \partial \dot{x}} J + \frac{\partial^2 L}{\partial \dot{x}^2} \dot{J} ,$$

$$\frac{d}{dt}\left(\frac{\partial}{\partial \dot{x}} \frac{\partial L}{\partial p} \right) = \frac{d}{dt}\left(\frac{\partial^2 L}{\partial x \partial \dot{x}} \right) J + \frac{\partial^2 L}{\partial x \partial \dot{x}} \dot{J} + \frac{d}{dt}\left(\frac{\partial^2 L}{\partial \dot{x}^2} \dot{J} \right) , \tag{4.62}$$

$$\frac{\partial}{\partial x} \frac{\partial L}{\partial p} = \frac{\partial^2 L}{\partial x^2} J + \frac{\partial^2 L}{\partial x \partial \dot{x}} \dot{J} . \tag{4.63}$$

If we subtract (4.63) from (4.62) we find that J satisfies the equation

$$\frac{d}{dt}\left(\frac{\partial^2 L}{\partial \dot{x}^2} \dot{J} \right) + \left[\frac{d}{dt}\left(\frac{\partial^2 L}{\partial x \partial \dot{x}} \right) - \frac{\partial^2 L}{\partial x^2} \right] J(x,p) = 0 . \tag{4.64}$$

But this is precisely the Jacobi equation. Using $x(p,0) = \text{const.} = x_1$, we add to (4.64) the initial condition

$$J(p,0) = 0 . \tag{4.65}$$

For a simple standard Lagrangian we may assume that the initial velocity $\dot{x}(p,0)$ and p are related by $p = m\dot{x}(p,0)$. Then $\partial \dot{x}(p,0)/\partial p = 1/m$ implies

$$\frac{\partial J(p,0)}{\partial t} = \frac{1}{m} . \tag{4.66}$$

Hence, although (4.65) tells us that J begins with zero, the derivative is nonzero, however. $\varepsilon J(p,t)$ is a measure for the distance between two neighboring paths. They meet again at conjugate points, where for some T at $x(p,T) = x_2$,

$$J(p,T) = 0 . \tag{4.67}$$

The action computed along the trajectory between $(x_1, 0)$ and (x_2, T) is denoted by $S(x_2, T; x_1, 0)$. Now we found in (1.34) that this implies for the initial momentum:

$$p(t_1 = 0) \equiv p = -\frac{\partial S}{\partial x(t_1 = 0)} \equiv -\frac{\partial S}{\partial x_1} . \tag{4.68}$$

Differentiating (4.68) again with respect to x_2 yields

$$\frac{\partial p}{\partial x_2} = \frac{1}{J} = -\frac{\partial^2 S}{\partial x_1 \partial x_2} . \tag{4.69}$$

The discussion so far can be extended to $N(>1)$ dimensions. Then, similarly, it holds that J_{ik} and $\partial^2 S/\partial x_{1i} \partial x_{2k}$, $i, k = 1, 2, \ldots, N$ are inverse matrices. The determinant,

$$D := \det\left[-\left(\frac{\partial^2 S}{\partial x_1 \partial x_2} \right)_{ik} \right] = \det[(J^{-1})_{ik}] \tag{4.70}$$

is known as the Van-Vleck determinant and plays an important role in semiclassical approximations in quantum mechanics.

At focal points, D becomes infinite ($J = 0$). At this point many paths which left x_1 at $t_1 = 0$ have come together again at $x_2(T)$. Using the explicit form of the action for the harmonic oscillator

$$S = \frac{m\omega}{2\sin(\omega T)}\left[(x_1^2 + x_2^2)\cos(\omega T) - 2x_1x_2\right] \tag{4.71}$$

we find

$$\frac{\partial^2 S}{\partial x_1 \partial x_2} = -\frac{m\omega}{\sin(\omega T)}, \tag{4.72}$$

and this is infinite at each half-period, as stated already in (4.58).

An important theorem exists, relating the conjugate points along a classical trajectory to the negative eigenvalues of $\delta^2 S$. If we call the index of $\delta^2 S$ the number of eigenvalues λ_n with $\lambda_n < 0$, then the Morse Index theorem makes roughly the following statement: Let $x(t),\ 0 \leq t \leq T$, be an extremum action-path of S. Then the index of $\delta^2 S$ is equal to the number of conjugate points to $x(0)$ along the curve $x(t),\ 0 \leq t \leq T$. In fact, from our earlier examples, we can read off immediately that once a curve traverses a conjugate point, $\delta^2 S$ picks up a negative eigenvalue.

5. Canonical Transformations

Let $q_1, q_2, \dots, q_N, p_1 . p_2, \dots p_N$ be $2N$ independent canonical variables, which satisfy Hamilton's equations:

$$\dot{q}_i = \frac{\partial H}{\partial p_i} , \quad \dot{p}_i = -\frac{\partial H}{\partial q_i} , \quad i = 1, 2, \dots, N . \tag{5.1}$$

We now transform to a new set of $2N$ coordinates $Q_1, \dots Q_N, P_1, \dots P_N$, which can be expressed as functions of the old coordinates:

$$Q_i = Q_i(q_i, p_i; t) , \quad P_i = P_i(q_i, p_i; t) . \tag{5.2}$$

These transformations should be invertible. The new coordinates Q_i, P_i are then exactly canonical if a new Hamiltonian $K(Q, P, t)$ exists with

$$\dot{Q}_i = \frac{\partial K}{\partial P_i} , \quad \dot{P}_i = -\frac{\partial K}{\partial Q_i} . \tag{5.3}$$

Our goal in using the transformations (5.2) is to solve a given physical problem in the new coordinates more easily. Canonical transformations are problem-independent; i.e., (Q_i, P_i) is a set of canonical coordinates for all dynamical systems with the same number of degrees of freedom, e.g., for the two-dimensional oscillator and the two-dimensional Kepler problem. Strictly speaking, for fixed N, the topology of the phase space can still be different, e.g., $\mathbb{R}^{2N}$, $\mathbb{R}^n \times (S^1)^m$, $n + m = 2N$ etc.

Using a canonical transformation, it is occasionally possible to attain a particularly simple form for the new Hamiltonian, e.g.,

$$K(Q_i, P_i, t) = 0 , \tag{5.4}$$

leading to

$$\dot{Q}_i = \frac{\partial K}{\partial P_i} = 0 , \quad \dot{P}_i = -\frac{\partial K}{\partial Q_i} = 0 . \tag{5.5}$$

The solutions are

$$Q_i(t) = \text{const.} = P_i(t) . \tag{5.6}$$

This manner of solving the problem is called "reduction to initial values".

Another simple solution results if the Q_i are ignorable:

$$K = K(P_1, P_2, \dots, P_N) . \tag{5.7}$$

Then it follows from

$$\dot{P}_i = -\frac{\partial K}{\partial Q_i} = 0 \tag{5.8}$$

that $P_i = \text{const.}$ for all $i = 1, \dots, N$ and, thus

$$\dot{Q}_i = \frac{\partial K}{\partial P_i} = \text{const.} . \tag{5.9}$$

This means that Q is linear in time:

$$Q_i(t) = \beta_i t + \gamma_i \tag{5.10}$$

with constants β_i and γ_i. This kind of procedure is called "reduction to an equilibrium problem".

One has to be able to derive the Hamiltonian equations (3.3) from Hamilton's principle:

$$\delta \int_{t_1}^{t_2} dt[P_i \dot{Q}_i - K(Q_i, P_i, t)] = 0 . \tag{5.11}$$

The integrands in (5.11) and in

$$\delta \int_{t_1}^{t_2} dt[p_i \dot{q}_i - H(q_i, p_i, t)] = 0 \tag{5.12}$$

differ only by a total differential:

$$p_i \dot{q}_i - H = P_i \dot{Q}_i - K + \frac{dF}{dt} \tag{5.13}$$

with

$$\delta \int_{t_1}^{t_2} dt \frac{dF}{dt} = \delta F(t_2) - \delta F(t_1) = 0 . \tag{5.14}$$

F is called the generating function of the canonical transformation of $(q_i, p_i, t) \to (Q_i, P_i, t)$. There are four possibilities for a generating function:

$$F_1 = F_1(q, Q, t) \ ; \quad F_2 = F_2(q, P, t) \ ; \quad F_3 = F_3(p, Q, t) \ ; \quad F_4 = F_4(p, P, t) \ .$$

It follows from (5.13) that

$$p_i \dot{q}_i - H = P_i \dot{Q}_i - K + \frac{\partial F_1}{\partial t} + \frac{\partial F_1}{\partial q_i} \dot{q}_i + \frac{\partial F_1}{\partial Q_i} \dot{Q}_i . \tag{5.15}$$

which means that

$$p_i = \frac{\partial F_1}{\partial q_i} , \tag{5.16}$$

$$P_i = -\frac{\partial F_1}{\partial Q_i} , \tag{5.17}$$

$$K = H + \frac{\partial F_1}{\partial t} . \tag{5.18}$$

There are similar equations for $F_2 = F_2(q, P, t)$.

Using the relation $F_1 = F_2(q, P, t) - Q_i P_i$ it follows from (5.15) that

$$p_i \dot{q}_i - H = P_i \dot{Q}_i - K - \dot{Q}_i P_i - Q_i \dot{P}_i + \frac{\partial F_2}{\partial t} + \frac{\partial F_2}{\partial q_i} \dot{q}_i + \frac{\partial F_2}{\partial P_i} \dot{P}_i$$

with which we get

$$p_i = \frac{\partial F_2}{\partial q_i} , \tag{5.19}$$

$$Q_i = \frac{\partial F_2}{\partial P_i} , \tag{5.20}$$

$$K = H + \frac{\partial F_2}{\partial t} . \tag{5.21}$$

The other cases, F_3 and F_4 can be dealt with in the same manner.

Useful simple examples of generating functions are given by:

(a) $F_2 = q_i P_i$. (5.22)

From

$$p_i = \frac{\partial F_2}{\partial q_i} = P_i , \quad Q_i = \frac{\partial F_2}{\partial P_i} = q_i , \quad K = H ,$$

it is clear that (5.22) generates the identity transformation. The choice of this generating function is, however, not unique, since $F_3 = -Q_i p_i$ accomplishes the same:

$$q_i = -\frac{\partial F_3}{\partial p_i} = Q_i , \quad P_i = -\frac{\partial F_3}{\partial Q_i} = p_i , \quad K = H .$$

(b) Generating function of an exchange transformation: $F_1 = q_i Q_i$.

$$p_i = \frac{\partial F_1}{\partial q_i} = Q_i , \quad P_i = -\frac{\partial F_1}{\partial Q_i} = -q_i , \quad K = H .$$

Here, "coordinates" and "momenta" are exchanged. Again, the choice of $F_1 = q_i Q_i$ is not unique for the generation of an exchange transformation. This is also accomplished by $F_4 = p_i P_i$:

$$q_i = -\frac{\partial F_4}{\partial p_i} = -P_i \ , \quad Q_i = \frac{\partial F_4}{\partial P_i} = p_i \ , \quad K = H \ .$$

(c) Point transformation:

$$F_2 = f_i(q_1, \ldots, q_N, t) P_i \ , \tag{5.23}$$

$$Q_i = \frac{\partial F_2}{\partial P_i} = f_i(q_1, \ldots, q_N, t) \ . \tag{5.24}$$

This is the generating function of a canonical transformation that affects a change of the coordinates; e.g., of $(x, y) \equiv (q_1, q_2) \to (r, \varphi) \equiv (Q_1, Q_2)$ with

$$q_1 = Q_1 \cos Q_2 \ , \quad q_2 = Q_1 \sin Q_2 \ . \tag{5.25}$$

If we invert this transformation, we then get the form (5.24):

$$Q_1 = \sqrt{q_1^2 + q_2^2} \ , \quad Q_2 = \arctan \frac{q_2}{q_1} \ . \tag{5.26}$$

As generating function of this transformation we choose

$$F_2 = \sqrt{q_1^2 + q_2^2} P_1 + \arctan\left(\frac{q_2}{q_1}\right) P_2 \ . \tag{5.27}$$

With this F_2, we can reproduce (5.26) immediately, since

$$Q_1 = \frac{\partial F_2}{\partial P_1} = \sqrt{q_1^2 + q_2^2} \ , \quad Q_2 = \frac{\partial F_2}{\partial P_2} = \arctan\left(\frac{q_2}{q_1}\right) \ . \tag{5.28}$$

The momenta are then given by

$$\begin{aligned} p_1 &= \frac{\partial F_2}{\partial q_1} = \frac{q_1 P_1}{\sqrt{q_1^2 + q_2^2}} - \frac{(q_2 P_2 / q_1^2)}{1 + (q_2/q_1)^2} = \frac{q_1 P_1}{\sqrt{q_1^2 + q_2^2}} - \frac{q_2 P_2}{q_1^2 + q_2^2} \ , \\ p_2 &= \frac{\partial F_2}{\partial q_2} = \frac{q_2 P_1}{\sqrt{q_1^2 + q_2^2}} + \frac{q_1 P_2}{q_1^2 + q_2^2} \ . \end{aligned} \tag{5.29}$$

If we now express the q_i as functions of the Q_i by means of (4.25, 26), we get

$$p_1 = P_1 \cos Q_2 - \frac{\sin Q_2}{Q_1} P_2 \ , \quad p_2 = P_1 \sin Q_2 + \frac{\cos Q_2}{Q_1} P_2 \ . \tag{5.30}$$

But we can also solve (5.29) for P_i:

$$P_1 = \frac{q_1 p_1 + q_2 p_2}{\sqrt{q_1^2 + q_2^2}} \ , \quad P_2 = q_1 p_2 - q_2 p_1 \ , \tag{5.31}$$

or

$$p_r = p_x \cos\varphi + p_y \sin\varphi\ , \qquad p_\varphi = -p_x r \sin\varphi + p_y r \cos\varphi\ .$$

At this point we select a simple Hamiltonian, e.g.,

$$H = \frac{1}{2m}(p_1^2 + p_2^2)\ . \tag{5.32}$$

Then, in the old coordinates, it holds that:

$$\dot{q}_1 = \frac{\partial H}{\partial p_1} = \frac{p_1}{m}\ ,\ \dot{q}_2 = \frac{\partial H}{\partial p_2} = \frac{p_2}{m}\ ,\ \dot{p}_1 = -\frac{\partial H}{\partial q_1} = 0\ ,\ \dot{p}_2 = -\frac{\partial H}{\partial q_2} = 0\ . \tag{5.33}$$

The Hamiltonian in the new coordinates follows from (5.30):

$$K = H = \frac{1}{2m}\left(P_1^2 + \frac{1}{Q_1^2}P_2^2\right), \tag{5.34}$$

and the canonical equations read:

$$\dot{Q}_1 = \frac{\partial K}{\partial P_1} = \frac{P_1}{m}\ , \qquad \dot{Q}_2 = \frac{\partial K}{\partial P_2} = \frac{P_2}{mQ_1^2}\ ,$$

$$\dot{P}_1 = -\frac{\partial K}{\partial Q_1} = \frac{P_2^2}{mQ_1^3}\ , \quad \dot{P}_2 = -\frac{\partial K}{\partial Q_2} = 0\ ,$$

or, in the familiar form:

$$K = \frac{1}{2m}\left(p_r^2 + \frac{1}{r^2}\,p_\varphi^2\right)\ , \tag{5.35}$$

$$p_r \equiv P_1 = m\dot{r}\ , \quad p_\varphi \equiv P_2 = mr^2\dot{\varphi} = \text{const.}\ , \quad \dot{P}_1 = \frac{P_2^2}{mr^3} = m\ddot{r} = \dot{p}_r\ .$$

The change of Cartesian coordinates to spherical coordinates is not much more complicated: $(x, y, z) \equiv (q_1, q_2, q_3) \to (r, \varphi, \vartheta) \equiv (Q_1, Q_2, Q_3)$:

$$\begin{aligned} q_1 &= Q_1 \cos Q_2 \sin Q_3\ , \\ q_2 &= Q_1 \sin Q_2 \sin Q_3\ , \\ q_3 &= Q_1 \cos Q_3\ . \end{aligned} \tag{5.36}$$

Inversion of these equations yields

$$Q_1 = \sqrt{q_1^2 + q_2^2 + q_3^2}\ , \quad Q_2 = \arctan\frac{q_2}{q_1}\ , \quad Q_3 = \arctan\frac{\sqrt{q_1^2 + q_2^2}}{q_3}\ . \tag{5.37}$$

Again, it is convenient to choose the following expression for the generating function of this point transformation:

$$F_2(q_i, P_i) = \sqrt{q_1^2 + q_2^2 + q_3^2}\, P_1 + \arctan\left(\frac{q_2}{q_1}\right) P_2 + \arctan\left(\frac{\sqrt{q_1^2 + q_2^2}}{q_3}\right) P_3 . \tag{5.38}$$

Because $Q_i = \partial F_2/\partial P_i$, it is clear that (5.37) is reproduced.

Now we come to the calculation of the momenta p_i:

$$\begin{aligned}
p_1 &= \frac{\partial F_2}{\partial q_1} = \frac{q_1 P_1}{\sqrt{q_1^2 + q_2^2 + q_3^2}} - \frac{q_2 P_2}{q_1^2 + q_2^2} + \frac{q_1 q_3 P_3}{(q_1^2 + q_2^2 + q_3^2)\sqrt{q_1^2 + q_2^2}} , \\
p_2 &= \frac{\partial F_2}{\partial q_2} = \frac{q_2 P_1}{\sqrt{q_1^2 + q_2^2 + q_3^2}} + \frac{q_1 P_2}{q_1^2 + q_2^2} + \frac{q_1 q_3 P_3}{(q_1^2 + q_2^2 + q_3^2)\sqrt{q_1^2 + q_2^2}} , \\
p_3 &= \frac{\partial F_2}{\partial q_3} = \frac{q_3 P_1}{\sqrt{q_1^2 + q_2^2 + q_3^2}} - \frac{\sqrt{q_1^2 + q_2^2}}{q_1^2 + q_2^2 + q_3^2} P_3 .
\end{aligned} \tag{5.39}$$

The equations (5.39) can be rewritten with the expressions for Q_i from (5.37) and (5.37):

$$\begin{aligned}
p_1 &= P_1 \cos Q_2 \sin Q_3 - P_2 \frac{\sin Q_2}{Q_1 \sin Q_3} + P_3 \frac{\cos Q_2 \cos Q_3}{Q_1} , \\
p_2 &= P_1 \sin Q_2 \sin Q_3 + P_2 \frac{\cos Q_2}{Q_1 \sin Q_3} + P_3 \frac{\sin Q_2 \cos Q_3}{Q_1} , \\
p_3 &= P_1 \cos Q_3 - P_3 \frac{\sin Q_3}{Q_1} .
\end{aligned} \tag{5.40}$$

On the other hand, we also can invert (5.39), with the result:

$$\begin{aligned}
P_1 &= \frac{q_1 p_1 + q_2 p_2 + q_3 p_3}{\sqrt{q_1^2 + q_2^2 + q_3^2}} , \\
P_2 &= q_1 p_2 - q_2 p_1 , \\
P_3 &= \frac{q_1 q_3 p_1 + q_2 q_3 p_2 - (q_1^2 + q_2^2) p_3}{\sqrt{q_1^2 + q_2^2}} .
\end{aligned} \tag{5.41}$$

The Hamilton for a free particle,

$$H = \frac{1}{2m}(p_1^2 + p_2^2 + p_3^2)$$

with

$$\dot{q}_i = \frac{\partial H}{\partial p_i} = \frac{p_i}{m}\,, \qquad \dot{p}_i = -\frac{\partial H}{\partial q_i} = 0\,,$$

is transformed, taking (5.40) into account, into

$$K = H = \frac{1}{2m}\left(P_1^2 + \frac{1}{Q_1^2}P_3^2 + \frac{1}{Q_1^2\sin^2 Q_3}\,P_2^2\right), \tag{5.42}$$

or, the more familiar form:

$$K = \frac{1}{2m}\left(p_r^2 + \frac{1}{r^2}p_\vartheta^2 + \frac{1}{r^2\sin^2\vartheta}\,p_\varphi^2\right). \tag{5.43}$$

So, we have convincingly demonstrated that every point transformation is a canonical transformation.

Also useful is the generator of a canonical transformation of an inertial system (x, y, z) to a coordinate system (X, Y, Z) rotating around the $z(Z)$ axis. Let the angular velocity be $\boldsymbol{\omega}$. Then, it holds that, with $\tau := \omega t$,

$$X = x\cos\tau + y\sin\tau\,, \qquad Y = -x\sin\tau + y\cos\tau\,, \qquad Z = z\,. \tag{5.44}$$

The invariant is

$$r^2 = x^2 + y^2 = X^2 + Y^2 = R^2\,.$$

We should like to again attempt to get the transformation (5.44) with the help of the following generating function:

$$F_2 = (x\cos\tau + y\sin\tau)P_1 + (-x\sin\tau + y\cos\tau)P_2 + zP_3\,. \tag{5.45}$$

Of course we have

$$\frac{\partial F_2}{\partial P_i} = Q_i = (X, Y, Z)\,.$$

Note that zP_3 generates the identity $Z = z$. Now the calculation of the momenta is brought in:

$$\begin{aligned} p_x &= \frac{\partial F_2}{\partial x} = P_1\cos\tau - P_2\sin\tau\,,\\ p_y &= \frac{\partial F_2}{\partial y} = P_1\sin\tau + P_2\cos\tau\,,\\ p_z &= \frac{\partial F_2}{\partial z} = P_3\,. \end{aligned} \tag{5.46}$$

This system of equations can be solved for P_1 and P_2:

$$\begin{aligned} &P_1 = p_x\cos\tau + p_y\sin\tau\,, \qquad P_2 = -p_x\sin\tau + p_y\cos\tau\,,\\ &P_3 = p_z\,; \quad P_1^2 + P_2^2 + P_3^2 = p_1^2 + p_2^2 + p_3^2\,. \end{aligned}$$

One must not forget here that F_2 is time-dependent, so that ($\tau = \omega t$):

$$\begin{aligned}\frac{\partial F_2}{\partial t} &= \omega\,[(-x\sin\tau + y\cos\tau)P_1 + (-x\cos\tau - y\sin\tau)P_2] = \omega\,[YP_1 - XP_2] \\ &= -\omega L_3\,.\end{aligned}$$

So the new Hamiltonian reads:

$$K = H + \frac{\partial F_2}{\partial t} = H - \omega L_3 = \frac{1}{2m}(P_1^2 + P_2^2 + P_3^2) + V(R) - \omega L_3\,. \tag{5.47}$$

Here we have assumed that the original Hamiltonian had been given for a particle in the potential $V(r, z)$, which is axial symmetric:

$$H = \frac{1}{2m}p_i^2 + V(r, z)\,.$$

Since $\partial H/\partial t = 0$, H is a constant of motion. Furthermore, no torque acts around the z axis, so that $L_3 = \text{const}$. Hence it follows from this that K is also a constant of motion. K describes the time development relative to the moving system:

$$K = \frac{1}{2m}(P_1^2 + P_2^2 + P_3^2) + V(R) + \omega(YP_1 - XP_2)\,.$$

The corresponding canonical equations read:

$$\dot{X} = \frac{\partial K}{\partial P_1} = \frac{P_1}{m} + \omega Y\,, \quad \dot{Y} = \frac{\partial K}{\partial P_2} = \frac{P_2}{m} - \omega X\,, \quad \dot{Z} = \frac{P_3}{m}$$

or

$$P_1 = m(\dot{X} - \omega Y)\,, \quad P_2 = m(\dot{Y} + \omega X)\,, \quad P_3 = m\dot{Z}\,.$$

The other half of the Hamiltonian equations gives

$$\begin{aligned}\dot{P}_1 &= -\frac{\partial K}{\partial X} = \omega P_2 - \frac{\partial V}{\partial X} = m\omega\dot{Y} + m\omega^2 X - \frac{\partial V}{\partial X}\,, \\ \dot{P}_2 &= -\frac{\partial K}{\partial Y} = -\omega P_1 - \frac{\partial V}{\partial Y} = -m\omega\dot{X} + m\omega^2 Y - \frac{\partial V}{\partial Y}\,, \\ \dot{P}_3 &= -\frac{\partial K}{\partial Z} = -\frac{\partial V}{\partial Z}\,.\end{aligned}$$

If we here substitute $\dot{P}_1 = m(\ddot{X} - \omega\dot{Y})$, etc., we get for the equations of motion relative to the rotating coordinate system:

$$\begin{aligned}m\ddot{X} &= -\frac{\partial V}{\partial X} + 2m\omega\dot{Y} + m\omega^2 X\,, \\ m\ddot{Y} &= -\frac{\partial V}{\partial Y} - 2m\omega\dot{X} + m\omega^2 Y\,, \\ m\ddot{Z} &= -\frac{\partial V}{\partial Z}\,.\end{aligned}$$

Here the Coriolis and centrifugal forces appear relative to the rotating reference system – as is to be expected.

Now we consider $F_1 = (m/2)\omega q^2 \cot Q$ and look for the transformations $(q, p) \to (Q, P)$ which are generated by F_1. First we have

$$p = \frac{\partial F_1}{\partial q} = m\omega q \cot Q \ ,$$

$$P = -\frac{\partial F_1}{\partial Q} = \frac{m}{2}\omega q^2 \frac{1}{\sin^2 Q} \ .$$

This can also be written as

$$\cot^2 Q = \left(\frac{p}{m\omega q}\right)^2 , \tag{5.48}$$

$$\frac{1}{\sin^2 Q} = \frac{2P}{m\omega q^2} \ . \tag{5.49}$$

From this follows

$$\frac{\cos^2 Q}{\sin^2 Q} - \frac{1}{\sin^2 Q} = -1 = \frac{1}{m\omega q^2}\left(\frac{p^2}{m\omega} - 2P\right)$$

or

$$P = \frac{1}{2}\left(m\omega q^2 + \frac{p^2}{m\omega}\right). \tag{5.50}$$

Now let us rewrite (5.49) as

$$\frac{1}{\sin^2 Q} = \frac{1}{m\omega q^2}\left(m\omega q^2 + \frac{p^2}{m\omega}\right) = 1 + \frac{p^2}{(m\omega q)^2}$$

or

$$\frac{1}{\sin^2 Q} - 1 = (\tan^2 Q)^{-1} = \frac{p^2}{(m\omega q)^2} \ .$$

Solving for Q gives

$$Q = \arctan\left(\frac{m\omega q}{p}\right). \tag{5.51}$$

So we have from (5.49):

$$q = \sqrt{\frac{2P}{m\omega}} \sin Q \tag{5.52}$$

and (5.48) yields with q from (5.52)

$$\frac{\cos^2 Q}{\sin^2 Q} = \frac{p^2}{m^2\omega^2 q^2} = \frac{p^2 m\omega}{2m^2\omega^2 P \sin^2 Q}$$

or

$$p = \sqrt{2m\omega P}\cos Q\ . \tag{5.53}$$

At last we can rewrite H in the simple form

$$H = \frac{p^2}{2m} + \frac{m}{2}\omega^2 q^2 = \omega P \cos^2 Q + \omega P \sin^2 Q = \omega P = K(P)\ . \tag{5.54}$$

K is ignorable with respect to Q; therefore, P is a constant of motion:

$$P = \frac{E}{\omega}\ . \tag{5.55}$$

The canonical equations now simply read:

$$\dot{Q} = \frac{\partial K}{\partial P} = \omega$$

with the solution

$$Q(t) = \omega t + \alpha\ ,$$

and

$$\dot{P} = -\frac{\partial K}{\partial Q} = 0: \quad P = \text{const.} = \frac{E}{\omega}\ .$$

Finally, from (5.52), the solution for q follows:

$$q(t) = \sqrt{\frac{2E}{m\omega^2}}\sin(\omega t + \alpha)\ , \tag{5.56}$$

the usual solution for the harmonic oscillator.

The above choice of F_1 reduces our problem to an equilibrium problem. Contrary to this, the following F_2 reduces our problem to the initial conditions.

$$F_2(q, P) = -\frac{m\omega q^2}{2}\tan[\omega(t - P)]\ ,$$

$$p = \frac{\partial F_2}{\partial q} = -m\omega q \tan[\omega(t - P)]\ , \tag{5.57}$$

$$Q = \frac{\partial F_2}{\partial P} = \frac{m\omega^2 q^2}{2}\frac{1}{\cos^2[\omega(t - P)]} = -\frac{\partial F_2}{\partial t}\ . \tag{5.58}$$

$$K = H + \frac{\partial F_2}{\partial t} = H - \frac{m}{2}\omega^2 q^2 \frac{1}{\cos^2[\omega(t - P)]}\ .$$

It follows from (5.58) that

$$q(t) = \sqrt{\frac{2Q}{m\omega^2}} \cos[\omega(t-P)] , \tag{5.59}$$

and this, when inserted in (5.57), gives

$$p(t) = -\sqrt{2mQ}\, \sin[\omega(t-P)] . \tag{5.60}$$

Combining (5.59) and (5.60) yields

$$P(p,q) = \frac{1}{\omega} \arctan\left(\frac{p}{m\omega q}\right) + t .$$

Squaring (5.59) and (5.60) and adding gives

$$Q(p,q) = \frac{p^2}{2m} + \frac{m}{2}\omega^2 q^2 . \tag{5.61}$$

So we have found explicitly the canonical transformations which are generated by $F_2(q, P)$.

According to (5.61) we have

$$K = H - Q = \frac{p^2}{2m} + \frac{m}{2}\omega^2 q^2 - \left(\frac{p^2}{2m} + \frac{m}{2}\omega^2 q^2\right) = 0 .$$

Therefore the canonical equations are simply

$$\dot{Q} = \frac{\partial K}{\partial P} = 0 , \quad \dot{P} = -\frac{\partial K}{\partial Q} = 0 ,$$

with the solutions $Q, P =$ const. In a further example we look for the canonical transformation which is generated by the following F_2:

$$F_2(q_1, q_2, P_1, P_2) = \sqrt{2m(P_1 - P_2)}\, q_1 - \frac{2}{3}\sqrt{\frac{2}{m}} \frac{(P_2 - mgq_2)^{3/2}}{g} . \tag{5.62}$$

Furthermore, how does the new Hamiltonian read and what do the canonical equations look like in the new variables?

We begin with the set of equations

$$p_1 = \frac{\partial F_2}{\partial q_1} = \sqrt{2m(P_1 - P_2)} , \tag{5.63}$$

$$p_2 = \frac{\partial F_2}{\partial q_2} = \sqrt{2m(P_2 - mgq_2)} , \tag{5.64}$$

$$Q_1 = \frac{\partial F_2}{\partial P_1} = \frac{mq_1}{\sqrt{2m(P_1 - P_2)}} , \tag{5.65}$$

$$Q_2 = \frac{\partial F_2}{\partial P_2} = -\frac{mq_1}{\sqrt{2m(P_1 - P_2)}} - \sqrt{\frac{2}{m}} \frac{1}{g}(P_2 - mgq_2)^{1/2} . \tag{5.66}$$

From (5.64) we get

$$\frac{p_2^2}{2m} = P_2 - mgq_2$$

or

$$P_2 = \frac{p_2^2}{2m} + mgq_2 \,. \tag{5.67}$$

Using (5.67) in (5.63) results in

$$\frac{p_1^2}{2m} = P_1 - P_2 = P_1 - \frac{p_2^2}{2m} - mgq_2$$

or

$$P_1 = \frac{p_1^2 + p_2^2}{2m} + mgq_2 = H =: \alpha_1 = \text{const} \,. \tag{5.68}$$

The equations (5.65) and (5.66) lead to

$$Q_1 = m\frac{q_1}{p_1} \,; \quad q_1 = \frac{Q_1}{m}\sqrt{2m(P_1 - P_2)} \tag{5.69}$$

$$Q_2 = -m\frac{q_1}{p_1} - \frac{p_2}{mg} \,,$$

$$Q_1 + Q_2 = -\frac{p_2}{mg} \,; \quad p_2 = -mg(Q_1 + Q_2) \,.$$

Squaring the last equation and using (5.67) yields

$$(Q_1 + Q_2)^2 = \frac{2m}{m^2 g^2}(P_2 - mgq_2)$$

or

$$q_2 = \frac{P_2}{mg} - \frac{g}{2}(Q_1 + Q_2)^2 \,. \tag{5.70}$$

The new Hamiltonian is

$$K = H + \frac{\partial F_2}{\partial t} = H$$

or

$$K = \frac{p_1^2 + p_2^2}{2m} + mgq_2 = \frac{p_1^2}{2m} + \left(\frac{p_2^2}{2m} + mgq_2\right) = P_1 - P_2 + P_2 = P_1 \,.$$

With $K = P_1$ we can readily find the canonical equations:

$$\dot{Q}_1 = \frac{\partial K}{\partial P_1} = 1 \quad : \quad Q_1 = t + \beta_1 ,$$
$$\dot{Q}_2 = \frac{\partial K}{\partial P_2} = 0 \quad : \quad Q_2 = \beta_2 ,$$
$$\dot{P}_1 = -\frac{\partial K}{\partial Q_1} = 0 \quad : \quad P_1 = \text{const.} = \alpha_1 = H ,$$
$$\dot{P}_2 = -\frac{\partial K}{\partial Q_2} = 0 \quad : \quad P_2 = \text{const.} = \alpha_2 .$$

If we set $q_1 = x$, $q_2 = y$, then it follows from (5.69) and (5.70) that

$$x(t) = \frac{1}{m}(t + \beta_1)\sqrt{2m(\alpha_1 - \alpha_2)} = \text{const.}\, t + \text{const.}$$
$$y(t) = -\frac{g}{2}(t + \text{const.})^2 + \frac{\alpha_2}{mg} = -\frac{g}{2}t^2 + \text{const.}\, t + \text{const.}$$

The constants have to be fixed by the initial conditions.

In the usual calculation of the projectile motion in the x-y plane, one chooses at $t = 0 : x_0 = 0 = y_0$ and $\boldsymbol{v}_0 = (v_0 \cos\alpha, v_0 \sin\alpha)$ as the initial velocity. Our result is then given by

$$x = (v_0 \cos\alpha)t$$
$$y = (v_0 \sin\alpha)t - \frac{g}{2}t^2 .$$

The above procedure may seem like a very difficult way to solve an easy problem, and indeed it is. The following problem is along the same line; it concerns the damped harmonic oscillator. The equation of motion is given by

$$m\ddot{q} + b\dot{q} + kq = 0 , \tag{5.71}$$

where $F = -b\dot{q}$ denotes the frictional force and, as usual, $k = m\omega_0^2$.

The equation of motion (5.71) can be derived from a Lagrangian, which we define according to

$$L = e^{bt/m}\left(\frac{m}{2}\dot{q}^2 - \frac{k}{2}q^2\right) . \tag{5.72}$$

However, this choice of L is not unique! The canonical momentum is

$$p = \frac{\partial L}{\partial \dot{q}} = m\dot{q}e^{bt/m} . \tag{5.73}$$

So the Hamiltonian becomes

$$H = p\dot{q} - L = e^{-bt/m}\frac{p^2}{2m} + e^{bt/m}\frac{k}{2}q^2 . \tag{5.74}$$

This Hamiltonian is explicitly time-dependent and indicates the dissipation inherent in the system. The canonical equations belonging to (5.74) are:

$$\dot{q} = \frac{\partial H}{\partial q} = \frac{p}{m}\mathrm{e}^{-bt/m} \,, \quad \dot{p} = -\frac{\partial H}{\partial q} = -kq\mathrm{e}^{bt/m} \,. \tag{5.75}$$

With this we reproduce (5.71):

$$m\ddot{q} = \left(\dot{p} - \frac{bp}{m}\right)\mathrm{e}^{-bt/m} = -kq - b\dot{q} \,.$$

The form of (5.74) suggests that the following canonical transformation should simplify the Hamiltonian:

$$Q = q\mathrm{e}^{bt/2m} \,, \quad P = p\mathrm{e}^{-bt/2m} \,. \tag{5.76}$$

One can be easily convinced that the generating function of the canonical transformation (5.76) is given by

$$F_2(q, P, t) = \mathrm{e}^{bt/2m} qP \,. \tag{5.77}$$

Indeed:

$$p = \frac{\partial F_2}{\partial q} = \mathrm{e}^{bt/2m} P \,, \quad Q = \frac{\partial F_2}{\partial P} = \mathrm{e}^{bt/2m} q \,.$$

The new Hamiltonian therefore reads:

$$K(Q, P, t) = H(q, p, t) - \frac{\partial F_2}{\partial t} = \frac{P^2}{2m} + \frac{k}{2}Q^2 + \frac{b}{2m}QP \,. \tag{5.78}$$

Note that K does not explicitly depend on time and thus is conserved. If we then express (5.78) again as a function of the old canonical variables, we get:

$$\mathrm{e}^{-bt/m}\frac{p^2}{2m} + \mathrm{e}^{bt/m}\frac{k}{2}q^2 + \frac{b}{2m}pq \,, \tag{5.79}$$

and this expression is an integral of motion – a fact that would not have been seen so easily from the original form of H.

6. The Hamilton–Jacobi Equation

We already know that canonical transformations are useful for solving mechanical problems. We now want to look for a canonical transformation that transforms the $2N$ coordinates (q_i, p_i) to $2N$ constant values (Q_i, P_i), e.g., to the $2N$ initial values (q_i^0, p_i^0) at time $t = 0$. Then the problem would be solved, $q = q(q_0, p_0, t)$, $p = p(q_0, p_0, t)$.

We can now automatically make sure that the new variables are all constant by requiring that the new transformed Hamiltonian $K(Q, P, t)$ vanish identically, $K = 0$:

$$\dot{Q}_i = \frac{\partial K}{\partial P_i} = 0\,, \quad \dot{P}_i = -\frac{\partial K}{\partial Q_i} = 0\,. \tag{6.1}$$

Now, however, $0 = K = H + \partial F/\partial t$, and thus $H(q, p, t) + \partial F/\partial t = 0$ must be valid for F. At this point we choose F as a function which depends on the old coordinates q_i and the new constant momenta P_i, so that we are talking for a while about $F = F_2(q_i, P_i, t)$. If we add to this relation

$$p_i = \frac{\partial F_2}{\partial q_i} \tag{6.2}$$

then the differential equation for F_2 takes the form:

$$H\left(q_i, \frac{\partial F_2}{\partial q_i}; t\right) + \frac{\partial F_2}{\partial t} = 0\,. \tag{6.3}$$

This is the well-known Hamilton–Jacobi equation for finding $F_2(q_i, P_i, t)$ – the generating function of that canonical transformation which brings us to the constant values Q_i and P_i.

The Hamilton–Jacobi equation is a partial differential equation of first order in the $N + 1$ variables $(q_1, q_2, \ldots, q_N; t)$. Normally the solution to (6.3) is denoted by S: this is known as Hamilton's principal function.

Note that in (6.3) the derivatives of F appear, but not F itself. Thus, along with F_2, $F_2 + \text{const.}$ is also a solution. This additive constant can be arbitrarily chosen.

Now we want to assume the existence of a solution to (6.3):

$$F_2 \equiv S = S(q_1, q_2, \ldots, q_N; \alpha_1, \alpha_2, \ldots, \alpha_N, \alpha_{N+1}; t) \tag{6.4}$$

with $N + 1$ independent constants of integration $\alpha_1, \alpha_2, \dots, \alpha_N, \alpha_{N+1}$. Such a solution is called a complete solution. In contrast to this, general solutions also exist with arbitrary functions of the independent variables instead of constants. For our further considerations, it is only important that there be a complete solution (6.4) to (6.3).

The constant α_{N+1} in (6.4) plays a special role. We can call it an additive constant, as mentioned above, and, since only partial derivatives of the generating function S appear in the transformation equations, we can just omit it. Then we get, as a complete solution to (6.3):

$$S = S(q_1, \dots, q_N; \alpha_1, \dots, \alpha_N; t) , \tag{6.5}$$

where none of the constants α_i is additive. Since the α_i are now arbitrary, we can put the new (constant) P_i's in their place: $\alpha_i = P_i$. Then as transformation equation, we have:

$$p_i = \frac{\partial S(q, \alpha; t)}{\partial q_i} , \tag{6.6}$$

$$Q_i = \frac{\partial S(q, \alpha; t)}{\partial \alpha_i} = \text{const.} = \beta_i . \tag{6.7}$$

If we assume that (6.7) is invertible, then $q_i = q_i(\alpha, \beta, t)$ and thus, from (6.6), $p_i = p_i(\alpha, \beta, t)$. In principle our dynamical problem with given $H(q, p, t)$ is hereby solved.

Finally we wish to show that the letter S was not chosen purely by chance. There is indeed a relation between the action functional $\int L(t)dt$ and the generating function of the canonical transformation S which transforms on constant values (Q, P). This can be seen as follows:

$$\begin{aligned} \frac{dS(q_i, P_i = \alpha_i; t)}{dt} &= \sum_{i=1}^{N} \left(\frac{\partial S}{\partial q_i} \dot{q}_i + \frac{\partial S}{\partial \alpha_i} \underbrace{\dot{\alpha}_i}_{=0} \right) + \frac{\partial S}{\partial t} \\ &= \sum_{i=1}^{N} p_i \dot{q}_i - H = L , \end{aligned}$$

where we have used (6.3) and (6.6). So the principal function S is given by the time integral of the Lagrangian:

$$S = \int_{t_1}^{t_2} dt\, L . \tag{6.8}$$

We need, however, the complete solution of the problem $q_i(t)$, $\dot{q}_i(t)$ to calculate S. Exactly this way was chosen earlier when we calculated the classical action for the forced oscillator.

We should like to point out that a Hamilton–Jacobi equation also exists for generating functions of the $F_1(q, Q, t)$-type, which also reads

$$H\left(q_i, \frac{\partial F_1}{\partial q_i}; t\right) + \frac{\partial F_1}{\partial t} = 0 . \tag{6.9}$$

Only the transformation equations are changed:

$$p_i = \frac{\partial F_1(q, Q, t)}{\partial q_i} , \qquad P_i = -\frac{\partial F_1(q, Q, t)}{\partial Q_i} . \tag{6.10}$$

One generally prefers the Hamilton–Jacobi equation for $F_2 = S$, since, in particular, the identical transformation has a generator of the F_2- and not of the F_1-type.

As a first example of the application of a solution to the Hamilton–Jacobi equation, we consider a particle in a time-dependent potential $V(q, t) = -qFt$, where F is a constant. The Lagrangian then reads:

$$L = \frac{m}{2}\dot{q}^2 + qFt , \tag{6.11}$$

so that, with

$$\frac{\partial L}{\partial \dot{q}} = m\dot{q} = p , \qquad \frac{\partial L}{\partial q} = Ft ,$$

the equation of motion follows: $m\ddot{q} = Ft$.

The explicitly time-dependent Hamiltonian is, accordingly,

$$H = \frac{p^2}{2m} - qFt . \tag{6.12}$$

From this we get the Hamilton–Jacobi equation:

$$\frac{1}{2m}\left(\frac{\partial S}{\partial q}\right)^2 - qFt + \frac{\partial S}{\partial t} = 0 . \tag{6.13}$$

This partial differential equation for S can be solved using the following ansatz:

$$S = f(t)q + g(t) . \tag{6.14}$$

Then we have

$$\left(\frac{\partial S}{\partial q}\right)^2 = f^2 , \qquad \frac{\partial S}{\partial t} = \dot{f}q + \dot{g} .$$

Inserted in (6.13), this gives:

$$\left(\frac{1}{2m}f^2 + \dot{g}\right) + q(\dot{f} - Ft) = 0 .$$

So the following equations must be integrated:

$$\dot{f} = Ft \,, \quad \dot{g} = -\frac{1}{2m} f^2 \,.$$

We immediately get the solutions

$$f(t) = \frac{F}{2} t^2 + \alpha$$

and with

$$\dot{g} = -\frac{1}{2m}\left(\frac{F}{2} t^2 + \alpha\right)^2 = -\frac{1}{8m} F^2 t^4 - \frac{F}{2m}\alpha t^2 - \frac{1}{2m}\alpha^2$$

it follows that

$$g(t) = -\frac{1}{40m} F^2 t^5 - \frac{F}{6m}\alpha t^3 - \frac{\alpha^2}{2m} t + t_1 \,.$$

t_1 plays the role here of an additive constant α_{N+1} and can be omitted. So far we have found:

$$S(q, \alpha; t) = \left(\frac{F}{2} t^2 + \alpha\right) q - \left(\frac{1}{40m} F^2 t^5 + \frac{F}{6m}\alpha t^3 + \frac{\alpha^2}{2m} t\right). \tag{6.15}$$

For the transformation equations we get with this S

$$\beta = Q = \frac{\partial S}{\partial \alpha} = \frac{\partial f}{\partial \alpha} q + \frac{\partial g}{\partial \alpha} = q - \frac{F}{6m} t^3 - \frac{\alpha}{m} t \,. \tag{6.16}$$

Solved for q, this yields

$$q(t) = \frac{F}{6m} t^3 + \frac{\alpha}{m} t + \beta \,. \tag{6.17}$$

$\beta = 0$ follows from the initial condition $q(0) = 0$, so that

$$q(t) = \frac{F}{6m} t^3 + \frac{\alpha}{m} t \,. \tag{6.18}$$

From this we have $\dot{q}(0) = \alpha/m$, so $\alpha = p(0) = P$. This is in accord with

$$p(t) = \frac{\partial S}{\partial q} = \frac{F}{2} t^2 + \alpha \,. \tag{6.19}$$

If H is not explicitly time-dependent, then

$$H\left(q_i, \frac{\partial S}{\partial q_i}\right) + \frac{\partial S}{\partial t} = 0 \,. \tag{6.20}$$

We can separate off the time variable with the following ansatz:

$$S(q_i, \alpha_i; t) = W(q_i, \alpha_i) - \alpha_1 t \ . \tag{6.21}$$

If we substitute this ansatz in (6.20), then

$$H\left(q_i, \frac{\partial W}{\partial q_i}\right) = \alpha_1 \ . \tag{6.22}$$

constants in S, α_1, is thus equal to the constant value $H(= E)$. W is known as Hamilton's characteristic function.

We now show that W is the generator of a canonical transformation in which the new momenta are constants $\alpha_i = P_i$ (or: the Q_i are ignorable), and, in particular, that α_1 is identical to the conserved quantity H. If, with foresight, we denote the generator of this canonical transformation $W(q, P)$, then the following transformation equations are valid:

$$p_i = \frac{\partial W}{\partial q_i} \ , \qquad Q_i = \frac{\partial W}{\partial P_i} = \frac{\partial W}{\partial \alpha_i} \ . \tag{6.23}$$

In order to determine W, we require that the following should hold for the conserved quantity H:

$$H(q_i, p_i) = \alpha_1 = P_1 \ .$$

This requirement yields, via (6.23), a partial differential equation for W:

$$H\left(q_i, \frac{\partial W}{\partial q_i}\right) = \alpha_1 \ ,$$

which is identical with (6.22). Furthermore, since W is time-independent,

$$K = H + \frac{\partial W}{\partial t} = H = \alpha_1 \tag{6.24}$$

is valid for the new Hamiltonian. With this new $K = \alpha_1$, the canonical equations follow:

$$\dot{P}_i = -\frac{\partial K}{\partial Q_i} = 0 \ . \tag{6.25}$$

Thus $P_i = \alpha_i =$ const., as required, and

$$\dot{Q}_i = \frac{\partial K}{\partial P_i} = \frac{\partial K}{\partial \alpha_i} = \delta_{1i} = \begin{cases} 1 \ , i = 1 \ , \\ 0 \ , i \neq 1 \ . \end{cases} \tag{6.26}$$

The solutions are simply

$$Q_1 = t + \beta_1 \equiv \frac{\partial W}{\partial \alpha_1} \tag{6.27}$$

$$Q_i = \beta_i \equiv \frac{\partial W}{\partial \alpha_i} \ , \qquad i \neq 1 \ . \tag{6.28}$$

Only (6.27) contains the time. Equations (6.28), which contain no time, can be used to determine the space trajectory (orbit).

One need not identify α_1 with H and the other integration constants with the new constant P_i's. The N constants P_i can also be linear combinations of the α_i: $P_i = P_i(\alpha_1, \dots, \alpha_N)$, $i = 1, 2, \dots, N$; for example, $P_1 = \alpha_2 + \alpha_2$, $P_2 = \alpha_1 - \alpha_2$. Then it holds that

$$\dot{Q}_i = \frac{\partial K}{\partial P_i} = \sum_j \overbrace{\frac{\partial K}{\partial \alpha_j}}^{\delta_{1j}} \frac{\partial \alpha_j}{\partial P_i} = \frac{\partial \alpha_1}{\partial P_i} =: \nu_i \tag{6.29}$$

with

$$Q_i = \nu_i t + \beta_i \tag{6.30}$$

and

$$K = \alpha_1(P_1, \dots, P_N) \ . \tag{6.31}$$

Hereby W is shown to be the generator of a canonical transformation in which the new Hamiltonian depends only on the constant new momenta. The new Q_i are ignorable and move linearly in time.

The characteristic function W has the following physical significance:

$$\frac{dW}{dt} = \sum_i \left(\frac{\partial W}{\partial q_i} \dot{q}_i + \frac{\partial W}{\partial \alpha_i} \underbrace{\dot{\alpha}_i}_{=0} \right) = \sum_i p_i \dot{q}_i \ . \tag{6.32}$$

We designated the time integral of the right-hand side in Chap. 1 as action. It should be recalled that, contrary to (6.32), the equation $dS/dt = L$ is valid for S.

We now want to show how to solve the Hamilton–Jacobi equation and supply some examples. The method of separation of the variables is of prime importance here. If, as had been discussed above, $\partial H/\partial t = 0$, we separate off the time dependence according to

$$S = W - \alpha_1 t \ . \tag{6.33}$$

Note that $\alpha_1 = H = E$ applies for conservative systems, so that the variable $-H$ canonically conjugate to t appears as a factor next to t in (6.33). Let us assume, likewise, that for a given k, $\partial H/\partial q_k = 0$, i.e., q_k is ignorable; then, as in (6.33), we write:

$$S = (\text{const.})\, q_k + S'(q_1, \dots, q_{k-1}, q_{k+1}, \dots, q_N; t) \ . \tag{6.34}$$

The constant next to the ignorable coordinate q_k results from $\partial S/\partial q_k = p_k = \text{const.}$ Then

$$S = p_k q_k + S' \tag{6.35}$$

or

$$W = p_k q_k + W' \ .$$

If all q_i (and t) are ignorable except for q_k, then we obtain:

$$W = \sum_{i \neq k} p_i q_i + W_k(q_k) \ . \tag{6.36}$$

Here W_k is the solution of the reduced Hamilton–Jacobi equation

$$H\left(q_k, \frac{\partial W}{\partial q_k}; \alpha_1, \dots, \alpha_{k-1}, \alpha_{k+1}, \dots, \alpha_N\right) = \alpha_1 \ . \tag{6.37}$$

This is a normal first-order differential equation in the variable q_k and can be immediately reduced to quadratures.

A dynamical problem is solvable if it is completely separable. There are unfortunately no general rules which indicate when a system is separable. A system can be separable in one coordinate system and not in another. Thus we need a cleverly chosen coordinate system. Furthermore, even if certain coordinates are not ignorable, the Hamilton–Jacobi equation can nevertheless be separable. Ignorability of coordinates is therefore not a necessary but sufficient condition for separability. Moreover, if a system is separable in more than one coordinate system, then we are necessarily dealing with a degenerate system (e.g. the Kepler problem).

Since many examples for calculating S (or W) can be found in pertinent textbooks, we shall limit ourselves in the following to the computation of three cases.

As a first example we consider a particle in the gravitational field with the Hamiltonian

$$H = \frac{1}{2m}(p_x^2 + p_y^2) + mgy \ . \tag{6.38}$$

Since we are dealing with a conservative system, we have $H = \text{const.} = E = \alpha_1$. The Hamilton–Jacobi equation associated with (6.38) is

$$\frac{1}{2m}\left[\left(\frac{\partial W}{\partial x}\right)^2 + \left(\frac{\partial W}{\partial y}\right)^2\right] + mgy = \alpha_1 = E \ . \tag{6.39}$$

We recall that W is the generator of a canonical transformation to new constant momenta: $W = W(x, y; P_x, P_y)$ with $P_x = \alpha_1 = E$, $P_y = \alpha_2$.

The separation ansatz

$$W = W_x(x, E, \alpha_2) + W_y(y, E, \alpha_2) \tag{6.40}$$

makes (6.39) become:

$$\frac{1}{2m}\left(\frac{dW_x}{dx}\right)^2 + \frac{1}{2m}\left(\frac{dW_y}{dy}\right)^2 + mgy = \alpha_1 \tag{6.41}$$

or

$$\frac{1}{2m}\left(\frac{dW_y}{dy}\right)^2 + mgy = \alpha_1 - \frac{1}{2m}\left(\frac{dW_x}{dx}\right)^2 . \tag{6.42}$$

Since the right- and left-hand sides of (6.42) are functions of different variables, e.g., y and x, they have to be equal to a constant, α_2. Then

$$dW_x = \sqrt{2m(\alpha_1 - \alpha_2)}\, dx$$

or

$$W_x = \sqrt{2m(\alpha_1 - \alpha_2)}\, x = p_x x$$

since x is an ignorable variable. Furthermore we get

$$dW_y = \sqrt{2m(\alpha_2 - mgy)}\, dy ,$$

which, when integrated, yields the following expression:

$$W_y = -\frac{2}{3}\frac{\sqrt{2m}}{mg}\,(\alpha_2 - mgy)^{3/2} .$$

So for the entire characteristic function we have:

$$W(x, y; \alpha_1, \alpha_2) = \sqrt{2m(\alpha_1 - \alpha_2)}\, x - \frac{2}{3}\sqrt{\frac{2}{m}}\frac{1}{g}(\alpha_2 - mgy)^{3/2} . \tag{6.43}$$

With this we can write

$$Q_1 = \frac{\partial W}{\partial \alpha_1} = \frac{m}{\sqrt{2m(\alpha_1 - \alpha_2)}}\, x$$

or, solved for x:

$$x = \sqrt{\frac{2}{m}}\sqrt{\alpha_1 - \alpha_2}\, Q_1 = \frac{Q_1}{m}\sqrt{2m(P_1 - P_2)} . \tag{6.44}$$

This should be compared with the result (4.69). Similarly, we find:

$$\begin{aligned} Q_2 = \frac{\partial W}{\partial \alpha_2} &= -\frac{m}{\sqrt{2m(\alpha_1 - \alpha_2)}}\, x - \frac{1}{g}\sqrt{\frac{2}{m}}\sqrt{\alpha_2 - mgy} \\ &= -Q_1 - \frac{1}{g}\sqrt{\frac{2}{m}}\sqrt{\alpha_2 - mgy} \end{aligned}$$

or

$$Q_1 + Q_2 = -\frac{1}{g}\sqrt{\frac{2}{m}(\alpha_2 - mgy)} . \tag{6.45}$$

By squaring, we can solve this equation for y:

$$y = -\frac{g}{2}(Q_1 + Q_2)^2 + \frac{\alpha_2}{mg} = \frac{P_2}{mg} - \frac{g}{2}(Q_1 + Q_2)^2 = y(x, Q_2; \alpha_2) \,. \tag{6.46}$$

This corresponds exactly to the result (4.70).

We now recall the canonical equations with $K = H = E = \alpha_1 = P_1$:

$$\dot{Q}_1 = \frac{\partial K}{\partial \alpha_1} = 1 \quad : \quad Q_1 = t + \beta_1 \,,$$
$$\dot{Q}_2 = \frac{\partial K}{\partial \alpha_2} = 0 \quad : \quad Q_2 = \beta_2 \,.$$

In addition, we have the equations:

$$\dot{\alpha}_1 = \dot{P}_1 = -\frac{\partial K}{\partial Q_1} = 0 \,,$$
$$\dot{\alpha}_2 = \dot{P}_2 = -\frac{\partial K}{\partial Q_2} = 0 \,,$$

as is to be expected. Now

$$Q_1 = t + \beta_1 = \frac{m}{\sqrt{2m(E - \alpha_2)}} x$$

or

$$x(t) = \frac{1}{m}\sqrt{2m(E - \alpha_2)}(t + \beta_1) \,.$$

For the initial conditions, $x(t = 0) = 0$, it holds that $\beta_1 = 0$, so that

$$x(t) = \frac{1}{m}\sqrt{2m(E - \alpha_2)}\, t = \dot{x}(0)t = v_0 \cos(\alpha)t \tag{6.47}$$

since

$$\dot{x}(0) = \frac{1}{m}\sqrt{2m(E - \alpha_2)} \,. \tag{6.48}$$

Squaring (6.48) gives $\alpha_2 = E - m\dot{x}^2(0)/2 = mgy(0) + (m/2)\dot{y}^2(0)$. With the initial condition $y(0) = 0$ we therefore conclude:

$$\alpha_2 = \frac{m}{2}\dot{y}^2(0) = P_2 \,.$$

Now we insert this expression for α_2 in (6.45), $\beta_1 = 0$:

$$t + \beta_2 = -\frac{1}{g}\sqrt{\frac{2}{m}\left(\frac{m}{2}\dot{y}(0)^2 - mgy\right)} \,.$$

Using $y(0) = 0$, it follows from the last equation that $\beta_2 = -\dot{y}(0)/g$. Therefore

$$t - \frac{1}{g}\dot{y}(0) = -\frac{1}{g}\sqrt{\frac{2}{m}\left(\frac{m}{2}\dot{y}^2(0) - mgy\right)} .$$

Squaring and solving for y finally yields

$$y(t) = -\frac{g}{2}t^2 + \dot{y}(0)t = -\frac{g}{2}t^2 + v_0 \sin(\alpha)t . \tag{6.49}$$

Together with (6.47), these are the familiar kinematic equations for a particle in presence of a gravitational field.

As our next example we consider the damped harmonic oscillator. Here, we solve the equation of motion for the new Hamiltonian $K(Q, P)$ from (5.78) with the help of the Hamilton–Jacobi equation:

$$\frac{1}{2m}\left(\frac{\partial S}{\partial Q}\right)^2 + \frac{k}{2}Q^2 + \frac{b}{2m}Q\frac{\partial S}{\partial Q} + \frac{\partial S}{\partial t} = 0 . \tag{6.50}$$

Since K is not explicitly time-dependent, i.e., $K = \alpha_1 =$ const., we again write $S = W - \alpha_1 t$ and get for W:

$$\frac{1}{2m}\left(\frac{\partial W}{\partial Q}\right)^2 + \frac{k}{2}Q^2 + \frac{b}{2m}Q\frac{\partial W}{\partial Q} = \alpha_1 . \tag{6.51}$$

This is a quadratic equation in $\partial W/\partial Q$ and has, as its solution,

$$\frac{\partial W}{\partial Q} = -\frac{b}{2}Q \pm \frac{1}{2}\sqrt{(b^2 - 4mk)Q^2 + 8m\alpha_1} .$$

Integration then gives the Hamiltonian characteristic function:

$$W(Q, \alpha_1) = -\frac{b}{4}Q^2 \pm \frac{1}{2}\int dQ\sqrt{(b^2 - 4mk)Q^2 + 8m\alpha_1} .$$

We need the derivative

$$\frac{\partial W}{\partial \alpha_1} = t + \beta_1 = \pm 2m \int dQ \frac{1}{\sqrt{(b^2 - 4mk)Q^2 + 8m\alpha_1}} .$$

The integral is elementary and yields

$$t + \beta_1 = \pm\frac{2m}{\sqrt{4mk - b^2}} \text{arc sin}\left[\sqrt{\frac{4mk - b^2}{8m\alpha_1}}Q\right] .$$

Since we want to assume that $4mk > b^2$, then, with $\omega_0^2 := k/m - (b/2m)^2$:

$$t + \beta_1 = \pm\frac{1}{\omega_0} \text{arc sin}\left(\sqrt{\frac{m}{2\alpha_1}}\,\omega_0 Q\right)$$

or

$$Q(t) = \pm\frac{1}{\omega_0}\sqrt{\frac{2\alpha_1}{m}}\,\sin[\omega_0(t+\beta_1)]\,. \tag{6.52}$$

The $\pm$ sign is unimportant here. So we finally have as a solution for the original variable:

$$q(t) = Q(t)\mathrm{e}^{-bt/2m} = \frac{1}{\omega_0}\sqrt{\frac{2\alpha_1}{m}}\,\mathrm{e}^{-bt/2m}\,\sin[\omega_0(t+\beta_1)]\,. \tag{6.53}$$

The constants α_1 and β_1 still have to be determined by the initial conditions.

Our last example concerns the Coulomb problem with an in z-direction. It is certainly true that the $1/r$-Coulomb problem is spherical symmetric. However, in presence of a constant F-field in z-direction, it is more useful to employ a parabolic coordinate system which distinguishes a certain direction; here, the z-direction. The potential is given by

$$V = \frac{\lambda}{r} - Fz\,, \tag{6.54}$$

and we are now going to separate the associated Hamilton–Jacobi equation in parabolic coordinates (ξ, η, φ). These are related to the cartesian coordinates (x, y, z) in the following way:

$$\begin{aligned}
x &= \sqrt{\xi\eta}\,\cos\varphi\\
y &= \sqrt{\xi\eta}\,\sin\varphi\\
z &= \tfrac{1}{2}(\xi-\eta)\,, \quad \varrho \equiv (x^2+y^2)^{1/2} = \sqrt{\xi\eta}\,.
\end{aligned} \tag{6.55}$$

The invariant r^2 is then given by

$$r^2 = x^2+y^2+z^2 = \varrho^2+z^2 = \xi\eta + \tfrac{1}{4}(\xi^2+\eta^2-2\xi\eta) = \tfrac{1}{4}(\xi+\eta)^2$$

or

$$r = \tfrac{1}{2}(\xi+\eta)\,. \tag{6.56}$$

The kinematics is contained in

$$T = \frac{m}{2}(\dot\varrho^2 + \varrho^2\dot\varphi^2 + \dot z^2)\,. \tag{6.57}$$

So we need:

$$\varrho^2 = \xi\eta\,, \quad 2\varrho\dot\varrho = \dot\xi\eta + \xi\dot\eta\,, \quad \dot\varrho^2 = \frac{1}{4\varrho^2}(\dot\xi\eta+\xi\dot\eta)^2$$

or

$$\dot{\varrho}^2 = \frac{1}{4\xi\eta}(\dot{\xi}^2\eta^2 + \xi^2\dot{\eta}^2 + 2\dot{\xi}\eta\xi\dot{\eta}) = \frac{1}{4}\left(\dot{\xi}^2\frac{\eta}{\xi} + \dot{\eta}^2\frac{\xi}{\eta} + 2\dot{\xi}\dot{\eta}\right),$$

$$\varrho^2\dot{\varphi}^2 + \dot{z}^2 = \xi\eta\dot{\varphi}^2 + \tfrac{1}{4}(\dot{\xi} - \dot{\eta})^2 = \xi\eta\dot{\varphi}^2 + \tfrac{1}{4}(\dot{\xi}^2 + \dot{\eta}^2 - 2\dot{\xi}\dot{\eta}).$$

The kinetic energy can then be expressed in the form

$$\begin{aligned} T &= \frac{m}{2}\left(\frac{1}{4}\dot{\xi}^2\frac{\eta}{\xi} + \frac{1}{4}\dot{\eta}^2\frac{\xi}{\eta} + \cancel{\frac{1}{2}\dot{\xi}\dot{\eta}} + \xi\eta\dot{\varphi}^2 + \frac{1}{4}\dot{\xi}^2 + \frac{1}{4}\dot{\eta}^2 - \cancel{\frac{1}{2}\dot{\xi}\dot{\eta}}\right) \\ &= \frac{m}{2}\frac{1}{4}(\xi + \eta)\left(\frac{\dot{\xi}^2}{\xi} + \frac{\dot{\eta}^2}{\eta}\right) + \frac{m}{2}\xi\eta\dot{\varphi}^2 . \end{aligned}$$

Therefore the Lagrangian reads, in parabolic coordinates:

$$L = \frac{m}{8}(\xi + \eta)\left(\frac{\dot{\xi}^2}{\xi} + \frac{\dot{\eta}^2}{\eta}\right) + \frac{m}{2}\xi\eta\dot{\varphi}^2 - \frac{2\lambda}{\xi + \eta} + \frac{F}{2}(\xi - \eta) . \tag{6.58}$$

From (6.58) we obtain the canonical momenta:

$$\begin{aligned} p_\xi &= \frac{\partial L}{\partial \dot{\xi}} = \frac{m}{4}(\xi + \eta)\frac{\dot{\xi}}{\xi} , \\ p_\eta &= \frac{\partial L}{\partial \dot{\eta}} = \frac{m}{4}(\xi + \eta)\frac{\dot{\eta}}{\eta} , \\ p_\varphi &= \frac{\partial L}{\partial \dot{\varphi}} = m\xi\eta\dot{\varphi} . \end{aligned} \tag{6.59}$$

The equations (6.59) are needed to build the Hamiltonian:

$$\begin{aligned} H &= p_\xi\dot{\xi} + p_\eta\dot{\eta} + p_\varphi\dot{\varphi} - L \\ &= \frac{2}{m}\frac{\xi p_\xi^2 + \eta p_\eta^2}{\xi + \eta} + \frac{p_\varphi^2}{2m\xi\eta} + \frac{2\lambda}{\xi + \eta} - \frac{F}{2}(\xi - \eta) . \end{aligned} \tag{6.60}$$

Since $\partial H/\partial t = 0 = \partial H/\partial\varphi$ and therefore $H = E = P_1 = \alpha_1 = \text{const.}$ and likewise,

$$p_\varphi = \frac{\partial W}{\partial \varphi} = \frac{\partial S}{\partial \varphi} = P_2 = \alpha_2 = \text{const.}$$

we can write

$$S = W - \alpha_1 t = W_\xi(\xi) + W_\eta(\eta) + p_\varphi\varphi - Et , \tag{6.61}$$

and thus obtain for Hamilton's characteristic function W the time-independent partial differential equation

$$\begin{aligned} &\frac{2}{m(\xi + \eta)}\left[\xi\left(\frac{dW_\xi(\xi)}{d\xi}\right)^2 + \eta\left(\frac{dW_\eta(\eta)}{d\eta}\right)^2\right] + \frac{p_\varphi^2}{2m\xi\eta} + \frac{2\lambda}{\xi + \eta} \\ &\quad - \frac{F}{2}(\xi - \eta) = \alpha_1 = E . \end{aligned}$$

Multiplying both sides by $m(\xi + \eta)$, we get

$$2\xi \left(\frac{dW_\xi}{d\xi}\right)^2 + \frac{p_\varphi^2}{2\xi} + 2m\lambda - \frac{mF}{2}\xi^2 - mE\xi$$
$$= -2\eta \left(\frac{dW_\eta}{d\eta}\right)^2 - \frac{p_\varphi^2}{2\eta} - \frac{mF}{2}\eta^2 + mE\eta \,. \tag{6.62}$$

The left-hand side of (6.62) depends only on ξ while the right-hand side is only η-dependent. Hence, in an obvious notation we set $[\cdot]_\xi = -\alpha_3 = -[\cdot/.]_\eta$. Now we take the left-hand side of (6.62) and solve for $dW_\xi/d\xi$; similarly for $[\cdot/.]_\eta$. Then we have the result:

$$S(\xi, \eta, \varphi, t; E, p_\varphi, \alpha_3) = \int^\xi d\xi' \sqrt{\frac{mE}{2} + \frac{\alpha_3 - 2m\lambda}{2\xi'} + \frac{mF}{4}\xi' - \frac{p_\varphi^2}{4\xi'^2}}$$
$$+ \int^\eta d\eta' \sqrt{\frac{mE}{2} - \frac{\alpha_3}{2\eta'} - \frac{mF}{4}\eta' - \frac{p_\varphi^2}{4\eta'^2}}$$
$$+ p_\varphi \varphi - Et \,. \tag{6.63}$$

If the calculation were to be continued, we would have

$$Q_1 = \frac{\partial W}{\partial P_1} = \frac{\partial W}{\partial E}\,, \quad Q_2 = \frac{\partial W}{\partial P_2} = \frac{\partial W}{\partial p_\varphi}\,, \quad Q_3 = \frac{\partial W}{\partial P_3} = \frac{\partial W}{\partial \alpha_3}\,. \tag{6.64}$$

The new Hamiltonian would depend only on the constants $(P_1, P_2, P_3) = (E, p_\varphi, \alpha_3)$ and $\dot{Q}_i = \nu_i = \text{const.}$ or $Q_i = \nu_i t + \beta_i$ with six further constants ν_i, β_i.

In the middle of this chapter [equations (6.11) – (6.19)], we discussed a rather elementary problem with external time dependence. We now consider a slightly more elaborate problem, namely, a harmonically bound particle driven by an arbitrary time-dependent prescribed force as expresssed by the Lagrangian

$$L = \frac{m}{2}\dot{q}^2 - \frac{m}{2}\omega^2 q^2 + qF(t) \tag{6.65}$$

or the Hamiltonian

$$H = \frac{p^2}{2m} + \frac{m}{2}\omega^2 q^2 - qF(t). \tag{6.66}$$

Here, the real-valued "source" $F(t)$ is coupled to the dynamical variable $q(t)$. Obviously, equation (6.66) could be augmented by an additional source coupling $pG(t)$, so as to write (6.66) symetrically. For our purposes, equation (6.66) will suffice. Associated with this equation is the Hamiltonian–Jacobi equation

$$\frac{1}{2m}\left(\frac{\partial S}{\partial q}\right)^2 + \frac{m}{2}\omega^2 q^2 - qF(t) + \frac{\partial S}{\partial t} = 0, \tag{6.67}$$

which we solve with the ansatz (we put $m = 1$)

$$S(q,t) = a(t)q^2 + b(t)q + c(t). \tag{6.68}$$

In equation (6.67) we need

$$\left(\frac{\partial S}{\partial q}\right)^2 = 4a^2q^2 + 4abq + b^2 \tag{6.69}$$

and

$$\frac{\partial S}{\partial t} = \dot{a}q^2 + \dot{b}q + \dot{c}. \tag{6.70}$$

We then obtain a set of equations for a, b and c:

$$\dot{a} + 2a^2 + \frac{1}{2}\omega^2 = 0, \tag{6.71}$$

$$\dot{b} + 2ab - F(t) = 0, \tag{6.72}$$

$$\dot{c} + \frac{b^2}{2} = 0. \tag{6.73}$$

Equation (6.71) can easily be solved with the result:

$$a(t) = -\frac{\omega}{2}\tan\omega(t - t_0). \tag{6.74}$$

Here, t_0 is our first constant of integration, $a(t_0) = 0$.
The solution for (6.72) is also standard and yields

$$\begin{aligned} b(t) &= \mathrm{e}^{-\int^t dt' 2a(t')} \int_\alpha^t \mathrm{e}^{\int^{t'} dt'' 2a(t'')} F(t')dt' \\ &= \frac{1}{\cos\omega(t - t_0)} \int_\alpha^t dt' F(t') \cos\omega(t' - t_0). \end{aligned} \tag{6.75}$$

Here, we meet another constant of integration, α, with the property $b(\alpha) = 0$.
Finally, equation (6.73) is solved by

$$c(t) = -\frac{1}{2}\int_{t_1}^t \frac{dt'}{\cos^2\omega(t' - t_0)} \left[\int_\alpha^{t'} dt'' F(t'') \cos\omega(t'' - t_0)\right]^2. \tag{6.76}$$

The constant t_1 is like α_{n+1}, our former additive constant which can either be dropped or choosen at will. The constant of integration t_0 is connected with the solution of the homogenous problem, i.e., without the driving F-term. What remains is the constant $\alpha(\equiv P)$, here a time variable, which occurs in the coefficients (6.75) and (6.76). Our generator of transformation is, correspondingly, $S = F_2(q, P \equiv \alpha, t)$, so that

$$
\begin{aligned}
p &= \frac{\partial S}{\partial q}, \\
\beta &= \text{const.} = Q = \frac{\partial S}{\partial P} \equiv \frac{\partial S}{\partial \alpha} = q\frac{\partial b}{\partial \alpha} + \frac{\partial c}{\partial \alpha} \\
&= -\frac{q}{\cos\omega(t-t_0)} F(\alpha)\cos\omega(\alpha - t_0) \\
&\quad + \int_{t_1}^{t} \frac{dt'}{\cos^2\omega(t'-t_0)} \left[\int_{\alpha}^{t'} dt'' F(t'')\cos\omega(t''-t_0) \right] F(\alpha)\cos(\alpha - t_0).
\end{aligned}
$$

If we solve this equation for q we obtain

$$
\begin{aligned}
\frac{\beta}{F(\alpha)\cos\omega(\alpha - t_0)} = &-\frac{q}{\cos\omega(t-t_0)} \\
&+ \int_{t_1}^{t} \frac{dt'}{\cos^2\omega(t'-t_0)} \int_{\alpha}^{t'} dt'' F(t'')\cos\omega(t''-t_0)
\end{aligned}
$$

or

$$
\begin{aligned}
q(t) = \Bigg[&-\frac{\beta}{F(\alpha)\cos\omega(\alpha - t_0)} \\
&+ \int_{t_1}^{t} dt' \frac{1}{\cos^2\omega(t'-t_0)} \int_{\alpha}^{t'} dt'' F(t'')\cos\omega(t''-t_0) \Bigg] \cos\omega(t-t_0).
\end{aligned}
$$

The first term in the square brackets is simply a constant, C. Since t_1 was arbitrary, we put it equal to α. Thus we end up with

$$
q(t) = \cos\omega(t-t_0) \left[C + \int_{\alpha}^{t} dt' \int_{\alpha}^{t'} dt'' \frac{F(t'')\cos\omega(t''-t_0)}{\cos^2\omega(t'-t_0)} \right]. \tag{6.77}
$$

The contribution of the integral can also be written as

$$
\begin{aligned}
&\frac{\cos\omega(t-t_0)}{\omega} \int_{\alpha}^{t} dt'' \int_{t''}^{t} \omega dt' \frac{F(t'')\cos\omega(t''-t_0)}{\cos^2\omega(t'-t_0)} \\
&= \frac{\cos\omega(t-t_0)}{\omega} \int_{\alpha}^{t} dt'' F(t'') \left[\tan\omega(t-t_0) - \tan\omega(t''-t_0) \right] \cos\omega(t''-t_0) \\
&= \frac{1}{\omega} \int_{\alpha}^{t} dt'' F(t'') \left[\sin\omega(t-t_0)\cos\omega(t''-t_0) - \sin\omega(t''-t_0)\cos\omega(t-t_0) \right] \\
&= \frac{1}{\omega} \int_{\alpha}^{t} dt' F(t') \sin\omega(t-t').
\end{aligned}
$$

Here, then, is our final result for the amplitude of a harmonically bound particle driven by a time-dependent external force $F(t)$:

$$
q(t) = C\cos\omega(t-t_0) + \int_{\alpha}^{t} dt G(t,t') F(t'), \tag{6.78}
$$

where the Greens's function is given by $(m = 1)$

$$G(t, t') = \frac{1}{\omega} \sin \omega(t - t'). \tag{6.79}$$

The associated generator F_2 as defined in (6.68) can now be identified and takes the form

$$\begin{aligned} S\{[q(t)], \alpha; t, t_0\} &= -\frac{\omega}{2} \tan \omega(t - t_0) q^2(t) \\ &+ \frac{1}{\cos \omega(t - t_0)} \int_\alpha^t dt' F(t') \cos \omega(t' - t_0) q(t) \\ &- \frac{1}{2} \int_\alpha^t \frac{dt'}{\cos^2 \omega(t' - t_0)} \left[\int_\alpha^{t'} dt'' F(t'') \cos \omega(t'' - t_0) \right]^2 . \end{aligned} \tag{6.80}$$

7. Action-Angle Variables

In the following we will assume that the Hamiltonian does not depend explicitly on time; $\partial H/\partial t = 0$. Then we know that the characteristic function $W(q_i, P_i)$ is the generator of a canonical transformation to new constant momenta P_i (all Q_i are ignorable), and the new Hamiltonian depends only on the P_i: $H = K = K(P_i)$. Besides, the following canonical equations are valid:

$$\dot{Q}_i = \frac{\partial K}{\partial P_i} = \nu_i = \text{const.} \tag{7.1}$$

$$\dot{P}_i = -\frac{\partial K}{\partial Q_i} = 0 \,. \tag{7.2}$$

The P_i are N independent functions of the N integration constants α_i, i.e., are not necessarily $P_i = \alpha_i$. But the P_i are, like the α_i, constants. On the other hand, Q_i develops linear with time:

$$Q_i = \nu_i t + \beta_i \,, \tag{7.3}$$

with constants $\nu_i = \nu_i(P_j)$ and β_i. The transformation equations which are associated with the above canonical transformation generated by $W(q_i, P_i)$ are given by

$$p_i = \frac{\partial W}{\partial q_i} \,, \qquad Q_i = \frac{\partial W}{\partial P_i} \,. \tag{7.4}$$

Before we come to the action-angle variables, the following canonical transformation may serve as an introduction. It is clear that $F_2 = qP/\alpha$, $\alpha = \text{const.}$ is the generator of a canonical transformation which causes a scale change (extension or stretching) of a canonical pair (q, p):

$$p = \frac{\partial F_2}{\partial q} = \frac{P}{\alpha} \,, \qquad P = \alpha p \,, \tag{7.5}$$

$$Q = \frac{\partial F_2}{\partial P} = \frac{q}{\alpha} \,, \qquad Q = \frac{1}{\alpha} q \,. \tag{7.6}$$

Along with (q, p), $(q/\alpha, \alpha p) = (Q, P)$ are also canonical variables. The two pairs are said to possess the same canonicity. The area of the phase plane remains unchanged when we go from $(q, p) \to (Q, P) : QP = (q/\alpha)\alpha p = qp$ in the simplest case.

In order to go from (q_i, p_i) to action-angle variables caused by canonical transformation, we have to require the system to be periodic; furthermore, it should be completely separable. The latter means that there exist functions

$$p_i = \frac{\partial W(q_j; \alpha_1, \dots, \alpha_N)}{\partial q_i}, \qquad i = 1, 2, \dots, N$$

$$p_i = p_i(q_j; \alpha_1, \dots, \alpha_N) \,. \tag{7.7}$$

The above equation gives the phase space trajectory. From these trajectories we have to require of all (q_i, p_i) pairs that they be either closed curves (libration: pendulum, harmonic oscillator) or that the p_i be periodic functions of the q_i (rotation: rotating pendulum). So if one of the q_i runs while the remaining $q_j (j \neq i)$ are "frozen", the system should, after a certain time, return to its original state (in the case of libration) in the (q_i, p_i) phase space. In addition to the above periodicity requirement, the canonical transformation should be of such a kind that for the transformation

$$(q_i, p_i) \to (Q_i, P_i) \mathrel{\hat{=}} (w_i, J_i)$$

the new $Q_i \mathrel{\hat{=}} w_i$ (angle variable) increases by one unit if q_i runs through one complete cycle so that the integral has to be performed over one period in q_i:

$$\oint dw_i(q_j) = 1 \,. \tag{7.8}$$

The new canonical variables corresponding to the angle variables $w_i = Q_i$ we shall call $J_i \mathrel{\hat{=}} P_i$. Then, according to $Q_i = \partial W / \partial P_i$, we have as transformation equation:

$$w_i = \frac{\partial W}{\partial J_i} \,. \tag{7.9}$$

(Actually, one ought to use a different letter, $\hat{W}$, for W, since W refers to the canonical transformation $(q, p) \to (Q, P)$.)

According to (7.8), it holds that in the case of a complete cycle by q_i while the others are held fixed, we have for the corresponding change of $w_j = w_j(q_i, J_i)$, $J_i = \text{const.}$, i.e., $(\partial w_j / \partial J_i) dJ_i = 0$:

$$1 = \oint dw_j = \oint \frac{\partial w_j}{\partial q_i} dq_i = \oint \frac{\partial^2 W}{\partial J_j \partial q_i} dq_i = \frac{\partial}{\partial J_j} \oint \frac{\partial W}{\partial q_i} dq_i = \frac{\partial}{\partial J_j} \oint p_i \, dq_i \,.$$

From this, the important relation

$$\oint p_i \, dq_i = J_i \,, \tag{7.10}$$

follows, with $p_i = p_i(q_j; \alpha_1, \dots, \alpha_N)$.

After performing the q_i integration in (7.10), each J_i is a function of the N integration constants α_i, which appeared upon integration of the Hamilton–Jacobi equation. Then it holds that

$$J_i = J_i(\alpha_1, \ldots, \alpha_N) . \tag{7.11}$$

As a result, the J_i are indeed constants of motion. In the following we shall assume that the system of equations (7.11) is invertible. Then the J_i are N independent functions of the integration constants $\alpha_i = P_i$ and can thus be considered to be our new momenta.

The canonical transformation $(q, p) \to (w, J)$ generated by $W(q, J)$ with the transformation equations

$$p = \frac{\partial W}{\partial q} , \qquad w = \frac{\partial W}{\partial J} , \tag{7.12}$$

is an area (volume)-preserving transformation – similar to the canonical transformation $(q, p) \to (Q, P)$ introduced earlier with the generating function $F_2 = qP/\alpha$.

We could have introduced the action-angle variables in this manner: we are looking for transformations $(q, p) \to (w, J)$, $J =$ const., which are volume preserving (in phase space), whereby we require that when q completes a single period in the (q, p) phase space (in the case of libration), the corresponding new variable w must change by one unit.

With the canonical transformation $(q, p) \to (w, J)$ and its transformation equations (7.12), we have simultaneously succeeded in making the new Hamiltonian dependent only on the new constant "momenta" J_i; the w_i are, like the Q_i, ignorable coordinates. This was exactly the intention of the generating function $W(q, J)$, which is of the F_2 type. So we can write:

$$K = H = \alpha_1 = H(J_1, \ldots, J_N) .$$

The w_i's, like the Q_i's, are linear functions of time:

$$w_i = \nu_i t + \beta_i , \tag{7.13}$$

$$\dot{w}_i = \frac{\partial H(J)}{\partial J} = \nu_i \quad \left(\mathrel{\hat{=}} \dot{Q}_i = \frac{\partial K}{\partial P_i} , \quad K = H \right) . \tag{7.14}$$

The constant ν_i now proves to be a real frequency, since if we go through a period, $t \to t + T$, then according to (7.13),

$$\Delta w_i = \nu_i T_i = 1$$

is valid; or

$$\nu_i = \frac{1}{T_i} .$$

Thus ν_i is the frequency of motion with which the path will run through the (q_i, p_i) phase space. The rule

$$\nu_i = \frac{\partial H(J)}{\partial J_i} , \qquad i = 1, 2, \ldots, N \tag{7.15}$$

supplies us with an extremely useful method of calculating frequencies without prior knowledge of the time development of the system as contained in the equations of motion. But we need $H = H(J_1, \dots, J_N)$.

Here, again, briefly, the "recipe" for finding the frequencies ν_i:
(a) calculate $\oint p_i \, dq_i$ and call this expression J_i;
(b) determine $H = H(J_1, \dots, J_N)$;
(c) construct $\nu_i = \partial H / \partial J_i$.

As our first example, let us take the linear harmonic oscillator:

$$H(q, p) = \frac{p^2}{2m} + \frac{m}{2} \omega^2 q^2 = \alpha = E \, .$$

Then

$$p(q, \alpha = E) = \sqrt{2m\alpha - (m\omega q)^2} \, .$$

Now we follow the above scheme:

$$\text{(a)} \oint p \, dq = \oint \sqrt{2m\alpha - (m\omega q)^2} dq = \oint \sqrt{1 - \frac{m\omega^2 q^2}{2\alpha}} \sqrt{2m\alpha} \, dq$$

$$= \frac{2\alpha}{\omega} \oint \sqrt{1 - \frac{m\omega^2 q^2}{2\alpha}} \sqrt{\frac{m\omega^2}{2\alpha}} dq \, ;$$

$$\varrho := \sqrt{\frac{m\omega^2}{2\alpha}} q \, , \quad d\varrho = \sqrt{\frac{m\omega^2}{2\alpha}} dq$$

$$= \frac{2\alpha}{\omega} \oint \sqrt{1 - \varrho^2} \, d\varrho \, , \quad \varrho = \sin\varphi \, , \quad d\varrho = \cos\varphi \, d\varphi$$

$$= \frac{2\alpha}{\omega} \underbrace{\int_0^{2\pi} \cos^2\varphi \, d\varphi}_{=\pi} = \alpha \frac{2\pi}{\omega} =: J \, .$$

$$\text{(b)} \, \alpha = H(J) = \frac{\omega}{2\pi} J \, . \tag{7.16}$$

$$\text{(c)} \, \frac{\partial H}{\partial J} = \frac{\omega}{2\pi} = \nu \, . \tag{7.17}$$

As our next example we consider the cycloid pendulum, i.e., the motion of a particle m, that is constrained to swing back and forth in the gravitational field along a prescribed curve (cylcoid). The lowest point through which the particle swings is to be the origin of an x-y coordinate system. Then the parametric form of the cycloid is:

$$x = a(\theta + \sin\theta) \, ,$$
$$y = a(1 - \cos\theta) \, , \quad -\pi \leq \theta \leq \pi \, .$$

The square of the line element is given by

$$(ds)^2 = (dx)^2 + (dy)^2 = (vdt)^2$$
$$= a^2[(1+\cos\theta)^2 + \sin^2\theta](d\theta)^2 = 2a^2(1+\cos\theta)(d\theta)^2 \ .$$

With this, we get for the kinetic energy

$$T = \frac{m}{2}v^2 = ma^2(1+\cos\theta)\dot{\theta}^2 \ .$$

Using $V = mgy = mga(1-\cos\theta)$, we then have for L:

$$L = T - V = ma^2\left[(1+\cos\theta)\dot{\theta}^2 - \frac{g}{a}(1-\cos\theta)\right] \ .$$

This gives us the canonical momentum:

$$p_\theta = \frac{\partial L}{\partial \dot{\theta}} = 2ma^2(1+\cos\theta)\dot{\theta}$$

or

$$\dot{\theta} = \frac{p_\theta}{2ma^2(1+\cos\theta)} \ .$$

So the Hamiltonian can be written

$$H(\theta, p_\theta) = T + V = \frac{1}{4ma^2}\frac{p_\theta^2}{1+\cos\theta} + mga(1-\cos\theta) \ . \tag{7.18}$$

According to our "recipe" we need p_θ to build $\oint p_\theta \, d\theta$. Now, H is conserved; then it holds, in particular for $\theta = \pi$: $p_\theta = 0$ and therefore

$$H = \alpha = 2mga \ .$$

If we set this equal to the right-hand side of (7.18) and solve for p_θ^2, we get

$$p_\theta^2 = 4m^2a^3g(1+\cos\theta)^2 \ .$$

Now we can calculate successively:

$$\text{(a)} \quad \oint p_\theta \, d\theta = 4ma\sqrt{ag}\int_{-\pi}^{+\pi} d\theta(1+\cos\theta)$$
$$= 4\pi(2mag)\sqrt{\frac{a}{g}} = 4\pi\sqrt{\frac{a}{g}}H =: J \ .$$

$$\text{(b)} \quad H = \frac{1}{4\pi}\sqrt{\frac{g}{a}}\,J = \frac{1}{2\pi}\sqrt{\frac{g}{4a}}\,J \ .$$

$$\text{(c)} \quad \nu = \frac{\partial H}{\partial J} = \frac{1}{2\pi}\sqrt{\frac{g}{4a}} \ .$$

The frequency is thus independent of the amplitude. This conforms to a simple pendulum of the length $l = 4a$ – but for small amplitudes. The above system corresponds to the famous Huygens cycloid pendulum discovered in 1673.

Before we go on to the next example, we still want to express $q(t)$ and $p(t)$ for the simple linear harmonic oscillator as function of the action-angle variables. First of all it holds that

$$H(q,p) = \frac{p^2}{2m} + \frac{m}{2}\omega^2 q^2 = E \ ,$$
$$q(t) = \sqrt{\frac{2E}{m\omega^2}} \sin(\omega t + \alpha) \ , \quad p(t) = \sqrt{2mE} \cos(\omega t + \alpha) \ .$$

With the result (7.16),

$$H(J) = \frac{\omega}{2\pi} J = E$$

as well as

$$w = \nu t + \beta \quad : \quad 2\pi w = \omega t + \alpha$$

it follows that

$$q(t) = \sqrt{\frac{J}{\pi m \omega}} \sin(2\pi w) \ , \tag{7.19}$$
$$p(t) = \sqrt{\frac{m\omega J}{\pi}} \cos(2\pi w) \ . \tag{7.20}$$

The transformation of (q, p) to the action-angle variables (w, J) is generated by $W(q, J)$, which is of the $F_2(q, P)$ type. For a period of motion in q it now holds that $W(J)$ changes by

$$\Delta W(J) = \oint dq \frac{\partial W(q,J)}{\partial q} = \oint p\, dq = J \ . \tag{7.21}$$

For the generating function $F_1(q, Q)$ we found the Legendre transformation:

$$F_1(q,Q) = F_2(q,P) - QP$$
$$p = \frac{\partial F_1}{\partial q} \ , \quad P = -\frac{\partial F_1}{\partial Q} \ .$$

In correspondence to these equations, we set

$$W'(q,w) = W(q,J) - wJ \ ,$$
$$p = \frac{\partial W'}{\partial q} \ , \quad J = -\frac{\partial W'}{\partial w} \ .$$

Here, W' generates the same canonical transformation as W, which is nothing new. But contrary to (7.21), W' is a periodic function in w with the period 1:

$$\begin{aligned}\Delta W'(w) &= \Delta W - \Delta(wJ)\,, \qquad J = \text{const.}\\ &= \Delta W - J\underbrace{\Delta w}_{=1} = 0\,,\end{aligned}$$

where we have used (7.21). According to this result, $W'(q, w)$ returns to its initial value after one period in q, whereas $W(q, J)$ increases by J. Thus, W' is indeed periodic in w with the period $w = 1$, whereas W is not. (But W' does not satisfy the Hamilton–Jacobi equation!)

We recall that for the individual periodic coordinates $q_k = q_k(w, J)$, it holds that:

$$\text{Libration:} \quad q_k(w_k + 1) = q_k(w_k)\,. \tag{7.22}$$

Thus, q_k is periodic in w_k with period 1, and we can, therefore, expand q_k in a Fourier series:

$$q_k = \sum_{m=-\infty}^{+\infty} a_m^{(k)} \mathrm{e}^{2\pi i m w_k}$$

or

$$q_k(t) = \sum_{m=-\infty}^{+\infty} a_m^{(k)} \mathrm{e}^{2\pi i m(\nu_k t + \beta_k)}\,, \tag{7.23}$$

where the Fourier coefficients are determined in the usual manner:

$$a_m^{(k)} = \int_0^1 dw_k\, q_k(w_k)\, \mathrm{e}^{-2\pi i m w_k}\,. \tag{7.24}$$

In the case of rotation, we have

$$\text{Rotation:} \quad q_k(w_k + 1) = q_k(w_k) + q_{k0}\,, \tag{7.25}$$

whereby q_{k0} usually is a constant angle value, like 2π for a rotating pendulum. Now since $w_k q_{k0}$ increases by one unit for each q_k period, the following expression is periodic and can thus also be expanded in a Fourier series:

$$q_k - w_k q_{k0} = \sum_{m=-\infty}^{+\infty} a_m^{(k)} \mathrm{e}^{2\pi i m w_k} \tag{7.26}$$

with

$$a_m^{(k)} = \int_0^1 dw_k (q_k - w_k q_{k0})\, \mathrm{e}^{-2\pi i m w_k}\,. \tag{7.27}$$

If the motion is not limited to the subspace (q_k, p_k), then because of the periodicity in all coordinates q_k generally, it holds that

$$q_k(w_1+1, w_2+1, \dots, w_N+1) = q_k(w_1, \dots, w_N) \,.$$

For this reason, every q_i (or p_i) and every function of it can be expanded in an N-fold Fourier series. In particular, for the trajectory in phase space and its time-dependence, it holds that

$$q_i = \sum_{m_1=-\infty}^{+\infty} \dots \sum_{m_N=-\infty}^{+\infty} a^{(i)}_{m_1 \dots m_N} \mathrm{e}^{2\pi \mathrm{i}(m_1 w_1 + \dots + m_N w_N)}$$

or

$$q_i(t) = \sum_{m_1=-\infty}^{+\infty} \dots \sum_{m_N=-\infty}^{+\infty} b^{(i)}_{m_1 \dots m_N} \mathrm{e}^{2\pi \mathrm{i}(m_1 \nu_1 + \dots + m_N \nu_N)t} \tag{7.28}$$

with

$$b^{(i)}_{m_1 \dots m_N} = a^{(i)}_{m_1 \dots m_N} \mathrm{e}^{2\pi \mathrm{i}(m_1 \beta_1 + \dots + m_N \beta_N)} \,. \tag{7.29}$$

The various frequencies $\nu_i = 1/T_i$ in (7.28) are generally different, so that the motion of the whole system (in time) does not return to its initial state. But if the fundamental frequencies are commensurate, i.e., are rationally related so that N integers $r, s, \dots, t$ exist with

$$\frac{\nu_1}{r} = \frac{\nu_2}{s} = \dots = \frac{\nu_N}{t} = \nu \,, \tag{7.30}$$

then for the coordinates $q_i(t)$, we have

$$q_i(t) = \sum_{m_1, \dots, m_N=-\infty}^{+\infty} b^{(i)}_{m_1 \dots m_N} \mathrm{e}^{2\pi \mathrm{i}(m_1 r + m_2 s + \dots + m_N t)\nu t} \,. \tag{7.31}$$

After the time $T = 1/\nu$, all separation coordinates return to their initial positions. In the process, they have completed $r, s, \dots t$ cycles, since

$$\Delta w_i = \nu_i T = \frac{\nu_i}{\nu} = \{r, s, \dots, t\} \,. \tag{7.32}$$

If this is not the case, i.e., no commensurability prevails, the motion is called conditionally or multiply periodic.

We continue, however, to be interested in commensurate frequencies and say that a system with N degrees of freedom is m-fold degenerate, if relations exist between the frequencies of the kind:

$$\sum_{i=1}^{N} j_{ki} \nu_i = 0 \,, \quad k = 1, 2, \dots, m \,; \quad j_{ki} \in \mathbb{Z} \,. \tag{7.33}$$

Simple examples are $N = 2$, $m = k = 1$:

$$\underbrace{j_{11}}_{=r}\nu_1 + \underbrace{j_{12}}_{=-s}\nu_2 = 0\,, \qquad \frac{\nu_2}{\nu_1} = \frac{r}{s}\,.$$

Another example is $N = 3$, $m = 2 = N(= 3) - 1$:

$$\begin{aligned} j_{11}\nu_1 + j_{12}\nu_2 + j_{13}\nu_3 &= 0\,, \\ j_{21}\nu_1 + j_{22}\nu_2 + j_{23}\nu_3 &= 0\,. \end{aligned}$$

In the Kepler problem with $r \equiv 1, \theta \equiv 2, \phi \equiv 3$, it holds that

$$\nu_r = \nu_\theta = \nu_\phi =: \nu$$

or

$$\nu_\phi - \nu_\theta = 0\,, \quad \nu_\theta - \nu_r = 0\,, \tag{7.34}$$

which can also be written as

$$\begin{aligned} (0)\nu_r + (-1)\nu_\theta + (1)\nu_\phi &= 0\,, \\ (-1)\nu_r + (1)\nu_\theta + (0)\nu_\phi &= 0\,, \end{aligned}$$

with

$$j_{11} = j_{23} = 0\,, \quad j_{13} = j_{22} = 1\,, \quad j_{21} = j_{12} = -1\,. \tag{7.35}$$

In the last example (Kepler problem) we have $m = 2$ commensurability relations. Here, the ratios of all frequencies are rational ($= 1$). A system like this is called completely degenerate if $m = N - 1$ equations of the form (7.33) exist between the frequencies. If only $m < N - 1$ such equations exist, the system is called m-fold degenerate. The Kepler problem is thus completely degenerate. In general we can say that every system with a closed path is completely degenerate ($V(r) \sim 1/r, r^2$).

We mention incidentally that $H(J_i)$ and the frequencies of the Kepler problem ($V = -k/r$) are given by

$$H = E = -\frac{2\pi^2 mk^2}{(J_r + J_\theta + J_\phi)^2}\,, \tag{7.36}$$

$$\nu = \frac{\partial H}{\partial J_r} = \frac{\partial H}{\partial J_\theta} = \frac{\partial H}{\partial J_\phi} = \frac{4\pi^2 mk^2}{(J_r + J_\theta + J_\phi)^3}\,. \tag{7.37}$$

For the cyclotron motion (charged particle in a homogeneous magnetic field B) with $N = 2$ and plane polar coordinates r, θ, it similarly holds that

$$H = \nu_c(J_r + J_\theta)\,,$$

with $\nu_c = \omega_c/2\pi = (1/2\pi)eB/mc$ and due to $\nu_r = \nu_\theta \equiv \nu_c$, we have the commensurability condition

$$-\nu_r + \nu_\theta = 0\,. \tag{7.38}$$

So this system is also completely degenerate.

If a system is m-fold degenerate, one can, via a canonical transformation to new action-angle variables (w', J'), make the new frequencies vanish. For example, for the Kepler problem, it holds for the transition from $(w_r, w_\theta, w_\phi; J_r, J_\theta, J_\phi) \to (w_1, w_2, w_3; J_1, J_2, J_3)$, with two new vanishing frequencies, that the associated generating function reads:

$$F_2(w_r, w_\theta, w_\phi; J_1, J_2, J_3) = (-w_\theta + w_\phi)J_1 + (-w_r + w_\theta)J_2 + w_r J_3\,,$$

since

$$\frac{\partial F_2}{\partial J_1} = w_\phi - w_\theta = w_1 \quad : \quad \nu_1 = \dot{w}_1 = 0$$

because of

$$w_\phi = \nu t + \beta_\phi\,, \qquad w_\theta = \nu t + \beta_\theta\,.$$

Likewise,

$$\frac{\partial F_2}{\partial J_2} = w_\theta - w_r = w_2 \quad : \quad \nu_2 = \dot{w}_2 = 0$$

with $w_r = \nu t + \beta_r$. Finally we have

$$\frac{\partial F_2}{\partial J_3} = w_r = w_3 \quad : \quad \nu_3 = \dot{w}_3 = \nu\,.$$

The new action variables follow from the transformation equations

$$\frac{\partial F_2(w_r, \ldots; J_1, \ldots)}{\partial w_r, \ldots} = J_{r,\ldots}$$

$$J_3 - J_2 = J_r\,, \quad J_2 - J_1 = J_\theta\,, \quad J_1 = J_\phi$$

or

$$J_1 = J_\phi\,, \quad J_2 = J_\theta + J_\phi\,, \quad J_3 = J_r + J_\theta + J_\phi\,.$$

The new (only) action variable with nonvanishing frequency is J_3. Therefore we find

$$H = H(J_3) = -\frac{2\pi^2 m k^2}{J_3^2}\,.$$

J_3 is called the "proper" action variable. (Only these become multiples of h in the older quantum theory!)

For the example of a charged particle in a homogeneous magnetic field B, it holds similarly that $(N = 2, m = 1)$

$$-\nu_r + \nu_\theta = 0\,, \quad \nu_r = \nu_\theta = \nu_c$$
$$j_{11} = -1\,, \quad j_{12} = 1\,; \quad w_r = \nu_c t + \beta_r\,, \quad w_\theta = \nu_c t + \beta_\theta\,.$$

The generating function which brings us to a single nonvanishing frequency is given by

$$F_2(w_r, w_\theta; J_1, J_2) = (-w_r + w_\theta)J_1 + w_\theta J_2\,;$$
$$\frac{\partial F_2}{\partial J_1} = w_\theta - w_r = w_1 \quad : \quad \nu_1 = \dot{w}_1 = 0$$
$$\frac{\partial F_2}{\partial J_2} = w_\theta = w_2 \quad : \quad \nu_2 = \dot{w}_2 = \nu_c\,.$$

The transformation equations $\partial F_2(w_r, \ldots; J_1, \ldots)/\partial w_r, \ldots = J_r, \ldots$ yield

$$-J_1 = J_r\,, \quad J_1 + J_2 = J_\theta\,,$$

or

$$J_1 = -J_r\,, \quad J_2 = J_r + J_\theta\,.$$

The new Hamiltonian again contains only the action variable with nonvanishing frequency:

$$H = H(J_2) = \nu_c J_2 = \frac{1}{2\pi}\frac{eB}{mc}\,J_2\,.$$

The general form of the generating function for the canonical transformation from (w, J) to (w', J') in which we want to obtain zero-frequencies for m of the new actions, is given by

$$F_2 = F_2(w, J') = \sum_{k=1}^{m}\sum_{i=1}^{N} J'_k j_{ki} w_i + \sum_{k=m+1}^{N} J'_k w_k\,.$$

The transformed coordinates are:

$$w'_k = \frac{\partial F_2}{\partial J'_k} = \sum_{i=1}^{N} j_{ki} w_i\,, \quad k = 1, 2, \ldots, m$$
$$w'_k = w_k\,, \quad k = m+1, \ldots, N\,.$$

The corresponding new frequencies result from (cf. (7.33))

$$\nu'_k = \dot{w}'_k = \sum_{i=1}^{N} j_{ki}\nu_i = 0\,, \quad k = 1, 2, \ldots, m$$
$$= \nu_k\,, \quad k = m+1, \ldots, N\,.$$

The associated new constant action variables follow from the solution of $\partial F_2(w_j, J'_k)/\partial w_i = J_i$:

$$J_i = \sum_{k=1}^{m} J'_k j_{ki} + \sum_{k=m+1}^{N} J'_k \delta_{ki} .$$

This then yields: $H = H(J'_k)$ with

$$\nu'_k = \frac{\partial H}{\partial J'_k} \neq 0 .$$

The results concerning the Coulomb or Kepler problem are well known (Born, Goldstein). We want to still prove the formulae used above for a particle in a magnetic field. To this end, we begin with the Lagrangian for a particle with charge e and mass m in a magnetic field:

$$\begin{aligned} L(r, \varphi, z; \dot{r}, \dot{\varphi}, \dot{z}) &= \frac{m}{2} v^2 + \frac{e}{c} \boldsymbol{v} \cdot \boldsymbol{A}(\boldsymbol{r}) \\ &= \frac{m}{2}(\dot{r}^2 + r^2\dot{\varphi}^2 + \dot{z}^2) + \frac{e}{c}(\dot{r}A_r + r\dot{\varphi}A_\varphi + \dot{z}A_z) . \end{aligned}$$

Here we have used cylindrical coordinates (r, φ, z). The vector potential is given by

$$\boldsymbol{A}(\boldsymbol{r}) = \left(0, A_\varphi(r) = \frac{B_0}{2} r, 0\right) .$$

If we then express $\nabla \times \boldsymbol{A}$ in cylindrical coordinates, only

$$\boldsymbol{B} \cdot \boldsymbol{e}_3 = \frac{1}{r}\frac{\partial}{\partial r}(rA_\varphi) = \frac{1}{r}\frac{\partial}{\partial r}\left(B_0 \frac{r^2}{2}\right) = B_0$$

remains for the third component of $\boldsymbol{B}$. Thus we are dealing with a time-independent magnetic field in z direction – as desired. In the following we suppress the uninteresting z-part in L and therefore write:

$$L = \frac{m}{2}(\dot{r}^2 + r^2\dot{\varphi}^2) + \frac{e}{c} r\dot{\varphi}A_\varphi . \tag{7.39}$$

From this L we get the canonical momenta:

$$p_r = \frac{\partial L}{\partial \dot{r}} = m\dot{r} , \quad p_\varphi = \frac{\partial L}{\partial \dot{\varphi}} = mr^2\dot{\varphi} + \frac{e}{c} rA_\varphi$$

or

$$\dot{r} = \frac{p_r}{m} , \quad \dot{\varphi} = \frac{1}{mr^2}\left(p_\varphi - \frac{e}{c} rA_\varphi\right) . \tag{7.40}$$

We use these equations in

$$H = p_r\dot{r} + p_\varphi\dot{\varphi} - \frac{m}{2}(\dot{r}^2 + r^2\dot{\varphi}^2) - \frac{e}{c}\, r\dot{\varphi}A_\varphi$$

and thus obtain

$$H = \frac{1}{2m}\left[p_r^2 + \left(\frac{1}{r}\, p_\varphi - \frac{e}{c}\, A_\varphi\right)^2\right] . \tag{7.41}$$

Since in our gauge it holds that $A_\varphi = B_0 r/2$, (7.41) becomes

$$H = \frac{1}{2m}\left[p_r^2 + \left(\frac{p_\varphi}{r} - \frac{eB_0}{2c}\, r\right)^2\right] . \tag{7.42}$$

The canonical equations of motion are then given by

$$\begin{aligned}
\dot{r} &= \frac{\partial H}{\partial p_r} = \frac{p_r}{m} , \\
\dot{\varphi} &= \frac{\partial H}{\partial p_\varphi} = \frac{1}{mr^2}\left(p_\varphi - \frac{eB_0}{2c}\, r^2\right), \qquad (7.43)\\
\dot{p}_r &= -\frac{\partial H}{\partial r} = \frac{1}{mr^3}\left(p_\varphi - \frac{eB_0}{2c}\, r^2\right)\left(p_\varphi + \frac{eB_0}{2c}\, r^2\right), \\
\dot{p}_\varphi &= 0 . \qquad (7.44)
\end{aligned}$$

For a circular motion, $p_r = 0$, $\dot{p}_r = 0$ is valid. Then (7.44) yields

$$\begin{aligned}
&\text{(a)} \quad p_\varphi = \frac{eB_0}{2c}\, r^2 \quad \text{and with (7.43),} \quad \dot{\varphi} = 0 ; \quad \dot{r} = 0 . \\
&\text{(b)} \quad p_\varphi = -\frac{eB_0}{2c}\, r^2 < 0! \quad \text{for} \quad (eB_0) > 0 . \qquad (7.45)
\end{aligned}$$

When (b) is inserted in (7.43), it gives

$$mr^2\dot{\varphi} + \frac{eB_0}{c}\, r^2 = 0$$

or

$$\dot{\varphi} = -\frac{eB_0}{mc} =: -\omega_c , \quad \omega_c \equiv \frac{eB_0}{mc} . \tag{7.46}$$

With the cyclotron frequency (7.46), H from (7.42) can be written as

$$H = \frac{1}{2m}\left[p_r^2 + \left(\frac{p_\varphi}{r} - \frac{m\omega_c r}{2}\right)^2\right] , \quad p_\varphi < 0 . \tag{7.47}$$

The Hamilton–Jacobi equation reads, accordingly:

$$\frac{1}{2m}\left\{\left(\frac{\partial W}{\partial r}\right)^2 + \left[\frac{1}{r}\,\frac{\partial W}{\partial \varphi} - \frac{m\omega_c}{2}\, r\right]^2\right\} = E = \alpha_1 .$$

With $\alpha_1 = E, \alpha_2 = p_\varphi$ and the separation ansatz,

$$W = W(r, \varphi; \alpha_1, \alpha_2) = \varphi p_\varphi + W_r(r) \ , \tag{7.48}$$

we get

$$\frac{dW_r}{dr} = \left[2mE - \left(\frac{p_\varphi}{r} - \frac{m\omega_c}{2} r\right)^2\right]^{1/2}$$

so that (7.48) can be written as

$$W = \varphi p_\varphi + \int^r dr' \left[2mE - \left(\frac{p_\varphi}{r'} - \frac{m\omega_c}{2} r'\right)^2\right]^{1/2} .$$

The action variables J_φ and J_r must be calculated next:

$$J_\varphi = \oint_{p_\varphi < 0} p_\varphi \, d\varphi = 2\pi |p_\varphi| \ . \tag{7.49}$$

$$\begin{aligned} J_r &= \oint p_r \, dr = \oint \frac{\partial W}{\partial r} dr \\ &= \oint dr \sqrt{2mE - \left(\frac{p_\varphi^2}{r^2} - p_\varphi m\omega_c + \frac{m^2\omega_c^2}{4} r^2\right)} \\ &= \frac{m\omega_c}{2} \oint \frac{dr}{r} \sqrt{\frac{2mE + p_\varphi m\omega_c}{(m\omega_c/2)^2} r^2 - \frac{p_\varphi^2}{(m\omega_c/2)^2} - r^4} \ , \qquad (r^2 = x) \\ &= \frac{m\omega_c}{4} \oint \frac{dx}{x} \sqrt{-a + 2bx - x^2} = \frac{m\omega_c}{4} 2\pi (b - \sqrt{a}) \ . \end{aligned} \tag{7.50}$$

Here we have used the following abbreviations:

$$a = \frac{p_\varphi^2}{(m\omega_c/2)^2} \ , \qquad b = \frac{mE + p_\varphi (m\omega_c/2)}{(m\omega_c/2)^2} \ .$$

If we now use

$$\sqrt{a} = \frac{|p_\varphi|}{(m\omega_c/2)}$$

as well as

$$b = \frac{mE}{(m\omega_c/2)^2} + p_\varphi \frac{1}{m\omega_c/2} = \frac{4E}{m\omega_c^2} - \frac{2|p_\varphi|}{m\omega_c} \ , \qquad |p_\varphi| = \frac{J_\varphi}{2\pi} \ ,$$

then it follows from (7.50) that

$$J_r = \frac{m\omega_c}{4} 2\pi \left[\left(\frac{4E}{m\omega_c^2} - \frac{2J_\varphi}{2\pi m\omega_c}\right) - \frac{J_\varphi}{2\pi (m\omega_c/2)}\right] = \frac{2\pi}{\omega_c} E - J_\varphi \ .$$

Thus we get

$$H(J_r, J_\varphi) = \nu_c(J_r + J_\varphi) \tag{7.51}$$

and from this,

$$\nu_r = \nu_\varphi = \nu_c \equiv \frac{1}{2\pi}\frac{eB}{mc} . \tag{7.52}$$

As a further example we determine the action variables and frequencies of the plane mathematical pendulum. We begin with the Lagrangian

$$L = T - V = \frac{m}{2}(l^2\dot{\varphi}^2) + mgl\cos\varphi . \tag{7.53}$$

φ is the angle of deviation from the lower (stable) equilibrium position. l is the length of the pendulum. From (7.53) it follows that

$$p_\varphi = \frac{\partial L}{\partial\dot{\varphi}} = ml^2\dot{\varphi} ,$$

so that

$$H(\varphi, p_\varphi) = p_\varphi\dot{\varphi} - L = \frac{1}{2ml^2}p_\varphi^2 - mgl\cos\varphi . \tag{7.54}$$

Since the system is conservative, $\partial H/\partial t = 0$, we set $H = E = \alpha$. The Hamilton–Jacobi equation is, accordingly:

$$\frac{1}{2ml^2}\left(\frac{dW}{d\varphi}\right)^2 - mgl\cos\varphi = E , \tag{7.55}$$

from which follows

$$W(\varphi; E) = \int^{\varphi} d\varphi' \underbrace{[2ml^2E + 2m^2gl^3\cos\varphi']^{1/2}}_{=[\cdot/.]^{1/2}} . \tag{7.56}$$

The action variable J then follows from

$$J_\varphi = \oint p_\varphi\, d\varphi = \oint \frac{dW}{d\varphi}\, d\varphi = \oint d\varphi[2ml^2E + 2m^2gl^3\cos\varphi]^{1/2} . \tag{7.57}$$

The limits of integration are determined in the case of libration from $\dot{\varphi} = 0$ at $p_\varphi = 0$; i.e., they result from setting the expression in parentheses in (7.57) equal to zero. At this point we have to distinguish between two cases:

(a) Libration: $|H| < mgl$; then φ is always smaller than π. If we start at $\varphi = 0$, then the angle $0 \to \varphi_{\max} \to 0 \to (-\varphi_{\max}) \to 0$ will be covered in one period: $T = 1/\nu_\varphi$. Then we can write:

$$J_\varphi = \oint_{1\,\text{period}} d\varphi[\cdot/.]^{1/2} = 4\int_0^{\varphi_{\max}} d\varphi[\cdot/.]^{1/2} . \tag{7.58}$$

So we have to integrate four times over a fourth of one period.

(b) Rotation: $H > mgl$; here, $\dot{\varphi}$ always has the same sign, and for the action variable it now holds that

$$J_\varphi = \oint d\varphi[\cdot/.]^{1/2} = \int_0^{2\pi} d\varphi[\cdot/.]^{1/2} = 2\int_0^{\pi} d\varphi[\cdot/.]^{1/2} . \tag{7.59}$$

One should note the discontinuity in the definition of a period (of factor 2) when going with $H \to mgl$ from below (libration: $-mgl < H \overset{\rightarrow}{<} mgl$) or above (rotation) with $H \to mgl$. This is, however, only a matter of definition of where to start a period.

We now come to the determination of the frequencies associated with the above two cases. First of all, it generally holds that

$$\nu_\varphi = \frac{1}{\partial J_\varphi/\partial H} = \left[\oint d\varphi \frac{ml^2}{\sqrt{2ml^2H + 2m^2gl^3\cos\varphi}}\right]^{-1} \tag{7.60}$$

or

$$\frac{1}{\nu_\varphi} = \sqrt{\frac{l}{g}} \oint d\varphi \frac{1}{\sqrt{2(\cos\varphi + (H/mgl))}} . \tag{7.61}$$

From here on it is convenient to introduce the parameter

$$h = \frac{H}{mgl} .$$

(a′) Here, as in the case (a) above, it holds for $|h| < 1$ and $\varphi_{\max} = \operatorname{arc}\cos(-h)$ that

$$\frac{1}{\nu_\varphi} = 4\sqrt{\frac{l}{g}} \int_0^{\varphi_{\max}} d\varphi \frac{1}{\sqrt{2(\cos\varphi - \cos\varphi_{\max})}}$$

or, with $\cos\varphi = 1 - 2\sin^2\varphi/2$

$$\frac{1}{\nu_\varphi} = 4\sqrt{\frac{l}{g}} \int_0^{\varphi_{\max}} d\varphi \frac{1}{\sqrt{4(\sin^2(\varphi_{\max}/2) - \sin^2(\varphi/2))}} . \tag{7.62}$$

At this point we introduce the following additional variables:

$$\sin\frac{\varphi}{2} = \sin\frac{\varphi_{\max}}{2}\sin\xi =: k\sin\xi \tag{7.63}$$

with

$$k = \sin\frac{\varphi_{\max}}{2} = \sqrt{\frac{1+h}{2}} . \tag{7.64}$$

The last formula is valid because

$$2k^2 = 2\sin^2\frac{\varphi_{\max}}{2} = 1 - \cos\varphi_{\max} = 1 + h \equiv \left(1 + \frac{H}{mgl}\right).$$

In this manner we get the expression

$$\frac{1}{\nu_\varphi} = 4\sqrt{\frac{l}{g}}\int_0^{\pi/2} d\xi \frac{1}{\sqrt{1 - k^2\sin^2\xi}} = 4\sqrt{\frac{l}{g}}\,K(k)\,. \tag{7.65}$$

Here, the complete elliptic integral of the first kind appears:

$$K(k) = \int_0^{\pi/2} d\xi \frac{1}{\sqrt{1 - k^2\sin^2\xi}}\,, \qquad 0 \le k < 1 \tag{7.66}$$

$$= \frac{\pi}{2}\left\{1 + \left(\frac{1}{2}\right)^2 k^2 + \left(\frac{1\cdot 3}{2\cdot 4}\right)^2 k^4 + \left(\frac{1\cdot 3\cdot 5}{2\cdot 4\cdot 6}\right)^2 k^6 + \dots\right\}. \tag{7.67}$$

For

$$\varphi_{\max} \to 0\,(k \to 0) \quad : \quad K(0) = \frac{\pi}{2}\,. \tag{7.68}$$

In this case (small angle), the familiar amplitude-independent frequency follows:

$$\nu_0 = \frac{1}{2\pi}\sqrt{\frac{g}{l}}\,, \qquad T_0 = 2\pi\sqrt{\frac{l}{g}}\,. \tag{7.69}$$

As normalized frequency we thus find

$$\frac{\nu_\varphi(k)}{\nu_0} = \frac{\pi}{2}\,\frac{1}{K(k)}\,, \qquad 0 \le k < 1\,. \tag{7.70}$$

(b′) Here we have $h > 1$ (complete rotation):

$$\frac{1}{\nu_\varphi} = 2\sqrt{\frac{l}{g}}\int_0^{\pi} d\varphi \frac{1}{\sqrt{2(\cos\varphi + h)}} = 2\sqrt{\frac{l}{g}}\int_0^{\pi} d\varphi \frac{1}{\sqrt{2h + 2 - 4\sin^2(\varphi/2)}}$$
$$= \sqrt{\frac{l}{g}}\int_0^{\pi} d\varphi \frac{1}{\sqrt{(1/k'^2) - \sin^2(\varphi/2)}}\,, \qquad k' := \frac{1}{k} = \sqrt{\frac{2}{1+h}}\,, \quad \varphi = 2\xi\,, \tag{7.71}$$

$$= 2k'\sqrt{\frac{l}{g}}\int_0^{\pi/2} d\xi \frac{1}{\sqrt{1 - k'^2\sin^2\xi}} = 2k'\sqrt{\frac{l}{g}}K(k')\,. \tag{7.72}$$

Using

$$k' = \sqrt{\frac{2}{1+h}} = \sqrt{\frac{2mgl}{H+mgl}} \tag{7.73}$$

we immediately get

$$\frac{1}{\nu_\varphi} = 2\sqrt{\frac{2ml^2}{H+mgl}}\,K(k') \tag{7.74}$$

and the normalized frequency:

$$\frac{\nu_\varphi}{\nu_0} = \frac{\pi}{2}\left(\frac{2k}{K(1/k)}\right), \quad k > 1(k' < 1)\,. \tag{7.75}$$

In both cases (a′) and (b′) the value of K goes to infinity, $K \to \infty$, as $H \to mgl$ ($k \to 1$) – but only slowly; namely, logarithmically:

$$\lim_{k\to 1} \frac{\nu_\varphi}{\nu_0} = \begin{cases} \dfrac{\pi}{2}\dfrac{1}{\ln[4/(1-k^2)^{1/2}]}, & \text{k} < 1 \quad (7.76) \\ \pi\dfrac{1}{\ln[4/(k^2-1)^{1/2}]}, & \text{k} > 1\,. \quad (7.77) \end{cases}$$

For $\nu_\varphi \to 0$, the period $T = 1/\nu_\varphi$ is then infinite; the mass m is at the upper, unstable equilibrium point.

Finally we summarize the most important results for the mathematical pendulum:

$$\text{Libration}: \quad |H| < mgl\,, \quad h = \frac{H}{mgl}\,, \quad -1 < h < 1\,,$$

$$J = 16ml\sqrt{lg}\,[E(k) - (1-k^2)K(k)]\,, \quad k^2 = \frac{1+h}{2}\,. \tag{7.78}$$

$K(k)$ and $E(k)$ are the complete elliptic integrals of the first and second kind:

$$K(k) = \int_0^{\pi/2} \frac{d\xi}{\sqrt{1-k^2\sin^2\xi}}\,, \quad 0 \le k < 1\,, \tag{7.79}$$

$$E(k) = \int_0^{\pi/2} d\xi\sqrt{1-k^2\sin^2\xi}\,, \quad 0 \le k \le 1\,. \tag{7.80}$$

For $h \to 1$ ($k \to 1$), we have $(1-k^2)K(k) \to 0$. Furthermore, $E(1) = 1$. For this limiting case it follows from (7.78) that

$$J \underset{k\to 1}{\to} 16ml\sqrt{lg}\,. \tag{7.81}$$

$$\text{Rotation:} \quad h > 1\,, \quad k'^2 = \frac{1}{k^2} = \frac{2}{1+h}$$

$$J = 4\sqrt{2ml^2(H+mgl)}\,E(k') \tag{7.82}$$

and

$$J \underset{k\to 1}{\to} 8ml\sqrt{lg}\ . \tag{7.83}$$

When comparing (7.81) with (7.83), we again meet the factor 2, which was mentioned earlier when defining the frequency: the jump in J results from the inconsistency of the definition of the period.

The energy and phase diagram of the plane pendulum is well known. The phase trajectory for the separatrix can be obtained from $H = p_\varphi^2/2ml^2 - mgl\cos\varphi = E$ for $\varphi = \pm\pi$ with $E = E_{\rm sx} = mgl$. Then

$$\frac{p_{\varphi_{\rm sx}}^2}{2ml^2} = mgl(1+\cos\varphi_{\rm sx})$$

or

$$\begin{aligned} p_{\varphi_{\rm sx}} &= \pm\sqrt{2}\,ml^2(2\pi\nu_0 \underbrace{(1+\cos\varphi_{\rm sx})^{1/2}}_{=\sqrt{2}\cos\varphi_{\rm sx}/2} \\ &= \pm 2ml^2(2\pi\nu_0)\cos\frac{\varphi_{\rm sx}}{2}\ . \end{aligned} \tag{7.84}$$

The two signs refer to the upper and lower branch. From $\dot\varphi = \partial H/\partial p_\varphi$ we have along the separatrix $\dot\varphi_{\rm sx} = p_{\varphi_{\rm sx}}/ml^2$, and with (7.84) we get

$$\frac{d\varphi_{\rm sx}}{dt} = \pm 2(2\pi\nu_0)\cos\frac{\varphi_{\rm sx}}{2}\ . \tag{7.85}$$

If we integrate the differential equation with $\varphi(t=0) = 0$, we obtain

$$2\pi\nu_0 t = \int_0^{\varphi_{\rm sx}} \frac{d(\varphi/2)}{\cos(\varphi/2)} = \ln\tan\left(\frac{\varphi_{\rm sx}}{4} + \frac{\pi}{4}\right)$$

or, solved for $\varphi_{\rm sx}$:

$$\varphi_{\rm sx}(t) = 4\arctan\left(e^{2\pi\nu_0 t}\right) - \pi\ . \tag{7.86}$$

This expression clearly shows the asymptotic behavior along the separatrix: $\varphi_{\rm sx} \to \pm\pi,\ t \to \pm\infty$.

As a final example we consider the "Toda molecule" . Here we are dealing with a system of three degrees of freedom, which will, surprisingly, prove to be completely integrable. The Hamiltonian is given by

$$H(\varphi_i, p_i) = \tfrac{1}{2}(p_1^2 + p_2^2 + p_3^2) + e^{-(\varphi_1-\varphi_3)} + e^{-(\varphi_2-\varphi_1)} + e^{-(\varphi_3-\varphi_2)} - 3\ , \tag{7.87}$$

and describes three particles that are moving on a circle and between which exponentially decreasing repulsive forces are acting.

A first integral is obviously the energy. In addition, the total momentum (= angular momentum) P_3 is conserved:

$$P_3 = (p_1 + p_2 + p_3) = \text{const} . \tag{7.88}$$

This results from the fact that H is invariant under rigid rotation: $\varphi_i \to \varphi_i + \varphi_0, p_i \to p_i$. Of course we could also prove this with the help of a canonical transformation, by transforming H to the new momenta $P_1 = p_1$, $P_2 = p_2$, and $P_3 = p_1 + p_2 + p_3$. The generating function that generates this momentum transformation is

$$F_2(\varphi_i, P_i) = \varphi_1 P_1 + \varphi_2 P_2 + \varphi_3(P_3 - P_1 - P_2) , \tag{7.89}$$

since it holds that

$$\left.\begin{aligned} p_1 &= \frac{\partial F_2}{\partial \varphi_1} = P_1 , \\ p_2 &= \frac{\partial F_2}{\partial \varphi_2} = P_2 , \\ p_3 &= \frac{\partial F_2}{\partial \varphi_3} = P_3 - P_1 - P_2 \end{aligned}\right\} : \ p_1 + p_2 + p_3 = P_3 .$$

If we designate the new variable canonically conjugate to P_i by ϕ_i, then it holds further that

$$\left.\begin{aligned} \phi_1 &= \frac{\partial F_2}{\partial P_1} = \varphi_1 - \varphi_3 , \\ \phi_2 &= \frac{\partial F_2}{\partial P_2} = \varphi_2 - \varphi_3 , \end{aligned}\right\} : \ \phi_1 - \phi_2 = \varphi_1 - \varphi_2 ,$$

$$\phi_3 = \frac{\partial F_2}{\partial P_3} = \varphi_3 .$$

Now inserting the newly found variables into (7.87), we have, as new Hamiltonian,

$$\begin{aligned} \mathcal{H}(\phi_i, P_i) = \tfrac{1}{2}[P_1^2 + P_2^2 + (P_3 - P_1 - P_2)^2] \\ + \mathrm{e}^{-\phi_1} + \mathrm{e}^{-(\phi_2 - \phi_1)} + \mathrm{e}^{\phi_2} - 3. \end{aligned} \tag{7.90}$$

Since $\mathcal{H}$ is independent of ϕ_3 (ϕ_3 is ignorable), P_3 is indeed conserved. Without loss of generality we set $P_3 = 0$, which represents a transition onto the rotating system with vanishing angular momentum. Note that at this point we have reduced our problem with three degrees of freedom to one with only two:

$$\mathcal{H} = P_1^2 + P_2^2 + P_1 P_2 + \mathrm{e}^{-\phi_1} + \mathrm{e}^{-(\phi_2 - \phi_1)} + \mathrm{e}^{\phi_2} - 3 . \tag{7.91}$$

We now want to demonstrate that we are dealing here with the dynamics of a particle moving in a two-dimensional potential. To do so, we introduce a second canonical transformation with the generating function

$$F_2'(\phi_i, p_i') = \frac{1}{4\sqrt{3}} \left[(p_x' - \sqrt{3}\, p_y')\phi_1 + (p_x' + \sqrt{3}\, p_y')\phi_2 \right] . \tag{7.92}$$

From this follow the transformation equations:

$$P_1 = \frac{\partial F_2'}{\partial \phi_1} = \frac{1}{4\sqrt{3}}(p_x' - \sqrt{3}\,p_y') ,$$
$$P_2 = \frac{\partial F_2'}{\partial \phi_2} = \frac{1}{4\sqrt{3}}(p_x' + \sqrt{3}\,p_y') .$$

Let the conjugate variables to p_x', p_y' be x', y':

$$x' = \frac{\partial F_2'}{\partial p_x'} = \frac{1}{4\sqrt{3}}(\phi_1 + \phi_2) ,$$
$$y' = \frac{\partial F_2'}{\partial p_y'} = \frac{1}{4}(\phi_2 - \phi_1) .$$

From here follow the equations necessary for (7.91):

$$\phi_1 = 2\sqrt{3}\left(x' - \frac{y'}{\sqrt{3}}\right), \quad \phi_2 = 2\sqrt{3}\left(x' + \frac{y'}{\sqrt{3}}\right), \quad \phi_2 - \phi_1 = 4y' .$$

Accordingly, our new Hamiltonian is written:

$$\begin{aligned}\mathcal{H} \to H' &= \frac{1}{48}\left[(p_x' - \sqrt{3}\,p_y')^2 + (p_x' + \sqrt{3}\,p_y')^2 + (p_x'^2 - 3p_y'^2)\right] \\ &\quad + e^{-2\sqrt{3}\,(x' - y'/\sqrt{3})} + e^{-4y'} + e^{2\sqrt{3}(x' + y'/\sqrt{3})} - 3 \\ &= \frac{1}{48}(3p_x'^2 + 3p_y'^2) + \sum e^{\cdots} - 3 .\end{aligned}$$

The following noncanonical trivial transformations then supply the Toda Hamiltonian:

$$x' = x , \quad p_x' = 8\sqrt{3}\,p_x , \quad y' = y , \quad p_y' = 8\sqrt{3}\,p_y ; \tag{7.93}$$
$$H_\mathrm{T} = \frac{H'}{24} , \quad H_\mathrm{T} = \tfrac{1}{2}(p_x^2 + p_y^2) + \tfrac{1}{24}\left[e^{2(y+\sqrt{3}\,x)} + e^{2(y-\sqrt{3}\,x)} + e^{-4y}\right] - \tfrac{1}{8} .$$

This Hamiltonian describes the motion of a particle in a potential $U(x, y)$ with threefold symmetry.

For small $H_\mathrm{T} = E$ (also for small x and y), one can expand (7.93) to get, up to cubic terms,

$$\bar{H}' = \tfrac{1}{2}(p_x^2 + x^2) + \tfrac{1}{2}(p_y^2 + y^2) + x^2 y - \tfrac{1}{3}\,y^3 . \tag{7.94}$$

Whereas (7.93) proves to be integrable, (7.94), a two-dimensional oscillator with the perturbation term $x^2y - y^3/3$, is not. We shall return to this and similar systems later when considering stochastic systems.

If the Hamiltonian for the Toda molecule is to be completely integrable, then, in addition to the energy H and angular momentum P_3, still another conserved quantity I must exist. This has in fact been found and reads:

$$
\begin{aligned}
I(x, y, p_x, p_y) = & \, 8p_x(p_x^2 - 3p_y^2) + (p_x + \sqrt{3}\,p_y)\mathrm{e}^{2(y-\sqrt{3}\,x)} \\
& + (p_x - \sqrt{3}p_y)\mathrm{e}^{2(y+\sqrt{3}\,x)} - 2p_x \mathrm{e}^{-4y} = \text{const}\,.
\end{aligned}
\tag{7.95}
$$

Discovering that this is a conserved quantity is, of course, no trivial task. Nevertheless, it is relatively simple to confirm that $\dot{I} = 0$.

In order to prove this explicitly, we begin with (7.93) and get, as equations of motion:

$$
\dot{x} = \frac{\partial H_\mathrm{T}}{\partial p_x} = p_x\,, \quad \dot{y} = \frac{\partial H_\mathrm{T}}{\partial p_y} = p_y\,, \tag{7.96}
$$

$$
\begin{aligned}
\dot{p}_x &= -\frac{\partial H_\mathrm{T}}{\partial x} = -\frac{1}{24}\left[2\sqrt{3}\,\mathrm{e}^{2(y+\sqrt{3}\,x)} - 2\sqrt{3}\,\mathrm{e}^{2(y-\sqrt{3}\,x)}\right] \\
&= -\frac{\mathrm{e}^{2y}}{2\sqrt{3}} \sinh(2\sqrt{3}\,x)\,,
\end{aligned}
\tag{7.97}
$$

$$
\dot{p}_y = -\frac{\partial H_\mathrm{T}}{\partial y} = \frac{1}{6}\left[\mathrm{e}^{-4y} - \mathrm{e}^{2y}\cosh(2\sqrt{3}\,x)\right]. \tag{7.98}
$$

Now I can be written in the form:

$$
\begin{aligned}
I = & \, 8p_x(p_x^2 - 3p_y^2) + p_x\left(\mathrm{e}^{2(y-\sqrt{3}\,x)} + \mathrm{e}^{2(y+\sqrt{3}\,x)}\right) \\
& + \sqrt{3}\,p_y\left(\mathrm{e}^{2(y-\sqrt{3}\,x)} - \mathrm{e}^{2(y+\sqrt{3}\,x)}\right) - 2p_x\,\mathrm{e}^{-4y} \\
= & \, 8p_x(p_x^2 - 3p_y^2) - 2p_x\left[\mathrm{e}^{-4y} - \mathrm{e}^{2y}\cosh(2\sqrt{3}\,x)\right] - 2\sqrt{3}\,p_y\mathrm{e}^{2y}\sinh(2\sqrt{3}\,x) \\
= & \, 8p_x(p_x^2 - 3p_y^2) - 12p_x\dot{p}_y + 12p_y\dot{p}_x\,.
\end{aligned}
\tag{7.99}
$$

The time derivative of I is, accordingly:

$$
\dot{I} = 24p_x^2\dot{p}_x - 24\dot{p}_x p_y^2 - 48p_x p_y\dot{p}_y - \cancel{12\dot{p}_x\dot{p}_y} - 12p_x\ddot{p}_y + \cancel{12\dot{p}_y\dot{p}_x} + 12p_y\ddot{p}_x
$$

or

$$
\dot{I} = 24(p_x^2 - p_y^2)\dot{p}_x - 48p_x p_y\dot{p}_y - 12p_x\ddot{p}_y + 12p_y\ddot{p}_x\,. \tag{7.100}
$$

If we insert (7.97, 98) and the time derivatives of these into (7.100), we indeed obtain $\dot{I} = 0$.

8. The Adiabatic Invariance of the Action Variables

We shall first use an example to explain the concept of adiabatic invariance. Let us consider a "super ball" of mass m, which bounces back and forth between two walls (distance l) with velocity $\boldsymbol{v}_0$. Let gravitation be neglected, and the collisions with the walls be elastic. If F_m denotes the average force onto each wall, then we have

$$F_m T = -\int_{\text{coll. time}} f\,dt\,. \tag{8.1}$$

f is the force acting on the ball during one collision, and T is the time between collisions. Now according to the law of conservation of momentum we have

$$\int_{1\ \text{coll.}} f\,dt = P_\text{f} - P_\text{i} = -mv_0 - mv_0 = -2mv_0\,. \tag{8.2}$$

Here, $P_{\text{i, f}}$ are the initial and final momenta of the ball. Equations (8.1) and (8.2), taken together, yield

$$F_m T = 2mv_0\,. \tag{8.3}$$

Since the ball travels the distance $2l$ between collisions, with the velocity $\boldsymbol{v}_0$, the corresponding time interval is

$$T = \frac{2l}{v_0}\,, \tag{8.4}$$

so that the average force on each wall follows from (8.3):

$$F_m = \frac{2mv_0}{T} = \frac{mv_0^2}{l}\,. \tag{8.5}$$

Now let the right wall move toward the left one with the velocity $V \ll \boldsymbol{v}$. Here, too, it is valid that

$$F_m T = -\int_{1\ \text{coll.}} f\,dt = -(P_\text{f} - P_\text{i})\,. \tag{8.6}$$

In order to determine the right-hand side, we go from the laboratory system into the comoving system, i.e., place ourselves into a system that moves with the constant velocity V toward the left wall. This system is also an inertial system, since $V =$ const. Then it is clear that

$$P_\mathrm{f} - P_\mathrm{i} = -m(v+V) - m(v+V) = -2m(v+V)$$

and, accordingly,

$$F_m T = 2m(v+V)\,. \tag{8.7}$$

We still need T. To get it, we take advantage of the fact that $V \ll v$. Then the ball moves very rapidly, whereas the wall hardly moves between collisions:

$$T \cong \frac{2x}{v}\,. \tag{8.8}$$

Here, x indicates the present distance between the walls. Now, because of $v \gg V$, $(v+V) \cong v$ is valid, so that from (8.7) it follows that

$$F_m T \cong F_m \frac{2x}{v} \cong 2mv$$

and thus

$$F_m \cong \frac{mv^2}{x}\,. \tag{8.9}$$

Now we still need the velocity as a function of the distance x. To find this, we again go back into the comoving system and find for the change of velocity $(v+2V)-v = 2V = \Delta v$. This change occurs at each collision or once within every $2x/v$ seconds (cf. (8.8)); thus, it holds that

$$\frac{\Delta v}{\Delta t} = \frac{2V}{2x/v} = \frac{vV}{x}\,, \qquad \frac{dv}{dt} = \frac{vV}{x}$$

or

$$\frac{dv}{v} = \frac{V}{x}\,dt\,. \tag{8.10}$$

Now $x = (l - Vt)$, so that $dx = -Vdt$. With this we find for (8.10):

$$\frac{dv}{v} = -\frac{dx}{x}\,.$$

The integration is simple:

$$\int_{v_0}^{v} \frac{dv'}{v'} = -\int_{l}^{x} \frac{dx'}{x'}$$

or

$$\ln\left(\frac{v}{v_0}\right) = -\ln\left(\frac{x}{l}\right) \quad : \quad \frac{v}{v_0} = \frac{l}{x}\,.$$

Thus, we find for v as a function of x:

$$v = v(x) = \frac{v_0 l}{x} . \tag{8.11}$$

As was to be expected, the velocity of the ball increases as the distance between the walls decreases. Moreover, as the distance decreases, the number of collisions per unit time (collision rate) increases.

Finally, the average force on the walls can be given as a function of the momentary distance x:

$$F_m \cong \frac{mv^2(x)}{x} = \frac{m}{x}\frac{v_0^2 l^2}{x^2} = \frac{mv_0^2 l^2}{x^3} . \tag{8.12}$$

With (8.12) it is easy to show that the work performed on the ball by the wall is equal to the increase of kinetic energy of the ball. For, according to the work-energy theorem, it holds for the work performed on the ball using (8.12)

$$W = -\int_l^x F_m dx' = \frac{m}{2}v^2(x) - \frac{m}{2}v_0^2 ,$$

thus

$$-W = \frac{m}{2}v^2(l) - \frac{m}{2}v^2(x) \equiv T(l) - T(x) .$$

Although the distance and thus the kinetic energy of the particle now change, the action J is practically constant. This can be seen as follows: first, it holds that $(p = mv)$

$$J = \oint p\,dx = m\int_0^T v^2\,dt .$$

If it were true that $l =$ const., then, with $v = v_0$ we could write $T = 2l/v_0$ and therefore

$$J = mv_0^2\frac{2l}{v_0} = 2mlv_0 \equiv 2mlv(l) . \tag{8.13}$$

We can easily confirm that the action variable J practically does not change when the distance between the walls is slowly changed: J is an adiabatic invariant; i.e., if the walls are at a distance of x apart, then, from (8.11) and (8.13) it follows that

$$J_x = 2mxv(x) . \tag{8.14}$$

For the change in time of J it therefore holds that

$$\frac{dJ_x}{dt} = 2m\left(\underbrace{\dot{x}}_{=-V}v + x\underbrace{\dot{v}}_{=\frac{vV}{x}}\right) = 2m(-Vv + Vv) = 0 . \tag{8.15}$$

At a distance x, (8.14) is valid. After the occurrence of the collision onto the right wall and shortly prior to the next collision, the following changes apply:

$$x - \Delta x = x - TV = x - \frac{2x}{v} V = \left(1 - \frac{2V}{v}\right) x \,,$$
$$v(x - \Delta x) = v(x) + 2V \,,$$

$$\begin{aligned} J_{x-\Delta x} &= 2m(x - \Delta x)v(x - \Delta x) \\ &= 2m\left[\left(1 - \frac{2V}{v}\right) x(v + 2V)\right] = 2mxv\left[1 - \frac{4}{v^2} V^2\right] . \end{aligned} \tag{8.16}$$

A comparison of (8.14) and (8.16) shows that the action variable J has changed after one period by only a small amount of the order of $V^2 \sim \dot{x}^2$.

As our next example we consider the harmonic oscillator with a slowly changing restoring force or frequency ω. We want to assume that the change in time of $\omega(t)$ within the time of one period $1/\omega$ is small compared to ω, i.e., $\dot{\omega}/\omega \ll \omega$. Thus our assumption is

$$\frac{1}{\omega}\frac{\dot{\omega}}{\omega} \ll 1 \,. \tag{8.17}$$

$1/\omega = T/2\pi$ is the oscillation period, and $(\dot{\omega}/\omega)^{-1}$ corresponds to the time scale during which the restoring force changes; this is very large compared to T. The fact that $\omega(t)$ is supposed to be slowly changing during one period T is expressed by the differential equation,

$$\left[\frac{d^2}{dt^2} + \omega^2(\varepsilon t)\right] x(t) = 0 \,. \tag{8.18}$$

Again: the argument of ω^2 emphasizes the slow change of the "coupling constant" ω^2; it does not mean ω^2 is small; after a certain (long) time the coupling ω^2 will reach its maximal strength, which need not be small. We now introduce the new variable τ:

$$\tau(t) = \varepsilon t \; ; \quad \varepsilon \text{ dimensionless, small} \,. \tag{8.19}$$

Then

$$\frac{d}{dt} = \frac{d\tau}{dt}\frac{d}{d\tau} = \varepsilon \frac{d}{d\tau} \,, \quad \left(\frac{d}{dt}\right)^2 = \varepsilon^2 \left(\frac{d}{d\tau}\right)^2$$

and (8.18) can be written as:

$$\left[\frac{d^2}{d\tau^2} + \left(\frac{1}{\varepsilon}\right)^2 \omega^2(\tau)\right] x(\tau) = 0 \,. \tag{8.20}$$

We try to solve (8.20) with a WKB ansatz:

$$x(\tau) = f(\tau)\mathrm{e}^{\mathrm{i}g(\tau)}\,; \quad f, g \quad \text{real}\,. \tag{8.21}$$

Then

$$\begin{aligned}
\frac{d}{d\tau}x(\tau) =: x'(\tau) &= f'(\tau)\mathrm{e}^{\mathrm{i}g(\tau)} + fig'(\tau)\mathrm{e}^{\mathrm{i}g(\tau)} \\
x''(\tau) &= f''\mathrm{e}^{\mathrm{i}g} + f'\mathrm{i}g'\mathrm{e}^{\mathrm{i}g} + f'\mathrm{i}g'\mathrm{e}^{\mathrm{i}g} + fig''\mathrm{e}^{\mathrm{i}g} - fg'^2\mathrm{e}^{\mathrm{i}g} \\
&= \left[f'' + 2\mathrm{i}f'g' + fig'' - fg'^2\right]\mathrm{e}^{\mathrm{i}g}\,.
\end{aligned}$$

Our oscillator equation is thus

$$\left[f'' + 2\mathrm{i}f'g' + \mathrm{i}fg'' - fg'^2 + \frac{1}{\varepsilon^2}\omega^2 f\right]\mathrm{e}^{\mathrm{i}g} = 0$$

or

$$\left\{\left[f'' - fg'^2 + \frac{1}{\varepsilon^2}\omega^2 f\right] + \mathrm{i}\left[2f'g' + fg''\right]\right\}\mathrm{e}^{\mathrm{i}g} = 0\,.$$

So we have to solve the differential equations

$$f'' - fg'^2 + \frac{\omega^2}{\varepsilon^2}f = 0\,, \tag{8.22}$$

$$2f'g' + fg'' = 0\,. \tag{8.23}$$

We shall soon need the last equation in the following form:

$$\frac{f'}{f} = -\frac{1}{2}\frac{g''}{g'}\,. \tag{8.24}$$

But first we multiply (8.23) by f:

$$2f'fg' + f^2g'' = \frac{d}{d\tau}(f^2g') = 0$$

so that

$$f^2g' = C^2 \quad \text{or} \quad f = \frac{C}{\sqrt{g'}}\,. \tag{8.25}$$

We now write (8.22) as

$$fg'^2 = \frac{\omega^2}{\varepsilon^2}f + f''$$

or

$$g'^2 = \frac{\omega^2}{\varepsilon^2} + \frac{f''}{f} = \frac{\omega^2}{\varepsilon^2} + \frac{d}{d\tau}\left(\frac{f'}{f}\right) + \left(\frac{f'}{f}\right)^2\,.$$

At this point we use (8.24) to obtain

$$g'^2 = \frac{\omega^2}{\varepsilon^2} - \frac{1}{2}\frac{d}{d\tau}\left(\frac{g''}{g'}\right) + \frac{1}{4}\left(\frac{g''}{g'}\right)^2 . \tag{8.26}$$

The last equation is in a form that allows us to set up a perturbation series.
0th approximation:

$$g' = \pm\frac{1}{\varepsilon}\,\omega .$$

1st approximation:

$$\begin{aligned}
g' &= \pm\sqrt{\left(\frac{1}{\varepsilon}\,\omega\right)^2 - \frac{1}{2}\frac{d}{d\tau}\left(\frac{\omega'}{\omega}\right) + \frac{1}{4}\left(\frac{\omega'}{\omega}\right)^2} \\
&\cong \pm\left(\frac{1}{\varepsilon}\,\omega - \frac{1}{4}\frac{\varepsilon}{\omega}\underbrace{\frac{d}{d\tau}\left(\frac{\omega'}{\omega}\right)}_{=\frac{\omega''\omega-\omega'^2}{\omega^2}} + \frac{1}{8}\frac{\varepsilon}{\omega}\left(\frac{\omega'}{\omega}\right)^2\right) \\
&\cong \pm\left(\frac{1}{\varepsilon}\,\omega + \frac{3}{8}\frac{\varepsilon}{\omega}\left(\frac{\omega'}{\omega}\right)^2 - \frac{1}{4}\frac{\varepsilon}{\omega}\frac{\omega''}{\omega}\right) .
\end{aligned}$$

If we re-introduce the normal time derivative, we get

$$\frac{1}{\varepsilon}\dot{g} \cong \pm\left(\frac{1}{\varepsilon}\,\omega + \frac{3}{8}\frac{1}{\varepsilon}\frac{\dot{\omega}^2}{\omega^3} - \frac{1}{4}\frac{1}{\varepsilon}\frac{\ddot{\omega}}{\omega^2}\right)$$

or

$$\dot{g} \cong \pm\left(\omega + \frac{3}{8}\left(\frac{\dot{\omega}}{\omega}\right)\underbrace{\left(\frac{\dot{\omega}}{\omega^2}\right)}_{(17):\ll 1} - \frac{1}{4}\frac{\ddot{\omega}}{\omega^2}\right) .$$

Thus we set

$$g' \cong \pm\frac{1}{\varepsilon}\,\omega(\tau) , \quad \dot{g} \cong \pm\omega(t) \tag{8.27}$$

in first WKB approximation and obtain, according to our solution ansatz (8.21) with (8.25),

$$x(\tau) \cong \frac{C}{\sqrt{g'}}\exp\left[\pm \mathrm{i}\,\frac{1}{\varepsilon}\int^{\tau} d\tau'\,\omega(\tau')\right]$$

or

$$x(t) \cong \frac{C}{\sqrt{\omega(t)}}\exp\left[\pm \mathrm{i}\int^{t} dt'\,\omega(t')\right] . \tag{8.28}$$

The real part of (8.28) reads:

$$x(t) \cong a\sqrt{\frac{\omega_0}{\omega(t)}}\sin\left[\int^t \omega(t')dt' + \alpha\right]\,; \qquad a, \alpha = \text{const}\,.$$
$$\equiv a\sqrt{\frac{\omega_0}{\omega(t)}}\sin\varphi(t)\,. \tag{8.29}$$

In the following we shall need the time derivative of (8.29):

$$\dot{x}(t) \cong a\sqrt{\omega_0}\left[\sqrt{\omega}\cos\varphi - \frac{\dot{\omega}}{2\omega^{3/2}}\sin\varphi\right]\,.$$

Using $p = m\dot{x}$, $dq = dx = \dot{x}dt$ we get for the action variable $\left(\omega \equiv \frac{d\varphi}{dt}\right)$

$$J = \oint_{1\,\text{per.}} p\,dq = \oint m\dot{x}^2 dt$$
$$= m\omega_0 a^2 \oint \left[\omega\cos^2\varphi + \frac{\dot{\omega}^2}{4\omega^3}\sin^2\varphi - \frac{\dot{\omega}}{\omega}\overbrace{\sin\varphi\cos\varphi}^{=\frac{1}{2}\sin(2\varphi)}\right]dt$$
$$= m\omega_0 a^2 \oint d\varphi \left[\cos^2\varphi + \left(\frac{\dot{\omega}}{2\omega^2}\right)^2\sin^2\varphi - \frac{\dot{\omega}}{2\omega^2}\sin(2\varphi)\right]\,.$$

With the familiar integrals $\oint \sin^2\varphi\, d\varphi = \pi = \oint \cos^2\varphi\, d\varphi$, $\oint \sin(2\varphi)d\varphi = 0$, we obtain the result,

$$J = \pi m\omega_0 a^2\left[1 + \left(\frac{1}{2}\frac{\dot{\omega}}{\omega^2}\right)^2\right]\,. \tag{8.30}$$

Our adiabatic invariant is thus

$$J = \pi m\omega_0 a^2 = \pi\sqrt{mk}\,a^2\,. \tag{8.31}$$

The correction term is of the order $O(\dot{\omega}^2)$.

If we use (8.31) in the form $a\sqrt{\omega_0} = \sqrt{J/\pi m}$, then (8.29) can be written as

$$x(t) \cong \sqrt{\frac{J}{\pi m\omega(t)}}\sin\left[\int^t \omega(t')dt' + \alpha\right] \tag{8.32}$$

or

$$p(t) \cong \sqrt{\frac{m\omega(t)J}{\pi}}\cos\left[\int^t \omega(t')dt' + \alpha\right]\,. \tag{8.33}$$

These results should be compared with the formulae (7.19) and (7.20). During the time interval $(t, t + 2\pi/\omega)$, (8.32, 33) represents (approximately) an ellipse in (x, p)-phase space.

Next we again consider the problem of a charged particle in an external homogeneous magnetic field $\boldsymbol{B}$ which points in the z-direction. The force acting on the moving particle is the Lorentz force: $m(d\boldsymbol{v}/dt) = (e/c)\boldsymbol{v} \times \boldsymbol{B}$ or

$$\frac{d\boldsymbol{v}}{dt} = \left(-\frac{e\boldsymbol{B}}{mc}\right) \times \boldsymbol{v} = \boldsymbol{\omega}_c \times \boldsymbol{v} . \tag{8.34}$$

Equation (8.34) says that the velocity vector precesses around the direction of the $\boldsymbol{B}$-field with the angular frequency (cyclotron frequency) $\omega_c = -eB/mc$. Of course, the Lorentz force also follows from the Lagrange formulation of the problem:

$$L = T - V = \frac{m}{2}v^2 + \boldsymbol{\mu} \cdot \boldsymbol{B} . \tag{8.35}$$

The last term in (8.35) is the potential energy of a magnetic dipole in presence of a magnetic field: $V = -\boldsymbol{\mu} \cdot \boldsymbol{B}$; more precisely: $\boldsymbol{\mu}$ is the orbital magnetic dipole moment: $\boldsymbol{\mu} = (e/2mc)\boldsymbol{L}$. Then the z-component of $\boldsymbol{\mu}$ is given in cylindrical coordinates by

$$\mu_z = \frac{e}{2mc}(\boldsymbol{r} \times \boldsymbol{p})_z = \frac{e}{2c}(\boldsymbol{r} \times \boldsymbol{v})_z = \frac{e}{2c}r^2\dot{\varphi} \tag{8.36}$$

so that, from (8.35), it follows for L that

$$L = \frac{m}{2}(\dot{r}^2 + r^2\dot{\varphi}^2 + \dot{z}^2) + \frac{eB}{2c}r^2\dot{\varphi} . \tag{8.37}$$

Obviously, φ is an ignorable variable; thus the canonically conjugate momentum p_φ is conserved:

$$p_\varphi = \frac{\partial L}{\partial \dot{\varphi}} = mr^2\dot{\varphi} + \frac{eB}{2c}r^2 = \text{const.} \tag{8.38}$$

The radial equation can be obtained from

$$\frac{d}{dt}\frac{\partial L}{\partial \dot{r}} - \frac{\partial L}{\partial r} = m\ddot{r} - mr\dot{\varphi}^2 - \frac{eB}{c}r\dot{\varphi} = 0$$

or

$$m\ddot{r} - r\dot{\varphi}\left(m\dot{\varphi} + \frac{eB}{c}\right) = 0 . \tag{8.39}$$

The regular circular motion follows then from $\dot{r} = 0$, $\dot{\varphi} = \text{const.}$, where $\dot{\varphi} = \omega_c = -eB/mc$ (as above). For this we get from (8.38)

$$p_\varphi = -\frac{eB}{2c}r^2 = \text{const.}$$

and the action variable becomes

$$J_\varphi = \oint p_\varphi \, d\varphi = -\frac{eB}{2c}r^2 2\pi = -\frac{\pi eB}{c}r^2 . \tag{8.40}$$

If we use (8.36) in the form $er^2/c = 2\mu_z/\omega_c$ then (8.40) can be written as

$$J_\varphi = -\frac{2\pi\mu_z B}{\omega_c} = \frac{2\pi mc}{e}\,\mu_z\,. \tag{8.41}$$

The magnetic moment is thus an adiabatic invariant: in the case of sufficiently small changes of the external magnetic field, $\mu_z(J_\varphi)$ remains constant. If we look at (8.40), we can say that B times the encompassed area of the circular orbit (flux) remains constant.

Finally we consider once again the problem of the linear harmonic oscillator with time-dependent frequency: $\ddot{x}+\omega^2(t)x = 0$. Many physical problems can be reduced to this equation, e.g., the motion of a charged particle in a time-dependent magnetic field. Also the treatment of small oscillations of a pendulum, whose length changes constantly with time, belongs to this realm of problems. Here we are interested in the remarkable fact that the harmonic oscillator with time-dependent frequency possesses an exact invariant which reduces to the action variable J in case of an adiabatic change of ω.

The equation of motion for the harmonic oscillator is known to be derived from the Lagrangian or Hamiltonian:

$$L = \frac{m}{2}\dot{x}^2 - \frac{m}{2}\,\omega^2(t)x^2\,; \quad H = \frac{p^2}{2m} + \frac{m}{2}\,\omega^2(t)x^2\,. \tag{8.42}$$

The conserved quantity $I(t)$ with $\dot{I}(t) = 0$ is given by

$$I(t) = \frac{1}{2}\left[\frac{x^2}{\varrho^2} + (\varrho\dot{x} - x\dot{\varrho})^2\right]\,, \tag{8.43}$$

where $\varrho(t)$ satisfies the following differential equation:

$$\ddot{\varrho} + \omega^2(t)\varrho - \frac{1}{\varrho^3} = 0\,. \tag{8.44}$$

If we use this equation and, in the following, take advantage of the fact that $\ddot{x} = -\omega^2(t)x$, then it is easy to show that indeed $\dot{I} = 0$:

$$2\frac{dI}{dt} = \frac{d}{dt}\left(\frac{x^2}{\varrho^2}\right) + \frac{d}{dt}(\varrho\dot{x} - x\dot{\varrho})^2$$

or

$$\begin{aligned}\frac{dI}{dt} &= \frac{x\dot{x}}{\varrho^2} - \frac{\dot{\varrho}x^2}{\varrho^3} + \varrho x \overbrace{\ddot{x}}^{=-\omega^2 x} - \varrho\ddot{\varrho}x\dot{x} - \varrho\dot{\varrho}x \overbrace{\ddot{x}}^{=-\omega^2 x} + x^2\dot{\varrho}\ddot{\varrho} \\ &= \varrho x\dot{x}\left(-\ddot{\varrho} - \omega^2\varrho + \frac{1}{\varrho^3}\right) + x^2\dot{\varrho}\left(\ddot{\varrho} + \omega^2\varrho - \frac{1}{\varrho^3}\right) = 0\,. \end{aligned} \tag{8.45}$$

In order to better understand the physical significance of the invariant (8.43), we consider the motion of the one-dimensional harmonic oscillator as a projection of the motion of a plane two-dimensional oscillator on the x axis. This kind of consideration of a linear harmonic oscillator is also valid when the frequency ω is time-dependent. So we shall first study as an auxiliary problem a central force problem with time-dependent potential

$$V = \frac{m}{2}\omega^2(t)\varrho^2 , \quad \varrho^2 = x^2 + y^2 .$$

In plane polar coordinates $x = \varrho\cos\varphi$, $y = \varrho\sin\varphi$, L reads for our auxiliary problem:

$$L = \frac{m}{2}\left[(\dot{x}^2 - \omega^2 x^2) + (\dot{y}^2 - \omega^2 y^2)\right] = \frac{m}{2}(\dot{\varrho}^2 + \varrho^2\dot{\varphi}^2 - \omega^2(t)\varrho^2) . \tag{8.46}$$

In cartesian coordinates we have two linear uncoupled harmonic oscillators. In plane polar coordinates, (8.46) tells us that φ is ignorable and therefore the angular momentum is conserved:

$$L_z \equiv p_\varphi = \frac{\partial L}{\partial\dot{\varphi}} = m\varrho^2\dot{\varphi} = \text{const.}$$

or

$$\varrho^2\dot{\varphi} = \frac{L_z}{m} = h = \text{const.} \tag{8.47}$$

In the following we need the radial equation; it follows from $d(\partial L/\partial\dot{\varrho})/dt - \partial L/\partial\varrho = 0$:

$$m\ddot{\varrho} + m\omega^2\varrho - m\dot{\varphi}^2\varrho = 0 \tag{8.48}$$

or, with (8.47):

$$\ddot{\varrho} + \omega^2\varrho - \frac{h^2}{\varrho^3} = 0 . \tag{8.49}$$

For $h = 1$, this equation becomes (8.44). But since ϱ now satisfies (8.49), the invariant at first reads – with the help of the conserved quantity h –

$$I' = \frac{1}{2}\left[\frac{h^2x^2}{\varrho^2} + (\varrho\dot{x} - x\dot{\varrho})^2\right] . \tag{8.50}$$

One can immediately write down the value for this invariant if one takes into consideration that, with $x = \varrho\cos\varphi$, the following holds:

$$\varrho\dot{x} - x\dot{\varrho} = \varrho(\dot{\varrho}\cos\varphi - \varrho\dot{\varphi}\sin\varphi) - \varrho\dot{\varrho}\cos\varphi = -\varrho^2\dot{\varphi}\sin\varphi = -h\sin\varphi$$

and thus, it follows from (8.50) that

$$I' = \frac{1}{2}(h^2\cos^2\varphi + h^2\sin^2\varphi) = \frac{h^2}{2} . \tag{8.51}$$

Thus the constancy of I' is proven equivalent to the conservation of angular momentum in the associated two-dimensional oscillator problem.

Now we want to explain why it is always possible to choose the initial value for the y-amplitude and the phase between the x- and y-motion in such a way that h takes the value 1. In order to do so, we begin with an initial frequency ω_0, which is to be constant: $x = x_0 \sin(\omega_0 t)$, $y = y_0 \cos(\omega_0 t + \alpha)$. Then, by definition, h becomes:

$$\frac{L_z}{m} = h = x\dot{y} - y\dot{x} = -\omega_0 x_0 y_0 \cos\alpha \ . \tag{8.52}$$

If we allow ω to change in time, h of course maintains its value as conserved quantity. We can make this 1 because of the free choice of y_0, α for every x_0. Thus we can relate (embed) each linear harmonic oscillator with time-dependent frequency to a plane isotropic oscillator with $h = 1$. At that moment (8.49) becomes (8.44), and I' from (8.50) with $h = 1$ becomes (8.43). The existence of the exact invariant I in the case of the one-dimensional harmonic oscillator thus results from the fact that the angular momentum is conserved in the associated problem of the two-dimensional oscillator.

Now we finally come to the relation between J and I for the harmonic oscillator. If ω is constant, then a possible solution of (8.44) is $\varrho = \sqrt{1/\omega}$ (plane circular orbit). Since $\dot{\varrho} = 0$, we obtain directly from (8.43):

$$I(t) = \frac{1}{2}\left[x^2\omega + \frac{\dot{x}^2}{\omega}\right] = \frac{1}{2\omega}[\dot{x}^2 + \omega^2 x^2] \tag{8.53}$$

$$= \frac{1}{m\omega}\left[\frac{m}{2}\dot{x}^2 + \frac{m}{2}\omega^2 x^2\right] = \frac{1}{m}\frac{E}{\omega} = \frac{J}{2\pi m} \ . \tag{8.54}$$

So, in lowest order ($\omega =$ const.), the exact invariant I is proportional to the action variable J.

9. Time-Independent Canonical Perturbation Theory

First we consider the perturbation calculation only to first order, limiting ourselves to only one degree of freedom. Furthermore, the system is to be conservative, $\partial H/\partial t = 0$, and periodic in both the unperturbed and perturbed case. In addition to periodicity, we shall require the Hamilton-Jacobi equation to be separable for the unperturbed situation. The unperturbed problem $H_0(J_0)$ which is described by the action-angle variables J_0 and w_0 will be assumed to be solved. Thus we have, for the unperturbed frequency:

$$\nu_0 = \frac{\partial H_0}{\partial J_0} \tag{9.1}$$

and

$$w_0 = \nu_0 t + \beta_0 \,. \tag{9.2}$$

Then the new Hamiltonian reads, up to a perturbation term of first order:

$$H = H_0\big(J_0\big) + \varepsilon H_1\big(w_0, J_0\big) \,, \tag{9.3}$$

where ε is a small parameter. Our goal now is to find a canonical transformation from the action-angle variables (J_0, w_0) of the unperturbed problem $H_0(J_0)$ to action-angle variables (J, w) of the total problem $H(J)$; this canonical transformation should make the perturbed problem become solvable. If we can achieve this, then it holds that $H = E(J)$, where $J = \text{const.}$, and now (1) and (2) are replaced by

$$\nu = \frac{\partial H(J)}{\partial J} \tag{9.4}$$

and

$$w = \nu t + \beta \,. \tag{9.5}$$

The canonical transformation in question can be generated with the help of the generating function of the type $F_2(q, P) : W = W(w_0, J)$. w_0 stands for the old coordinate and J for the new momentum. Since we are limiting ourselves to $\partial F_2/\partial t = 0$, it holds that $H_{\text{old}} = H_{\text{new}}$. Then the Hamilton-Jacobi equation reads:

$$H\big(w_0, J_0\big) = H\left(w_0, \frac{\partial W}{\partial w_0}\right) = E(J) \tag{9.6}$$

with

$$J_0 = \frac{\partial W(w_0, J)}{\partial w_0}, \quad w = \frac{\partial W(w_0, J)}{\partial J}. \tag{9.7}$$

This corresponds to the familiar transformation equations $p = \partial F_2/\partial q$ and $Q = \partial F_2/\partial P$.

It is important to emphasize that for the perturbed problem, the (w, J) are "good" action-angle variables, while the (w_0, J_0) "basis" no longer plays the role of action-angle variables. w_0 is angle variable for the unperturbed case and is related to the original coordinate q by

$$q = \sum_{k=-\infty}^{+\infty} a_k\big(J_0\big)\,\mathrm{e}^{2\pi i k w_0} \quad \text{(libration)}, \tag{9.8}$$

or

$$q - q_0 w_0 = \sum_{k=-\infty}^{+\infty} a_k\big(J_0\big)\,\mathrm{e}^{2\pi i k w_0} \quad \text{(rotation)}. \tag{9.9}$$

Certainly (w_0, J_0) remain canonical variables for the perturbed situation, since they are, according to the above, related to the original canonical variables (q, p) by a canonical transformation. J_0 is now, however, no longer constant $[\dot{J}_0 = -\partial H/\partial w_0 = -\varepsilon(\partial H_1(w_0, J_0)/\partial w_0)]$ and w_0 is no longer a linear function in time $[\dot{w}_0 = \partial H/\partial J_0 = \partial H_0(J_0)/\partial J_0 + \varepsilon(\partial H_1(w_0, J_0)/\partial J_0) \neq \text{const.}]$. Since (w, J) are action-angle variables, w increases by one unit when q runs through one period. This also applies, however, to w_0, because q is, according to (9.8), a periodic function of w_0 with period 1. The canonical transformation (9.8) expresses q in dependence of (w_0, J_0), and has nothing to do with the particular form of the Hamiltonian.

We now return to (9.6) and treat this equation perturbatively, i.e., we expand both sides:

$$H\big(w_0, J_0\big) = H_0\big(J_0\big) + \varepsilon H_1\big(w_0, J_0\big) + \ldots \tag{9.10}$$

$$E(J; \varepsilon) = E_0(J) + \varepsilon E_1(J) + \varepsilon^2 E_2(J) + \ldots. \tag{9.11}$$

We apply the same procedure to the generating function $W(w_0, J)$ of the canonical transformation (9.7), which transforms (w_0, J_0) to (w, J):

$$W\big(w_0, J\big) = \underbrace{W_0(w_0, J)}_{=w_0 J} + \varepsilon W_1\big(w_0, J\big) + \varepsilon^2 W_2\big(w_0, J\big) + \ldots. \tag{9.12}$$

For $\varepsilon = 0$, only the identity transformation $w_0 J$ remains. The transformation equations (9.7) take on the following form:

$$J_0 = \frac{\partial W(w_0, J)}{\partial w_0} = J + \varepsilon \frac{\partial W_1(w_0, J)}{\partial w_0} + \ldots$$

$$w = \frac{\partial W(w_0, J)}{\partial J} = w_0 + \varepsilon \frac{\partial W_1(w_0, J)}{\partial J} + \ldots. \tag{9.13}$$

The Hamilton-Jacobi equation (9.6) can then be written in first-order perturbation theory:

$$H\left(w_0, \frac{\partial W}{\partial w_0}\right) = E(J) :$$

$$H_0\big(J_0\big) + \varepsilon H_1\Bigg(w_0, \underbrace{\frac{\partial W}{\partial w_0}}_{(13):J+\varepsilon\partial W_1/\partial w_0}\Bigg) = E_0(J) + \varepsilon E_1(J) \tag{9.14}$$

and with

$$\begin{aligned} H_0\big(J_0\big) &\underset{(13)}{=} H_0\left(J + \varepsilon\frac{\partial W_1}{\partial w_0}\right) \\ &= H_0(J) + \varepsilon\frac{\partial W_1}{\partial w_0}\frac{\partial H_0(J)}{\partial J} + \mathrm{O}\big(\varepsilon^2\big) \end{aligned}$$

we get

$$\begin{aligned} &H_0(J) + \varepsilon\left[H_1\big(w_0, J\big) + \frac{\partial H_0(J)}{\partial J}\frac{\partial W_1(w_0, J)}{\partial w_0}\right] \\ &\quad = E_0(J) + \varepsilon E_1(J) \; . \end{aligned} \tag{9.15}$$

Here, only w_0 and the constant J still appear. Comparison of coefficients in ε finally yields:

$$\varepsilon^0 : \; H_0(J) = E_0(J) \; , \tag{9.16}$$

$$\varepsilon^1 : \; H_1\big(w_0, J\big) + \frac{\partial H_0(J)}{\partial J}\frac{\partial W_1(w_0, J)}{\partial w_0} = E_1(J) \; . \tag{9.17}$$

Equation (9.17) contains the two unknown functions $W_1(w_0, J)$ and $E_1(J)$. Two assumptions permit us to solve (9.17). First of all, we set

$$\frac{\partial H_0(J)}{\partial J} = \left.\frac{\partial H_0(J_0)}{\partial J_0}\right|_{J_0=J} = \nu_0 \; . \tag{9.18}$$

ν_0 is the frequency of the solved problem! Then (9.17) becomes

$$H_1\big(w_0, J\big) + \nu_0\frac{\partial W_1(w_0, J)}{\partial w_0} = E_1(J) \; . \tag{9.19}$$

The inhomogeneous term H_1 is given, $E_1(J)$ is unknown. Thus, (9.19) is a linear partial differential equation with constant coefficient (ν_0) for W_1.

Next we take advantage of the fact that the function W_1 is a periodic function of w_0. In this respect we recall that the function $W^*(w_0, w) \equiv W(w_0, J) - w_0 J$ is a periodic function of w_0; since J is an action variable here, it holds that $J = \oint p\,dq = \oint(\partial W/\partial q)dq$, so that for a single rotation in q, the action increases by J. Simultaneously, w_0 increases by one unit, (9.8). Then it holds that

$$W^*(w_0+1, w) = W(w_0+1, J) - (w_0+1)J = W(w_0, J) + J - w_0 J - J \\ = W(w_0, J) - w_0 J = W^*(w_0, w) \,. \tag{9.20}$$

Because

$$W^* \underset{(12)}{=} \varepsilon W_1(w_0, J) + \varepsilon^2 W_2(w_0, J) + \ldots \tag{9.21}$$

every W_i, in particular, W_1, is also a periodic function in w_0:

$$W_1(w_0, J) = \sum_{k=-\infty}^{+\infty} C_k(J)\, e^{2\pi i k w_0} \,. \tag{9.22}$$

Consequently, $\partial W_1/\partial w_0$ in (9.19) contains no constant term. If one now averages (9.19) over one period w_0 of the unperturbed problem, one gets

$$E_1(J) = \overline{H_1(w_0, J)} \,, \tag{9.23}$$

because the average over the derivative of the periodic function (9.22) vanishes. If we then insert the expression for $E_1(J)$ in (9.23) into (9.19), we have

$$\nu_0 \frac{\partial W_1(w_0, J)}{\partial w_0} = -\left[H_1(w_0, J) - \overline{H_1(w_0, J)}\right] =: -\{H_1\} \,. \tag{9.24}$$

Here, the right-hand side is known, and we thus get a linear partial differential equation with constant coefficients for W_1. Note that averaging the right-hand side of (9.24) indeed yields zero.

If we now are interested in the new frequency, the knowledge of W_1 is superfluous, since we only need (9.23) in

$$\nu = \frac{\partial E(J)}{\partial J} = \nu_0 + \varepsilon \frac{\partial E_1(J)}{\partial J} = \nu_0 + \varepsilon \frac{\partial \bar{H}_1}{\partial J} \,. \tag{9.25}$$

We now come to a few simple illustrative examples and begin by determining the dependence of the frequency on the amplitude in first-order perturbation theory for a perturbed oscillator potential,

$$V(q) = \frac{k}{2} q^2 + \frac{1}{6} \varepsilon m q^6 \,, \tag{9.26}$$

where $k = m\omega_0^2$, and ω_0 is the small-amplitude frequency of the unperturbed oscillator. The Hamiltonian of the problem is given by

$$H = T + V = \frac{p^2}{2m} + \frac{m}{2}\omega_0^2 q^2 + \varepsilon \frac{m q^6}{6} = H_0 + \varepsilon H_1 \,. \tag{9.27}$$

For the unperturbed Hamiltonian we already have found that

$$H_0 = \nu_0 J_0 = \frac{\omega_0}{2\pi} J_0 \,, \qquad w_0 = \nu_0 t + \beta_0 \,, \tag{9.28}$$

$$q = \sqrt{\frac{J_0}{\pi m \omega_0}} \sin(2\pi w_0) \,, \quad p = \sqrt{\frac{m\omega_0 J_0}{\pi}} \cos(2\pi w_0) \,. \tag{9.29}$$

According to (9.23), we have to compute

$$E_1(J) = \bar{H}_1 = \frac{m}{6}\overline{q^6} = \frac{m}{6}\left(\frac{J}{\pi m \omega_0}\right)^3 \overline{\sin^6(2\pi w_0)} \,. \tag{9.30}$$

In order to determine the average value in (9.30), we recall that

$$\begin{aligned}\sin^6\alpha &= \left(\frac{1}{2\mathrm{i}}\right)^6 \left(\mathrm{e}^{\mathrm{i}\alpha} - \mathrm{e}^{-\mathrm{i}\alpha}\right)^6 = \left(-\frac{1}{4}\right)^3 \\ &\quad \times \left[1 \cdot \mathrm{e}^{6\mathrm{i}\alpha} - 6\mathrm{e}^{4\mathrm{i}\alpha} + 15\mathrm{e}^{2\mathrm{i}\alpha} - 20 + 15\mathrm{e}^{-2\mathrm{i}\alpha} + 6\mathrm{e}^{-4\mathrm{i}\alpha} + 1 \cdot \mathrm{e}^{-6\mathrm{i}\alpha}\right] \\ &= -\frac{2}{64}[\cos(6\alpha) - 6\cos(4\alpha) + 15\cos(2\alpha) - 10] \,.\end{aligned}$$

Thus we have

$$\overline{\sin^6\alpha} = \left(-\tfrac{2}{64}\right)(-10) = \tfrac{5}{16} \,. \tag{9.31}$$

For the energy correction E_1 in $E = E_0 + \varepsilon E_1$, it therefore follows from (9.30) that

$$E_1(J) = \frac{m}{6}\,\frac{5}{16}\left(\frac{J}{\pi m \omega_0}\right)^3 \,. \tag{9.32}$$

We have been looking for the new frequency,

$$\nu = \frac{\partial E(J)}{\partial J} = \nu_0 + \varepsilon \frac{5m}{32}\,\frac{J^2}{(\pi m \omega_0)^3} \,. \tag{9.33}$$

If A is the maximum amplitude of the unperturbed harmonic oscillator, then $J = J_0 = \pi m \omega_0 A^2$ in first-order perturbation theory. Then (9.33) becomes

$$\nu = \nu_0\left[1 + \frac{5}{32}\,\frac{\varepsilon m}{\nu_0}\,\frac{A^4}{\pi m \omega_0}\right] = \nu_0 + \frac{5}{64\pi^2}\,\frac{\varepsilon A^4}{\nu_0}$$

or

$$\Delta\nu = \nu - \nu_0 = \frac{5}{64\pi^2}\,\frac{\varepsilon A^4}{\nu_0} \,, \tag{9.34}$$

$$\frac{\Delta\nu}{\nu_0} = \frac{5}{64\pi^2}\,\frac{\varepsilon A^4}{\nu_0^2} \,; \qquad \frac{\Delta\omega}{\omega_0} = \frac{5}{16}\,\frac{\varepsilon A^4}{w_0^2} \,. \tag{9.35}$$

A further example with one degree of freedom is the plane mathematical pendulum with small amplitude. If l is the length of the pendulum and the origin of the coordinate system is assumed to be in the suspension point, then the Hamiltonian reads:

$$H = \frac{p^2}{2ml^2} + mgl(1 - \cos\varphi) \tag{9.36}$$
$$\cong \frac{p^2}{2ml^2} + mgl\left(\frac{\varphi^2}{2} - \frac{\varphi^4}{24}\right) .$$

Introducing $I = ml^2$, $\omega_0 = \sqrt{g/l}$ we have

$$H = \frac{p^2}{2I} + \frac{I\omega_0^2}{2}\varphi^2 - \frac{1}{24}I\omega_0^2\varphi^4 = H_0(\text{H.O.}) + \varepsilon H_1 . \tag{9.37}$$

We now substitute $m \to I$ and $q \to \varphi$ in (9.29):

$$\varphi = \sqrt{\frac{J_0}{I\pi\omega_0}}\sin(2\pi w_0) , \quad p = \sqrt{\frac{I\omega_0 J_0}{\pi}}\cos(2\pi w_0) . \tag{9.38}$$

Now we can express H in terms of action-angle variables and in this manner gain access to a perturbative treatment:

$$H = \frac{\omega_0}{2\pi}J_0 - \frac{1}{24}\frac{J_0^2}{I\pi^2}\sin^4(2\pi w_0) , \tag{9.39}$$

with

$$\varepsilon H_1 = -\frac{J_0^2}{24I\pi^2}\sin^4(2\pi w_0) . \tag{9.40}$$

For ε we choose φ_1^2, the maximum angle of the harmonically swinging pendulum (with small amplitude). Then (9.23) tells us that

$$E_1(J) = \overline{H_1(w_0, J)} = -\frac{J^2}{24I\pi^2\varphi_1^2}\overline{\sin^4(2\pi w_0)} .$$

Here, we have

$$\overline{\sin^4(2\pi w_0)} = \int_0^1 dw_0 \sin^4(2\pi w_0) = \frac{3}{8}$$

since

$$\sin^4\alpha = \left(\frac{1}{2i}\right)^4 (e^{i\alpha} - e^{-i\alpha})^4 = \frac{1}{16}\left[e^{4i\alpha} - 6e^{2i\alpha} + 6 - 6e^{-2i\alpha} + e^{-4i\alpha}\right]$$
$$= \frac{1}{8}[\cos(4\alpha) - 6\cos(2\alpha) + 3]$$

so that

$$\overline{\sin^4\alpha} = \tfrac{3}{8} .$$

Up until now we have

$$E_1(J) = -\frac{J^2}{64I\pi^2\varphi_1^2} \,.$$

The frequency change results from this as

$$\Delta\nu = \varepsilon\frac{\partial E_1(J)}{\partial J} = -\frac{J}{32I\pi^2} \,. \tag{9.41}$$

Since we are determining $\Delta\nu$ in first order, we can replace J by J_0 here: $J_0 = (2\pi/\omega_0)E_0$ with $E_0 = I\omega_0^2\varphi_1^2/2$. Then J_0 becomes

$$J_0 = \pi I\omega_0\varphi_1^2 = 2\pi^2\varphi_1^2\nu_0 I \,.$$

We insert this into (9.41) and get

$$\Delta\nu = -\frac{2\pi^2 I\varphi_1^2\nu_0}{32I\pi^2} = -\frac{\varphi_1^2}{16}\nu_0 \tag{9.42}$$

or

$$\frac{\Delta\nu}{\nu_0} = \frac{\nu-\nu_0}{\nu_0} = -\frac{\varphi_1^2}{16} \,. \tag{9.43}$$

10. Canonical Perturbation Theory with Several Degrees of Freedom

We extend the perturbation theory of the previous chapter by going one order further and permitting several degrees of freedom. So let the unperturbed problem $H_0(J_k^0)$ be solved. Then we expand the perturbed Hamiltonian in the (w_k^0, J_k^0)-"basis" according to

$$H(w_k^0, J_k^0) = H_0(J_k^0) + \varepsilon H_1(w_k^0, J_k^0) + \varepsilon^2 H_2(w_k^0, J_k^0) + \dots . \tag{10.1}$$

We are looking for the generating function of the canonical transformation which will lead us from the variables (J_k^0, w_k^0) to the new variables (J_k, w_k). This generating function is the solution to the Hamilton–Jacobi equation

$$H\left(w_k^0, \frac{\partial W}{\partial w_k^0}\right) = E(J_k) \tag{10.2}$$

with

$$J_k^0 = \frac{\partial W(w_k^0, J_k)}{\partial w_k^0}, \qquad w_k = \frac{\partial W(w_k^0, J_k)}{\partial J_k}. \tag{10.3}$$

Since we want to solve (10.2) perturbatively, we expand both sides as follows:

$$\begin{aligned} H_0\left(\frac{\partial W}{\partial w_k^0}\right) + \varepsilon H_1\left(w_k^0, \frac{\partial W}{\partial w_k^0}\right) + \varepsilon^2 H_2\left(w_k^0, \frac{\partial W}{\partial w_k^0}\right) + \dots \\ = E_0(J_k) + \varepsilon E_1(J_k) + \varepsilon^2 E_2(J_k) + \dots . \end{aligned} \tag{10.4}$$

With the expansion of $W(w_k^0, J_k)$,

$$W\left(w_k^0, J_k\right) = \sum_k w_k^0 J_k + \varepsilon W_1\left(w_k^0, J_k\right) + \varepsilon^2 W_2\left(w_k^0, J_k\right) + \dots \tag{10.5}$$

$$J_k^0 = \frac{\partial W(w_k^0, J_k)}{\partial w_k^0} = J_k + \varepsilon \frac{\partial W_1}{\partial w_k^0} + \varepsilon^2 \frac{\partial W_2}{\partial w_k^0} + \dots \tag{10.6}$$

is valid. We use this in (10.4), where we write

$$\begin{aligned} H_0\left(J_k^0\right) = & H_0(J_k) + \left(\varepsilon \frac{\partial W_1}{\partial w_k^0} + \varepsilon^2 \frac{\partial W_2}{\partial w_k^0}\right) \frac{\partial H_0(J_k)}{\partial J_k} \\ & + \frac{1}{2}\left(\varepsilon \frac{\partial W_1}{\partial w_k^0}\right)\left(\frac{\partial^2 H_0(J_k)}{\partial J_k^0 \partial J_l^0}\right)\left(\varepsilon \frac{\partial W_1}{\partial w_l^0}\right) + \dots . \end{aligned} \tag{10.7}$$

We again set

$$\frac{\partial H_0(J_k)}{\partial J_k} = \left.\frac{\partial H_0(J_k^0)}{\partial J_k^0}\right|_{J_k^0=J_k} = \nu_k^0 \, .$$

Furthermore, in (10.4) we need

$$\varepsilon H_1\left(w_k^0, J_k^0\right) = \varepsilon H_1\left(w_k^0, J_k\right) + \varepsilon^2 \frac{\partial W_1}{\partial w_k^0}\frac{\partial H_1(J_k)}{\partial J_k} \tag{10.8}$$

$$\varepsilon^2 H_2\left(w_k^0, J_k^0\right) = \varepsilon^2 H_2\left(w_k^0, J_k\right) \, . \tag{10.9}$$

The results (10.7–9) are now inserted into the left-hand side of (10.4); by comparing coefficients in ε we get

$$\varepsilon^0 : \quad H_0(J_k) = E_0(J_k) \, , \tag{10.10}$$

$$\varepsilon^1 : \quad H_1\left(w_k^0, J_k\right) + \sum_k \nu_k^0 \frac{\partial W_1}{\partial w_k^0} = E_1(J_k) \, , \tag{10.11}$$

$$\varepsilon^2 : \quad K_2\left(w_k^0, J_k\right) + \sum_k \nu_k^0 \frac{\partial W_2}{\partial w_k^0} = E_2(J_k) \, , \tag{10.12}$$

with

$$\begin{aligned} K_2\left(w_k^0, J_k\right) :=& H_2\left(w_k^0, J_k\right) + \frac{\partial W_1}{\partial w_k^0}\frac{\partial H_1(J_k)}{\partial J_k} \\ &+ \frac{1}{2}\frac{\partial W_1}{\partial w_k^0}\frac{\partial^2 H_0(J_k)}{\partial J_k \partial J_l}\frac{\partial W_1}{\partial w_l^0} \, . \end{aligned} \tag{10.13}$$

As in the previous chapter it can be shown that every W_i is a periodic function in every w_k^0 argument:

$$W_i\left(w_k^0, J_k\right) = \sum_{j_1=-\infty}^{+\infty} \cdots \sum_{j_N=-\infty}^{+\infty} B_{j_1 \dots j_N}^{(i)}(J_k) \mathrm{e}^{2\pi \mathrm{i}(j_1 w_1^0 + \dots + j_N w_N^0)} \, . \tag{10.14}$$

The derivatives of W_i with respect to w_k^0 have no constant term, so that, after averaging over a complete period $(w_1^0 \dots w_N^0)$ of the unperturbed motion, we obtain the following system of equations:

$$H_0(J_k) = E_0(J_k) \, , \tag{10.15}$$

$$\overline{H_1\left(w_k^0, J_k\right)} = E_1(J_k) \, , \tag{10.16}$$

$$\overline{K_2\left(w_k^0, J_k\right)} = E_2(J_k) \, . \tag{10.17}$$

If we insert (10.16) into (10.11), and (10.17) into (10.12), we get

$$\sum_{k=1}^{N} \nu_k^0 \frac{\partial W_1}{\partial w_k^0} = -(H_1 - \bar{H}_1) =: -\{H_1\}, \tag{10.18}$$

or

$$\sum_{k=1}^{N} \nu_k^0 \frac{\partial W_2}{\partial w_k^0} - (K_2 - \bar{K}_2) =: -\{K_2\}. \tag{10.19}$$

From this, W_1 and W_2 can be determined. According to (10.13), we need to know W_1 in order to calculate K_2 or $\bar{K}_2$. Since W_1 is periodic in w_k^0, it holds that

$$W_1 = \sum_{j_1,\dots,j_N=-\infty}^{+\infty} B^{(1)}_{j_1\dots j_N}(J_k) e^{2\pi i(j_1 w_1^0 + \dots + j_N w_N^0)}. \tag{10.20}$$

Correspondingly,

$$\sum_{k=1}^{N} \nu_k^0 \frac{\partial W_1}{\partial w_k^0} = \sum_{\substack{j_1,\dots,j_N=-\infty \\ j_i\neq 0}}^{+\infty} C_{j_1\dots j_N}(J_k)\, e^{2\pi i(j_1 w_1^0 + \dots + j_N w_N^0)} \tag{10.21}$$

$$\underset{(20)}{=} \sum_{\substack{j_i\in\mathbb{Z} \\ j_i\neq 0}} B^{(1)}_{j_1\dots j_N}(J_k) \sum_{k=1}^{N} j_k \nu_k^0\, e^{2\pi i(j_1 w_1^0 + \dots + j_N w_N^0)} 2\pi i. \tag{10.22}$$

A comparison of the last two equations yields

$$B^{(1)}_{j_1\dots j_N}(J_k) = \frac{C_{j_1\dots j_N}(J_k)}{2\pi i \sum_{k=1}^{N} j_k \nu_k^0}, \qquad j_i \neq 0. \tag{10.23}$$

With this, (10.20) gives

$$W_1 = \frac{1}{2\pi i} \sum_{\substack{j_i\in\mathbb{Z} \\ j_i\neq 0}} \frac{C_{j_1\dots j_N}(J_k)}{\sum_{k=1}^{N} j_k \nu_k^0} \exp\left[2\pi i \sum_{k=1}^{N} j_k w_k^0\right]. \tag{10.24}$$

We have to limit ourselves here to nondegenerate frequencies. The $C_{j_1,\dots,j_N}$ should converge fast enough because for large j_k, the scalar product $(\boldsymbol{j} \cdot \boldsymbol{\omega}^0)$ can come arbitrarily close to zero. [It can be shown (cf. later: KAM theorem) that, under certain assumptions for the unperturbed frequencies – these must be "very irrational" – the series (10.24) converges.] The above series does not, in a strict sense, converge, and the perturbation theory becomes meaningless. This problem, "problem of small divisors," was first clearly recognized by Poincaré. Nevertheless, one gets in celestial mechanics, for example, very useful results by cutting off the perturbation series at an appropriate point. The motion of the system is then determined for finite times only.

Finally we present the frequencies of the quasiperiodic motions up to second order:

$$\nu_k\left(J_k\right) = \nu_k^0 + \varepsilon \frac{\partial \bar{H}_1}{\partial J_k} + \varepsilon^2 \frac{\partial \bar{K}_2}{\partial J_k} \,. \tag{10.25}$$

We now again consider a few examples, beginning with the linear harmonic oscillator with the perturbation term $H = (m\omega_0^2/2q_0)q^3$. Then the total Hamiltonian reads:

$$H = H_0 + \varepsilon H_1 = \frac{p^2}{2m} + \frac{m\omega_0^2}{2} q^2 + \varepsilon \frac{m\omega_0^2}{2q_0} q^3 \,. \tag{10.26}$$

In terms of action-angle variables we have

$$H_0 = \nu_0 J_0 \,, \quad q = \sqrt{\frac{J_0}{m\pi\omega_0}} \sin\left(2\pi w_0\right) ,$$

so that

$$H_1 = \frac{m\omega_0^2}{2q_0} \left(\frac{J_0}{m\pi\omega_0}\right)^{3/2} \sin^3\left(2\pi w_0\right) . \tag{10.27}$$

Equations (10.15–17) then tell us:

$$E_0(J) = \nu_0 J \,, \tag{10.28}$$

$$E_1(J) = \bar{H}_1 = 0 \,, \quad \left(\overline{\sin^3\alpha} = 0\right) , \tag{10.29}$$

$$E_2(J) = \bar{K}_2 \,, \tag{10.30}$$

with $K_2 = (\partial W_1/\partial w^0)\partial H_1/\partial J$, since $H_2 \equiv 0$; $\partial W_1/\partial w_0 = -H_1/\nu_0$, so that it holds for K_2 that

$$K_2 = -\frac{H_1}{\nu_0} \frac{\partial H_1}{\partial J} = -\frac{1}{2\nu_0} \frac{\partial H_1^2}{\partial J} \,.$$

With the above expression for H_1, (10.27), we then obtain

$$\begin{aligned} K_2 &= -\frac{1}{2\nu_0} \frac{m^2\omega_0^4}{4q_0^2} \frac{1}{(m\pi\omega_0)^3} \sin^6\left(2\pi w_0\right) \frac{\partial}{\partial J} J^3 \\ &= -\frac{3J^2}{4\pi^2 m q_0^2} \sin^6\left(2\pi w_0\right) . \end{aligned} \tag{10.31}$$

We already found the average of $\sin^6(2\pi w_0)$ in the last chapter: $\overline{\sin^6\alpha} = 5/16$. Thus we have

$$E_2(J) = \bar{K}_2 = -\frac{15J^2}{64\pi^2 m q_0^2} \,, \tag{10.32}$$

from which, according to (10.25), we get the frequency,

$$\nu(J) = \nu_0 - \varepsilon^2 \frac{15J}{32\pi^2 m q_0^2} \,. \tag{10.33}$$

Now, with q_0 as maximum amplitude of the unperturbed oscillator, $E = (m\omega_0^2/2)q_0^2$, which we can rewrite (in lowest order) as

$$\frac{m\omega_0^2 q_0^2}{2} = \nu_0 J = \frac{\omega_0}{2\pi} J : \quad m q_0^2 = \frac{J}{\pi\omega_0}$$

we obtain from (10.33)

$$\nu = \nu_0 - \varepsilon^2 \frac{15 J\pi\omega_0}{32\pi^2 J} = \nu_0 - \varepsilon^2 \frac{15}{16}\nu_0$$

or

$$\frac{\Delta\nu}{\nu_0} = -\frac{15}{16}\varepsilon^2 \,. \tag{10.34}$$

More interesting is the following example with two degrees of freedom ($m = 1$):

$$H = \sum_{k=1}^{2} \frac{1}{2}\left(p_k^2 + \omega_k^{02} q_k^2\right) + \varepsilon\omega_1^{02}\omega_2^{02} q_1^2 q_2^2 \,. \tag{10.35}$$

In action-angle variables:

$$H_0 = \nu_1^0 J_1^0 + \nu_2^0 J_2^0 = E_0 \,,$$

$$q_k = \sqrt{\frac{J_k^0}{\pi\omega_k^0}} \sin\left(2\pi w_k^0\right) \,,$$

$$H_1\left(w_k^0, J_k\right) = \frac{\omega_1^0\omega_2^0}{\pi^2} J_1 J_2 \sin^2\left(2\pi w_1^0\right) \sin^2\left(2\pi w_2^0\right) \,, \tag{10.36}$$

$$\begin{aligned} E(J_k) &= E_0(J_k) + \varepsilon E_1(J_k) \,, \quad E_1(J_k) = \overline{H_1\left(w_k^0, J_k\right)} \\ &= \nu_1^0 J_1 + \nu_2^0 J_2 + \varepsilon \frac{\omega_1^0\omega_2^0}{4\pi^2} J_1 J_2 \\ &= \nu_1^0 J_1 + \nu_2^0 J_2 + \varepsilon\nu_1^0\nu_2^0 J_1 J_2 \,. \end{aligned} \tag{10.37}$$

The frequency spectrum of H_1 is ($\alpha := 2\pi w_1^0$, $\beta := 2\pi w_2^0$)

$$\begin{aligned} H_1 &= 4\nu_1^0\nu_2^0 J_1 J_2 \sin^2\left(2\pi w_1^0\right) \sin^2\left(2\pi w_2^0\right) \\ &= 4\nu_1^0\nu_2^0 J_1 J_2 \left(\frac{1}{2\mathrm{i}}\right)^4 \left(\mathrm{e}^{2\pi \mathrm{i} w_1^0} - \mathrm{e}^{-2\pi \mathrm{i} w_1^0}\right)^2 \left(\mathrm{e}^{2\pi \mathrm{i} w_2^0} - \mathrm{e}^{-2\pi \mathrm{i} w_2^0}\right)^2 \\ &= \frac{1}{4}\nu_1^0\nu_2^0 J_1 J_2 \left(\mathrm{e}^{2\mathrm{i}\alpha} - 2 + \mathrm{e}^{-2\mathrm{i}\alpha}\right)\left(\mathrm{e}^{2\mathrm{i}\beta} - 2 + \mathrm{e}^{-2\mathrm{i}\beta}\right) \\ &= \frac{1}{4}\nu_1^0\nu_2^0 J_1 J_2 \Big[\mathrm{e}^{2\mathrm{i}(\alpha+\beta)} - 2\mathrm{e}^{2\mathrm{i}\alpha} + \mathrm{e}^{2\mathrm{i}(\alpha-\beta)} \\ &\quad - 2\,\mathrm{e}^{2\mathrm{i}\beta} + 4 - 2\,\mathrm{e}^{-2\mathrm{i}\beta} \\ &\quad + \mathrm{e}^{2\mathrm{i}(\beta-\alpha)} - 2\,\mathrm{e}^{-2\mathrm{i}\alpha} + \mathrm{e}^{-2\mathrm{i}(\alpha+\beta)}\Big] \,. \end{aligned} \tag{10.38}$$

With

$$H_1 - \nu_1^0 \nu_2^0 J_1 J_2 = \frac{1}{4} \nu_1^0 \nu_2^0 J_1 J_2 \sum_{m,n} H_{1mn} \, e^{2\pi i (m w_1^0 + n w_2^0)} = \{H_1\} \tag{10.39}$$

in comparison to (10.38) we obtain

$$\begin{aligned} H_{122} &= H_{12-2} = H_{1-22} = H_{1-2-2} = 1 \,, \\ H_{120} &= H_{1-20} = H_{10-2} = H_{102} = -2 \,. \end{aligned} \tag{10.40}$$

All remaining H_{1mn} are equal to zero. Since $\bar{H}_1 = \nu_1^0 \nu_2^0 J_1 J_2$, (10.39) tells us that, in accord with (10.18), $H_1 - \bar{H}_1 = \{H_1\}$ is the oscillation part of H_1. The latter vanishes when averaging over one period.

According to (10.18), it holds that

$$\sum_{k=1}^{2} \nu_k^0 \frac{\partial W_1}{\partial w_k^0} = -\{H_1\}$$

so that

$$\sum_{k=1}^{2} \nu_k^0 \frac{\partial W_1}{\partial w_k^0} = -\frac{1}{4} \nu_1^0 \nu_2^0 J_1 J_2 \sum_{m,n} H_{1mn} \, e^{2\pi i (m w_1^0 + n w_2^0)} \,. \tag{10.41}$$

Then, with help of (10.21) and (10.24), it follows that

$$\begin{aligned} W_1 &= \frac{1}{2\pi i} \left(-\frac{1}{4} \nu_1^0 \nu_2^0 J_1 J_2 \right) \sum_{m,n} \frac{H_{1mn} \, e^{2\pi i (m w_1^0 + n w_2^0)}}{m \nu_1^0 + n \nu_2^0} \\ &= -\frac{\nu_1^0 \nu_2^0 J_1 J_2}{8\pi i} \left(\frac{e^{2\pi i (2 w_1^0 + 2 w_2^0)} - e^{-2\pi i (2 w_1^0 + 2 w_2^0)}}{2\nu_1^0 + 2\nu_2^0} + \ldots \right) \\ &= \frac{\nu_1^0 \nu_2^0 J_1 J_2}{8\pi} \left[-\frac{\sin[4\pi (w_1^0 + w_2^0)]}{\nu_1^0 + \nu_2^0} - \frac{\sin[4\pi (w_1^0 - w_2^0)]}{\nu_1^0 - \nu_2^0} \right. \\ &\quad \left. + \frac{2 \sin(4\pi w_1^0)}{\nu_1^0} + \frac{2 \sin(4\pi w_2^0)}{\nu_2^0} + \ldots \right] . \end{aligned} \tag{10.42}$$

At resonance $\nu_1^0 = \nu_2^0$ the procedure naturally fails. Incidentally, higher perturbation terms W_i, $i \geqslant 2$ produce even more critical denominators. But we know that the formula (10.24) diverges in general.

As another example of the application of canonical perturbation theory in more than one dimension, we consider a charged particle in a homogeneous magnetic field. In addition, an external "electrostatic" plane wave should act on the particle; i.e., we want to assume that the interaction of the particle with the magnetic part of the electromagnetic wave is smaller by v/c, so that we can neglect its contribution. Then the Hamiltonian of the unperturbed problem is:

$$H_0 = H_0(\boldsymbol{r}, \boldsymbol{p}) = \frac{1}{2m}\left(\boldsymbol{p} - \frac{e}{c}\boldsymbol{A}(\boldsymbol{r})\right)^2 . \tag{10.43}$$

The canonical momentum is $\boldsymbol{p} = m\boldsymbol{v} + (e/c)\boldsymbol{A}$. We choose the vector potential for a homogeneous magnetic field B_0 in z-direction in the form

$$\boldsymbol{A} = -B_0 y\hat{\boldsymbol{x}} , \quad \boldsymbol{B} = B_0\hat{\boldsymbol{z}} . \tag{10.44}$$

This choice of $\boldsymbol{A}$ emphasizes the y coordinate – contrary to $\boldsymbol{A} = B_0 x\hat{\boldsymbol{y}}$, which also leads to a homogeneous B_0 field in z-direction.

Our next goal is to transform the variables $q_1 = x$, $q_2 = y$, p_x, p_y into "guiding center coordinates." These are

$$\boldsymbol{Q} = (\phi, Y, Z = z) , \quad \boldsymbol{P} = \left(P_\phi, m\Omega X, P_z = p_z\right) .$$

Their dependence of the old cartesian coordinates is defined as follows:

$$\phi = \arctan\left[\frac{p_x + m\Omega y}{p_y}\right] , \quad Y = -\frac{p_x}{m\Omega} , \tag{10.45}$$

$$P_\phi = \frac{(p_x + m\Omega y)^2 + p_y^2}{2m\Omega} , \quad X = x + \frac{p_y}{m\Omega} .$$

That these variables are canonical can easily be proved by computing the Poisson brackets:

$$\left\{\phi, P_\phi\right\}_{q_i, p_i} = \sum_{i=x,y}\left(\frac{\partial\phi}{\partial q_i}\frac{\partial P_\phi}{\partial p_i} - \frac{\partial P_\phi}{\partial q_i}\frac{\partial\phi}{\partial p_i}\right) = 1 , \quad \{Y, m\Omega X\} = 1 ,$$

$$\{\phi, Y\} = \{\phi, X\} = \left\{P_\phi, Y\right\} = \left\{P_\phi, X\right\} = 0 .$$

The generating function for the canonical transformation $(q_i, p_i) \to (Q_i, P_i)$ is of type F_1:

$$F_1 = F_1\left(q_1, q_2, Q_1, Q_2\right) \equiv F_1(x, y, \phi, Y) ,$$

where $Q_1 \equiv \phi$ and $Q_2 \equiv Y$ are the new canonical coordinates. Here, too, we emphasize the Y coordinate in F_1, reflecting the presence of y in the gauge (10.44). The explicit form of F_1 is given by

$$F_1 = m\Omega\left[\tfrac{1}{2}\left(q_2 - Q_2\right)^2 \cot Q_1 - q_1 Q_2\right] . \tag{10.46}$$

Here, $\Omega = eB_0/mc$ is the cyclotron frequency, and (X, Y) specify the origin of the circular orbit of the electron. The direction of motion is clockwise if we choose $e > 0$, $B_0 > 0$. Then, the following formulae are obvious:

$$x = X + \varrho \sin\phi , \quad v_x = \dot{x} = -\varrho\Omega \sin\phi \equiv -v_\perp \sin\phi ,$$

$$y = Y - \varrho \sin\phi , \quad v_y = \dot{y} = -\varrho\Omega \sin\phi \equiv -v_\perp \sin\phi ,$$

$$z = Z , \quad v_z = \dot{z} = \dot{Z} .$$

From this, it follows that

$$\tan\phi = \frac{v_x}{v_y}\,, \quad v_\perp = \sqrt{\dot{x}^2 + \dot{y}^2} = \varrho\Omega$$
$$v_\perp^2 = v_x^2 + v_y^2 = v^2 - v_z^2 = \text{const.} \equiv (\varrho\Omega)^2\,.$$

The transformation equations associated with F_1 are:

$$p_i = \frac{\partial F_1}{\partial q_i}\,, \quad P_i = -\frac{\partial F_1}{\partial Q_i}\,, \quad i = 1, 2\,.$$

The various partial derivatives are then given by

$$p_1 = \frac{\partial F_1}{\partial q_1} = -m\Omega Q_2 :$$
$$p_x = -m\Omega Y \tag{10.47}$$
$$p_2 = \frac{\partial F_1}{\partial q_2} = m\Omega(q_2 - Q_2)\cot Q_1 :$$
$$p_y = m\Omega(y - Y)\cot\phi \tag{10.48}$$
$$P_1 = -\frac{\partial F_1}{\partial Q_1} = \frac{m\Omega}{2}(q_2 - Q_2)^2\frac{1}{\sin^2 Q_1} :$$
$$P_\phi = \frac{m\Omega}{2}(y - Y)^2\frac{1}{\sin^2\phi} \tag{10.49}$$
$$P_2 = -\frac{\partial F_1}{\partial Q_2} = m\Omega\left[(q_2 - Q_2)\cot Q_1 + q_1\right] :$$
$$P_Y = m\Omega[(y - Y)\cot\phi + x]\,. \tag{10.50}$$

We shall now use the equations (10.47–50) to express the new coordinates (ϕ, Y, P_ϕ, P_Y) as functions of the old coordinates (x, y, p_x, p_y) and in this way justify the formulae (10.45). It follows from (10.47) that

$$Y = -\frac{p_x}{m\Omega}\,. \tag{10.51}$$

If we insert (10.48) into (10.50), then we can write

$$P_Y = p_y + m\Omega x\,. \tag{10.52}$$

Finally, it follows from (10.48) that

$$\cot\phi = \frac{p_y}{m\Omega(y - Y)} \underset{(51)}{=} \frac{p_y}{m\Omega[y - (-p_x/m\Omega)]}$$

or

$$\cot\phi = \frac{p_y}{p_x + m\Omega y} : \quad \phi = \arctan\left[\frac{p_x + m\Omega y}{p_y}\right]\,. \tag{10.53}$$

The results (10.51) and (10.53) can now be inserted into (10.49):

$$P_\phi = \frac{m\Omega}{2}\left(y + \frac{p_x}{m\Omega}\right)^2 \left[1 + \left(\frac{p_y}{p_x + m\Omega y}\right)^2\right]$$

or

$$\begin{aligned} P_\phi &= \frac{(p_x + m\Omega y)^2 + p_y^2}{2m\Omega} = \frac{(mv_x)^2 + (mv_y)^2}{2m\Omega} \\ &= \frac{1}{2} m\Omega\varrho^2 = \frac{1}{2}\frac{mv_\perp^2}{\Omega} \equiv \frac{mc}{e}\mu \,. \end{aligned} \tag{10.54}$$

On the other hand, we also can express (x, y, p_x, p_y) as functions of (ϕ, Y, P_ϕ, P_Y):

$$(47): \qquad p_x = -m\Omega Y \,, \tag{10.55}$$

$$(49): \qquad y = Y \pm \sqrt{\frac{2P_\phi}{m\Omega}} \sin\phi \,. \tag{10.56}$$

Choosing the minus sign,

$$y = Y - \varrho \sin\phi \,. \tag{10.57}$$

$$\begin{aligned} (50): \qquad x &= \frac{P_Y}{m\Omega} - \underbrace{(y - Y)}_{=(57)} \cot\phi \\ &= \frac{P_Y}{m\Omega} + \varrho \sin\phi = X + \varrho \sin\phi \,. \end{aligned} \tag{10.58}$$

Let us keep in mind:

$$\varrho = \varrho(P_\phi) = \sqrt{\frac{2P_\phi}{m\Omega}} \,, \tag{10.59}$$

$$X = \frac{P_Y}{m\Omega} \underset{(52)}{=} x + \frac{p_y}{m\Omega} \,. \tag{10.60}$$

Finally, (10.48) and (10.57) yield

$$p_y = m\Omega\varrho \sin\phi \underset{(59)}{=} \sqrt{2m\Omega P_\phi} \sin\phi \,. \tag{10.61}$$

Now we return to the Hamiltonian (10.43) with the vector potential (10.44):

$$H_0 = \frac{1}{2m}\left[(p_x + m\Omega y)^2 + p_y^2 + p_z^2\right] . \tag{10.62}$$

The equations of motion read, accordingly, ($\dot{q}_i = \partial H_0/\partial p_i$, $\dot{p}_i = -\partial H_0/\partial q_i$):

$$\begin{aligned} &mv_x = p_x + m\Omega y \,, \quad \dot{p}_x = 0 \,, \ p_x = \text{const.} \,, \\ &mv_y = p_y \,, \\ &mv_z = p_z \,, \qquad\qquad \dot{p}_z = 0 \,, \ p_z = mv_z = \text{const.} = P_z \end{aligned}$$

Now, according to (10.54) it holds that

$$\Omega P_\phi = \frac{1}{2m}\left[\left(p_x + m\Omega y\right)^2 + p_y^2\right] .$$

So we have for the Hamiltonian in the new variables

$$H_0\left(P_\phi, P_z\right) = \Omega P_\phi + \frac{1}{2m}P_z^2 \equiv H_\perp + \frac{P_z^2}{2m} . \tag{10.63}$$

Here, no external time dependence appears; the momenta $J_1 \equiv P_z$ and $J_2 \equiv P_\phi$ $(= m\Omega\varrho^2/2)$ are our new action variables. With this new Hamiltonian we get for the canonical equations of the variables $(Y, P_Y, \phi, P_\phi, z, p_z)$:

$$\dot{Y} = \dot{P}_Y = \dot{P}_\phi = \dot{p}_z = \dot{P}_z = 0 ; \quad \dot{H}_0 = 0 ,$$

thus $\{Y, P_Y(= m\Omega X), P_\phi, P_z\}$ are all constant. But it holds that

$$\dot{\phi} = \frac{\partial H_0}{\partial P_\phi} = \Omega = \text{const.}$$

$$\dot{Z} = \dot{z} = \frac{p_z}{m} \equiv \frac{P_z}{m} .$$

So the guiding center coordinates of the particle trajectory (X, Y) with $X = P_Y/m\Omega$, the transverse energy ΩP_ϕ as well as $\dot{\phi}$ and the longitudinal energy $P_z^2/2m$ are all constants of motion. They are, therefore, the appropriate variables with which to set up a perturbation theory. A small perturbation term εH_1 will then make these quantities slowly change. As an example of a perturbation, let us consider a propagating "electrostatic" wave with the amplitude ϕ_0, frequently ω and wave vector $\boldsymbol{k}$, which lies in the y-z plane:

$$H_1 = e\phi(y, z, t) , \quad \phi = \phi_0 \sin\left(k_z z + k_\perp y - \omega t\right) , \qquad \boldsymbol{E} = -\nabla\phi . \tag{10.64}$$

With

$$y = Y - \varrho \sin\phi ,$$

$$\varrho\left(P_\phi\right) = \sqrt{\frac{2P_\phi}{m\Omega}} ,$$

we obtain with (10.63) and (10.64) the time-dependent Hamiltonian,

$$\begin{aligned} H = H_0 + \varepsilon H_1 &= \frac{p_z^2}{2m} + \Omega P_\phi + \varepsilon e\phi_0 \sin\left(k_z z + k_\perp Y - k_\perp\varrho \sin\phi - \omega t\right) \\ &= H\left(\phi, P_\phi, Y, z, p_z; t\right) . \end{aligned} \tag{10.65}$$

But $P_Y(= m\Omega X)$ does not appear in (10.65). Therefore, Y is a constant, as is $k_\perp Y$. We can then eliminate the $k_\perp Y$ in (10.65) by choosing the origin of z or t appropriately.

Since, furthermore, the two variables z and t only occur in the combination $k_z z - \omega t$, we shall try to eliminate the time by means of a transformation to the wave system. Thereby, the following generating function is of help:

$$F_2(q_1, q_2, P_1, P_2; t) = F_2(\phi, z, P_\phi, P_z; t) .$$

Here we put $q_1 = \phi$, $q_2 = z$, $P_1 = P_\phi$, $P_2 = P_z$, and begin with the generating function for the identity $q_1 P_1 = \phi P_\phi$:

$$F_2 = q_1 P_1 + (k_z z - \omega t) P_\psi$$

or

$$F_2 = \phi P_\phi + (k_z z - \omega t) P_\psi . \tag{10.66}$$

The transformation equations then yield, together with (10.66),

$$p_\phi = P_\phi = \frac{\partial F_2}{\partial \phi} , \quad p_z = \frac{\partial F_2}{\partial z} = k_z P_\psi \tag{10.67}$$

$$Q_\phi \equiv \phi = \frac{\partial F_2}{\partial P_\phi} , \quad Q_\psi \equiv \psi = \frac{\partial F_2}{\partial P_\psi} = k_z z - \omega t . \tag{10.68}$$

Thus the combination of two variables, z and t in $k_z z - \omega t$ is replaced by one variable, ψ.

Finally, our new transformed, time-dependent Hamiltonian reads:

$$H_{\text{new}} = H_{\text{old}} + \frac{\partial F_2}{\partial t} = H_{\text{old}} - \omega P_\psi$$

or

$$H = \frac{k_z^2 P_\psi^2}{2m} - \omega P_\psi + \Omega P_\phi + \varepsilon e \phi_0 \sin(\psi - k_\perp \varrho \sin \phi) = E = \text{const.} \tag{10.69}$$

The last term on the right-hand side with $\sin \phi$ and $\varrho(P_\phi)$ indeed represents a highly nonlinear perturbation. This causes many resonances, which is immediately clear, when one considers that (J_n Bessel functions)

$$\sin(\psi - k_\perp \varrho \sin \phi) = \sum_{n=-\infty}^{+\infty} J_n(k_\perp \varrho) \sin(\psi - n\phi) . \tag{10.70}$$

Accordingly, the Hamiltonian in question reads:

$$H = \frac{k_z^2 P_\psi^2}{2m} - \omega P_\psi + \Omega P_\phi + \varepsilon e \phi_0 \sum_n J_n(k_\perp \varrho) \sin(\psi - n\phi) . \tag{10.71}$$

We have already pointed out (cf. discussion on convergence of the perturbation series) that it is necessary to stay away from the unperturbed frequencies; the Fourier amplitudes ($J_n(k_\perp \varrho)$) of the nth frequency will then vanish more rapidly than the next resonating denominator.

In order to obtain the resonances between the unperturbed frequencies caused by H_1, we first need those frequencies. Now it holds for $\varepsilon \to 0$:

$$H_0 = \frac{(k_z P_\psi)^2}{2m} - \omega P_\psi + \Omega P_\phi \equiv H\big(P_\phi, P_\psi\big) . \tag{10.72}$$

Note the nonlinearity of H_0 in P_ψ. The P's are our new action variables; i.e., J_ϕ and J_ψ are constants. Their conjugate angles develop linearly in time:

$$\frac{\partial H_0}{\partial P_i} = \omega_i , \quad Q_i = \omega_i t + \beta_i . \tag{10.73}$$

(Note: At this point we re-define our action variable, i.e.,

$$J = \oint p\, dq , \quad w = \frac{\partial H}{\partial J} . \tag{10.74}$$

We hereby adopt Lichtenberg and Lieberman's notation.) Then it holds that

$$\omega_\phi = \frac{\partial H_0}{\partial P_\phi} = \Omega \tag{10.75}$$

and

$$\omega_\psi = \frac{\partial H_0}{\partial P_\psi} = \frac{k_z^2}{m} P_\psi - \omega \underset{(67)}{=} \frac{k_z p_z}{m} - \omega = k_z v_z - \omega . \tag{10.76}$$

The perturbation (εH_1) contains terms in the form $\sin(\psi - n\phi)$ and can thus lead to resonances between the frequencies ω_ψ (Doppler-shifted frequency of the incoming wave) and the various harmonics of ω_ϕ. This occurs when the following resonance condition is satisfied:

$$\dot{\psi} - n\dot{\phi} = \omega_\psi - n\omega_\phi = \omega_\psi - n\Omega = k_z v_z^{(n)} - \omega - n\Omega = 0 , \quad n \in \mathbb{Z} . \tag{10.77}$$

v_z is the particle velocity. So there is a set of resonant parallel velocities $\{v_z^{(n)}\}$ if $k_z \neq 0$. The resonance condition contains two interesting limiting cases:

$$\text{(a)} \quad k_z = 0 : \quad \omega_\psi = -\omega : \quad \omega + n\Omega = 0 , \quad \Omega = \frac{eB_0}{mc} . \tag{10.78}$$

The nonlinearity only enters via the perturbation. This case is called perturbation with intrinsic degeneracy. While (a) does not contain the particle velocity, it shows up in the case of an degeneracy:

$$\text{(b)} \quad k_z \neq 0 : \quad k_z v_z^{(n)} = \omega + n\Omega = \frac{k_z^2 P_\psi}{m}$$

or

$$J \equiv P_\psi = \frac{m}{k_z^2}(\omega + m\Omega) = \frac{m v_z^{(n)}}{k_z} = \frac{p_z^{(n)}}{k_z} \, . \tag{10.79}$$

We shall return to this case in a later section.

Right now we consider case (a), i.e., orthogonal wave propagation ($k_z = 0$), where it is assumed that we are staying away from the "primary" resonances as defined in (10.78). Now we use (10.18):

$$\sum_{k=1}^{2} \frac{\partial H_0}{\partial J_k} \frac{\partial W_1}{\partial w_k^0} = -\left(H_1 - \bar{H}_1\right) = -\left\{H_1\right\} , \tag{10.80}$$

where $\left\{H_1\right\}$ denotes the oscillating part. Now, according to (10.75, 76),

$$\frac{\partial H_0}{\partial P_\psi} = \omega_\psi = -\omega \, , \qquad \frac{\partial H_0}{\partial P_\phi} = \omega_\phi = \Omega$$

so that altogether it holds that $\left(W_1 = W_1(\psi, \phi, \bar{J}_\psi, \bar{J}_\phi)\right)$

$$-\omega \frac{\partial W_1}{\partial \psi} + \Omega \frac{\partial W_1}{\partial \phi} = -e\phi_0 \sum_n J_n\left(k_\perp \bar{\varrho}\right) \sin(\psi - n\phi) \, . \tag{10.81}$$

Here, $\bar{\varrho}$ is a function of the new constant action variables $\bar{P}_\phi \equiv \bar{J}_\phi$. The partial differential equation (10.81) for W_1 can be solved easily:

$$W_1 = -e\phi_0 \sum_n J_n\left(k_\perp \bar{\varrho}\right) \frac{\sin(\psi - n\phi)}{\omega + n\Omega} \tag{10.82}$$

because

$$-\omega \frac{\partial W_1}{\partial \psi} = -e\phi_0 \sum_n J_n\left(k_\perp \bar{\varrho}\right) \omega \frac{\sin(\psi - n\phi)}{\omega + n\Omega}$$

and

$$\Omega \frac{\partial W_1}{\partial \phi} = -e\phi_0 \sum_n J_n\left(k_\perp \bar{\varrho}\right) n\Omega \frac{\sin(\psi - n\phi)}{\omega + n\Omega} \, .$$

Adding the last two equations indeed yields (10.81).

W_1 is part of the generating function W, which takes us from the old action variables P_ϕ, P_ψ to the new constant ones $\bar{P}_\phi$, $\bar{P}_\psi$:

$$W = W_0 + \varepsilon W_1 + \dots \, .$$

Here, W_0 is the generator of the identity transformation; $p_i = \partial W(q_i, P_i)/\partial q_i$ becomes

$$P_\psi = \bar{P}_\psi + \varepsilon \frac{\partial W_1}{\partial \psi} + \ldots \tag{10.83}$$

$$P_\phi = \bar{P}_\phi + \varepsilon \frac{\partial W_1}{\partial \phi} + \ldots . \tag{10.84}$$

Thus, if we stay away from the primary resonances ($\omega + n\Omega = 0$), then we get in first order

$$\bar{P}_\psi \equiv \bar{J}_\psi = P_\psi - \varepsilon e \phi_0 \sum_n J_n(k_\perp \varrho) \frac{\sin(\psi - n\phi)}{\omega + n\Omega} = \text{const.}$$

$$\bar{P}_\phi \equiv \bar{J}_\phi = P_\phi + \varepsilon e \phi_0 \sum_n n J_n(k_\perp \varrho) \frac{\sin(\psi - n\phi)}{\omega + n\Omega} = \text{const.}$$

with $\varrho = \sqrt{2P_\phi/m\Omega}$. The $\bar{P}_\psi$, $\bar{P}_\phi$ are constants. This was the intention of the canonical transformation W.

$$\varrho^2 = \frac{2}{m\Omega} P_\phi$$

describes the change (oscillation) of the Larmor radius in the vicinity of a resonant trajectory.

11. Canonical Adiabatic Theory

In the present chapter we are concerned with systems, the change of which – with the exception of a single degree of freedom – should proceed slowly. (Compare the pertinent remarks about ε as slow parameter in Chap. 7.) Accordingly, the Hamiltonian reads:

$$H = H_0\big(J, \varepsilon p_i, \varepsilon q_i; \varepsilon t\big) + \varepsilon H_1\big(J, \theta, \varepsilon p_i, \varepsilon q_i; \varepsilon t\big) . \tag{11.1}$$

Here, (J, θ) designates the "fast" action-angle variables for the unperturbed, solved problem $H_0(\varepsilon = 0)$, and the (p_i, q_i) represent the remaining "slow" canonical variables, which do not necessarily have to be action-angle variables. Naturally, we again wish to eliminate the fast variable θ in (11.1). In zero-th order, the quantity which is associated to θ is denoted by J. In order to then calculate the effect of the perturbation εH_1, we look for a canonical transformation $(J, \theta, p_i, q_i) \to (\bar{J}, \bar{\theta}, \bar{p}_i, \bar{q}_i)$ which makes the new Hamiltonian $\tilde{H}$ independent of the new fast variable $\bar{\theta}$.

It is only logical to now begin with a generating function,

$$W\big(\bar{J}, \theta, \bar{p}_i, q_i; t\big) = \bar{J}\theta + \bar{p}_i q_i + \varepsilon W_1\big(\bar{J}, \theta, \bar{p}_i, q_i; t\big) + \dots . \tag{11.2}$$

To this belong the transformation equations (in first order)

$$J = \frac{\partial W}{\partial \theta} = \bar{J} + \varepsilon \frac{\partial W_1}{\partial \theta} = \bar{J} + \varepsilon \frac{\partial W_1}{\partial \bar{\theta}} , \tag{11.3}$$

$$\theta = \frac{\partial W}{\partial \bar{J}} = \theta + \varepsilon \frac{\partial W_1}{\partial \bar{J}} = \theta + \varepsilon \frac{\partial W_1}{\partial J} , \tag{11.4}$$

$$p_i = \frac{\partial W}{\partial q_i} = \bar{p}_i + \varepsilon \frac{\partial W_1}{\partial q_i} = \bar{p}_i + \varepsilon \frac{\partial W_1}{\partial \bar{q}_i} , \tag{11.5}$$

$$\bar{q}_i = \frac{\partial W}{\partial \bar{p}_i} = q_i + \varepsilon \frac{\partial W_1}{\partial \bar{p}_i} = q_i + \varepsilon \frac{\partial W_1}{\partial p_i} . \tag{11.6}$$

We insert these expressions into H_0 and expand up to first order in ε:

$$\begin{aligned} H_0 &= H_0(J, \varepsilon p_i, \varepsilon q_i; \varepsilon t) \\ &= H_0\left(\bar{J} + \varepsilon \frac{\partial W_1}{\partial \theta}\ ,\ \varepsilon\left(\bar{p}_i + \varepsilon \frac{\partial W_1}{\partial q_i}\right)\ ,\ \varepsilon\left(\bar{q}_i - \varepsilon \frac{\partial W_1}{\partial p_i}\right)\ ;\ \varepsilon t\right) \\ &= H_0(\bar{J}, \varepsilon \bar{p}_i, \varepsilon \bar{q}_i; \varepsilon t) + \underbrace{\left.\frac{\partial H_0}{\partial J}\right|_{J=\bar{J}}}_{=\omega} \varepsilon \frac{\partial W_1}{\partial \theta} + \mathrm{O}(\varepsilon^2)\,. \end{aligned} \tag{11.7}$$

ω is the fast frequency corresponding to θ. Note that in (11.7) we have omitted the following terms, since they appear with ε^2:

$$-\frac{\partial H_0}{\partial \bar{q}_i} \underbrace{\frac{\partial W_1}{\partial p_i}}_{=\partial W_1/\partial \bar{p}_i}\ ,\quad \frac{\partial H_0}{\partial \bar{p}_i} \underbrace{\frac{\partial W_1}{\partial q_i}}_{=\partial W_1/\partial \bar{q}_i}\ . \tag{11.8}$$

Now it holds that

$$\tilde{H} = H + \frac{\partial W}{\partial t}$$

or

$$\tilde{H}(\bar{J}, \bar{\theta}, \varepsilon \bar{p}_i, \varepsilon \bar{q}_i; \varepsilon t) = H(J, \theta, \varepsilon p_i, \varepsilon q_i; \varepsilon t) + \varepsilon \frac{\partial W(\bar{J}, \theta, \varepsilon \bar{p}_i, \varepsilon q_i; \varepsilon t)}{\partial(\varepsilon t)}$$

or

$$\begin{aligned} &\tilde{H}_0(\bar{J}, \varepsilon \bar{p}_i, \varepsilon \bar{q}_i; \varepsilon t) + \varepsilon \tilde{H}_1(\bar{J}, \bar{\theta}, \varepsilon \bar{p}_i, \varepsilon \bar{q}_i; \varepsilon t) \\ &\underset{(7)}{=} H_0(\bar{J}, \varepsilon \bar{p}_i, \varepsilon \bar{q}_i; \varepsilon t) + \varepsilon \omega \frac{\partial W_1}{\partial \theta} + \varepsilon H_1 \\ &\quad + \varepsilon \frac{\partial [\overbrace{W_0}^{=\bar{J}\theta} + \varepsilon W_1 + \ldots]}{\partial(\varepsilon t)}\,. \end{aligned} \tag{11.9}$$

Since W_0 has no external time dependence, the last term on the right-hand side is second order in ε and is thus neglected. A comparison of coefficients of ε in (11.9) then yields:

$$\varepsilon^0 :\quad \tilde{H}_0(\bar{J}, \varepsilon \bar{p}_i, \varepsilon \bar{q}_i; \varepsilon t) = H_0(\bar{J}, \varepsilon \bar{p}_i, \varepsilon \bar{q}_i; \varepsilon t) \tag{11.10}$$

$$\begin{aligned} \varepsilon^1 :\quad \tilde{H}_1(\bar{J}, \bar{\theta}, \varepsilon \bar{p}_i, \varepsilon \bar{q}_i; \varepsilon t) &= \omega \frac{\partial W_1(\bar{J}, \overbrace{\theta}^{=\bar{\theta}+\ldots}, \ldots)}{\partial \theta} + H_1 \\ &= \omega \frac{\partial W_1(\bar{J}, \bar{\theta}, \varepsilon \bar{p}_i, \varepsilon \bar{q}_i; \varepsilon t)}{\partial \bar{\theta}} + H_1(\bar{J}, \bar{\theta}, \varepsilon \bar{p}_i, \varepsilon \bar{q}_i; \varepsilon t)\,. \end{aligned} \tag{11.11}$$

We now write (11.11) in the form

$$\tilde{H}_1(\bar{J},\bar{\theta},\dots) = \omega\frac{\partial W_1(\bar{J},\bar{\theta},\dots)}{\partial\bar{\theta}} + \overbrace{\left(H_1 - \langle H_1\rangle_{\bar{\theta}}\right)}^{=\{H_1\}_{\bar{\theta}}} + \langle H_1\rangle_{\bar{\theta}} \tag{11.12}$$

with

$$\langle H_1\rangle_{\bar{\theta}} = \frac{1}{2\pi}\int_0^{2\pi} H_1 d\bar{\theta} \,.$$

Then $\tilde{H}_1$ becomes a function which only depends on $\bar{J}$ and not on $\bar{\theta}$, if we choose W_1 so that

$$\omega\frac{\partial W_1}{\partial\bar{\theta}} + \{H_1\}_{\bar{\theta}} = 0 \,. \tag{11.13}$$

This differential equation is immediately solvable and gives for W_1:

$$W_1 = -\frac{1}{\omega}\int_0^{\theta}\{H_1\}_{\theta'}d\theta' \,. \tag{11.14}$$

Here, we have replaced $\bar{\theta}$ by θ. So we finally get from (11.12)

$$\tilde{H}_1(\bar{J},\,\dots) = \langle H_1\rangle_\theta \tag{11.15}$$

and, altogether:

$$\tilde{H}\left(\bar{J},\varepsilon\bar{p}_i,\varepsilon\bar{q}_i;\varepsilon t\right) = H_0 + \varepsilon\langle H_1\rangle_\theta \,. \tag{11.16}$$

The old adiabatic invariant was J. Now the new (constant) adiabatic invariant is called $\bar{J}$ and is related with J to first order according to

$$\bar{J}\left(J,\theta,\varepsilon p_i,\varepsilon q_i;\varepsilon t\right) = J - \varepsilon\frac{\partial W_1}{\partial\theta} \,. \tag{11.17}$$

J varies slowly now with θ:

$$J = \underbrace{\bar{J}}_{=\text{const.}} + \varepsilon\frac{\partial W_1}{\partial\theta} \underset{(13)}{=} \bar{J} - \varepsilon\frac{\{H_1\}_\theta}{\omega} \,. \tag{11.18}$$

At this point we return once again to Poincaré's "small divisors". These were responsible for the fact that our perturbation series for W did not converge. These small denominators are present in the problem under discussion as well, which can be seen immediately if we do not neglect (11.8) and $\partial W_1/\partial t$ in (11.9). Then, instead of (11.13), we get

$$\omega\frac{\partial W_1}{\partial\bar{\theta}} + \varepsilon\omega_1\frac{\partial W_1}{\partial(\varepsilon\bar{q}_1)} + \varepsilon\omega_2\frac{\partial W_1}{\partial(\varepsilon\bar{q}_2)} + \dots + \varepsilon\frac{\partial W_1}{\partial t} + \{H_1\}_{\bar{\theta}} = 0 \,. \tag{11.19}$$

The (p_i, q_i) are to be understood here as action-angle variables (J_i, θ_i). The solution to (11.19) can be immediately written in the form of a Fourier series if we take into account the fact that W_1 and $\{H_1\}_{\bar{\theta}}$ are periodic in the $\bar{\theta}$'s and Ωt:

$$W_1 = \mathrm{i} \sum_{\substack{k,m,n,l \\ k \neq 0}} \frac{H_{1klmn\ldots}(\bar{J}, \bar{p}_i)}{k\omega + \varepsilon(m\omega_1 + n\omega_2 + \ldots + l\Omega)} \times \mathrm{e}^{\mathrm{i}[k\bar{\theta} + \varepsilon(m\bar{q}_1 + n\bar{q}_2 + \ldots + l\Omega t)]} \,. \tag{11.20}$$

One can tell by the denominator that even for small ω_i, Ω, which belong to the slow variables, resonance behavior can occur between the slow and fast oscillation (ω) if the integers $m, n \ldots$ are large enough. We are not permitted to neglect the terms of the order ε in (11.19) in sufficient proximity to the resonances. It is thus not surprising that the adiabatic perturbation series for W, which neglects these resonance effects, can only be asymptotically correct and thus formally diverges.

To illustrate the above perturbative procedure, we calculate in first order the adiabatic invariant of the slowly changing harmonic oscillator,

$$H = \tfrac{1}{2} G(\tau) p^2 + \tfrac{1}{2} F(\tau) q^2 \,, \tag{11.21}$$

with $\tau = \varepsilon t$. We again switch to action-angle variables, in order to make the system accessible to an adiabatic perturbation. To this end we use a generating function of the F_1-type:

$$F_1 = F_1(q, \theta, \tau) = \tfrac{1}{2} R(\tau) q^2 \cot\theta \tag{11.22}$$

with

$$R(\tau) = \left(\frac{F(\tau)}{G(\tau)} \right)^{1/2} .$$

Now the following transformation equations are valid:

$$p = \frac{\partial F_1}{\partial q} = Rq \cot\theta \,, \tag{11.23}$$

$$P \equiv J = -\frac{\partial F_1}{\partial Q} \equiv -\frac{\partial F_1}{\partial \theta} = \frac{1}{2} Rq^2 \frac{1}{\sin^2\theta} \,, \tag{11.24}$$

or, solved for the trigonometric functions:

$$\cot^2\theta = \left(\frac{p}{Rq} \right)^2 , \tag{11.25}$$

$$\frac{1}{\sin^2\theta} = \frac{2J}{Rq^2} \,. \tag{11.26}$$

Subtraction yields

$$\frac{\cos^2\theta}{\sin^2\theta} - \frac{1}{\sin^2\theta} = -1 = \frac{1}{Rq^2}\left(\frac{p^2}{R} - 2J\right)$$

from which we obtain

$$Rq^2 = 2J - \frac{p^2}{R} ,$$

or, finally:

$$P \equiv J = \frac{1}{2}\left(Rq^2 + \frac{p^2}{R}\right) . \tag{11.27}$$

Equation (11.26) then reads, with (11.27):

$$\frac{1}{\sin^2\theta} = \frac{1}{Rq^2}2J = \frac{1}{Rq^2}\left(Rq^2 + \frac{p^2}{R}\right) = 1 + \frac{p^2}{(Rq)^2}$$

or

$$\frac{1}{\sin^2\theta} - 1 = \frac{1}{\tan^2\theta} = \frac{p^2}{(Rq)^2}$$

and thus

$$\theta = \arctan\left(\frac{Rq}{p}\right) ,$$

or, solved for q:

$$q = \sqrt{\frac{2J}{R}}\sin\theta . \tag{11.28}$$

If we now use (11.25), then it follows that

$$\frac{\cos^2\theta}{\sin^2\theta} = \frac{p^2}{R^2q^2} \underset{(28)}{=} \frac{p^2R}{R^2 2J\sin^2\theta}$$

or

$$p = \sqrt{2RJ}\cos\theta . \tag{11.29}$$

Finally, it holds for the new Hamiltonian that

$$H_{\text{new}} = H_{\text{old}} + \frac{\partial F_1}{\partial t} = \omega_0 J + \frac{\partial F_1}{\partial t} . \tag{11.30}$$

So we still need $\partial F_1/\partial t$:

$$\frac{\partial F_1}{\partial t} = \frac{1}{2}q^2\cot(\theta)\varepsilon \underbrace{\partial R(\tau)/\partial\tau}_{=:R'(\tau)} \underset{(25)}{=} \frac{1}{2}q^2\frac{p}{Rq}\varepsilon R' = \frac{1}{2}\varepsilon\frac{R'}{R}\underbrace{qp}_{(28/29)}$$

$$= \frac{1}{2}\varepsilon\frac{R'}{R}2J\sin\theta\cos\theta = \frac{1}{2}\varepsilon\frac{R'}{R}J\sin(2\theta) . \tag{11.31}$$

So the Hamiltonian transformed to action-angle variables reads:

$$H = \omega_0 J + \varepsilon \frac{1}{2} \frac{R'}{R} J \sin(2\theta) \tag{11.32}$$

with

$$\omega_0(\tau) = (FG)^{1/2} \, . \tag{11.33}$$

In lowest order the adiabatic invariant is simply

$$J = \frac{H_0}{\omega_0} = \text{const.} \tag{11.34}$$

This result is familiar to us from Chap. 7. In order to see how this quantity changes if we use adiabatic perturbation theory to order ε, we use (11.18):

$$\begin{aligned}\bar{J} &= J + \varepsilon \frac{\{H_1\}_\theta}{\omega} = J + \varepsilon \frac{1}{2\omega_0} \frac{R'}{R} J \sin(2\theta) \\ &= J\big(1 + \varepsilon P(\tau) \sin(2\theta)\big) = \text{const.}\end{aligned} \tag{11.35}$$

with

$$P(\tau) = \frac{R'}{2\omega_0 R} \, .$$

Accordingly, J changes in first order with a small component, which oscillates with twice the frequency of the fast variable.

We now want to verify that $\bar{J}$ indeed is constant. In order to do so, we take the time derivative ($\dot{\theta} = \omega_0$)

$$\frac{d}{dt}\bar{J} = \dot{J} + \varepsilon \dot{P} J \sin(2\theta) + 2\omega_0 \varepsilon P J \cos(2\theta) + \mathrm{O}\big(\varepsilon^2\big) \, . \tag{11.36}$$

On the other hand, it follows from (11.32) that

$$\dot{J} = -\frac{\partial H}{\partial \theta} = -\varepsilon \frac{1}{2} \frac{R'}{R} J 2 \cos(2\theta) = -\varepsilon \omega_0 P 2 J \cos(2\theta)$$

so that (11.36) reduces to

$$\dot{\bar{J}} = \varepsilon \dot{P} J \sin 2\theta \, . \tag{11.37}$$

However, since $P(t)$ is supposed to change slowly (adiabatically), i.e., $\dot{P} \sim \varepsilon P$, it follows from (11.37) that $\dot{\bar{J}}$ is of the order ε^2; then $\bar{J}$ is indeed an adiabatic invariant of first order.

12. Removal of Resonances

From the perturbative procedure in the last chapter we have learned that in the proximity of resonances of the unperturbed system, resonant denominators appear in the expression for the adiabatic invariants. We now wish to begin to locally remove such resonances by trying, with the help of a canonical transformation, to go to a coordinate system which rotates with the resonant frequency.

Let the unperturbed, solved problem with two degrees of freedom be given by

$$H_0 = \frac{1}{2}\sum_{k=1}^{2}\left(p_k^2 + \omega_k^2 q_k^2\right) . \tag{12.1}$$

The transition to action-angle variables J_i, θ_i is achieved with the transformation

$$q_i = \sqrt{\frac{2J_i}{\omega_i}} \cos\theta_i , \tag{12.2}$$

$$p_i = -\sqrt{2\omega_i J_i} \sin\theta_i . \tag{12.3}$$

These formulae agree with (10.28/29) in so far as we have replaced θ by $\theta + \pi/2$ there. This corresponds to a simple phase change in $\theta_i = \omega_i t + \beta_i$. Furthermore, it holds that $J_i = (1/2\pi) \oint p_i \, dq_i$. Thus we can write (12.1) as

$$H_0(J_i) = \omega_1 J_1 + \omega_2 J_2 , \quad \omega_i = \frac{\partial H_0}{\partial J_i} . \tag{12.4}$$

Let the perturbation term be given by

$$H_1 = q_1^2 q_2 - \tfrac{1}{3} q_2^3 , \tag{12.5}$$

and let us assume a 1:2 resonance between ω_1 and ω_2, i.e., that oscillator 1 is slower than oscillator 2. Then our complete Hamiltonian reads

$$H = \tfrac{1}{2}\left(p_1^2 + p_2^2\right) + \tfrac{1}{2} q_1^2 + 2q_2^2 + q_1^2 q_2 - \tfrac{1}{3} q_2^3 \tag{12.6}$$

with

$$\omega_1 = 1 , \quad \omega_2 = 2 .$$

The resonance of the unperturbed frequencies,

$$\omega_2/\omega_1 = r/s = 2/1 \,, \tag{12.7}$$

leads to divergent expressions in the perturbative solution of the problem. We shall therefore attempt to eliminate the commensurability (12.7),

$$r\omega_1 - s\omega_2 = 0 \,, \tag{12.8}$$

by making a canonical transformation to new action-angle variables $\hat{J}_i, \hat{\theta}_i$, so that only one of the two actions $\hat{J}_i$ appears in the new, unperturbed Hamiltonian. In order to do so, we choose the generating function

$$F_2 = \left(r\theta_1 - s\theta_2\right)\hat{J}_1 + \theta_2\hat{J}_2 \,. \tag{12.9}$$

The corresponding transformation equations then read

$$\left.\begin{aligned} J_1 &= \tfrac{\partial F_2}{\partial \theta_1} = r\hat{J}_1 = 2\hat{J}_1 \\ J_2 &= \tfrac{\partial F_2}{\partial \theta_2} = \hat{J}_2 - s\hat{J}_1 = \hat{J}_2 - \hat{J}_1 \end{aligned}\right\} : \qquad \begin{aligned} \hat{J}_1 &= \tfrac{1}{2}J_1 \\ \hat{J}_2 &= \tfrac{1}{2}J_1 + J_2 \end{aligned} \tag{12.10}$$

$$\left.\begin{aligned} \hat{\theta}_1 &= \tfrac{\partial F_2}{\partial \hat{J}_1} = r\theta_1 - s\theta_2 = 2\theta_1 - \theta_2 \\ \hat{\theta}_2 &= \tfrac{\partial F_2}{\partial \hat{J}_2} = \theta_2 \end{aligned}\right\} : \qquad \begin{aligned} \theta_1 &= \tfrac{\hat{\theta}_1 + \hat{\theta}_2}{2} \\ \theta_2 &= \hat{\theta}_2 \,. \end{aligned} \tag{12.11}$$

This choice of coordinates puts the observer into a coordinate system in which the change of $\hat{\theta}_1$,

$$\dot{\hat{\theta}}_1 = r\dot{\theta}_1 - s\dot{\theta}_2 = r\omega_1 - s\omega_2 \,, \tag{12.12}$$

measures small deviations from the resonance (12.8). For $\dot{\hat{\theta}}_1 = 0$, the system is in resonance. The variable $\hat{\theta}_1$ changes slowly and is, in the resonant case, a constant. Thus $\hat{\theta}_2$ is the fast variable, and we shall average over it.

One should note that the new Hamiltonian is now actually only dependent on a single action variable, i.e., $\hat{J}_2$; $\hat{J}_1$ does not appear:

$$H_0 \underset{(10)}{=} \omega_1\left(2\hat{J}_1\right) + \omega_2\left(\hat{J}_2 - \hat{J}_1\right) \underset{(7)}{=} \frac{\omega_2}{2}\left(2\hat{J}_1\right) + \omega_2\left(\hat{J}_2 - \hat{J}_1\right) = \omega_2\hat{J}_2 \,. \tag{12.13}$$

The perturbation term is then

$$\begin{aligned} \varepsilon H_1 &= \varepsilon q_1^2 q_2 - \frac{\varepsilon}{3}q_2^3 \\ &\underset{(2)}{=} \varepsilon\left(\frac{2J_1}{\omega_1}\right)\left(\frac{2J_2}{\omega_2}\right)^{1/2}\cos^2\theta_1\cos\theta_2 - \frac{\varepsilon}{3}\left(\frac{2J_2}{\omega_2}\right)^{3/2}\cos^3\theta_2 \,. \end{aligned}$$

The product of the cosines is

$$\begin{aligned}
\cos^2\theta_1 \cos\theta_2 &= \frac{1}{2^3}\left(e^{i\theta_1} + e^{-i\theta_1}\right)^2\left(e^{i\theta_2} + e^{-i\theta_2}\right) \\
&= \frac{1}{8}\left(e^{2i\theta_1} + 2 + e^{-2i\theta_1}\right)\left(e^{i\theta_2} + e^{-i\theta_2}\right) \\
&= \frac{1}{8}\Big[e^{i(2\theta_1+\theta_2)} + e^{-i(2\theta_1+\theta_2)} + e^{i(2\theta_1-\theta_2)} + e^{-i(2\theta_1-\theta_2)} \\
&\quad + 2\left(e^{i\theta_2} + e^{-i\theta_2}\right)\Big] \\
&\underset{(11)}{=} \frac{1}{8}\left[e^{i(\hat{\theta}_1+2\hat{\theta}_2)} + e^{-i(\hat{\theta}_1+2\hat{\theta}_2)} + e^{i\hat{\theta}_1} + e^{-i\hat{\theta}_1} + 2\left(e^{i\hat{\theta}_2} + e^{-i\hat{\theta}_2}\right)\right] \\
&= \frac{1}{4}\left[\cos(\hat{\theta}_1 + 2\hat{\theta}_2) + \cos\hat{\theta}_1 + 2\cos\hat{\theta}_2\right] .
\end{aligned}$$

Likewise, for $\cos^3\theta_2$, it holds that (recall $\theta_2 = \hat{\theta}_2$):

$$\begin{aligned}
\cos^3\theta_2 &= \tfrac{1}{8}\left(e^{i3\theta} + e^{-i3\theta_2} + 3e^{i\theta_2} + 3e^{-i\theta_2}\right) \\
&= \tfrac{1}{4}\left[\cos 3\hat{\theta}_2 + 3\cos\hat{\theta}_2\right] .
\end{aligned}$$

We then have for the frequency spectrum of H_1:

$$\begin{aligned}
H_1 = &\left(\frac{2J_1}{\omega_1}\right)\left(\frac{2J_2}{\omega_2}\right)^{1/2} \frac{1}{8}\sum_{l,m} H_{lm}\, e^{i(l\theta_1+m\theta_2)} \\
&- \frac{1}{3}\left(\frac{2J_2}{\omega_2}\right)^{3/2} \frac{1}{8}\sum_{l,m} \tilde{H}_{lm}\, e^{i(l\theta_1+m\theta_2)}
\end{aligned}$$

with

$$\begin{aligned}
&H_{21} = H_{-2-1} = H_{2-1} = H_{-21} = \tilde{H}_{03} = \tilde{H}_{0-3} = 1 , \\
&H_{01} = H_{0-1} = 2 , \quad \tilde{H}_{01} = \tilde{H}_{0-1} = 3 .
\end{aligned}$$

The complete Hamiltonian can be written in terms of the new hat variables as ($J_1 = 2\hat{J}_1$, $J_2 = \hat{J}_2 - \hat{J}_1$):

$$\begin{aligned}
\hat{H}(\hat{J}_i, \hat{\theta}_i) = \omega_2\hat{J}_2 + \varepsilon&\left(\frac{4}{\omega_2}\right)\left(\frac{2}{\omega_2}\right)^{1/2} 2\hat{J}_1\left(\hat{J}_2 - \hat{J}_1\right)^{1/2} \\
&\times\frac{1}{4}\left[\cos(\hat{\theta}_1 + 2\hat{\theta}_2) + \cos\hat{\theta}_1 + 2\cos\hat{\theta}_2\right] \\
&-\frac{\varepsilon}{3}\left(\frac{2}{\omega_2}\right)^{3/2}\left(\hat{J}_2 - \hat{J}_1\right)^{3/2}\frac{1}{4}\left[\cos(3\hat{\theta}_2) + 3\cos\hat{\theta}_2\right] .
\end{aligned}$$

If we now average over $\hat{\theta}_2$, we obtain the dependence on $\hat{\theta}_1$:

$$\varepsilon \bar{H}_1 = \frac{\varepsilon}{4}\left(\frac{8\hat{J}_1}{\omega_2}\right)\left(\frac{2(\hat{J}_2 - \hat{J}_1)}{\omega_2}\right)^{1/2} \cos\hat{\theta}_1 \,. \tag{12.14}$$

Because $\partial \bar{H}_1/\partial\hat{\theta}_2 = 0$ and of course $\partial \bar{H}_0/\partial\hat{\theta}_2 = 0$, $\hat{J}_2$ is proved to be a constant – up to averaging over the fast angle:

$$\hat{J}_2 = \hat{J}_{20} = \text{const.} \tag{12.15}$$

Altogether our new Hamiltonian now reads

$$\bar{H} = \bar{H}_0(\hat{J}_2) + \varepsilon \bar{H}_1(\hat{J}_1, \hat{J}_2, \hat{\theta}_1) \tag{12.16}$$

$$= \omega_2 \hat{J}_2 + 2\varepsilon \frac{\hat{J}_1}{\omega_2}\sqrt{\frac{2(\hat{J}_2 - \hat{J}_1)}{\omega_2}} \cos\hat{\theta}_1 \,. \tag{12.17}$$

Because of (12.10), it holds that

$$\hat{J}_2 = J_2 + \frac{s}{r} J_1 = \text{const.} \tag{12.18}$$

If, therefore, $s \gg r$ – which means a primary resonance of high order – then J_1 also becomes (almost) a constant.

After having transformed away the original (2:1) primary resonance, we can now study, in the usual manner, the motion that $\bar{H}$ (12.17) implies: one determines the singular points, analyzes their characteristics, etc. It is remarkable that since $\hat{J}_2 = $ const., the problem has become one-dimensional and therefore integrable: $\bar{H} = \bar{H}(\hat{J}_1, \hat{\theta}_1)$. Closed (periodic) trajectories can occur, etc. In the generic form

$$\bar{H} = \text{const.} + 2\varepsilon h(\hat{J}_1) \cos\hat{\theta}_1 \tag{12.19}$$

the stationary (fixed) points can be located in the $\hat{J}_1 - \hat{\theta}_1$ phase plane:

$$\dot{\hat{\theta}}_1 = \frac{\partial \bar{H}}{\partial \hat{J}_1} = 0\,, \quad \dot{\hat{J}}_1 = -\frac{\partial \bar{H}}{\partial \hat{\theta}_1} = 0 \,. \tag{12.20}$$

or

$$2\varepsilon \frac{\partial h}{\partial \hat{J}_1} \cos\hat{\theta}_1 = 0\,, \quad -2\varepsilon h(\hat{J}_1) \sin\hat{\theta}_1 = 0\,. \tag{12.21}$$

The "elliptic" fixed point is given by $\hat{J}_{10}$, $\hat{\theta}_{10} = 0$, while the "hyperbolic" fixed point is given by $\hat{J}_{10} = 0$, $\hat{\theta}_{10} = \pm\pi$. These singular points then determine the topology in phase space.

For (12.19) we write

$$\Delta \bar{H} = 2\varepsilon h(\hat{J}_1) \cos\hat{\theta}_1 \tag{12.22}$$

and expand around the elliptic fixed point $(\hat{J}_{10}, \hat{\theta}_{10} = 0)$:

$$\begin{aligned}&\Delta\hat{J}_1 = \hat{J}_1 - \hat{J}_{10}\,, \quad \Delta\hat{\theta}_1 = \hat{\theta}_1 - \hat{\theta}_{10} = \hat{\theta}_1\\ &\cos\hat{\theta}_1 = 1 - \tfrac{1}{2}\big(\Delta\hat{\theta}_1\big)^2 + \ldots\,.\end{aligned}$$

Then

$$\begin{aligned}\Delta\bar{H} &= 2\varepsilon h\big(\hat{J}_1\big)\left(1 - \tfrac{1}{2}\big(\Delta\hat{\theta}_1\big)^2\right)\\ &= 2\varepsilon h\big(\hat{J}_1\big) + \tfrac{1}{2}\big(-2\varepsilon h(\hat{J}_1)\big)\big(\Delta\hat{\theta}_1\big)^2\end{aligned}$$

and with

$$h\big(\hat{J}_1\big) = h\big(\hat{J}_{10}\big) + \underbrace{\frac{\partial h}{\partial \hat{J}_{10}}}_{=0}\big(\Delta\hat{J}_1\big) + \frac{1}{2}\frac{\partial^2 h}{\partial \hat{J}_{10}^2}\big(\Delta\hat{J}_1\big)^2$$

neglecting the unimportant constant term $h(\hat{J}_{10})$, we get the standard Hamiltonian for the harmonic oscillator,

$$\Delta\bar{H} = \tfrac{1}{2}G\big(\Delta\hat{J}_1\big)^2 + \tfrac{1}{2}F\big(\Delta\hat{\theta}_1\big)^2 \tag{12.23}$$

with

$$G = 2\varepsilon\frac{\partial^2 h(\hat{J}_{10})}{\partial \hat{J}_{10}^2}\,, \quad F = -2\varepsilon h\big(\hat{J}_{10}\big)\,. \tag{12.24}$$

The frequency of the $\hat{J}_1 - \hat{\theta}_1$-oscillation in the proximity of the elliptic fixed point is [compare (11.21, 33)]

$$\hat{\omega}_1 = (FG)^{1/2} = \mathrm{O}(\varepsilon) \tag{12.25}$$

and the ratio of the semiaxes of the ellipse reads

$$\frac{\Delta\hat{J}_1}{\Delta\hat{\theta}_1} = \left(\frac{F}{G}\right)^{1/2} = \mathrm{O}(1)\,.$$

We conclude this chapter by again considering the resonant particle-wave interaction which was introduced at the end of Chap. 10. However, we first begin with the case in which no external magnetic field is present; i.e., we consider the one-dimensional motion of a charged particle in presence of a plane wave field. Let the direction of the particle and the direction of propagation of the wave be the positive z-axis:

$$H = \frac{p_z^2}{2m} + \varepsilon e\phi_0 \sin\big(k_z z - \omega t\big)\,.$$

With

$$F_2(z, P_\psi) = (k_z z - \omega t) P_\psi$$
$$p_z = \frac{\partial F_2}{\partial z} = k_z P_\psi \,, \qquad \frac{\partial F_2}{\partial t} = -\omega P_\psi \,,$$
$$\psi = \frac{\partial H}{\partial P_\psi} = k_z z - \omega t \,,$$

we get a new time-independent Hamiltonian:

$$H \to \mathcal{H} = H + \frac{\partial F_2}{\partial t} = \frac{(k_z P_\psi)^2}{2m} - \omega P_\psi + \varepsilon e \phi_0 \sin \psi$$
$$=: \mathcal{H}_0 + \varepsilon e \phi_0 \sin \psi \,.$$

From the nonlinear free Hamiltonian $\mathcal{H}_0 = (k_z P_\psi)^2/2m - \omega P_\psi$ we find for the frequency

$$\dot{\psi} = \omega_\psi = \frac{\partial \mathcal{H}_0}{\partial P_\psi} = \frac{k_z^2 P_\psi}{m} - \omega = \frac{p_z}{m} k_z - \omega$$

or

$$\omega_\psi(v_z) = v_z k_z - \omega \,,$$

where v_z is the particle velocity. Resonance occurs if the particle velocity is equal to the phase velocity of the wave:

$$v_z^{(\mathrm{r})} = \frac{\omega}{k_z} \,.$$

From the equation of motion

$$\dot{P}_\psi = -\frac{\partial \mathcal{H}}{\partial \psi} = -\varepsilon e \phi_0 \cos \psi$$

and with

$$\ddot{\psi} = \frac{k_z^2}{m} \dot{P}_\psi = -\frac{k_z^2}{m} \varepsilon e \phi_0 \cos \psi$$

and a simple change of phase, $\psi \to \psi - \pi/2$, the pendulum equation follows:

$$\ddot{\psi} + \frac{k_z^2}{m} \varepsilon e \phi_0 \sin \psi = 0$$

or

$$\ddot{\psi} + \omega_b^2 \sin \psi = 0 \,, \qquad \omega_b = k_z \left(\frac{\varepsilon e \phi_0}{m} \right)^{1/2} \,.$$

In linearized form we get $\ddot{\psi} + \omega_b^2 \psi = 0$. Here we can see that ω_b is the frequency (for small amplitudes) of the electron trapped in the wave. For this reason, ω_b is called the "bounce" frequency. The amplitude of the wave must, however, be high enough. We can determine just how high by calculating the width of the separatrix. In order to do so, we linearize $\mathcal{H}$ in the neighborhood of the resonance $v_z^{(r)}$:

$$v_z = v_z^{(r)} + \Delta v_z = \frac{\omega}{k_z} + \Delta v_z \,,$$

$$\dot{\psi} = \omega_\psi = v_z k_z - \omega = \left(\frac{\omega}{k_z} + \Delta v_z\right) k_z - \omega = (\Delta v_z) k_z \,.$$

The Hamiltonian then becomes

$$\mathcal{H} \to \Delta\bar{\mathcal{H}} = \tfrac{1}{2}\dot{\psi}^2 - \omega_b^2 \cos\psi$$

or

$$\Delta\bar{\mathcal{H}} = \tfrac{1}{2}k_z^2 (\Delta v_z)^2 - \omega_b^2 \cos\psi \equiv \tfrac{1}{2} G (\Delta v_z)^2 - F\cos\psi \,.$$

From this we get for the width of the separatrix:

$$(\Delta v_z)_{\max} = 2\left(\frac{F}{G}\right)^{1/2} = 2\frac{\omega_b}{k_z} = 2\left(\frac{\varepsilon e \phi_0}{m}\right)^{1/2} .$$

$(\Delta v_z)_{\max}$ is the maximum "oscillation amplitude" of the particle trapped in the wave. Particles whose velocity differs from the phase velocity of the wave, ω/k_z, by less than the trapping velocity, $2(\varepsilon e\phi_0/m)^{1/2}$, may be trapped into orbits and then oscillate with bounce frequency $k_z(\varepsilon e\phi_0/m)^{1/2}$ around the phase velocity $\omega/k_z = v_z^{(r)}$. This periodic colliding of the particle with the potential wall of the wave (Landau damping) limits the energy transfer of the wave to the resonant particle.

The situation changes drastically, however, if we now apply a magnetic field. Let us recall that it was not possible to find adiabatic invariants with the aid of perturbation theory because resonant denominators appeared. We therefore now wish to apply the procedure developed at the beginning of this chapter to remove the resonances locally. In doing so, we must distinguish between two cases: (1) oblique wave propagation $k_z \neq 0$ (accidental degeneracy) and (2) right-angle propagation $k_z = 0$ – with respect to the direction of the magnetic field. The latter case corresponds to intrinsic degeneracy.

(1) For $k_z \neq 0$, accidental degeneracy occurs if the resonance condition (10.79) is satisfied; this is satisfied for a series of n values for particles with different z-momentum. We now choose a special resonance $n = l$ and transform again to the comoving system with the following generating function:

$$F_2 = (\psi - l\phi)\hat{P}_\psi + \phi\hat{P}_\phi \,. \tag{12.26}$$

With this F_2 we obtain the following transformation equations:

$$\hat{\psi} = \frac{\partial F_2}{\partial \hat{P}_\psi} = \psi - l\phi\,, \quad \hat{\phi} = \frac{\partial F_2}{\partial \hat{P}_\phi} = \phi\,, \tag{12.27}$$

$$P_\psi = \frac{\partial F_2}{\partial \psi} = \hat{P}_\psi\,, \quad P_\phi = \frac{\partial F_2}{\partial \phi} = \hat{P}_\phi - l\hat{P}_\psi\,. \tag{12.28}$$

Then (10.71) yields the new Hamiltonian:

$$\begin{aligned}\hat{H} =& \frac{k_z^2}{2m}\hat{P}_\psi^2 + \Omega\big(\hat{P}_\phi - l\hat{P}_\psi\big) - \omega\hat{P}_\psi \\ &+ \varepsilon e\phi_0 \sum_n J_n\big(k_\perp\varrho\big)\sin[\hat{\psi} - (n-l)\hat{\phi}]\,,\end{aligned} \tag{12.29}$$

where

$$\varrho = \varrho(\hat{P}_\phi, \hat{P}_\psi) = \left(\frac{2P_\phi}{m\Omega}\right)^{1/2} = \left(\frac{2}{m\Omega}\right)^{1/2}\big(\hat{P}_\phi - l\hat{P}_\psi\big)^{1/2}\,. \tag{12.30}$$

$\hat{\psi}$ is slowly changing, so we average over the fast phase $\hat{\phi}$. According to (12.29), only the term $n = l$ survives. Then the averaged Hamiltonian reads:

$$\begin{aligned}\bar{H} &= \frac{k_z^2}{2m}P_\psi^2 + \Omega\big(\hat{P}_\phi - l\hat{P}_\psi\big) - \omega\hat{P}_\psi + \varepsilon e\phi_0 J_l\big(k_\perp\varrho\big)\sin \underbrace{\hat{\psi}}_{\to\hat{\psi}+\pi/2} \\ &= \frac{k_z^2}{2m}P_\psi^2 + \Omega\big(\hat{P}_\phi - l\hat{P}_\psi\big) - \omega\hat{P}_\psi + \varepsilon e\phi_0 J_l\big(k_\perp\varrho\big)\cos\hat{\psi}\,.\end{aligned} \tag{12.31}$$

Since $\bar{H}$ is independent of $\hat{\phi}$, the associated action is constant (up to averaging):

$$\hat{P}_\phi = P_\phi + lP_\psi = \hat{P}_{\phi 0}\,. \tag{12.32}$$

The fixed points can be found, as in (12.20),

$$\psi_0 = 0\,, \quad \pm\pi\,, \tag{12.33}$$

and if we replace $\hat{J}_1$ by $\hat{P}_\psi$ in (12.20), we obtain with $\bar{H}$ of (12.31)

$$\frac{\partial \bar{H}}{\partial \hat{P}_\psi} = 0: \quad \frac{k_z^2}{m}\hat{P}_\psi^2 - l\Omega - \omega = \underset{\hat{\psi}_0 = \pm\pi}{\overset{\hat{\psi}_0 = 0}{\mp}} \varepsilon e\phi_0 \frac{\partial J_l(k_\perp\varrho)}{\partial \hat{P}_\psi}\,, \tag{12.34}$$

where, according to (12.30), ϱ depends on the actions $\hat{P}_\phi$ and $\hat{P}_\psi$. Equation (12.34) implicity determines $\hat{P}_{\psi 0}$. If we now linearize again in $\hat{P}_\psi$ but not in $\hat{\psi}$, we get the standard Hamiltonian of a pendulum (12.23) with

$$G = \frac{k_z^2}{m} \tag{12.35}$$

and

$$F = -\varepsilon e\phi_0 J_l\big(k_\perp \varrho_0\big)\,. \tag{12.36}$$

In the proximity of the elliptic singular point, it holds for the (slow) frequency of the perturbed motion that

$$\hat{\omega}_\psi = (FG)^{1/2} = \left|\frac{\varepsilon e\phi_0 J_l k_z^2}{m}\right|^{1/2}\,. \tag{12.37}$$

The maximum $\Delta\hat{P}_\psi$ (the separatrix) is given by

$$\Delta\hat{P}_{\psi\,\max} = \frac{2\hat{\omega}_\psi}{G}\,. \tag{12.38}$$

Both $\hat{\omega}_\psi$ and $\Delta\hat{P}_\psi$ are proportional to $\sqrt{\varepsilon}$. From (9.79) follows the distance between neighboring resonances:

$$\delta\hat{P}_\psi = \frac{m\Omega}{k_z^2}\,. \tag{12.39}$$

Finally, the ratio of oscillation width (twice the half-width) to the distance between neighboring oscillations is, according to (11.38/39),

$$\frac{2\Delta\hat{P}_{\psi\,\max}}{\delta\hat{P}_\psi} = \frac{4\hat{\omega}_\psi}{\Omega}\,. \tag{12.40}$$

(2) We now come to the intrinsic degeneracy. For this case, $k_z = 0$ in (12.31). We now expand again in $\Delta\hat{P}_\psi$ and $\Delta\hat{\psi}$ around the elliptical point and get in the standard Hamiltonian of the pendulum the parameters (12.24) without 2ε and $h \to \bar{H}$:

$$G = \varepsilon e\phi_0 \frac{\partial^2 J_l(k_\perp \varrho_0)}{\partial \hat{P}_{\psi 0}^2} \tag{12.41}$$

$$F = -\varepsilon e\phi_0 J_l\big(k_\perp \varrho_0\big)\,. \tag{12.42}$$

The corresponding (slow) frequency and half-width are given by

$$\hat{\omega}_\psi = (FG)^{1/2} = \varepsilon e\phi_0 \left| J_l \frac{\partial^2 J_l}{\partial \hat{P}_{\psi 0}^2}\right|^{1/2}$$

$$\Delta\hat{P}_{\psi\,\max} = \frac{2\hat{\omega}_\psi}{G}\,.$$

Compared with (11.37/38), $\hat{\omega}_\psi$ is of order ε for the present intrinsic oscillation, i.e., $\varepsilon^{1/2}$ slower than in the case of accidental degeneracy, whereas the deviation $\Delta\hat{P}_\psi$ for the intrinsic degeneracy is of order unity, i.e., $\varepsilon^{-1/2}$ larger than for the case of accidental degeneracy.

13. Superconvergent Perturbation Theory, KAM Theorem (Introduction)

Here we are dealing with an especially fast converging perturbation series, which is of particular importance for the proof of the KAM theorem (cf. below).

Until now we have transformed the Hamiltonian $H = H_0 + \varepsilon H_1$ by successive canonical transformations in such a manner that the order of the perturbation grows by one power in ε with every step. After the nth transformation we therefore obtain

$$\varepsilon H_1 \to \varepsilon^2 H_2 \to \ldots \to \varepsilon^n H_n . \tag{13.1}$$

Following Kolmogorov, we can find a succession of canonical transformations for which the order of the perturbation series increases much faster:

$$\varepsilon H_1 \to \varepsilon^2 H_2 \to \varepsilon^4 H_3 \to \ldots \to \varepsilon^{2^{n-1}} H_n . \tag{13.2}$$

We should now like to establish an analogy between the two procedures. It is based on the two following methods of finding the zero of a function $f(x)$. We begin by assuming that the zero is at x_0 (unperturbed value of the action J_0). The next improved approximation x_1 is obtained from a Taylor expansion around x_0:

$$f(x) = f(x_0) + \left.\frac{df(x)}{dx}\right|_{x=x_0} (x - x_0) + \text{rem.} = 0 .$$

If we neglect the remainder, then we obtain as our first approximation

$$x_1 = x_0 - \frac{f(x_0)}{f'(x_0)} . \tag{13.3}$$

In order to establish the error, we consider the first neglected term in the Taylor series: for this reason, let us define $\varepsilon := (x - x_0)$, and write

$$f(x) = f(x_0) + f'(x_0)\varepsilon + \tfrac{1}{2} f''(x_0)\varepsilon^2 = 0 .$$

If we subtract from this expression $f(x_1) = f(x_0) + f'(x_0)(x_1 - x_0) = 0$, then we get as error

$$e_1 := x - x_1 = -\frac{1}{2!}\frac{f''(x_0)}{f'(x_0)}\varepsilon^2 . \tag{13.4}$$

If we are considering n terms, then we would have to solve the following polynomial (of nth degree) for x_n, in order to determine x_n:

$$\sum_{m=0}^{n} \frac{1}{m!} f^{(m)}(x_0)(x_n - x_0)^m = 0 \, .$$

If we were to now subtract this result from the Taylor series around x_0, we would obtain, after the nth step, an error of

$$e_n := x - x_n \sim \frac{1}{(n+1)!} \frac{f^{(n+1)}(x_0)}{f'(x_0)} \varepsilon^{n+1} \, , \tag{13.5}$$

i.e.,

$$e_n \sim \varepsilon^{n+1} \, . \tag{13.6}$$

One should note that in the denominator, the derivative is always taken at x_0.

Matters look completely different when using Newton's procedure. The first step is identical to that of the foregoing procedure and gives

$$x_1 = x_0 - \frac{f(x_0)}{f'(x_0)} \tag{13.7}$$

with the error

$$e_1 := x - x_1 =: \alpha(x_0)\varepsilon^2 \, ,$$

where

$$\alpha(x_0) = -\frac{1}{2} \frac{f''(x_0)}{f'(x_0)} \, . \tag{13.8}$$

However, the second step consists of an expansion around x_1 (not x_0!) which we have just found:

$$f(x) = f(x_1) + f'(x_1)(x - x_1) + \ldots = 0 \, .$$

Thus, we obtain as solution for x_2 [unlike in (13.3)]:

$$x_2 = x_1 - \frac{f(x_1)}{f'(x_1)} \, .$$

The error in the second step can now be determined by subtracting the equation

$$f(x_1) + f'(x_1)(x_2 - x_1) = 0$$

from

$$f(x_1) + f'(x_1)(x - x_1) + \frac{1}{2!} f''(x_1)(x - x_1)^2 = 0 \, .$$

Thus

$$\begin{aligned} x - x_2 =: e_2 &= -\frac{1}{2!}\frac{f''(x_1)}{f'(x_1)}\big(x - x_1\big)^2 \\ &= \alpha\big(x_1\big)\big(\alpha(x_0)\varepsilon^2\big)^2 = \alpha\big(x_1\big)\alpha\big(x_0\big)^2\varepsilon^4 \,. \end{aligned} \tag{13.9}$$

Hence, it follows for the error e_n after the nth step:

$$e_n \sim \prod_{i=1}^{n}\left(\alpha^{2^{n-i}}\big(x_{i-1}\big)\right)\varepsilon^{2^n} \,. \tag{13.10}$$

The fast convergence of the Newtonian iteration procedure is evident if we write

$$e_1 \sim \varepsilon^2 \,, \quad e_2 \sim \varepsilon^4 \,, \quad e_3 \sim \varepsilon^8 \,, \quad e_4 \sim \varepsilon^{16} \,, \ldots \,. \tag{13.11}$$

The reason for this fast convergence (superconvergence) is the fact that at each step, $f(x)$ is taken at the just previously calculated approximation x_n, rather than at x_0.

This is precisely the procedure used to prove the KAM theorem. Each new "torus" which was generated by the preceding approximation becomes the basis of the next approximation itself. Thus we do not generate all successive approximations – as in the canonical perturbation series –, always starting with the unperturbed torus (J_i, θ_i), with the Hamiltonian $H_0(J_i)$.

We now look a bit ahead: the crux of the KAM theorem (according to Kolmogorov, Arnol'd and Moser) is that the process of generating "perturbed tori" indeed "almost always" converges for small but finite ε. Thus most of the phase space trajectories remain for all times on tori $\mathcal{M}$ of N dimensions and do not migrate into the entire $2N-1$ dimensional energy hyperplane. But "almost all" those unperturbed tori found in the proximity of tori whose orbits are closed will be destroyed. These orbits lie on tori with commensurate frequencies:

$$\boldsymbol{\omega}_0 \cdot \boldsymbol{m} = 0 \,. \tag{13.12}$$

These destroyed tori are precisely the ones which give rise to the famous small denominators. But we have already seen that for every arbitrary $\boldsymbol{\omega}_0$, there are "rational tori" which satisfy (13.12). So if we destroy all rational tori and their close neighbors, are there any at all which remain intact – although somewhat deformed? Indeed! In order to understand this, and to specify the width of the destroyed regions, we must concern ourselves briefly with rational and irrational numbers. These are necessary for the arithmetic of torus destruction.

In the following we consider two degrees of freedom. This is the most simple nontrivial case. Then for closed orbits it holds for the frequency ratio that

$$\frac{\omega_{01}}{\omega_{02}} =: \sigma = \frac{r}{s} \,. \tag{13.13}$$

r and s are integers, and σ is therefore rational. A torus with incommensurate frequencies possesses irrational σ and cannot be represented in the form of (13.13).

But one can approximate its frequency ratio arbitrarily precisely by rational σ's. Let us take, for example, the number π:

$$\sigma = \pi = 3.1415926535\ldots$$

with

$$\frac{r}{s} = \frac{3}{1}, \frac{31}{10}, \frac{314}{100}, \frac{3142}{1000}, \frac{31426}{10000}, \ldots .$$

The better approximations contain larger values for r and s. In fact, for each of these approximations, it holds that

$$\left|\sigma - \frac{r}{s}\right| < \frac{1}{s} . \tag{13.14}$$

But we can approximate irrational tori even better (faster): namely, with the help of continued fractions. Here are a few examples:

$$\frac{747}{61} = 12 + \frac{15}{61} = 12 + \frac{1}{61/15} = 12 + \frac{1}{4 + 1/15}$$

$$\frac{7}{10} = \frac{1}{10/7} = \frac{1}{1 + 3/7} = \frac{1}{1 + \frac{1}{7/3}} = \frac{1}{1 + \frac{1}{2+1/3}} .$$

For the number π, matters are more complicated:

$$\begin{aligned}
\pi \approx 3.1415926534 &= 3 + \frac{141592654}{10^9} \\
&= 3 + \frac{1}{\frac{10^9}{141592654}} = 3 + \frac{1}{7 + \frac{8851436}{141592654}} \\
&= 3 + \frac{1}{7 + \frac{1}{141592654/8851436}} = 3 + \frac{1}{7 + \frac{1}{15+8821114/8851436}} \\
&= 3 + \frac{1}{7 + \frac{1}{15+\frac{1}{8851436/8821114}}} = 3 + \frac{1}{7 + \frac{1}{15+\frac{1}{1+30322/8821114}}}
\end{aligned}$$

$$\pi = 3 + \frac{1}{7 + \frac{1}{15+\frac{1}{1+1/291+\ldots}}} .$$

Thus the approximands of the continued fractions read

$$\pi = \sigma = \frac{r}{s} = \frac{3}{1}, \underbrace{\frac{22}{7}}_{3.14}, \underbrace{\frac{333}{106}}_{3.141}, \underbrace{\frac{355}{113}}_{3.141592}, \ldots .$$

The rational numbers r_n/s_n appearing here are alternatively larger and smaller than σ and approximate σ with quadratic convergence:

$$\left|\sigma - \frac{r_n}{s_n}\right| < \frac{1}{s_n s_{n-1}} . \tag{13.15}$$

The slowest convergence, i.e., the most irrational number which one is least able to approximate using continued fractions, is given by the golden mean:

$$\sigma = \frac{1}{1+\frac{1}{1+1/1+\dots}} = 0.618033989 \tag{13.16}$$

$$= \frac{\sqrt{5}-1}{2} = \text{“golden mean."} \tag{13.17}$$

Apparently σ satisfies the equation

$$\sigma = \frac{1}{1+\sigma} \tag{13.18}$$

of which (13.16) is the iteration. Let us write (13.18) in the form

$$\sigma(1+\sigma) = 1 : \ \sigma^2 + \sigma - 1 = 0 .$$

Then one solution is indeed

$$\sigma = \tfrac{1}{2}(\sqrt{5}-1) .$$

Another famous irrational number is e = 2.7182818285 . . .

$$\mathrm{e} = 2 + \cfrac{1}{1+\cfrac{1}{2+\cfrac{1}{1+\cfrac{1}{1+\cfrac{1}{4+\cfrac{1}{1+\dots}}}}}} .$$

Having completed these mathematical preliminaries, let us return to physics. We know that a system with rational frequency ratios is not integrable – perturbatively speaking. It looks as if the system would at the most be integrable for irrational values of ω_1/ω_2, and that convergence of the perturbation series in ε would exist in this case. We shall therefore first answer the question as to what happens to an integrable unperturbed system $H_0(J_1, J_2)$ whose unperturbed frequency ratio ω_1/ω_2 lies in the neighborhood of an irrational number, and which is perturbed by εH_1. What happens to the rational ω_1/ω_2 in the case of a perturbation will be answered later on.

The KAM theorem now says that if, in addition to other assumptions (cf. below), in particular the functional determinant of the (action-dependent) frequencies does not vanish,

$$\left|\frac{\partial \omega_i}{\partial J_j}\right| \neq 0 , \tag{13.19}$$

for those tori, whose frequency ratio ω_1/ω_2 is “sufficiently” irrational,

$$\left|\frac{\omega_1}{\omega_2} - \frac{r}{s}\right| > \frac{K(\varepsilon)}{s^{2.5}} , \tag{13.20}$$

with r and s relatively prime, the iterated (according to Kolmogorov) perturbation series for the generator $W(\theta_k^0, J_k)$ converges (for small enough ε), and therefore the invariant tori are not destroyed.

One should note that the set of frequency ratios or space for the KAM curves, i.e., curves for which condition (13.20) holds, indeed amounts to a finite part of, e.g., the interval $0 \leqslant \omega_1/\omega_2 \leqslant 1$, because we obtain for the total length L of the intervals for which (13.20) is not valid, i.e., $|\omega_1/\omega_2 - r/s| < K(\varepsilon)/s^{2.5}$,

$$L < \sum_{s=1}^{\infty} s \frac{K(\varepsilon)}{s^{2.5}} = K(\varepsilon)\zeta(1.5) \approx 2.6K(\varepsilon) < 1 . \tag{13.21}$$

Here, $K(\varepsilon)/s^{2.5}$ is the width of an interval around the rational value r/s, for which (13.20) does not hold, and s is the number of r-values with $r/s < 1$.

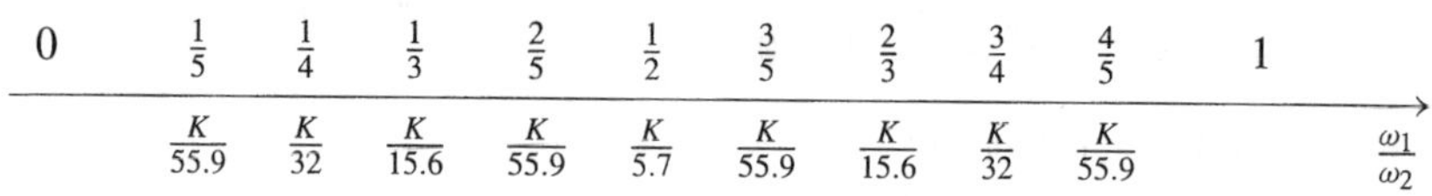

For sufficiently large ε, εH_1 destroys all tori. The last KAM torus which is destroyed is the one whose frequency ratio is the most irrational; namely,

$$\frac{\omega_1}{\omega_2} = \frac{1}{2}(\sqrt{5} - 1) .$$

By way of illustration of the KAM theorem, we consider a system from the dynamics of the solar system. (Admittedly, the following treatment will be somewhat oversimplified; nevertheless, it contains much truth.) Let three bodies move under the influence of mutual gravitation, e.g., a very massive main body M (sun or Saturn), a perturbing body m (Jupiter or moon of Saturn) and a test body of mass μ, whose long-term behavior we wish to study. Furthermore, let us assume that $\mu \ll m \ll M$; second, that all these bodies are moving in a fixed plane; and third, that m is rotating on a circle (rather than on an ellipse) around the common center of mass of M and m.

The Hamiltonian for the motion of the test body μ in presence of the gravitational fields of m and M is thus

$$H\big(\boldsymbol{q}, \boldsymbol{p}; t\big) = \frac{\boldsymbol{p}^2}{2\mu} - \frac{GM\mu}{r} - \frac{Gm\mu}{|\boldsymbol{q} - \boldsymbol{r}_m(t)|} . \tag{13.22}$$

It is obvious that (13.22) is a time-dependent Hamiltonian. Now, according to our assumption, m is supposed to move around the center of mass at the angular velocity Ω, so that it seems logical to eliminate the time dependence of H by making a transformation to the comoving, rotating system. In this system, m is at rest. Then the new conservative Hamiltonian reads

$$
\begin{aligned}
H(\boldsymbol{q},\boldsymbol{p}) &= \frac{\boldsymbol{p}^2}{2\mu} - \Omega p_\phi - \frac{GM\mu}{r} - \frac{Gm\mu}{|\boldsymbol{q}-\boldsymbol{r}_m|} && (13.23)\\
&= H_0(\boldsymbol{q},\boldsymbol{p}) + \varepsilon H_1\,, && (13.24)
\end{aligned}
$$

where

$$
H_{m=0} \equiv H_0(\boldsymbol{q}\equiv\boldsymbol{r},\boldsymbol{p}) = \frac{1}{2\mu}\left(p_r^2 + \frac{1}{r^2}p_\phi^2\right) - \Omega p_\phi - \frac{GM\mu}{r} \tag{13.25}
$$

and

$$
\varepsilon H_1 = -\frac{Gm\mu}{|\boldsymbol{q}-\boldsymbol{r}_m|}\,. \tag{13.26}
$$

Note that we have replaced $\boldsymbol{q}\to\boldsymbol{r}$ as $m\to 0$ in (13.25). Neither t nor ϕ appears in (13.25), so that the two constants of motion are p_ϕ and H_0. Equation (13.26) gives the nonintegrable perturbation term in which m plays the role of the small parameter.

Specifying the conserved quantities p_ϕ and H_0, then a certain torus is defined. The action variables J_ϕ and J_r in terms of p_ϕ and H_0 are given as follows:

$$
J_\phi = \frac{1}{2\pi}\int_0^{2\pi} p_\phi\, d\phi = p_\phi \tag{13.27}
$$

$$
\begin{aligned}
J_r &= \frac{1}{2\pi}\oint p_r\, dr = \frac{1}{2\pi}\oint dr\sqrt{2\mu\left(H_0 - \Omega J_\phi + \frac{GM\mu}{r}\right) - \frac{J_\phi^2}{r^2}} && (13.28)\\
&= -J_\phi + \frac{GM\mu^2}{\sqrt{-2\mu(H_0+\Omega J_\phi)}}\,. && (13.29)
\end{aligned}
$$

The new Hamiltonian – relative to the rotating system – thus reads, as a function of the action variables J_ϕ and J_r:

$$
H_0(J_r, J_\phi) = -\Omega J_\phi - \frac{G^2M^2\mu^3}{2(J_r+J_\phi)^2}\,. \tag{13.30}
$$

Hence, the unperturbed Hamiltonian is a nonlinear function of the actions. For the unperturbed frequencies we find, using $\omega_{0i} = \partial H_0/\partial J_i$:

$$
\omega_{0r} = \frac{G^2M^2\mu^3}{(J_r+J_\phi)^3}\,, \quad \omega_{0\phi} = -\Omega + \frac{G^2M^2\mu^3}{(J_r+J_\phi)^3}\,. \tag{13.31}
$$

Here,

$$
\omega_\mu := \frac{G^2M^2\mu^3}{(J_r+J_\phi)^3} \tag{13.32}
$$

is the frequency of the Kepler motion relative to the nonrotating coordinate system in which the r- and ϕ-motion have the same frequency (accidental degeneracy in the $1/r$-potential). Then we finally obtain

$$\omega_{0r} = \omega_\mu \, , \quad \omega_{0\phi} = -\Omega + \omega_\mu \, . \tag{13.33}$$

So the decision as to the regular or stochastic behavior of the motion of the problem perturbed by εH_1 depends on the following frequency ratio:

$$\frac{\omega_{0\phi}}{\omega_{0r}} = 1 - \frac{\Omega}{\omega_\mu} \, . \tag{13.34}$$

The invariant tori are thus destroyed if the frequency ratio, Ω/ω_μ, of the m- and μ-motion is rational. In fact, there are distributions of test bodies in the solar system in which gaps between tori can be observed. This is the case for the asteroid belt between Mars and Jupiter. Here, the sun is the main body, and Jupiter, the perturbing body. The test mass μ is any asteroid. According to the KAM theorem, one should expect gaps (instabilities) in the asteroid belt if the frequency of the asteroids and the Jupiter frequency Ω_J are commensurate. These gaps were observed by Kirkwood in 1866 and are therefore called Kirkwood gaps. They occur at $\omega_\mu/\Omega_J = 2, 3, 4$ especially clearly, at $\omega_\mu/\Omega_J = 3/2, 5/2, 7/2$, less so.

14. Poincaré Surface of Sections, Mappings

We consider a system with two degrees of freedom, which we describe in four-dimensional phase space. In this (finite) space we define an (oriented) two-dimensional surface. If we then consider the trajectory in phase space, we are interested primarily in its piercing points through this surface. This piercing can occur repeatedly in the same direction. If the motion of the trajectory is determined by the Hamiltonian equations, then the $n+1$-th piercing point depends only on the nth. The Hamiltonian thus induces a mapping $n \rightarrow n+1$ in the "Poincaré surface of section" (P.S.S.). The mapping transforms points of the P.S.S. into other (or the same) points of the P.S.S. In the following we shall limit ourselves to autonomous Hamiltonian systems, $\partial H/\partial t = 0$, so that because of the canonicity (Liouville's theorem) the mapping is area-preserving (canonical mapping).

So let $H(q_1, q_2, p_1, p_2) = E = \text{const}$. Then the motion is reduced to a three-dimensional energy hypersurface. Now we can, for example, delete p_2, since this quantity can be expressed by the remaining three (q_1, q_2, p_1). In this three-dimensional space we now construct the two-dimensional P.S.S., $q_2 = \text{const}$. Here it is advantageous to choose action variables rather than the original q_i, p_i. At this moment, we are dealing with tori which are nested within each other. We may say that phase space becomes foliated with different tori. Closed circles in the P.S.S. are indicative of integrable systems, because if there is another constant of motion (in addition to the energy), then the piercing points always lie on a curve. Conversely, a closed curve in the P.S.S. indicates an additional conserved quantity. If that conserved quantity does not – or no longer – exists (the torus is destroyed!) the points wander around chaotically (stochastically) in the P.S.S.

The above approach (after Poincaré) of the study of mappings replaces the integration of equations of motion. The development in time of the Hamiltonian system will now be treated on the basis of a series of piercing points in the P.S.S. This is an algebraic iteration method which, in the age of fast computers, has proven especially advantageous in the iteration of nonlinear algebraic mappings.

We return to our oscillator system with two degrees of freedom and a time-independent Hamiltonian. Let the system be integrable:

$$H_0 = H_0(J_1, J_2) = E \; . \tag{14.1}$$

The J_i are constants of motion. Equation (14.1) reduces our motion to a three-dimensional space, and fixing of one of the actions finally reduces the motion process to a two-dimensional surface (torus). We parametrize this torus with the help of the angle variables θ_i:

$$\theta_1 = \omega_1 t + \beta_1 \, , \quad \theta_2 = \omega_2 t + \beta_2 \, , \tag{14.2}$$

where the frequencies are determined according to

$$\omega_i = \frac{\partial H_0(J_1, J_2)}{\partial J_i} \, . \tag{14.3}$$

The trajectory then runs on the torus. We are particularly interested in the frequency ratio α:

$$\alpha = \frac{\omega_1}{\omega_2} \, . \tag{14.4}$$

If it holds that $\alpha = r/s$, with r and s relatively prime, then there is a common frequency: after r rotations in θ_1, and s in θ_2, the trajectory returns to its initial point of departure. If, however, α is irrational, then the surface of the torus eventually becomes densely filled with arbitrarily close-lying trajectories – like Lissajous figures. The system is then only conditionally periodic. The condition for a periodic solution (commensurability) is

$$m_1\omega_1 + m_2\omega_2 = 0 \equiv \boldsymbol{m} \cdot \boldsymbol{\omega}(\boldsymbol{J}) \, . \tag{14.5}$$

The time for one complete rotation in θ_2, i.e., the time between two piercings in the P.S.S. (J_1, θ_1) is

$$\Delta t = \frac{2\pi}{\omega_2} \, . \tag{14.6}$$

Meanwhile, the angle in the P.S.S. has progressed by $\Delta\theta_1$:

$$\Delta\theta_1 = \omega_1 \Delta t \underset{(6)}{=} \frac{\omega_1}{\omega_2} 2\pi =: 2\pi\alpha\big(J_1\big) \, . \tag{14.7}$$

$\alpha = \alpha(J_1)$ is termed the rotation or winding number. Since E is given and J_1 was chosen fixed (P.S.S.), $J_2 = J_2(E, J_1)$ is also determined. Hence, everything, i.e., J_2, ω_1, ω_2, α, can be expressed in terms of J_1. In the following we shall suppress the subscript 1 in J_1. Then we obtain for our unperturbed problem (no angle in H_0) the mapping (J is conserved)

$$J_{n+1} = J_n \, , \tag{14.8}$$

$$\theta_{n+1} = \theta_n + 2\pi\alpha\big(J_{n+1}\big) \, , \tag{14.9}$$

where it is useful to write J_{n+1} instead of J_n in the argument of α. The mapping defined by (14.8,9) is called twist mapping. The motion proceeds on the torus defined by $J(\equiv J_1)$. The twist mapping is area-preserving; i.e., the Jacobian of the transformation matrix is equal to one:

$$\frac{\partial(J_{n+1},\theta_{n+1})}{\partial(J_n,\theta_n)} = \begin{vmatrix} \frac{\partial J_{n+1}}{\partial J_n} & \frac{\partial J_{n+1}}{\partial \theta_n} \\ \frac{\partial \theta_{n+1}}{\partial J_n} & \frac{\partial \theta_{n+1}}{\partial \theta_n} \end{vmatrix} = 1 \ . \tag{14.10}$$

Equation (14.10) is written as Poisson bracket in the (θ_n, J_n)-basis as

$$\frac{\partial J_{n+1}}{\partial J_n}\frac{\partial \theta_{n+1}}{\partial \theta_n} - \frac{\partial J_{n+1}}{\partial \theta_n}\frac{\partial \theta_{n+1}}{\partial J_n} = \{\theta_{n+1},\ J_{n+1}\}_{\theta_n, J_n} = 1 \ .$$

The proof that (14.8,9) is indeed area-preserving is given upon insertion into the determinant (14.10):

$$\begin{vmatrix} 1 & 0 \\ 2\pi\alpha' & 1 \end{vmatrix} = 1 \ .$$

Now we are going to study systems which are almost integrable; i.e., we add a perturbation term to H_0:

$$H = H_0(\boldsymbol{J}) + \varepsilon H_1(\boldsymbol{J},\boldsymbol{\theta}) \ . \tag{14.11}$$

In this case we have to modify the twist mapping. In the $J_1 - \theta_1$ P.S.S. $\theta_2 = \text{const.}$, we get, instead, the perturbed twist mapping

$$J_{n+1} = J_n + \varepsilon f(J_{n+1},\theta_n) \tag{14.12}$$

$$\theta_{n+1} = \theta_n + 2\pi\alpha(J_{n+1}) + \varepsilon g(J_{n+1},\theta_n) \ , \tag{14.13}$$

where f and g are supposed to be periodic in θ; i.e., the modified torus remains periodic in θ $(\equiv \theta_1)$.

The following generating function $F_2 = F_2(q \equiv \theta_n,\ P \equiv J_{n+1})$ yields the transformation equations (14.12,13):

$$F_2 = \underbrace{J_{n+1}\theta_n}_{\text{ident.}} + 2\pi\mathfrak{A}(J_{n+1}) + \varepsilon\mathfrak{G}(J_{n+1},\theta_n) \tag{14.14}$$

with

$$J_n = \frac{\partial F_2}{\partial \theta_n} = J_{n+1} + \varepsilon\frac{\partial \mathfrak{G}}{\partial \theta_n}$$

or

$$J_{n+1} = J_n - \varepsilon\frac{\partial \mathfrak{G}}{\partial \theta_n} \ . \tag{14.15}$$

Comparing (14.12) with (14.15) we obtain

$$f = -\frac{\partial \mathfrak{G}}{\partial \theta_n} \ . \tag{14.16}$$

Finally, $\theta = \partial F_2/\partial P$ also belongs to the transformation equations:

$$\theta_{n+1} = \frac{\partial F_2}{\partial J_{n+1}} = \theta_n + 2\pi \frac{d\mathfrak{A}}{dJ_{n+1}} + \varepsilon \frac{\partial \mathfrak{G}}{\partial J_{n+1}} . \tag{14.17}$$

If we compare (14.17) with (14.13), we obtain

$$\alpha = \frac{d\mathfrak{A}}{dJ_{n+1}} , \tag{14.18}$$

$$g = \frac{\partial \mathfrak{G}}{\partial J_{n+1}} . \tag{14.19}$$

Since we are still dealing with a Hamiltonian system, (14.12,13) is naturally area-preserving, since it holds that

$$\frac{\partial f}{\partial J_{n+1}} + \frac{\partial g}{\partial \theta_n} = \frac{\partial}{\partial J_{n+1}} \left(-\frac{\partial \mathfrak{G}}{\partial \theta_n} \right) + \frac{\partial}{\partial \theta_n} \left(\frac{\partial \mathfrak{G}}{\partial J_{n+1}} \right) = 0 .$$

For many interesting mappings, $g = 0$ is valid and the function f is independent of J: $\partial f/\partial J = 0$. Then (14.12,13) takes on the form of the "radial" twist mapping:

$$J_{n+1} = J_n + \varepsilon f(\theta_n) , \tag{14.20}$$

$$\theta_{n+1} = \theta_n + 2\pi\alpha(J_{n+1}) . \tag{14.21}$$

Next we assume that we have a fixed point (J_0, θ_0) with period 1, i.e., corresponding to one rotation in θ_2. Per definition, a fixed point is a point that is mapped into itself. Let the winding number $\alpha(J_0)$ be an integer p, and $f(\theta_0) = 0$. The radial twist mapping then becomes simply

$$\begin{aligned} J_{n+1} &= J_n = J_0 \\ \theta_{n+1} &= \theta_n + 2\pi p , \quad p \in \mathbb{Z} , \end{aligned} \tag{14.22}$$

or

$$\begin{aligned} \theta_{n+1} &= \theta_n \pmod{2\pi} \\ &= \theta_0 . \end{aligned} \tag{14.23}$$

In the $J_1 - \theta_1$ Poincaré S.S., we shall now linearize (14.21) near the fixed point (J_0, θ_0); it then holds that, in direct proximity of the action J_0 at the nth step,

$$J_n = J_0 + \Delta J_n . \tag{14.24}$$

Then

$$\Delta J_{n+1} \underset{(20)}{=} \Delta J_n + \varepsilon f(\theta_n) , \quad f(\theta_0) = 0$$

$$\begin{aligned}\theta_{n+1} &= \theta_n + \underbrace{2\pi[\alpha(J_0)}_{\text{multiple of } 2\pi} + \left.\frac{d\alpha}{dJ}\right|_{J_0} \Delta J_{n+1}] \\ &= \theta_n + 2\pi \underbrace{\alpha'(J_0)}_{=\text{const.}} \Delta J_{n+1} \quad (\text{mod } 2\pi) \ .\end{aligned} \tag{14.25}$$

Here we have assumed that $\alpha(J_{n+1})$ is a slowly changing function of the action. We now define a new action I_n, according to

$$I_n := 2\pi\alpha' \Delta J_n \ . \tag{14.26}$$

Then, from (14.24,25) we obtain

$$I_{n+1} = I_n + 2\pi\alpha'\varepsilon f(\theta_n) \ , \tag{14.27}$$

$$\theta_{n+1} = \theta_n + I_{n+1} \quad (\text{mod } 2\pi) \ . \tag{14.28}$$

This is the desired linearized radial twist mapping in the neighborhood of the fixed point (J_0, θ_0), period 1, in the direction of θ_2. Instead of $2\pi\alpha'\varepsilon f(\theta_n)$ in (14.27), we now use the product $K f^*(\theta_n)$:

$$K = 2\pi\alpha'\varepsilon f_{\max} \tag{14.29}$$

$$f^*(\theta_n) = \frac{f(\theta_n)}{f_{\max}} \ . \tag{14.30}$$

K is called the "stochasticity parameter" and $f^*(\theta_n)$ measures the jump in the action $I_n \to I_{n+1}$. The maximum value of $f^*(\theta_n)$ is 1 because of the normalization (14.30): $f^*(\theta_n) \leqslant 1$. The thus defined mapping is called the "generalized standard mapping":

$$I_{n+1} = I_n + K f^*(\theta_n) \ , \tag{14.31}$$

$$\theta_{n+1} = \theta_n + I_{n+1} \ . \tag{14.32}$$

Until now, $f^*(\theta_n)$ has been a fairly general function. If we now simply choose $\sin\theta_n$ for $f^*(\theta_n)$, then we obtain the "standard mapping":

$$I_{n+1} = I_n + K \sin\theta_n \ , \tag{14.33}$$

$$\theta_{n+1} = \theta_n + I_{n+1} \ . \tag{14.34}$$

To conclude, we want to show how to transform the Hamiltonian development of a system into a mapping and, vice versa, how a certain class of mappings can be re-written into Hamiltonian form.

Let us begin by finding a mapping from a Hamiltonian. In particular, we again consider the two-dimensional $J_1 - \theta_1$ P.S.S. with $\theta_2 = \text{const.}$ Furthermore, let the unperturbed problem ($\varepsilon = 0$) with $H_0 = H_0(J_1, J_2)$ be solved. Since $H_0 = \text{const.} = E$, when fixing one of the actions, for example, J_1, the other, J_2, can be expressed as $J_2 = J_2(J_1, E)$. So we find ourselves once again in the P.S.S. of a 2-torus.

Now in analogy to $\dot{p} = -\partial H/\partial q$ with $H = H_0 + \varepsilon H_1(J_1, J_2, \theta_1, \theta_2)$ and $\partial H_0/\partial\theta_i = 0$:

$$\frac{dJ_1}{dt} = -\varepsilon \frac{\partial H_1}{\partial \theta_1} . \tag{14.35}$$

If we go around the torus once in the direction θ_2, period T, then $J_1(\equiv J_{n+1})$ changes by the amount

$$\Delta J_1 = -\varepsilon \int_0^T dt \frac{\partial}{\partial \theta_1} H_1\big(J_{n+1} \equiv J_1, J_2, \theta_n + \omega_1 t, \theta_{20} + \omega_2 t\big) . \tag{14.36}$$

Let us recall that J_2, ω_1 and ω_2 are all determined by $J_1 \equiv J_{n+1}$. Since ΔJ_1 is of order ε, we shall replace the arguments of H_1 by those of H_0; i.e., we integrate $\partial H_1/\partial\theta$ along the unperturbed orbit. The jump in the action J_1 (during a single rotation around the torus in θ_2-direction) is thus

$$\Delta J_1\big(J_{n+1}, \theta_n\big) = J_{n+1} - J_n \underset{(12)}{=} \varepsilon f\big(J_{n+1}, \theta_n\big) .$$

So we have determined εf in

$$J_{n+1} = J_n + \varepsilon f\big(J_{n+1}, \theta_n\big) . \tag{14.37}$$

We still need the phase jump in the phase given by εg in

$$\theta_{n+1} \underset{(13)}{=} \theta_n + 2\pi\alpha\big(J_{n+1}\big) + \varepsilon g\big(J_{n+1}, \theta_n\big) . \tag{14.38}$$

Here, α is given. g can be obtained most conveniently from the requirement that the perturbed twist mapping (14.37,38) be area-preserving:

$$\left| \frac{\partial(J_{n+1}, \theta_{n+1})}{\partial(J_n, \theta_n)} \right| = 1 . \tag{14.39}$$

If we apply (14.39) to (14.37,38), we again obtain

$$\frac{\partial f}{\partial J_{n+1}} + \frac{\partial g}{\partial \theta_n} = 0 ,$$

so that, by simple integration, we get

$$g\big(J_{n+1}, \theta_n\big) = -\int^{\theta_n} \frac{\partial f}{\partial J_{n+1}} d\theta_n' , \tag{14.40}$$

where f already has been identified. In this manner we can, in principle, generate a mapping for any given Hamiltonian.

Now, conversely, we want to determine a Hamiltonian from a given mapping. For this reason, let us imagine an infinite series of sharp spikes (kicks) along the $t(\equiv n)$-axis, represented by δ-functions at $n = t = 0, \pm 1, \pm 2, \ldots$. Since this periodic δ-function [period 1:$\delta_1(n)$] is even, its Fourier series reads:

$$\delta_1(n) \equiv \sum_{m=-\infty}^{+\infty} \delta(n-m) = \frac{a_0}{2} + \sum_{q=1}^{\infty} a_q \cos(2\pi qn) ,$$

or, since

$$\frac{a_0}{2} = \int_{-1}^{+1} \delta_1(n)dn = 1 , \quad a_q = 2\int_{-1}^{+1} \delta_1(n)\cos(2\pi n)dn = 2 ,$$

$$\delta_1(n) = 1 + 2\sum_{q=1}^{\infty} \cos(2\pi qn) . \tag{14.41}$$

The iteration number n takes over the role of the time here. As mapping we take the radial twist mapping (14.20,21): $\Delta J = J_{n+1} - J_n = \varepsilon f(\theta_n)$, and because of $\Delta n = 1$, it holds that

$$\frac{\Delta J}{\Delta n} = \Delta J : \quad \frac{dJ}{dn} = \varepsilon f(\theta)\delta_1(n) . \tag{14.42}$$

The jumps in $J = J(n)$ are measured by $\varepsilon f(\theta_n)$. Accordingly, it holds that

$$\frac{\Delta\theta}{\Delta n} = \Delta\theta : \quad \frac{d\theta}{dn} = 2\pi\alpha(J) , \tag{14.43}$$

where J_n and θ_n are $J(n)$ and $\theta(n)$ at $n - 0$, i.e., just before the "time" n. We now write the equations (14.42,43) in the form of Hamilton's equations of motion:

$$\frac{dJ}{dt} \equiv \frac{dJ}{dn} = -\frac{\partial H}{\partial\theta} , \quad \frac{d\theta}{dt} \equiv \frac{d\theta}{dn} = \frac{\partial H}{\partial J} .$$

These equations can be integrated and yield

$$H = H(J, \theta; n) = 2\pi \int^{J} \alpha(J')dJ' - \varepsilon\delta_1(n) \int^{\theta} f(\theta')d\theta' . \tag{14.44}$$

Note that H is nonautonomous with one degree of freedom In this manner we have reached our goal of constructing the appropriate Hamiltonian from a given mapping – here, the radial twist mapping.

Another example is provided by the standard mapping (14.33,34) that corresponds to a Hamiltonian which we can construct, again with the help of the periodic δ-function $\delta_1(n)$. For this reason, let us first replace J by I in (14.44) and put

$$\Delta I = I_{n+1} - I_n = f(\theta) = K \sin\theta , \tag{14.45}$$

and

$$\Delta\theta = \theta_{n+1} - \theta_n = 2\pi\alpha(I) = I . \tag{14.46}$$

Then we obtain for (14.44) ($n \equiv$ time)

$$\begin{aligned} H &= \int^I I'\,dI' - \delta_1(n)K \int^\theta \sin\theta'\,d\theta' \\ &= \frac{I^2}{2} + K\cos\theta \sum_{m=-\infty}^{+\infty} \mathrm{e}^{2\pi imn} \\ &= \frac{I^2}{2} + K \sum_{m=-\infty}^{+\infty} \cos(\theta - 2\pi mn)\ . \end{aligned} \tag{14.47}$$

At this point we want to draw attention to the fact that we are dealing here with a periodically driven pendulum which displays both regular and stochastic behavior. We can thus write (cf. 7.54)

$$H = \frac{p_\varphi^2}{2ml^2} + V(\varphi)\delta_1(n)$$

with

$$V(\varphi) = -mgl\cos\varphi = -m\omega_0^2 l^2\cos\varphi\ , \quad \omega_0^2 = \frac{g}{l}\ ,$$

so that

$$H = \frac{p_\varphi^2}{2ml^2} - m\omega_0^2 l^2\cos\varphi\,\delta_1(n)\ . \tag{14.48}$$

This Hamiltonian is that of a free rotator that is perturbed every second by a δ-type kick. If we now set

$$\frac{H}{ml^2} = \tilde{H}\ , \quad \frac{p_\varphi}{ml^2} = I\ ,$$

then we obtain for (14.48)

$$\begin{aligned} \tilde{H} &= \frac{1}{2}I^2 - \omega_0^2\cos\varphi\delta_1(n) \\ &= 2\pi\int^I \alpha(I')dI' + \omega_0^2\delta_1(n)\int^\varphi \sin\varphi'\,d\varphi'\ . \end{aligned} \tag{14.49}$$

From here, it obviously follows that (cf. 14.45,46)

$$I_{n+1} = I_n - K\sin\varphi_n\ , \quad K = \omega_0^2\ , \tag{14.50}$$

$$\varphi_{n+1} = \varphi_n + I_{n+1} \quad (\mathrm{mod}\ 2\pi)\ . \tag{14.51}$$

Replacing $\varphi_n \to \theta_n = \varphi_n - \pi$ we again make it possible for (14.50) to be written in the form (14.45):

$$I_{n+1} = I_n + K\sin\theta_n\ .$$

Finally, we again use

$$\delta_1(n) = \sum_{q=-\infty}^{+\infty} e^{2\pi i q n} = 1 + 2\sum_{q=1}^{\infty} \cos(2\pi q n)$$

and thus obtain

$$\begin{aligned}\tilde{H} = \tilde{H}(I, \varphi; n) &= \frac{1}{2}I^2 - \omega_0^2 \cos\varphi \sum_{q=-\infty}^{+\infty} e^{2\pi i q n} \\ &= \left(\frac{1}{2}I^2 - \omega_0^2 \cos\varphi\right) - 2\omega_0^2 \cos\varphi \sum_{q=1}^{\infty} \cos(2\pi q n) .\end{aligned} \tag{14.52}$$

This is indeed the Hamiltonian of a mathematical pendulum that is driven by an external periodically acting force. $K = \omega_0^2$ is the stochasticity parameter. If the time n becomes increasingly shorter between the δ-kicks, i.e., goes to zero (n is the fast parameter), then the graviational potential will be continuously turned on and $\tilde{H}$ becomes $H = p_\varphi^2/2ml^2 - mgl\cos\varphi$, i.e., the integrable Hamiltonian of the mathematical pendulum.

If the time interval of the δ-kick gradually grows out of zero, the phase space orbits of the new Hamiltonian deviate more and more from the integrable Hamiltonian H and ultimately exhibit stochastic behavior. In the case of sufficiently small $K = \omega_0^2$, however, the mapping (14.50,51) is, according to the KAM theorem, almost integrable; most of the orbits are still lying on invariant KAM curves.

We now proceed with (14.47), and want to assume that θ is a slow variable. We again retain only the most important terms, $m = 0, \pm 1$, and obtain to this order

$$H = \frac{I^2}{2} + K\cos\theta + 2K\cos\theta\cos(2\pi n) .$$

We have used the fact that

$$\cos(\theta - 2\pi n) + \cos(\theta + 2\pi n) = 2\cos\theta\cos(2\pi n) .$$

Assuming that the third term on the right-hand side of H is a perturbation term whose averaging over n vanishes, we obtain for the unperturbed Hamiltonian

$$H = \frac{I^2}{2} + K\cos\theta ,$$

i.e., the pendulum motion once again. The elliptic fixed point lies at $I = 0$, $\theta = \pi$, and the separatrix trajectory runs through $\theta = 0$, $\theta = 2\pi$. The libration frequency in the vicinity of the elliptic point $\theta = \pi$ is

$$\omega_0 = K^{1/2}$$

and the separatrix width is

$$\Delta I_{\max} = 2K^{1/2} .$$

Since the distance between the primary resonances δI is equal to the period 2π in the case of the standard mapping, it holds for the ratios of the whole separatrix width to the distance between the resonances that

$$\frac{2\Delta I_{\max}}{\delta I} = \frac{4K^{1/2}}{2\pi} .$$

15. The KAM Theorem

This theorem guarantees that, under certain assumptions, in the case of a perturbation $\varepsilon H_1(\boldsymbol{J}, \boldsymbol{\theta})$ with small enough ε, the iterated series for the generator $W(\theta_i^0, J_i)$ converges (according to Newton's procedure) and thus the invariant tori are not destroyed. The KAM theorem is valid for systems with two and more degrees of freedom. However, in the following, we shall deal exclusively with the case of two degrees of freedom.

Thus, let an integrable Hamiltonian system $H_0(J_1, J_2)$ be perturbed by a term $\varepsilon H_1(J_1, J_2, \theta_1, \theta_2)$ which depends on the angle variables θ_i. Then we know that the convergence of the various perturbation series is destroyed by the presence of the resonance condition $\sum_{i=1}^{2} m_i \omega_i = 0$ in the denominators (Poincaré's problem of small divisors). Nevertheless, under certain conditions concerning the ratio of the unperturbed frequencies, numerous invariant tori (J_i, θ_i) survive a "moderate" perturbation, albeit somewhat deformed.

Let us imagine these tori to be parametrized by η, where the relation between the unperturbed and perturbed tori is given by

$$\begin{aligned} \boldsymbol{J} &= \boldsymbol{J}_0 + \boldsymbol{v}(\boldsymbol{\eta}, \varepsilon) , \\ \boldsymbol{\theta} &= \boldsymbol{\eta} + \boldsymbol{u}(\boldsymbol{\eta}, \varepsilon) . \end{aligned} \tag{15.1}$$

η is a generalized angle variable. The perturbation terms $\boldsymbol{u}$ and $\boldsymbol{v}$ are periodic in η and vanish with $\varepsilon \to 0$. $\dot{\boldsymbol{\eta}} = \boldsymbol{\omega}$ for $\varepsilon \to 0$ are the unperturbed frequencies on the torus.

The conditions which must be fulfilled in order for invariant tori to survive a perturbation are:

(1) Linear independence of the frequencies,

$$\sum_i m_i \omega_i(\boldsymbol{J}) \neq 0 , \quad m_i \in \mathbb{Z}/\{0\} , \tag{15.2}$$

over a certain region of $\boldsymbol{J}$. $\omega_i(J_i)$ are the components of $\boldsymbol{\omega} = \nabla_J H_0(\boldsymbol{J})$, and the m_i are the components of the vector $\boldsymbol{m}$; thus, it should hold that $\boldsymbol{m} \cdot \boldsymbol{\omega}(\boldsymbol{J}) \neq 0$.

(2) Existence of sufficiently numerous derivatives of H_1; i.e., we require a certain "softness" of the perturbation.

(3) "Sufficiently large distance" from the resonance:

$$|\boldsymbol{m} \cdot \boldsymbol{\omega}| \geqslant \gamma(\varepsilon) |\boldsymbol{m}|^{-\tau} , \quad \forall \boldsymbol{m} . \tag{15.3}$$

Here, τ depends on the number of degrees of freedom and the softness of the perturbation term. γ depends on ε and the "nonlinearity" G of the unperturbed Hamiltonian H_0.

Since condition (3) cannot be fulfilled if $\gamma(\varepsilon)$ is too large, and – as we shall see – grows with ε, the smallness of the perturbation is a condition for the existence of KAM tori.

In order to elucidate the terms "linear independence" and "moderate nonlinearity", the following examples may prove helpful.

We are familiar with the Hamiltonian for the uncoupled harmonic oscillator (in two dimensions),

$$H_0(J_1, J_2) = H_0(J_1) + H_0(J_2) = \omega_1 J_1 + \omega_2 J_2 \,.$$

$H_0(J_1, J_2)$ is a linear function of the action variables, and ω_i is independent of the actions J_i, so that the frequencies are indeed constants:

$$\frac{\partial H_0}{\partial J_1} = \omega_1 = \text{const.}\,, \qquad \frac{\partial H_0}{\partial J_2} = \omega_2 = \text{const.}$$

Furthermore, it holds that

$$\frac{\partial^2 H_0}{\partial J_i^2} = \frac{\partial^2 H_0}{\partial J_i \partial J_k} = \frac{\partial \omega_i}{\partial J_k} = 0\,, \quad i, k = 1, 2\,.$$

The situation is different in the Kepler problem. There it is well known that for the motion of a particle in a plane with polar coordinates (r, θ), we have

$$H_0 = -\frac{\text{const.}}{(J_r + J_\theta)^2}\,.$$

Here, $H_0(J_r, J_\theta)$ is obviously nonlinear in the action variables. Since H_0 depends only on the sum $(J_r + J_\theta)$ (degeneracy), the frequencies are equal:

$$\omega_i = \omega = \frac{\partial H_0}{\partial J_i} = \frac{\text{const.}}{(J_r + J_\theta)^3}\,, \quad i = r, \theta\,.$$

Furthermore, the second derivatives are equal – but unequal to zero:

$$\frac{\partial^2 H_0}{\partial J_i^2} = \frac{\partial^2 H_0}{\partial J_i \partial J_k} = \frac{\partial \omega_i}{\partial J_k} = -\frac{\text{const.}}{(J_r + J_\theta)^4} \neq 0\,.$$

Again, we shall limit ourselves in the following to a two-dimensional system and shall assume from now on that the (unperturbed) frequencies of the two degrees of freedom,

$$\omega_i = \frac{\partial H_0(J_1, J_2)}{\partial J_i}\,, \quad i = 1, 2$$

are functions of the action variables:

$$\omega_i = \omega_i\big(J_1, J_2\big) \ .$$

Let us assume that between the two degrees of freedom of our so far unperturbed system commensurable frequencies $\sum_i m_i\omega_i(J_1, J_2) = 0$, $m_i \in \mathbb{Z}/\{0\}$ exist for a certain pair (J_1, J_2); e.g., $\omega_1(J_1, J_2) = \omega_2(J_1, J_2)$, i.e., $m_1\omega_1 + m_2\omega_2 = 0$, with $m_1 = -m_2 = 1$. Then, according to the KAM theorem, the invariant torus will be destroyed after turning on the interaction εH_1. Only those invariant tori for which the ω_i are linear independent in the sense of (1) come into further consideration and have a chance (if ε is small enough) to survive a perturbation.

So let $f(\omega_1, \omega_2) = 0$ be a general relation between the frequencies. We now wish to establish which requirements have to be made regarding $\omega_i(J_i)$ in order to be led to noncommensurate frequencies. We do not want a relation of the kind $\sum_i m_i\omega_i(\boldsymbol{J}) = 0$ for all J_i. In other words, we are interested in finding a condition for the linear independence of the frequencies,

$$\sum_i m_i\omega_i\big(J_1, J_2\big) \neq 0 \ . \tag{15.4}$$

To achieve this, we recall the functional dependence $f = f[\omega_1(J_1, J_2), \omega_2(J_1, J_2)] = 0$ and construct

$$\begin{aligned} df &= \frac{\partial f}{\partial\omega_1}\left[\frac{\partial\omega_1}{\partial J_1}dJ_1 + \frac{\partial\omega_1}{\partial J_2}dJ_2\right] + \frac{\partial f}{\partial\omega_2}\left[\frac{\partial\omega_2}{\partial J_1}dJ_1 + \frac{\partial\omega_2}{\partial J_2}dJ_2\right] \\ &= \left(\frac{\partial f}{\partial\omega_1}\frac{\partial\omega_1}{\partial J_1} + \frac{\partial f}{\partial\omega_2}\frac{\partial\omega_2}{\partial J_1}\right)dJ_1 + \left(\frac{\partial f}{\partial\omega_1}\frac{\partial\omega_1}{\partial J_2} + \frac{\partial f}{\partial\omega_2}\frac{\partial\omega_2}{\partial J_2}\right)dJ_2 = 0 \ . \\ &\quad \forall dJ_1, dJ_2 \end{aligned}$$

From this, we obtain the following pair of equations:

$$\begin{aligned} \frac{\partial f}{\partial\omega_1}\frac{\partial\omega_1}{\partial J_1} + \frac{\partial f}{\partial\omega_2}\frac{\partial\omega_2}{\partial J_1} &= 0 \ , \\ \frac{\partial f}{\partial\omega_1}\frac{\partial\omega_1}{\partial J_2} + \frac{\partial f}{\partial\omega_2}\frac{\partial\omega_2}{\partial J_2} &= 0 \ , \end{aligned}$$

or, written in matrix form:

$$\begin{pmatrix} \dfrac{\partial\omega_1}{\partial J_1} & \dfrac{\partial\omega_2}{\partial J_1} \\ \dfrac{\partial\omega_1}{\partial J_2} & \dfrac{\partial\omega_2}{\partial J_2} \end{pmatrix} \begin{pmatrix} \dfrac{\partial f}{\partial\omega_1} \\ \dfrac{\partial f}{\partial\omega_2} \end{pmatrix} = 0 \tag{15.5}$$

or

$$\begin{aligned} \sum_{k=1}^{2}\left(\frac{\partial}{\partial J_i}\omega_k\right)\frac{\partial f}{\partial\omega_k} &=: \left(\frac{\partial}{\partial J}\omega\right)_{ik}\frac{\partial f}{\partial\omega_k} \\ &= \big(\nabla_{\boldsymbol{J}}\boldsymbol{\omega}\big)\cdot\nabla_{\boldsymbol{\omega}}f(\boldsymbol{\omega}) = 0 \ . \end{aligned} \tag{15.6}$$

From $\det(\nabla_J \boldsymbol{\omega}) \neq 0$ follows, as only solution, $\nabla_\omega f = 0$; i.e., $\partial f/\partial\omega_1 = 0$ and $\partial f/\partial\omega_2 = 0$. Therefore, there is no valid relation for all J_i of the kind $f(\omega_1, \omega_2) = m_1\omega_1 + m_2\omega_2 = 0$, $m_i \neq 0$. Consequently, as necessary condition for the nonlinear dependence of the frequencies, we obtain

$$\det\bigl(\nabla_J \boldsymbol{\omega}(\boldsymbol{J})\bigr) = \det\left(\frac{\partial\omega_i}{\partial J_k}\right) = \det\left(\frac{\partial^2 H_0}{\partial J_i \partial J_k}\right) \neq 0 \,. \tag{15.7}$$

Now let us turn on the interaction εH_1 and assume a particular resonance behavior $f(\omega_1, \omega_2) \equiv r\omega_1 - s\omega_2 = 0$ in the Fourier decomposition of εH_1. This kind of commensurability also leads to the destruction of the torus. Thus it is necessary to formulate a nonlinearity condition for this case, too, in order not to completely destroy the invariant torus. Here we can show that a weaker condition than (15.7) prevails; required is only that the frequency $\boldsymbol{\omega}(\boldsymbol{J})$ not be zero along the direction of the actual change of $\boldsymbol{J}$. For proof, we again consider the Hamiltonian

$$H = H_0\bigl(J_1, J_2\bigr) + \varepsilon \sum_{l,m} H_{lm}^{(1)}\bigl(J_1, J_2\bigr)\, \mathrm{e}^{\mathrm{i}(l\theta_1 - m\theta_2)} \tag{15.8}$$

and choose a particular resonance $l = r$, $m = s$ and $\omega_2/\omega_1 = r/s$; in other words,

$$f\bigl(\omega_1, \omega_2\bigr)\big|_{\text{res.}} = r\dot{\theta}_1 - s\dot{\theta}_2 = r\omega_1 - s\omega_2 = 0 \,,$$

from which we obtain

$$\frac{\partial f}{\partial\omega_1} = r \,, \qquad \frac{\partial f}{\partial\omega_2} = -s \,. \tag{15.9}$$

At the point of resonance we get, with the help of the canonical equations,

$$\begin{aligned}
\dot{J}_1 &= -\frac{\partial H}{\partial\theta_1} = -\mathrm{i}r\varepsilon H_{rs}^{(1)}\, \mathrm{e}^{\mathrm{i}(r\theta_1 - s\theta_2)} \,, \\
\dot{J}_2 &= -\frac{\partial H}{\partial\theta_2} = -\mathrm{i}s\varepsilon H_{rs}^{(1)}\, \mathrm{e}^{\mathrm{i}(r\theta_1 - s\theta_2)} \,,
\end{aligned}$$

from which we derive

$$\frac{\dot{J}_1}{\dot{J}_2} = \left.\frac{dJ_1}{dJ_2}\right|_{\text{res}} = -\frac{r}{s} = -\frac{\omega_2}{\omega_1} \,. \tag{15.10}$$

Thus in a (J_1, J_2)-diagram, the direction of $\boldsymbol{J}$ at the location of the resonance is parallel to $\boldsymbol{m} = \{-r, s\} \sim \hat{\boldsymbol{J}}$, and since according to our assumption $\boldsymbol{m} \cdot \boldsymbol{\omega} = 0$, it also holds that $\hat{\boldsymbol{J}} \cdot \boldsymbol{\omega} = 0$. The unit vector $\hat{\boldsymbol{J}}$ is normalized according to

$$\hat{\boldsymbol{J}} = -\frac{r}{\sqrt{r^2 + s^2}}\hat{J}_1 + \frac{s}{\sqrt{r^2 + s^2}}\hat{J}_2 \,.$$

So we obtain

$$0 = \hat{\boldsymbol{J}} \cdot \boldsymbol{\omega} = \hat{\boldsymbol{J}} \cdot \nabla_J H_0(\boldsymbol{J})$$
$$= \frac{(-r)}{\sqrt{r^2+s^2}} \frac{\partial H_0}{\partial J_1} + \frac{s}{\sqrt{r^2+s^2}} \frac{\partial H_0}{\partial J_2} .$$

Together with $0 = f(\omega_1, \omega_2) = r\omega_1 - s\omega_2$, we then get for the frequencies:

$$\omega_1 = -\frac{1}{\sqrt{r^2+s^2}} \frac{\partial H_0}{\partial J_1} , \quad \omega_2 = -\frac{1}{\sqrt{r^2+s^2}} \frac{\partial H_0}{\partial J_2} . \tag{15.11}$$

We now use these expressions together with (15.9) in (15.5) and obtain, after some trivial changes,

$$r \frac{\partial^2 H_0}{\partial J_1^2} - s \frac{\partial^2 H_0}{\partial J_1 \partial J_2} = 0 ,$$
$$r \frac{\partial^2 H_0}{\partial J_1 \partial J_2} - s \frac{\partial^2 H_0}{\partial J_2^2} = 0 .$$

By multiplying the first equation by r, the second by $(-s)$ and then adding the two equations, we finally get as sufficient condition for the *nonlinearity*:

$$r^2 \frac{\partial^2 H_0}{\partial J_1^2} - 2rs \frac{\partial^2 H_0}{\partial J_1 \partial J_2} + s^2 \frac{\partial^2 H_0}{\partial J_2^2} \neq 0 . \tag{15.12}$$

We recall that in canonical perturbation theory, the Hamiltonian describing the motion in the vicinity of a resonance is given by

$$\Delta \bar{H} = \tfrac{1}{2} G \big(\Delta \hat{J}_1\big)^2 - F \cos \hat{\theta}_1 .$$

Here, G is the nonlinearity parameter, i.e., our measure of the degree of linear independence,

$$G\big(\hat{J}_{i0}\big) = \frac{\partial^2 \hat{H}_0}{\partial \hat{J}_{10}^2} ,$$

and F is the product of the strength of the perturbation, ε, and the Fourier amplitude:

$$F = -2\varepsilon H_{rs}^{(1)}\big(\hat{J}_{i0}\big) .$$

We still want to show that the nonlinearity condition can also be derived from canonical perturbation theory. To that end, we apply the generating function

$$F_2 = \big(r\theta_1 - s\theta_2\big) \hat{J}_1 + \theta_2 \hat{J}_2 \tag{15.13}$$

to the Hamiltonian

$$H = H_0\big(J_1, J_2\big) + \varepsilon \sum_{l,m} H_{lm}^{(1)}\big(J_1, J_2\big) \, \mathrm{e}^{\mathrm{i}(l\theta_1 - m\theta_2)} , \tag{15.14}$$

and obtain (in lowest order ε) in the new variables (12.10) $r\hat{J}_1 = J_1, \ldots$, after having expanded around the resonant value of the action and averaged over the fast variables:

$$\Delta\bar{H} = \frac{1}{2}\frac{\partial^2 \hat{H}_0}{\partial \hat{J}_1^2}(\Delta\hat{J}_1)^2 + 2\varepsilon H_{\mathrm{rs}}^{(1)} \cos\hat{\theta}_1 . \tag{15.15}$$

(Here, again, we consider the case of accidental degeneracy, $\partial^2\hat{H}_0/\partial\hat{J}_1^2 \neq 0$). Then we find

$$\begin{aligned}
0 \neq \frac{\partial^2 \hat{H}_0}{\partial \hat{J}_1^2} &= \frac{\partial}{\partial \hat{J}_1}\left[\frac{\partial \hat{H}_0}{\partial J_1}\overbrace{\frac{\partial J_1}{\partial \hat{J}_1}}^{=r} + \frac{\partial \hat{H}_0}{\partial J_2}\overbrace{\frac{\partial J_2}{\partial \hat{J}_1}}^{=-s}\right] \\
&= \frac{\partial}{\partial \hat{J}_1}\left[r\frac{\partial H_0}{\partial J_1} - s\frac{\partial H_0}{\partial J_2}\right] \\
&= r\left[\frac{\partial^2 H_0}{\partial J_1^2}\frac{\partial J_1}{\partial \hat{J}_1} + \frac{\partial^2 H_0}{\partial J_1 \partial J_2}\frac{\partial J_2}{\partial \hat{J}_1}\right] \\
&\quad - s\left[\frac{\partial^2 H_0}{\partial J_1 \partial J_2}\frac{\partial J_1}{\partial \hat{J}_1} + \frac{\partial^2 H_0}{\partial J_2^2}\frac{\partial J_2}{\partial \hat{J}_1}\right] \\
&= r^2\frac{\partial^2 H_0}{\partial J_1^2} - 2rs\frac{\partial^2 H_0}{\partial J_1 \partial J_2} + s^2\frac{\partial^2 H_0}{\partial J_2^2} \neq 0 .
\end{aligned} \tag{15.16}$$

The question that now arises is: "How far from zero" must we stay? For a fixed ε-value, one can estimate the required nonlinearity in G by assuming that the deviation in the action ΔJ_1 is much smaller than the unperturbed action $J_0(\varepsilon = 0)$. Now $\Delta J_1 = r\Delta\hat{J}_1$. For the half-width of the separatrix, we found $|\Delta\hat{J}_1| = 2(2\varepsilon H_{rs}/G)^{1/2}$, so that for the total width we estimate

$$4r\left(\frac{2\varepsilon H_{rs}}{G}\right)^{1/2} \ll J_0 \tag{15.17}$$

or

$$G \gg \frac{32r^2(\varepsilon H_{rs})}{J_0^2} . \tag{15.18}$$

In this case, we find KAM curves.

We conclude by considering the fate of the tori with rational frequency ratios $\alpha = r/s$ after a perturbation has been switched on. For tori with sufficiently irrational frequency ratio and small enough perturbation, we have made it plausible that these are only deformed and remain otherwise stable (KAM curves). It would thus be natural to suspect that for rational $\alpha = r/s$, where the KAM theorem fails, all the tori would be destroyed. The circumstances of just how this happens are, however, more complicated, as a theorem going back to Poincaré and Birkhoff shows.

For the unperturbed twist mapping (14.8,9), we have seen that every point on the circle, with $\alpha(J) = r/s$ =rational, is a fixed point of the twist mapping with period s (s =number of rotations along θ_2). Now the Poincaré–Birkhoff Theorem states that even after switching on the perturbation, the now perturbed twist mapping (14.12,13) still has fixed points, namely $2ks$ in number with $k \in \mathbb{N}$. Half of these are elliptic (stable); the other half are hyperbolic (unstable) fixed points. The simple proof of this theorem can be found in, for example, Lichtenberg and Lieberman.

16. Fundamental Principles of Quantum Mechanics

There are two alternative methods of quantizing a system:

a) quantization via the *Feynman Path Integral* (equivalent to Schwinger's Action Principle);
b) canonical quantization.

We shall favor the first method, which Feynman followed. Feynman, on his part, was put on the right track by – none other, of course, than – Dirac.

The first step on the way to quantizing a system entails rewriting the problem in Lagrangian form. We know from classical mechanics that this is a compact method with which to derive equations of motion. Let us refresh our memory by considering the one-dimensional motion of a particle along, say, the x-axis. Let the particle move from the point in space-time $A \equiv (x_1, t_1)$ to $B \equiv (x_2, t_2)$. In classical mechanics, the motion of a particle between A and B is described by the classical path $x = \bar{x}(t)$, which makes the action functional (for short: action) an extremum. We thus assign a number, the action S, to each path leading from A to B:

$$S = S\{[x(t)]; t_1, t_2\} = \int_{t_1}^{t_2} dt\, L\big(x(t), \dot{x}(t); t\big) \,. \tag{16.1}$$

Then the neighboring paths of the classical trajectory $\bar{x}(t)$ are given by

$$x(t) = \bar{x}(t) + \varepsilon y(t) \,. \tag{16.2}$$

The "perturbation" $y(t)$ around the classical path is arbitrary except for the boundary conditions at the terminal times,

$$y(t_1) = 0 = y(t_2) \,. \tag{16.3}$$

Besides, the time is not going to be varied. Then the action – considered as a function of ε –

$$S(\varepsilon) = \int_{t_1}^{t_2} dt\, L\big(\bar{x}(t) + \varepsilon y(t), \dot{\bar{x}}(t) + \varepsilon \dot{y}(t); t\big) \tag{16.4}$$

becomes extremal for $\varepsilon = 0$. The necessary condition for S to become stationary is therefore

$$\begin{aligned} 0 = \left.\frac{\partial S}{\partial \varepsilon}\right|_{\varepsilon=0} &= \int_{t_1}^{t_2} dt \left[\frac{\partial L}{\partial x} y(t) + \frac{\partial L}{\partial \dot{x}} \dot{y}(t)\right]_{\varepsilon=0} \\ &= \left[\frac{\partial L}{\partial \dot{x}} y(t)\right]_{t_1}^{t_2} + \int_{t_1}^{t_2} dt \left[\frac{\partial L}{\partial x} - \frac{d}{dt}\frac{\partial L}{\partial \dot{x}}\right]_{\varepsilon=0} y(t) \,. \end{aligned} \tag{16.5}$$

Since the surface term in (16.5) does not contribute, and $y(t)$ was chosen to be arbitrary, we obtain the Euler–Lagrange equation for the classical motion of the particle along the actual path $\bar{x}(t)$:

$$\left.\frac{\partial L}{\partial x}\right|_{\bar{x}} - \left.\frac{d}{dt}\frac{\partial L}{\partial \dot{x}}\right|_{\bar{x}} = 0 \,. \tag{16.6}$$

of the *probability*, or *transition, amplitude.* The motion of a particle between x_1 and x_2 is described in Feynman's quantum mechanical formulation by a phase-carrying transition amplitude. Furthermore, all possible particle paths between x_1 and x_2 contribute to the transition amplitude.

One possibility of explaining the meaning of the complex-valued transition amplitudes is provided by the well-known double-slit experiment. A double slit is irradiated with a parallel beam of electrons. We designate the registration of an electron at a point x of the detector (screen) as an event. Each event is assigned a complex-valued transition amplitude $\phi(x) \equiv \langle x|\phi\rangle$. The probability, $W(x)$, that an electron will be found at point x is given by the square of the amplitude:

$$W(x) = |\phi(x)|^2 \,. \tag{16.7}$$

The electron may, on its way to the detector, have taken path (1) or path (2) through slit (1) or (2), respectively. However, one cannot order the particle to take a particular path – through slit (1) or slit (2). Thus, there are two alternative paths which can lead to event x. Each of them is characterized by a probability amplitude $\phi_1(x) = \langle x|1\rangle$ and $\phi_2(x) \equiv \langle x|2\rangle$. The total amplitude then yields, by addition,

$$\phi(x) \equiv \langle x|\phi\rangle = \langle x|1\rangle \underbrace{\langle 1|\phi\rangle}_{=:a_1} + \langle x|2\rangle \underbrace{\langle 2|\phi\rangle}_{=:a_2} = a_1\phi_1(x) + a_2\phi_2(x) \,. \tag{16.8}$$

$|a_1|^2$ is the probability for the particle to have been selected by slit (1); likewise, for slit (2). Equation (16.8) is the well-known superposition principle of quantum mechanics which allows for interference effects. If we successively put up various slit screens, we obtain a number of possible paths that the electron can take in order to reach x. To each of these connecting paths, an amplitude is assigned, and the total amplitude of the event x is given by multiplication, e.g.,

$$\begin{aligned} \langle x|\phi\rangle =& \langle x|1\rangle\langle 1| \left(\sum_n |n\rangle\langle n|\right) |\phi\rangle \\ &+ \langle x|2\rangle\langle 2| \left(\sum_m |m\rangle\langle m|\right) |\phi\rangle + \ldots \,. \end{aligned} \tag{16.9}$$

Now we have completed all preparations and can begin with the quantum mechanical description of a propagating particle.

At time t_1, we have a probability amplitude $\psi(\boldsymbol{r}_1, t_1)$ of finding the particle at the location $\boldsymbol{r}_1$. Similarly, $\psi(\boldsymbol{r}_2, t_2)$ is the probability amplitude of the particle at the location $\boldsymbol{r}_2$ at time t_2.

With $K(\boldsymbol{r}_2, t_2|\boldsymbol{r}_1, t_1)$ we want to denote the transition amplitude for a particle that is emitted at $\boldsymbol{r}_1$ at time t_1, and is being detected at $\boldsymbol{r}_2$ at time t_2.

If a particle is selected by a screen with openings $\boldsymbol{r}_1$ to be at $(\boldsymbol{r}_1, t_1)$ with the amplitude $\psi(\boldsymbol{r}_1, t_1)$, then propagates, [i.e., is emitted at $(\boldsymbol{r}_1, t_1)$ and goes to $(\boldsymbol{r}_2, t_2)$, which is described by the quantum mechanical amplitude $K(\boldsymbol{r}_2, t_2|\boldsymbol{r}_1, t_1)$], and then is detected at $(\boldsymbol{r}_2, t_2)$ – amplitude $\psi(\boldsymbol{r}_2, t_2)$ – then, according to (16.9), the total amplitude $\psi(\boldsymbol{r}_2, t_2)$ reads

$$\psi(\boldsymbol{r}_2, t_2) = \int d^3\boldsymbol{r}_1 K(\boldsymbol{r}_2, t_2|\boldsymbol{r}_1, t_1)\psi(\boldsymbol{r}_1, t_1) . \tag{16.10}$$

integral equation, we shall show later on that it is completely equivalent to the Schrödinger equation. Our main concern now is how to find K, the kernel of the integral equation. So we have to study $K(\boldsymbol{r}_2, t_2|\boldsymbol{r}_1, t_1)$ more closely. K is also called the Feynman propagator, and once we have found its explicit form, we can control the dynamical development of the Schrödinger wave function.

In order to get from $A(\boldsymbol{r}_1, t_1)$ to $B(\boldsymbol{r}_2, t_2)$, the particle must have taken some path C. Let $\phi_{BA}[C]$ be the amplitude for the path of the particle going from A to B along C. Then it holds that

$$K(B|A) = \int [dC]\phi_{BA}[C] , \tag{16.11}$$

where the integral (or the sum) has to be taken over all paths from A to B. Obviously, the integral is very complicated, as infinitely many paths exist between A and B. The right-hand side of (16.11) is called the Feynman path integral. A precise definition of the path integral is anything but easy. Since we are interested in explicit examples, we shall always write down in detail, whereever appropriate, the right-hand side of (16.11), without considering the finer details of the existence of the path integral in general. Thus, we shall continue to use (16.11) in its naive form; for example, when we allow all possible paths in the (x, t)-plane between two points a and b, then the path integral is written as

$$K(b, a) = \int_a^b [dx(t)]\phi_{ba}[x(t)] \tag{16.12}$$

and the integral is taken over all possible paths from a to b.

We have until now reduced our problem to finding the amplitude $\phi_{BA}[C]$. But one cannot determine this amplitude from a fundamental physical principle! We shall therefore postulate $\phi_{BA}[C]$ at first according to Dirac. Precisely here we again come into contact with the Lagrangian formulation of classical mechanics. Let us recall that we assigned a classical action to each path:

$$S[C] = \int_{t_1}^{t_2} dt\, L\big(\boldsymbol{r}, \dot{\boldsymbol{r}}; t\big) \,. \tag{16.13}$$

Following an idea of Dirac's (1933), Feynman uses the following expression for $\phi[C]$:

$$\phi_{BA}[C] = \mathrm{e}^{(\mathrm{i}/\hbar)S[C]} \,. \tag{16.14}$$

With this we obtain the following formula for the Feynman propagator:

$$K\big(\boldsymbol{r}_2, t_2; \boldsymbol{r}_1, t_1\big) = \int_{\boldsymbol{r}(t_1)=\boldsymbol{r}_1}^{\boldsymbol{r}(t_2)=\boldsymbol{r}_2} \big[d\boldsymbol{r}(t)\big] \times \exp\left[\frac{\mathrm{i}}{\hbar}\int_{t_1}^{t_2} dt\, L\big(\boldsymbol{r}(t), \dot{\boldsymbol{r}}(t); t\big)\right] . \tag{16.15}$$

We can see from this form of $K = \int [d\boldsymbol{r}(t)] \exp\big[\mathrm{i}S[\boldsymbol{r}(t)]/\hbar\big]$ that the phase is constructed in such a way that in the classical limit, $S \gg \hbar$, exactly the actual classical particle path results, for the classical path is constructed in such a manner that S does not change in first order in the vicinity of the classical trajectory; i.e., the phase $S/\hbar$ stays constant in an infinitesimal neighborhood of the classical path $\boldsymbol{r}_{\mathrm{cl}}(t)$. Outside of this vicinity of $\boldsymbol{r}_{\mathrm{cl}}(t)$, the phase, in case $S_{\mathrm{cl}}/\hbar \gg 1$, will change rapidly, so that the corresponding amplitudes will be washed out by destructive interference.

Since the main contribution to the propagator comes from the infinitesimal strip around the classical path, as first approximation it holds that in the classical limit $\hbar \to 0$:

$$K\big(\boldsymbol{r}_2, t_2; \boldsymbol{r}_1, t_1\big) \sim \exp\left[\frac{\mathrm{i}}{\hbar}\int_{t_1}^{t_2} dt\, L\big(\boldsymbol{r}_{\mathrm{cl}}(t), \dot{\boldsymbol{r}}_{\mathrm{cl}}(t); t\big)\right] . \tag{16.16}$$

For a typical classical problem, the strip is very "narrow", but for a typical quantum mechanical problem, the strip is very "wide". Consequently, the classical path loses its meaning in a typical quantum mechanical situation, like, for example, the case of an electron, in its orbit around the nucleus. The path of the electron is "smeared out".

Before using (16.15), let us point out another characteristic of the propagator K. To this end we now keep x_1 and t_1 fixed, and consider $K(x_2, t_2; x_1, t_1)$ as a function of $x_2 \equiv x$ and $t_2 \equiv t$:

$$K_{(x_1,t_1)}(x, t) := K\big(x, t; x_1, t_1\big) \,. \tag{16.17}$$

This form makes it clear that $K_{(x_1,t_1)}(x, t)$ is a (Schrödinger) probability amplitude (wave function) of finding the particle at (x, t). But we know very well where the particle was located at time $t = t_1$, namely at $x = x_1$; i.e., for $t = t_1$, the amplitude is not smeared out. This can also be seen immediately from our integral equation (16.10) for the Schrödinger wave function:

$$\psi(x_2, t_2) = \int dx_1\, K(x_2, t_2; x_1, t_1)\psi(x_1, t_1)\ . \tag{16.18}$$

For $t \equiv t_2 = t_1$, we therefore obtain

$$\psi(x, t_1) = \int dx_1\, K(x, t_1; x_1, t_1)\psi(x_1, t_1)$$

which yields,

$$K(x, t_1; x_1, t_1) = \delta(x - x_1)\ . \tag{16.19}$$

A comparison with (16.17) shows clearly that $K_{(x_1,t_1)}(x, t)$ reduces to a δ-function for $t = t_1$:

$$K_{(x_1,t_1)}(x, t)|_{t=t_1} = \delta(x - x_1)\ .$$

Since $K_{(x_1,t_1)}(x, t)$ is now a Schrödinger wave function itself, it has to satisfy the integral equation (16.10):

$$K(x_3, t_3; x_1, t_1) = \int_{-\infty}^{+\infty} dx_2\, K(x_3, t_3; x_2, t_2)K(x_2, t_2; x_1, t_1)\ . \tag{16.20}$$

Thus we have derived the important *group property* for propagators. In general we can write $[b := (x_b, t_b),\ a = (x_a, t_a)]$

$$\begin{aligned} K(b, a) = \int_{-\infty}^{+\infty} dx_{N-1} \dots \int_{-\infty}^{+\infty} dx_1\, K(b, N-1) \\ \times K(N-1, N-2) \dots K(2, 1)K(1, a)\ . \end{aligned} \tag{16.21}$$

Note that the intermediate times t_i are not integrated over.

17. Functional Derivative Approach

Let us now leave the path integral formalism temporarily and reformulate operatorial quantum mechanics in a way which will make it easy later on to establish the formal connection between operator and path integral formalism. Our objective is to introduce the generating functional into quantum mechanics. Naturally we want to generate transistion amplitudes. The problem confronting us is how to transcribe operator quantum mechanics as expressed in Heisenberg's equation of motion into a theory formulated solely in terms of c-numbers. This can be achieved either by Schwinger's action principle or with the aid of a generation functional defined as follows:

$$\langle q_2, t_2 \mid q_1, t_1 \rangle^{Q,P} = \langle q_2, t_2 \mid T\left(\mathrm{e}^{\frac{\mathrm{i}}{\hbar}\int_{t_1}^{t_2} dt (Q(t)p(t)+P(t)q(t))}\right) \mid q_1, t_1 \rangle . \tag{17.1}$$

Here $Q(t)$ and $P(t)$ stand for arbitrary c-number functions ("sources") and T denotes the time-ordering operation with respect to the Heisenberg operators $q(t)$ and $p(t)$. When acting on a string of operators with different time arguments it orders the operators in such a way that the time arguments increase from the right to the left. The exponential in (17.1) should be interpreted as a power series, and in each term T acts according to the above rule:

$$T\mathrm{e}^{\int_{\tau_1}^{\tau_2} dt A(t)} \equiv \sum_{n=0}^{\infty} \frac{1}{n!} \int_{\tau_1}^{\tau_2} dt_1 \ldots \int_{\tau_1}^{\tau_2} dt_n T A(t_1) \ldots A(t_n). \tag{17.2}$$

The advantage offered by the introduction of external sources is that any number of operators between brackets can be generated by a simple functional differentiation, e.g.,

$$\langle q_2, t_2 \mid q(t) \mid q_1, t_1 \rangle = \frac{\hbar}{\mathrm{i}} \frac{\delta}{\delta P(t)} \langle q_2, t_2 \mid q_1, t_1 \rangle^{Q,P} \big|_{Q=P=0}, \tag{17.3}$$

$$\langle q_2, t_2 \mid p(t) \mid q_1, t_1 \rangle = \frac{\hbar}{\mathrm{i}} \frac{\delta}{\delta Q(t)} \langle q_2, t_2 \mid q_1, t_1 \rangle^{Q,P} \big|_{Q=P=0} \tag{17.4}$$

or, somewhat more generally:

$$\langle q_2, t_2 \mid TF[q, p] \mid q_1, t_1 \rangle = F\left[\frac{\hbar}{\mathrm{i}} \frac{\delta}{\delta P}, \frac{\hbar}{\mathrm{i}} \frac{\delta}{\delta Q}\right] \langle q_2, t_2 \mid q_1, t_1 \rangle^{Q,P} \big|_{Q=P=0}, \tag{17.5}$$

where $F[q, p]$ is any functional of $q(t)$ and $p(t)$. To be more specific, let us consider the two-point function

$$\langle q_2, t_2 \mid T(q(t)q(t')) \mid q_1, t_1 \rangle = \frac{\hbar}{\mathrm{i}} \frac{\delta}{\delta P(t)} \frac{\hbar}{\mathrm{i}} \frac{\delta}{\delta P(t')} \langle q_2, t_2 \mid q_1, t_1 \rangle^{Q,P} \mid_{Q=P=0} . \tag{17.6}$$

The LHS contains the time-ordering operation, which is given by

$$\begin{aligned} \langle q_2, t_2 \mid T(q(t)q(t')) \mid q_1, t_1 \rangle &= \langle q_2, t_2 \mid q(t)q(t') \mid q_1, t_1 \rangle \theta(t - t') \\ &+ \langle q_2, t_2 \mid q(t')q(t) \mid q_1, t_1 \rangle \theta(t' - t), \end{aligned} \tag{17.7}$$

where the step function $\theta(t)$ is defined by

$$\theta(t) = \begin{cases} 1, & t > 0 \\ 0, & t < 0. \end{cases}$$

The functional derivative referred to above is the limiting process

$$\frac{\delta F[Q]}{\delta Q(t)} = \lim_{\varepsilon \to 0} \frac{1}{\varepsilon} \big(F[Q(t') + \varepsilon \delta(t - t')] - F[Q(t')] \big). \tag{17.8}$$

Here we list a few examples:

$$\text{(a)} \quad F[Q] = \int Q(t') f(t') dt', \tag{17.9}$$

$$\frac{\delta F[Q]}{\delta Q(t)} = f(t).$$

Note $$\frac{\delta Q(t')}{\delta Q(t)} = \delta(t' - t). \tag{17.10}$$

$$\text{(b)} \quad F[Q] = \mathrm{e}^{\int Q(t') f(t') dt'}, \tag{17.11}$$

$$\frac{\delta F[Q]}{\delta Q(t)} = f(t) \mathrm{e}^{\int Q(t') f(t') dt'}.$$

$$\text{(c)} \quad F[Q] = \mathrm{e}^{\int dt' dt'' Q(t') A(t', t'') Q(t'')} \tag{17.12}$$

$$\frac{\delta F[Q]}{\delta Q(t)} = \left[\int dt'' A(t, t'') Q(t'') + \int dt' Q(t') A(t', t) \right] \mathrm{e}^{\int QAQ}.$$

Our main goal is to rewrite Heisenberg's operator quantum mechanics equations into a coupled set of c-number functional differential equations. For this reason we start with the Hamiltonian operator $H = H\big(q(t), p(t)\big)$ and write Heisenberg's equations of motion for the operators $q(t)$ and $p(t)$ as

$$\begin{aligned} \dot{q}(t) &= \frac{\partial H(q(t), p(t))}{\partial p(t)}, \\ \dot{p}(t) &= -\frac{\partial H(q(t), p(t))}{\partial q(t)} . \end{aligned} \tag{17.13}$$

In the following we want to keep the c-number sources, i.e., we do not set them equal to zero after functional differentiation. Thus we write

$$\langle a_2, t_2 \mid a_1, t_1 \rangle^{Q,P} = \langle a_2, t_2 \mid T\big(\mathrm{e}^{\frac{\mathrm{i}}{\hbar}\int_{t_1}^{t_2} dt[Q(t)p(t)+P(t)q(t)]}\big) \mid a_1, t_1 \rangle \tag{17.14}$$

and obtain

$$\begin{aligned}
\langle a_2, t_2 \mid q(t) \mid a_1, t_1 \rangle^{Q,P} &= \frac{\hbar}{\mathrm{i}} \frac{\delta}{\delta P(t)} \langle a_2, t_2 \mid a_1, t_1 \rangle^{Q,P} \\
\langle a_2, t_2 \mid p(t) \mid a_1, t_1 \rangle^{Q,P} &= \frac{\hbar}{\mathrm{i}} \frac{\delta}{\delta Q(t)} \langle a_2, t_2 \mid a_1, t_1 \rangle^{Q,P}
\end{aligned} \tag{17.15}$$

or, more generally

$$\langle a_2, t_2 \mid TF[q(t), p(t)] \mid a_1, t_1 \rangle^{Q,P} = F\left[\frac{\hbar}{\mathrm{i}} \frac{\delta}{\delta P(t)}, \frac{\hbar}{\mathrm{i}} \frac{\delta}{\delta Q(t)}\right] \langle a_2, t_2 \mid a_1, t_1 \rangle^{Q,P}. \tag{17.16}$$

Now we operate with the time derivative upon the RHS of (17.15):

$$\begin{aligned}
\frac{d}{dt} \frac{\hbar}{\mathrm{i}} \frac{\delta}{\delta P(t)} \langle a_2, t_2 \mid a_1, t_1 \rangle^{Q,P} &= \frac{d}{dt} \langle a_2, t_2 \mid T\big(q(t) \mathrm{e}^{\frac{\mathrm{i}}{\hbar}\int_{t_1}^{t_2} dt'(Qp+Pq)}\big) \mid a_1, t_1 \rangle \\
&= \frac{d}{dt} \langle a_2, t_2 \mid T\big(\mathrm{e}^{\frac{\mathrm{i}}{\hbar}\int_{t}^{t_2} dt'(Qp+Pq)}\big) q(t) T\big(\mathrm{e}^{\frac{\mathrm{i}}{\hbar}\int_{t_1}^{t} dt'(Qp+Pq)}\big) \mid a_1, t_1 \rangle \\
&= \langle a_2, t_2 \mid T\big(\mathrm{e}^{\frac{\mathrm{i}}{\hbar}\int_{t}^{t_2} dt'(Qp+Pq)}\big) \underbrace{\dot{q}(t)}_{=\frac{\partial H}{\partial p}} T\big(\mathrm{e}^{\frac{\mathrm{i}}{\hbar}\int_{t_1}^{t} dt'(Qp+Pq)}\big) \mid a_1, t_1 \rangle \\
&\quad + \langle a_2, t_2 \mid T\big(\mathrm{e}^{\frac{\mathrm{i}}{\hbar}\int_{t}^{t_2} \cdot/\cdot}\big) \underbrace{\big[q(t), \tfrac{\mathrm{i}}{\hbar}(Q(t)p(t) + P(t)q(t))\big]}_{=-Q(t)} T\big(\mathrm{e}^{\frac{\mathrm{i}}{\hbar}\int_{t_1}^{t} \cdot/\cdot}\big) \mid a_1, t_1 \rangle \\
&= \langle a_2, t_2 \mid T\Big(\frac{\partial H(q(t)p(t))}{\partial p(t)} \mathrm{e}^{\frac{\mathrm{i}}{\hbar}\int_{t_1}^{t_2} dt(Qp+Pq)}\Big) \mid a_1, t_1 \rangle \\
&\quad - Q(t) \langle a_2, t_2 \mid T\big(\mathrm{e}^{\frac{\mathrm{i}}{\hbar}\int_{t_1}^{t_2} dt(Qp+Pq)}\big) \mid a_1, t_1 \rangle \\
&= \frac{\partial H}{\partial p}\left[\frac{\hbar}{\mathrm{i}} \frac{\delta}{\delta P(t)}, \frac{\hbar}{\mathrm{i}} \frac{\delta}{\delta Q(t)}\right] \langle a_2, t_2 \mid a_1, t_1 \rangle^{Q,P} - Q(t) \langle a_2, t_2 \mid a_1, t_1 \rangle^{Q,P}.
\end{aligned}$$

So we obtain with the aid of the equations of motion and the equal-time commutation relation $[q(t), p(t')]_{t'=t} = \mathrm{i}\hbar$ the functional differential equation

$$\begin{aligned}
\frac{d}{dt} \frac{\hbar}{\mathrm{i}} \frac{\delta}{\delta P(t)} \langle a_2, t_2 \mid a_1, t_1 \rangle^{Q,P} &= \frac{\partial H}{\partial p}\left[\frac{\hbar}{\mathrm{i}} \frac{\delta}{\delta P(t)}, \frac{\hbar}{\mathrm{i}} \frac{\delta}{\delta Q(t)}\right] \langle a_2, t_2 \mid a_1, t_1 \rangle^{Q,P} \\
&\quad - Q(t) \langle a_2, t_2 \mid a_1, t_1 \rangle^{Q,P}.
\end{aligned} \tag{17.17}$$

Similarly,

$$\begin{aligned}
\frac{d}{dt} \frac{\hbar}{\mathrm{i}} \frac{\delta}{\delta Q(t)} \langle a_2, t_2 \mid a_1, t_1 \rangle^{Q,P} &= -\frac{\partial H}{\partial q}\left[\frac{\hbar}{\mathrm{i}} \frac{\delta}{\delta P(t)}, \frac{\hbar}{\mathrm{i}} \frac{\delta}{\delta Q(t)}\right] \langle a_2, t_2 \mid a_1, t_1 \rangle^{Q,P} \\
&\quad + P(t) \langle a_2, t_2 \mid a_1, t_1 \rangle^{Q,P}.
\end{aligned} \tag{17.18}$$

This set of coupled functional differential equations for the transition amplitude $\langle a_2, t_2 \mid a_1, t_1\rangle^{Q,P}$ can obviously be obtained by replacing in Heisenberg's equations of motion

$$q(t) \to \frac{\hbar}{\mathrm{i}} \frac{\delta}{\delta P(t)}, \tag{17.19}$$

$$p(t) \to \frac{\hbar}{\mathrm{i}} \frac{\delta}{\delta Q(t)} \tag{17.20}$$

and adding the source terms as done in (17.17) and (17.18).

Now we turn to the solution of our system of equations as expressed in (17.17) and (17.18), and begin with the simplified free situation $H = 0$. The corresponding functional equations are then reduced to

$$\frac{d}{dt}\frac{\hbar}{\mathrm{i}}\frac{\delta}{\delta P(t)}\langle a_2, t_2 \mid a_1, t_1\rangle^{Q,P}_{H=0} = -Q(t)\langle a_2, t_2 \mid a_1, t_1\rangle^{Q,P}_{H=0}, \tag{17.21}$$

$$\frac{d}{dt}\frac{\hbar}{\mathrm{i}}\frac{\delta}{\delta Q(t)}\langle a_2, t_2 \mid a_1, t_1\rangle^{Q,P}_{H=0} = P(t)\langle a_2, t_2 \mid a_1, t_1\rangle^{Q,P}_{H=0} \tag{17.22}$$

or, if we divide both sides by $\langle a_2, t_2 \mid a_1, t_1\rangle^{Q,P}_{H=0}$,

$$\frac{d}{dt}\frac{\hbar}{\mathrm{i}}\frac{\delta}{\delta P(t)}\log\langle a_2, t_2 \mid a_1, t_1\rangle^{Q,P}_{H=0} = -Q(t), \tag{17.23}$$

$$\frac{d}{dt}\frac{\hbar}{\mathrm{i}}\frac{\delta}{\delta Q(t)}\log\langle a_2, t_2 \mid a_1, t_1\rangle^{Q,P}_{H=0} = P(t). \tag{17.24}$$

In a moment we will choose some particular quantum numbers for our brackets. Without any further specification we can show, however, that the interacting transformation amplitude can be solved in terms of the free theory by

$$\begin{aligned}
&\langle a_2, t_2 \mid a_1, t_1\rangle^{Q,P}_{H} \\
&= \exp\left\{-\frac{\mathrm{i}}{\hbar}\int_{t_1}^{t_2} dt H\left(\frac{\hbar}{\mathrm{i}}\frac{\delta}{\delta P(t)}, \frac{\hbar}{\mathrm{i}}\frac{\delta}{\delta Q(t)}\right)\right\}\langle a_2, t_2 \mid a_1, t_1\rangle^{Q,P}_{H=0}.
\end{aligned} \tag{17.25}$$

Let us prove this statement by constructing

$$\begin{aligned}
&\frac{d}{dt}\frac{\hbar}{\mathrm{i}}\frac{\delta}{\delta P(t)}\langle a_2, t_2 \mid a_1, t_1\rangle^{Q,P}_{H} \\
&= \exp\left\{-\frac{\mathrm{i}}{\hbar}\int_{t_1}^{t_2} dt H\left(\frac{\hbar}{\mathrm{i}}\frac{\delta}{\delta P}, \frac{\hbar}{\mathrm{i}}\frac{\delta}{\delta Q}\right)\right\}\frac{d}{dt}\frac{\hbar}{\mathrm{i}}\frac{\delta}{\delta P(t)}\langle a_2, t_2 \mid a_1, t_1\rangle^{Q,P}_{H=0} \\
&= \exp\left\{-\frac{\mathrm{i}}{\hbar}\int_{t_1}^{t_2} dt H\left(\frac{\hbar}{\mathrm{i}}\frac{\delta}{\delta P}, \frac{\hbar}{\mathrm{i}}\frac{\delta}{\delta Q}\right)\right\}\left(-Q(t)\langle a_2, t_2 \mid a_1, t_1\rangle^{Q,P}_{H=0}\right) \\
&= \left[-Q(t) + \frac{\partial H}{\partial p}\left[\frac{\hbar}{\mathrm{i}}\frac{\delta}{\delta P}, \frac{\hbar}{\mathrm{i}}\frac{\delta}{\delta Q}\right]\right]\exp\left\{\frac{\mathrm{i}}{\hbar}\int_{t_1}^{t_2} H(\ldots)\right\}\langle a_2, t_2 \mid a_1, t_1\rangle^{Q,P}_{H=0} \\
&= \left[\frac{\partial H}{\partial p}\left[\frac{\hbar}{\mathrm{i}}\frac{\delta}{\delta P}, \frac{\hbar}{\mathrm{i}}\frac{\delta}{\delta Q}\right] - Q(t)\right]\langle a_2, t_2 \mid a_1, t_1\rangle^{Q,P}_{H}.
\end{aligned} \tag{17.26}$$

In the last line of (17.26) we arrived at the RHS of (17.17), as it should be. Here we used the result of the following calculation:

$$\left(e^{-\frac{i}{\hbar}\int_{t_1}^{t_2}dt' H(t')}Q(t)e^{\frac{i}{\hbar}\int_{t_1}^{t_2}dt' H(t')}\right)\underbrace{e^{-\frac{i}{\hbar}\int_{t_1}^{t_2}dt' H(t')}\langle a_2,t_2\mid a_1,t_1\rangle_{H=0}^{Q,P}}_{=\,\%}$$

$$=\left[1-\frac{i}{\hbar}\int_{t_1}^{t_2}dt' H(t')+\ldots\right]Q(t)\left[1+\frac{i}{\hbar}\int_{t_1}^{t_2}dt' H(t')+\ldots\right]\%$$

$$=\left\{Q(t)-\frac{i}{\hbar}\int_{t_1}^{t_2}dt'\underbrace{\left[H\left(\frac{\hbar}{i}\frac{\delta}{\delta P(t')},\frac{\hbar}{i}\frac{\delta}{\delta Q(t')}\right),Q(t)\right]}_{=-i\hbar\frac{\partial}{\partial p(t')}H\left[\frac{\hbar}{i}\frac{\delta}{\delta P(t')},\frac{\hbar}{i}\frac{\delta}{\delta Q(t')}\right]\delta(t-t')}+0\right\}\%$$

$$=\left\{Q(t)-\frac{\partial H}{\partial p}\left[\frac{\hbar}{i}\frac{\delta}{\delta P(t)},\frac{\hbar}{i}\frac{\delta}{\delta Q(t)}\right]\right\}\underbrace{e^{-\frac{i}{\hbar}\int_{t_1}^{t_2}dtH(t)}\langle a_2,t_2\mid a_1,t_1\rangle_{H=0}^{Q,P}}_{=\langle a_2,t_2|a_1,t_1\rangle_H^{Q,P}}.$$

In arriving at this result we made use of

$$\left[\frac{\delta}{\delta P},Q\right]=0,\qquad \left[\frac{\delta}{\delta Q(t)},Q(t')\right]=\delta(t-t').$$

At this stage we return to the $H=0$ case as contained in the equations (17.23) and (17.24). These can be solved with the ansatz

$$\begin{aligned}&\log\langle a_2,t_2\mid a_1,t_1\rangle_{H=0}^{Q,P}\\ &=\frac{i}{\hbar}\left\{\int_{t_1}^{t_2}d\sigma Q(\sigma)\int_{t_1}^{\sigma}d\tau P(\tau)+A\int_{t_1}^{t_2}dtQ(t)+B\int_{t_1}^{t_2}dtP(t)+C\right\},\end{aligned}\tag{17.27}$$

where A, B, C are constants of integration which do not depend on the source functions $Q(t)$ and $P(t)$. We can fix the particular values for A, B, C by selecting specific base states for our brackets. Here we choose the mixed representation $\langle q_2,t_2\mid p_1,t_1\rangle^{Q,P}$ and formulate our initial conditions in terms of the equations

$$\frac{\hbar}{i}\frac{\delta}{\delta Q(t_1)}\log\langle q_2,t_2\mid p_1,t_1\rangle_{H=0}^{Q,P}=p(t_1)=p_1,\tag{17.28}$$

$$\frac{\hbar}{i}\frac{\delta}{\delta P(t_2)}\log\langle q_2,t_2\mid p_1,t_1\rangle_{H=0}^{Q,P}=q(t_2)=q_2,\tag{17.29}$$

$$\langle q_2,t_2\mid q_1,t_1\rangle_{H=0}^{Q=P=0}=\langle q_2,t_2\mid p_1,t_1\rangle=\frac{1}{(2\pi\hbar)^{\frac{1}{2}}}e^{\frac{i}{\hbar}q_2p_1}.\tag{17.30}$$

When these conditions are employed in the above ansatz (17.27), we obtain

$$A=p_1,\qquad B=q_2,$$

$$C=q_2p_1+\frac{\hbar}{i}\frac{1}{2}\log(2\pi\hbar).$$

This result enables us to rewrite (17.27) more specifically as

$$\langle q_2, t_2 \mid p_1, t_1\rangle^{Q,P}_{H=0} = \frac{1}{(2\pi\hbar)^{\frac{1}{2}}} \exp\left\{\frac{i}{\hbar}\left[\left[\int_{t_1}^{t_2} d\sigma Q(\sigma) \int_{t_1}^{\sigma} d\tau P(\tau) + \right.\right.\right.$$

$$\left.\left.\left. + p_1 \int_{t_1}^{t_2} dt Q(t) + q_2 \int_{t_1}^{t_2} dt P(t) + q_2 p_1\right]\right]\right\}. \tag{17.31}$$

Now let us consider the free particle as a simple example. To evaluate the transformation amplitude $\langle q_2, t_2 \mid p_1, t_1\rangle^{Q,P}_{H=p^2/2m}$, we have to operate with

$$\exp\left\{-\frac{i}{\hbar}\int_{t_1}^{t_2} dt \frac{1}{2m}\left(\frac{\hbar}{i}\frac{\delta}{\delta Q(t)}\right)^2\right\} \qquad \text{on } \langle q_2, t_2 \mid p_1, t_1\rangle^{Q,P}_{H=0}.$$

Now the functional operation $\frac{\hbar}{i}\frac{\delta}{\delta Q(t)}$ on $\langle q_2, t_2 \mid p_1, t_1\rangle^{Q,P}_{H=0}$ yields the shift

$$\frac{\hbar}{i}\frac{\delta}{\delta Q(t)}\langle q_2, t_2 \mid p_1, t_1\rangle^{Q,P}_{H=0} = \left(p_1 + \int_{t_1}^{t} d\tau P(\tau)\right)\langle q_2, t_2 \mid p_1, t_1\rangle^{Q,P}_{H=0} \tag{17.32}$$

so that the transformation amplitude for the free particle is given by

$$\langle q_2, t_2 \mid p_1, t_1\rangle^{Q,P}_{H=\frac{p^2}{2m}} = \exp\left\{-\frac{i}{\hbar}\int_{t_1} dt \frac{1}{2m}\left(\frac{\hbar}{i}\frac{\delta}{\delta Q(t)}\right)^2\right\}\langle q_2, t_2 \mid p_1, t_1\rangle^{Q,P}_{H=0}$$

$$= \exp\left\{-\frac{i}{\hbar}\int_{t_1}^{t_2} dt \frac{1}{2m}\left(p_1 + \int_{t_1}^{t} d\tau P(\tau)\right)^2\right\}\langle q_2, t_2 \mid p_1, t_1\rangle^{Q,P}_{H=0}$$

$$= \frac{1}{(2\pi\hbar)^{\frac{1}{2}}}\exp\left\{\frac{i}{\hbar}\int_{t_1}^{t_2} dt\left[Q(t)\left(\int_{t_1}^{t} d\tau P(\tau) + p_1\right) + q_2 P(t)\right.\right.$$

$$\left.\left. - \frac{1}{2m}\left(p_1 + \int_{t_1}^{t} d\tau P(\tau)\right)^2\right] + \frac{i}{\hbar} q_2 p_1\right\}. \tag{17.33}$$

Setting $Q(t) \equiv 0$, i.e., $H = p^2/2m - q(t)P(t)$, we obtain

$$\langle q_2, t_2 \mid p_1, t_1\rangle^{P}_{H=\frac{p^2}{2m}} = \frac{1}{(2\pi\hbar)^{\frac{1}{2}}}\exp\left\{\frac{i}{\hbar}q_2\left(\int_{t_1}^{t_2} dt P(t) + p_1\right)\right.$$

$$\left. - \frac{i}{\hbar}\frac{1}{2m}\int_{t_1}^{t_2} dt\left(p_1 + \int_{t_1}^{t} d\tau P(\tau)\right)^2\right\} \tag{17.34}$$

and for $Q \equiv 0 \equiv P$, we rediscover the well-known result,

$$\langle q_2, t_2 \mid p_1, t_1\rangle^{Q=P=0}_{H=\frac{p^2}{2m}} = \frac{1}{(2\pi\hbar)^{\frac{1}{2}}} e^{\frac{i}{\hbar}q_2 p_1 - \frac{i}{\hbar}\frac{p_1^2}{2m}(t_2 - t_1)}. \tag{17.35}$$

Let us now return to the general discussion. Our aim is to construct a formal solution of the functional differential equations (17.17) and (17.18) in terms of a path integral.

In order to be precise, we choose the labels a_1 and a_2 to represent position eigenstates, and we try to calculate the transition matrix element $\langle q_2, t_2 \mid p_1, t_1 \rangle_H^{Q,P}$ for an arbitrary Hamiltonian H. We shall start with $H \equiv 0$ and use equation (17.25) later on to obtain the matrix element for an arbitrary H. The argument consists of the following four steps:

(i) We convert the mixed q-p matrix element (17.31) to a q-q matrix element by inserting a complete set of momentum eigenstates and using (17.30):

$$\begin{aligned}
\langle q_2, t_2 \mid q_1, t_1 \rangle_{H=0}^{Q,P} &= \int_{-\infty}^{+\infty} dp_1 \langle q_2, t_2 \mid p_1, t_1 \rangle_{H=0}^{Q,P} \langle p_1, t_1 \mid q_1, t_1 \rangle \\
&= \frac{1}{(2\pi\hbar)^{\frac{1}{2}}} \int_{-\infty}^{+\infty} dp_1 \langle q_2, t_2 \mid p_1, t_1 \rangle_{H=0}^{Q,P} \, \mathrm{e}^{-\frac{\mathrm{i}}{\hbar} p_1 q_1} . \qquad (17.36)
\end{aligned}$$

Employing (17.31) in (17.36), the momentum integral leads to a δ-function, and the q-q matrix element reads:

$$\begin{aligned}
\langle q_2, t_2 \mid q_1, t_1 \rangle_{H=0}^{Q,P} &= \delta\left(q_2 - q_1 + \int_{t_1}^{t_2} d\tau Q(\tau) \right) \\
&\cdot \exp\left\{ \frac{\mathrm{i}}{\hbar} \left[\int_{t_1}^{t_2} d\sigma Q(\sigma) \int_{t_1}^{\sigma} d\tau P(\tau) + q_2 \int_{t_1}^{t_2} d\tau P(\tau) \right] \right\} . \qquad (17.37)
\end{aligned}$$

(ii) We reproduce the amplitude (17.37) by an integral over paths in phase space, $(q(t), p(t)), t \in [t_1, t_2]$. It reads:

$$\begin{aligned}
\langle q_2, t_2 \mid q_1, t_1 \rangle_{H=0}^{Q,P} &= \int_{q(t_1)=q_1}^{q(t_2)=q_2} [dq(t)] \int [dp(t)] \\
&\cdot \exp\left\{ \frac{\mathrm{i}}{\hbar} \int_{t_1}^{t_2} dt \, [p(t)\dot{q}(t) + Q(t)p(t) + P(t)q(t)] \right\} . \qquad (17.38)
\end{aligned}$$

Here we only give a formal proof of equation (17.38); for a more rigorous treatment we refer to the literature. Note that the integration over the momenta $p(t)$ is not subject to any restrictions. Therefore we can easily perform the integrations which are implied by the measure

$$\int [dp(t)] \equiv \prod_{t \in [t_1, t_2]} \int_{-\infty}^{+\infty} dp(t) \, . \qquad (17.39)$$

Because the exponential appearing in (17.38) contains $p(t)$ linearly, the integral factorizes into an infinite product of delta functions:

$$\begin{aligned}
\prod_t \left\{ \int_{-\infty}^{+\infty} dp(t) \mathrm{e}^{\frac{\mathrm{i}}{\hbar} \int_{t_1}^{t_2} dt [\dot{q}(t) + Q(t)] p(t)} \right\} &= \mathcal{N} \prod_t \delta(\dot{q}(t) + Q(t)) \\
&\equiv \mathcal{N} \delta[\dot{q} + Q] \, . \qquad (17.40)
\end{aligned}$$

As a consequence,

$$\langle q_2, t_2 \mid q_1, t_1 \rangle^{Q,P}_{H=0} = \mathcal{N} \int_{q(t_1)=q_1}^{q(t_2)=q_2} [dq(t)]\delta[\dot{q} + Q] \cdot \exp\left\{ \frac{\mathrm{i}}{\hbar} \int_{t_1}^{t_2} dt P(t) q(t) \right\}. \tag{17.41}$$

The normalization factor $\mathcal{N}$ is ill defined a priori. We shall fix it by the requirement $\langle q_2, t \mid q_1, t \rangle = \delta(q_2 - q_1)$ later on. The remaining functional integral (17.41) is over paths in coordinate space only. Because of the "delta functional" $\delta[\dot{q} + Q]$, only paths satisfying

$$\dot{q}(t) = -Q(t) \tag{17.42}$$

can contribute. At most one single path exists which satisfies (17.42) and is consistent with the boundary conditions. Solving (17.42) subject to $q(t_2) = q_2$, one finds

$$q(t) = q_2 + \int_t^{t_2} d\tau Q(\tau). \tag{17.43}$$

The second boundary condition $q(t_1) = q_1$ is satisfied only if the triple $(q_1, q_2, Q(\cdot))$ is such that

$$q_2 - q_1 + \int_{t_1}^{t_2} d\tau Q(\tau) = 0. \tag{17.44}$$

This implies that the path integral (17.41) is proportional to a delta function enforcing equation (17.44):

$$\langle q_2, t_2 \mid q_1, t_1 \rangle^{Q,P}_{H=0} = \widetilde{\mathcal{N}} \delta\left(q_2 - q_1 + \int_{t_1}^{t_2} d\tau Q(\tau) \right) \cdot \exp\left\{ \frac{\mathrm{i}}{\hbar} \int_{t_1}^{t_2} dt P(t) \left(q_2 + \int_t^{t_2} d\tau Q(\tau) \right) \right\}. \tag{17.45}$$

We fix the new normalization constant $\widetilde{\mathcal{N}}$ as $\widetilde{\mathcal{N}} = 1$, because this yields the correct normalization $\langle q_2, t \mid q_1, t \rangle^{Q,P}_{H=0} = \delta(q_2 - q_1)$. If we note that

$$\int_{t_1}^{t_2} d\sigma Q(\sigma) \int_{t_1}^{\sigma} d\tau P(\tau) = \int_{t_1}^{t_2} dt P(t) \int_t^{t_2} d\tau Q(\tau) \tag{17.46}$$

we see that the equation (17.45) is precisely the same expression as matrix element (17.37). This completes the proof that the phase space path integral (17.38) provides a representation of the transition matrix element (17.37), i.e., that this path integral is a formal solution of the functional differential equations (17.17) and (17.18) for $H = 0$.

(iii) So far we have dealt with the case $H \equiv 0$. Now we allow for a generic Hamiltonian H and derive the corresponding phase space path integral. This is most easily done by using equation (17.25). If we insert (17.38) on its RHS and interchange the functional integration with the differentiation with respect to the sources, we arrive at

$$\begin{aligned}\langle q_2, t_2 \mid q_1, t_1\rangle_H^{Q,P} = & \int_{q(t_1)=q_1}^{q(t_2)=q_2} [dq(t)] \int [dp(t)] \exp\left\{\frac{\mathrm{i}}{\hbar}\int_{t_1}^{t_2} dt\, p\dot{q}\right\} \\ & \cdot \exp\left\{-\frac{\mathrm{i}}{\hbar}\int_{t_1}^{t_2} dt H\left(\frac{\hbar}{\mathrm{i}}\frac{\delta}{\delta P(t)}, \frac{\hbar}{\mathrm{i}}\frac{\delta}{\delta Q(t)}\right)\right\} \\ & \cdot \exp\left\{\frac{\mathrm{i}}{\hbar}\int_{t_1}^{t_2} dt\,[Q(t)p(t) + P(t)q(t)]\right\}. \end{aligned} \tag{17.47}$$

When acting on the last exponential in (17.47), we may replace the derivatives $-\mathrm{i}\hbar\delta/\delta P(t)$ and $-\mathrm{i}\hbar\delta/\delta Q(t)$ by $q(t)$ and $p(t)$, respectively. This can be shown by a power series expansion of the exponential and the rules of functional differentiation. Hence we obtain as our final result for the matrix element in presence of the sources P and Q, and for an arbitrary Hamiltonian H:

$$\begin{aligned}\langle q_2, t_2 \mid q_1, t_1\rangle_H^{Q,P} = & \int_{q(t_1)=q_1}^{q(t_2)=q_2} [dq(t)] \int [dp(t)] \\ \cdot \exp & \left\{\frac{\mathrm{i}}{\hbar}\int_{t_1}^{t_2} dt\,[p(t)\dot{q}(t) - H(q(t), p(t)) + Q(t)p(t) + P(t)q(t)]\right\}. \end{aligned} \tag{17.48}$$

The propagation kernel K introduced earlier is a special case of (17.48). It is obtained by setting Q and P equal to zero: $K(q_2, t_2; q_1, t_1) \equiv \langle q_2, t_2 \mid q_1, t_1\rangle_H$. This leads us to the important result that K is given by the following phase space integral:

$$\begin{aligned}K(q_2, t_2; q_1, t_1) = & \int_{q(t_1)=q_1}^{q(t_2)=q_2} [dq(t)] \int [dp(t)] \\ & \cdot \exp\left\{\frac{\mathrm{i}}{\hbar}\int_{t_1}^{t_2} dt\,[p(t)\dot{q}(t) - H(q(t), p(t))]\right\}. \end{aligned} \tag{17.49}$$

(iv) What is the relation between the phase space integral (17.49) and the position space (or, more precisely, configuration space) path integral (15.15), which we discussed earlier? To answer this question we define a functional $\widetilde{S}[q(t)]$ by the following momentum path integral:

$$\exp\left(\frac{\mathrm{i}}{\hbar}\widetilde{S}[q(t)]\right) \equiv \int [dp(t)] \exp\left[\frac{\mathrm{i}}{\hbar}\int_{t_1}^{t_2} dt\,\{p(t)\dot{q}(t) - H(q(t), p(t))\}\right] \tag{17.50}$$

so that (17.49) becomes

$$K(q_2, t_2; q_1, t_1) = \int_{q(t_1)=q_1}^{q(t_2)=q_2} [dq(t)] \exp\left(\frac{\mathrm{i}}{\hbar}\widetilde{S}[q(t)]\right). \tag{17.51}$$

Obviously the path integral (17.51) coincides with (15.15) only if we can show that $\widetilde{S}$ equals the conventional action S, the one related to H by the usual Legendre transform. From (17.50) it is clear that the equality $\widetilde{S}[q(t)] = S[q(t)]$ cannot hold true, however, for an arbitrary Hamiltonian. It is true, however, for all Hamiltonians which have a "natural" kinetic term proportional to p^2. If we make the ansatz

$$H = \frac{p^2}{2m} + V(q), \tag{17.52}$$

the momentum integral (17.50) becomes Gaussian and can be solved by completing the square in the exponent:

$$\begin{aligned} &\int [dp(t)] \exp\left[\frac{i}{\hbar}\int_{t_1}^{t_2} dt \left\{p(t)\dot{q}(t) - \frac{p(t)^2}{2m}\right\}\right] \\ &= \text{const.} \cdot \exp\left[\frac{i}{\hbar}\int_{t_1}^{t_2} dt \frac{1}{2} m\dot{q}(t)^2\right]. \end{aligned} \tag{17.53}$$

This means that, up to an irrelevant constant,[1]

$$\widetilde{S}[q] = \int_{t_1}^{t_2} dt \left(\frac{m}{2}\dot{q}^2 + V(q)\right) = S[q]. \tag{17.54}$$

In the sequel we shall primarily consider systems with Hamiltonians of the type (17.52). For them the functional $\widetilde{S}$ in the configuration space path integral (17.51) is indeed the same as the classical action, and Feynman's formula (15.15) may be used. However, it is important to keep in mind that for a generic momentum dependence of H, $\widetilde{S}$ might differ from S.

[1] Rather than keeping track of naively divergent constants of this sort, we shall determine the overall normalization of K from the condition (15.19) when we discuss concrete examples.

18. Examples for Calculating Path Integrals

We now want to compute the kernel $K(b, a)$ for a few simple Lagrangians. We have already found for the one-dimensional case that

$$K\left(x_2, t_2; x_1, t_1\right) = \int_{x(t_1)=x_1}^{x(t_2)=x_2} [dx(t)]\, \mathrm{e}^{(\mathrm{i}/\hbar)S} \tag{18.1}$$

with

$$S = \int_{t_1}^{t_2} dt\, L(x, \dot{x}; t)\ .$$

First we consider a free particle,

$$L = m\dot{x}^2/2\ , \tag{18.2}$$

and represent an arbitrary path in the form,

$$x(t) = \bar{x}(t) + y(t)\ . \tag{18.3}$$

Here, $\bar{x}(t)$ is the actual classical path, i.e., solution to the Euler–Lagrange equation:

$$\left.\frac{\partial L}{\partial x}\right|_{\bar{x}} = -\frac{d}{dt}\left.\frac{\partial L}{\partial \dot{x}}\right|_{\dot{\bar{x}}} = 0 = \ddot{\bar{x}}\ . \tag{18.4}$$

For the deviation from the classical path, $y(t)$, it holds that

$$y\left(t_1\right) = 0 = y\left(t_2\right)\ . \tag{18.5}$$

Now let us substitute

$$x(t) = \bar{x}(t) + y(t)\ ,$$
$$\dot{x}(t) = \dot{\bar{x}}(t) + \dot{y}(t)$$

in $L(\dot{x}) = m\dot{x}^2/2$ and expand the Lagrangian around $\dot{\bar{x}}$:

$$L(\dot{x}) = L(\dot{\bar{x}} + \dot{y}) = L(\dot{\bar{x}}) + \left.\frac{\partial L}{\partial \dot{x}}\right|_{\dot{\bar{x}}} \dot{y} + \frac{1}{2}\left.\frac{\partial^2 L}{\partial \dot{x}\partial \dot{x}}\right|_{\dot{\bar{x}}} \dot{y}^2\ . \tag{18.6}$$

This expansion is exact, i.e., terminates with the term of second order, since L is quadratic in $\dot{x}$. Hence, we can write the action in the following form:

$$S = \int_{t_1}^{t_2} dt \left[L(\dot{\bar{x}}) + \left.\frac{\partial L}{\partial \dot{x}}\right|_{\dot{\bar{x}}} \dot{y} + \frac{1}{2} \left.\frac{\partial^2 L}{\partial \dot{x} \partial \dot{x}}\right|_{\dot{\bar{x}}} \dot{y}^2 \right] . \tag{18.7}$$

Using

$$\int_{t_1}^{t_2} dt \left.\frac{\partial L}{\partial \dot{x}}\right|_{\dot{\bar{x}}} \dot{y} = \left[\left.\frac{\partial L}{\partial \dot{x}}\right|_{\dot{\bar{x}}} y(t) \right]_{t_1}^{t_2} - \int_{t_1}^{t_2} dt \underbrace{\frac{d}{dt} \left(\left.\frac{\partial L}{\partial \dot{x}}\right|_{\dot{\bar{x}}} \right)}_{= m\ddot{\bar{x}} = 0} y$$

$$S_{\mathrm{cl}} = \int_{t_1}^{t_2} dt \, L(\dot{\bar{x}}) \, , \qquad \left.\frac{\partial^2 L}{\partial \dot{x} \partial \dot{x}}\right|_{\dot{\bar{x}}} = m = \text{const.}$$

we finally get

$$S = S_{\mathrm{cl}} + \frac{m}{2} \int_{t_1}^{t_2} dt \, \dot{y}^2 \tag{18.8}$$

and thus, for the kernel:

$$K(x_2, t_2; x_1, t_1) = \mathrm{e}^{\mathrm{i} S_{\mathrm{cl}}[\bar{x}]/\hbar} \int_{y(t_1)=0}^{y(t_2)=0} [dy(t)] \exp \left[\frac{\mathrm{i}}{\hbar} \int_{t_1}^{t_2} dt \frac{m}{2} \dot{y}^2 \right] . \tag{18.9}$$

Here we have used

$$[dx(t)] = \left| \frac{\delta x(t)}{\delta y(t)} \right| [dy(t)] \underset{(3)}{=} [dy(t)] \, .$$

The classical action for a free particle was worked out in (2.32):

$$S_{\mathrm{cl}} = \frac{m}{2} \frac{(x_2 - x_1)^2}{t_2 - t_1} \, . \tag{18.10}$$

Equation (18.9) thus yields for the kernel

$$K(x_2, t_2; x_1, t_1) = \exp \left[\frac{\mathrm{i}}{\hbar} \frac{m}{2} \frac{(x_2 - x_1)^2}{t_2 - t_1} \right] \int_{y(t_1)=0}^{y(t_2)=0} [dy(t)] \times \exp \left[\frac{\mathrm{i}}{\hbar} \int_{t_1}^{t_2} dt \frac{m}{2} \dot{y}^2(t) \right] . \tag{18.11}$$

Later we shall calculate the path integral in (18.11) explicitly. Here we want to apply a trick which makes use of the group property (16.20). First of all, the path integral over $y(t)$ in (18.11) is independent of x_1 and x_2. Its value can thus depend only on t_1 and t_2, and since the entire problem is time-translation invariant, (conservation of energy!), the value of the path integral is only a function of the time difference, i.e.,

$$A\big(t_2-t_1\big) := \int_0^0 [dy(t)] \exp\left[\frac{\mathrm{i}}{\hbar}\int_{t_1}^{t_2} dt \frac{m}{2}\dot{y}^2\right] . \tag{18.12}$$

So we get

$$K\big(x_2,t_2;x_1,t_1\big) = A\big(t_2-t_1\big)\exp\left[\frac{\mathrm{i}}{\hbar}\frac{m}{2}\frac{(x_2-x_1)^2}{t_2-t_1}\right] . \tag{18.13}$$

To determine $A(t)$ we make use of the group property (16.20), which reads for $t_1 = t_2 (\equiv t)$:

$$\begin{aligned}\delta\big(x_2-x_1\big) &= K\big(x_2,t;x_1,t\big)\\ &= \int_{-\infty}^{+\infty} dx\, K\big(x_2,t;x,0\big)K\big(x,0;x_1,t\big) , \qquad H=\frac{p^2}{2m}\\ &= \int_{-\infty}^{+\infty} dx\, K\big(x_2,t;x,0\big)K^*\big(x_1,t;x,0\big) .\end{aligned}$$

Here we substitute

$$\begin{aligned}K\big(x_2,t;x,0\big) &= A(t)\,\mathrm{e}^{(\mathrm{i}/\hbar)S_{\mathrm{cl}}(x_2,t;x,0)} ,\\ K^*\big(x_1,t;x,0\big) &= A^*(t)\,\mathrm{e}^{(-\mathrm{i}/\hbar)S_{\mathrm{cl}}(x_1,t;x,0)} .\end{aligned}$$

Hence we can continue to write

$$\delta\big(x_2-x_1\big) = \int_{-\infty}^{+\infty} dx |A(t)|^2\, \mathrm{e}^{(\mathrm{i}/\hbar)[S_{\mathrm{cl}}(x_2,t;x,0)-S_{\mathrm{cl}}(x_1,t;x,0)]} .$$

The exponential can also be written as $(x_2 = x_1 + \Delta x)$

$$S_{\mathrm{cl}}\big(x_1+\Delta x,t;x,0\big) - S_{\mathrm{cl}}\big(x_1,t;x,0\big) = \frac{\partial S_{\mathrm{cl}}(x_1,t;x,0)}{\partial x_1}\underbrace{\Delta x}_{=(x_2-x_1)} , \qquad \Delta x\to 0 ,$$

where

$$\frac{\partial S_{\mathrm{cl}}(x_1,t;x,0)}{\partial x_1} = \frac{\partial}{\partial x_1}\left[\frac{m}{2}\frac{(x_1-x)^2}{t}\right] = \frac{m}{t}\big(x_1-x\big)$$

or

$$\alpha(x) := \frac{\partial S_{\mathrm{cl}}(x_1,t;x,0)}{\partial x_1} = \frac{m}{t}\big(x_1-x\big) .$$

Note that $\alpha(x)$ is a linear function of x, so that $d\alpha/dx$ is independent of x. With this information, we can continue to write

$$\begin{aligned}\delta\big(x_2-x_1\big) &= \int_{-\infty}^{+\infty} d\alpha \left|\frac{dx}{d\alpha}\right| |A(t)|^2\, \mathrm{e}^{(\mathrm{i}/\hbar)\alpha(x)(x_2-x_1)}\\ &= \underbrace{\int_{-\infty}^{+\infty}\frac{d\alpha}{2\pi\hbar}\,\mathrm{e}^{(\mathrm{i}/\hbar)\alpha(x_2-x_1)}}_{=\delta(x_2-x_1)}\,\frac{2\pi\hbar|A(t)|^2}{|d\alpha/dx|} ,\end{aligned}$$

so that we obtain

$$|A(t)|^2 = \frac{1}{2\pi\hbar}\left|\frac{d\alpha}{dx}\right| = \frac{1}{2\pi\hbar}\left|-\frac{m}{t}\right|$$
$$= \frac{1}{2\pi\hbar}\left|\frac{\partial^2 S_{\mathrm{cl}}(x_1, t; x, 0)}{\partial x_1 \partial x}\right|$$

or

$$A(t) = \mathrm{e}^{-\mathrm{i}\pi/4}\sqrt{\frac{m}{2\pi\hbar t}} = \sqrt{\frac{m}{2\pi \mathrm{i}\hbar t}}\,, \qquad \frac{1}{\sqrt{\mathrm{i}}} = \mathrm{e}^{-\mathrm{i}\pi/4}\,. \tag{18.14}$$

Here we have chosen the phase in such a manner that

$$K\big(x_2, t; x_1, 0\big) = \sqrt{\frac{m}{2\pi \mathrm{i}\hbar t}}\exp\left[\frac{\mathrm{i}}{\hbar}\frac{m}{2}\frac{(x_2 - x_1)^2}{t}\right] \tag{18.15}$$

reduces to the δ-function when the limit $t \to 0$ is performed. To prove this, let us use the following representation of the δ-function:

$$\delta\big(x_2 - x_1\big) = \frac{1}{\sqrt{\pi}}\lim_{t\to 0}\left[\frac{1}{\sqrt{t}}\exp\left[-\frac{(x_2 - x_1)^2}{t}\right]\right].$$

Then the limit of (18.15) takes the value

$$\lim_{t\to 0} K\big(x_2, t; x_1, 0\big) = \sqrt{\frac{m}{2\pi \mathrm{i}\hbar}}\lim_{t\to 0}\frac{1}{\sqrt{t}}\exp\left[-\frac{m}{2\mathrm{i}\hbar t}\big(x_2 - x_1\big)^2\right]$$
$$= \sqrt{\frac{m}{2\pi \mathrm{i}\hbar}}\sqrt{\frac{2\pi \mathrm{i}\hbar}{m}}\delta\big(x_2 - x_1\big) = \delta\big(x_2 - x_1\big)\,, \quad \text{q.e.d.}$$

So we have determined in detail the propagator of the free particle,

$$K\big(x_2, t_2; x_1, t_1\big) = \sqrt{\frac{m}{2\pi \mathrm{i}\hbar(t_2 - t_1)}}\exp\left[\frac{\mathrm{i}}{\hbar}\frac{m}{2}\frac{(x_2 - x_1)^2}{t_2 - t_1}\right] \tag{18.16}$$
$$= \sqrt{\frac{m}{2\pi \mathrm{i}\hbar(t_2 - t_1)}}\mathrm{e}^{(\mathrm{i}/\hbar)S_{\mathrm{cl}}}\,.$$

As a side-result we have [cf. (18.11)]

$$\int_{y(t_1)=0}^{y(t_2)=0} [dy(t)]\exp\left[\frac{\mathrm{i}}{\hbar}\int_{t_1}^{t_2} dt\frac{m}{2}\dot{y}^2(t)\right] = \left(\frac{m}{2\pi \mathrm{i}\hbar(t_2 - t_1)}\right)^{1/2}. \tag{18.17}$$

In three dimensions we obtain instead

$$K\big(\boldsymbol{r}_2, t_2; \boldsymbol{r}_1, t_1\big) = \left(\frac{m}{2\pi \mathrm{i}\hbar(t_2 - t_1)}\right)^{3/2}\exp\left[\frac{\mathrm{i}}{\hbar}\frac{m}{2}\frac{(\boldsymbol{r}_2 - \boldsymbol{r}_1)^2}{t_2 - t_1}\right]. \tag{18.18}$$

For future purposes, let us keep the above boundary condition in mind:

$$\lim_{t\to 0} K(x_2, t; x_1, 0) = \delta(x_2 - x_1) . \tag{18.19}$$

To conclude we shall use (18.16) to establish contact with the Schrödinger wave function. We already know from Chap. 15 that $K(x, t; 0, 0)$ represents the Schrödinger wave function for a free particle which was emitted at $x_1 = 0$ at time $t_1 = 0$ and at (x, t) is described by the probability amplitude $\psi(x, t)$:

$$\psi(x, t) = K(x, t; 0, 0) = \sqrt{\frac{m}{2\pi \mathrm{i}\hbar t}} \exp\left[\frac{\mathrm{i}}{\hbar}\frac{m}{2}\frac{x^2}{t}\right] . \tag{18.20}$$

Let (x_0, t_0) be a special point. If the particle is then observed at $x = x_0$ at time t_0, then it has, classically speaking, the momentum

$$p_0 = mv_0 = m\frac{x_0}{t_0}$$

and the energy

$$E_0 = \frac{1}{2}mv_0^2 = \frac{1}{2}m\frac{x_0^2}{t_0^2} .$$

The change in phase $(m/2\hbar)x^2/t$ in (18.20) in the vicinity of (x_0, t_0) is then

$$\begin{aligned}
\psi(x, t) &= \sqrt{\frac{m}{2\pi \mathrm{i}\hbar t}} \exp\left[\frac{\mathrm{i}}{\hbar}\frac{m}{2}\frac{x^2}{t}\right] \\
&\cong \sqrt{\frac{m}{2\pi \mathrm{i}\hbar t}} \\
&\quad \times \exp\Bigg\{\frac{\mathrm{i}}{\hbar}\frac{m}{2}\underbrace{\left[\frac{x_0^2}{t_0} + \frac{\partial}{\partial x}\left(\frac{x^2}{t}\right)_{x_0,t_0}(x - x_0) + \frac{\partial}{\partial t}\left(\frac{x^2}{t}\right)_{x_0,t_0}(t - t_0) + \ldots\right]}_{\underbrace{\frac{x_0^2}{t_0} + \frac{2x_0}{t_0}(x - x_0) - \frac{x_0^2}{t_0^2}(t - t_0)}_{\frac{x_0^2}{t_0} + \frac{2x_0 x}{t_0} - \frac{2x_0^2}{t_0} - \frac{x_0^2}{t_0^2}t + \frac{x_0^2}{t_0} = \frac{2x_0 x}{t_0} - \frac{x_0^2}{t_0^2}t}}\Bigg\} \\
&= \sqrt{\frac{m}{2\pi \mathrm{i}\hbar t}} \exp\left\{\frac{\mathrm{i}}{\hbar}\left[\underbrace{m\left(\frac{x_0}{t_0}\right)}_{=p_0} x - \underbrace{\frac{m}{2}\frac{x_0^2}{t_0^2}}_{=E_0} t\right]\right\} .
\end{aligned} \tag{18.21}$$

Thus, the wave function varies in the immediate vicinity of (x_0, t_0) according to

$$\psi(x, t) \cong \sqrt{\frac{m}{2\pi \mathrm{i}\hbar t}} \exp\left[\frac{\mathrm{i}}{\hbar}\left(p_0 x - E_0 t\right)\right] . \tag{18.22}$$

This is the well-known *Einstein–de Broglie relation*, according to which a particle with momentum p and energy E is assigned a wave function with the wave length $\lambda = h/p$ and the frequency $\nu = E/h$:

$$\exp\left[\mathrm{i}\left(\frac{2\pi}{\lambda}x - \frac{\nu}{2\pi}t\right)\right] = \exp\left[\frac{\mathrm{i}}{\hbar}(px - Et)\right] . \tag{18.23}$$

With $\psi_{x_1=0,t_1=0}(x,t) = K(x,t;0,0)$ we have a space-time description of the freely moving particle. We now want to proceed to the momentum (energy) description of the particle with the aid of (16.10) and (18.16):

$$K(x,t;p,0) \equiv \chi_{p,0}(x,t) = \int_{x'=-\infty}^{+\infty} dx'\, K(x,t;x',0)\, \overbrace{\chi_{p,0}(x',0)}^{\equiv K(x',0;p,0)} . \tag{18.24}$$

Later on we shall prove in more detail the following ansatz for the transformation amplitude $\chi_p(x,0)$:

$$\big(K(x,0;p,0) \equiv\big)\chi_p(x,0) = \frac{1}{\sqrt{2\pi\hbar}}\mathrm{e}^{(\mathrm{i}/\hbar)xp} . \tag{18.25}$$

Hence, (18.24) can be written as

$$\chi_p(x,t) = \int_{-\infty}^{+\infty} dx' \sqrt{\frac{m}{2\pi\mathrm{i}\hbar t}} \exp\left[\frac{\mathrm{i}}{\hbar}\frac{m}{2}\frac{(x-x')^2}{t}\right] \frac{1}{\sqrt{2\pi\hbar}}\exp\left[\frac{\mathrm{i}}{\hbar}x'p\right] .$$

With the aid of the identity

$$x'p + \frac{m}{2}\frac{(x-x')^2}{t} = \frac{m}{2t}\left[x' - \left(x - \frac{pt}{m}\right)\right]^2 + xp - \frac{p^2}{2m}t$$

we get

$$\begin{aligned}\chi_p(x,t) =&\exp\left[\frac{\mathrm{i}}{\hbar}\left(xp - \frac{p^2}{2m}t\right)\right]\sqrt{\frac{m}{2\pi\mathrm{i}\hbar t}}\sqrt{\frac{1}{2\pi\hbar}} \\ &\times \int_{-\infty}^{+\infty} dx' \exp\left\{\frac{\mathrm{i}}{\hbar}\frac{m}{2t}\overbrace{\left[x' - \left(x - \frac{pt}{m}\right)\right]}^{u}{}^{2}\right\}\end{aligned}$$

or

$$\chi_p(x,t) = \exp\left[\frac{\mathrm{i}}{\hbar}\left(xp - \frac{p^2}{2m}t\right)\right]\frac{1}{\sqrt{2\pi\hbar}} \underbrace{\sqrt{\frac{m}{2\pi\mathrm{i}\hbar t}}\int_{-\infty}^{+\infty} du \exp\left[\frac{\mathrm{i}}{\hbar}\frac{m}{2t}u^2\right]}_{=1}$$

or

$$\chi_p(x,t) = \frac{1}{\sqrt{2\pi\hbar}}\exp\left[\frac{\mathrm{i}}{\hbar}xp - \frac{\mathrm{i}}{\hbar}\frac{p^2}{2m}t\right] . \tag{18.26}$$

So we are describing a particle with momentum p and energy $E(p) = p^2/2m$.

In three dimensions:

$$\chi_{\boldsymbol{p}}(\boldsymbol{r}, t) = \frac{1}{(2\pi\hbar)^{3/2}} \exp\left[\frac{\mathrm{i}}{\hbar}\boldsymbol{r} \cdot \boldsymbol{p} - \frac{\mathrm{i}}{\hbar}\frac{\boldsymbol{p}^2}{2m}t\right] . \tag{18.27}$$

Of course a propagator K can also have momentum arguments. Again with the help of the group property (16.10) we obtain

$$\begin{aligned} K\big(p_2, t; p_1, 0\big) &= \int_{-\infty}^{+\infty} \underbrace{\underbrace{K(p_2, t; x, t)}_{K(p_2, 0; x, 0)}}_{\chi^*_{p_2}(x)} \; dx \; \underbrace{K\big(x, t; p_1, 0\big)}_{\chi_{p_1}(x,t)} \\ &= \int_{-\infty}^{+\infty} dx \frac{1}{\sqrt{2\pi\hbar}} \exp\left[-\frac{\mathrm{i}}{\hbar}p_2 x\right] \frac{1}{\sqrt{2\pi\hbar}} \exp\left[\frac{\mathrm{i}}{\hbar}p_1 x - \frac{\mathrm{i}}{\hbar}\frac{p_1^2}{2m}t\right] \\ &= \underbrace{\frac{1}{2\pi\hbar}\int_{-\infty}^{+\infty} dx\, \mathrm{e}^{-(\mathrm{i}/\hbar)x(p_2-p_1)}}_{=\delta(p_2-p_1)} \exp\left[-\frac{\mathrm{i}}{\hbar}\frac{p_1^2}{2m}t\right] \\ &= \delta\big(p_2 - p_1\big) \exp\left[-\frac{\mathrm{i}}{\hbar}\frac{p_1^2}{2m}t\right] . \end{aligned}$$

So for the free propagator in momentum space we have

$$K\big(p_2, t; p_1, 0\big) = \delta\big(p_2 - p_1\big)\exp\left[-\frac{\mathrm{i}}{\hbar}\frac{p_1^2}{2m}t\right] . \tag{18.28}$$

With this form for K we can, conversely, return to real space:

$$\begin{aligned} &K\big(x_2, t; x_1, 0\big) \\ &= \int \underbrace{K\big(x_2, t; p, t\big)}_{\frac{1}{\sqrt{2\pi\hbar}}\exp\left[\frac{\mathrm{i}}{\hbar}x_2 p\right]} \; dp \; \underbrace{K(p, t; p', 0)}_{\delta(p-p')\exp\left[-\frac{\mathrm{i}}{\hbar}\frac{p'^2}{2m}t\right]} \; dp' \; \underbrace{K\big(p', 0; x_1, 0\big)}_{\frac{1}{\sqrt{2\pi\hbar}}\exp\left[-\frac{\mathrm{i}}{\hbar}x_1 p'\right]} \\ &= \int \frac{dp}{2\pi\hbar} \mathrm{e}^{(\mathrm{i}/\hbar)(x_2-x_1)p} \exp\left[-\frac{\mathrm{i}}{\hbar}\frac{p^2}{2m}t\right] \\ &= \int \frac{dp}{2\pi\hbar} \exp\left\{-\frac{\mathrm{i}}{\hbar}\frac{t}{2m}\left[p - \frac{m(x_2-x_1)}{t}\right]^2 + \frac{\mathrm{i}}{\hbar}\frac{m}{2}\frac{(x_2-x_1)^2}{t}\right\} \\ &= \sqrt{\frac{m}{2\pi\mathrm{i}\hbar t}} \exp\left[\frac{\mathrm{i}}{\hbar}\frac{m}{2}\frac{(x_2-x_1)^2}{t}\right] , \end{aligned} \tag{18.29}$$

where we have again used

$$\int_{-\infty}^{+\infty} dx\, \mathrm{e}^{-\mathrm{i}ax^2} = \sqrt{\frac{\pi}{a\mathrm{i}}} , \quad a > 0 .$$

Using (18.28), we can show that $K(p_2, t; p_1, 0)$ satisfies the Schrödinger equation:

$$
\begin{aligned}
\mathrm{i}\hbar\frac{\partial}{\partial t}K(p_2, t; p_1, 0) &= \delta(p_2 - p_1)\frac{p_1^2}{2m}\exp\left[-\frac{\mathrm{i}}{\hbar}\frac{p_1^2}{2m}t\right] \\
&= \frac{p_2^2}{2m}K(p_2, t; p_1, 0)
\end{aligned}
$$

or

$$
\left(\mathrm{i}\hbar\frac{\partial}{\partial t} - \frac{p_2^2}{2m}\right)K(p_2, t; p_1, 0) = 0\,.
$$

Similarly, it holds that

$$
\left(\mathrm{i}\hbar\frac{\partial}{\partial t} - \frac{p^2}{2m}\right)\chi_p(x, t) = 0
$$

and

$$
\left(\mathrm{i}\hbar\frac{\partial}{\partial t} + \frac{\hbar^2}{2m}\frac{\partial^2}{\partial x^2}\right)K(x, t; x', 0) = 0\,.
$$

By way of illustration we consider the path integral solution of a particle in a one-dimensional infinite square well (walls separated by a distance L):

$$
V(x) = \begin{cases} 0\,, & 0 < x < L\,, \\ \infty\,, & x \leqslant 0\,,\ x \geqslant L\,. \end{cases}
$$

path integral treatment, let us first clarify how the particle can travel from (x_i, t_i) to (x_f, t_f). As in the double-slit experiment, we again have to allow all paths along which the particle can travel from $i \to f$. There is an infinite number of possibilities: the direct path (free propagation from $i \to f$), while on a second path, the particle bounces against the rigid wall one time before travelling on to the final point. Geometrically, one can say that the end point (x_f, t_f) is reached via the image point $(-x_f, t_f)$, starting at (x_i, t_i). If we now turn once again to the superposition principle, we expect the propagator to consist of the sum of the two contributions of the classical paths:

$$
\begin{aligned}
K_L(x_f, t_f; x_i, t_i) &= \sqrt{\frac{m}{2\pi\mathrm{i}\hbar(t_f - t_i)}}\left\{\exp\left[\frac{\mathrm{i}}{\hbar}\frac{m}{2}\frac{(x_f - x_i)^2}{(t_f - t_i)}\right]\right. \\
&\qquad \left. - \exp\left[\frac{\mathrm{i}}{\hbar}\frac{m}{2}\frac{(-x_f - x_i)^2}{(t_f - t_i)}\right]\right\} \\
&= K(x_f, t_f; x_i, t_i) - K(-x_f, t_f; x_i, t_i)\,. \qquad (18.30)
\end{aligned}
$$

The K's are free propagators, while the subscript L on K_L reminds us that the particle is trapped between the walls of the box. The minus sign between the two free propagators is due to the action of the elastic wall. We cannot keep the particle from bouncing off the wall arbitrarily often, i.e., an infinite number of times to the right and the left, while on its way from $i \to f$. Thus it is natural to construct the propagator by forming the sum of all these infinitely many classical paths. The contribution of each classical path is the free particle propagator, multiplied by (-1) for every collision against the wall. Thus it is clear that the generalization of the above formula can only read

$$K_L\big(x_f, t_f; x_i, t_i\big) = \sum_{r=-\infty}^{+\infty} (-1)^r K\big(x_r, t_f; x_i, t_i\big) \,. \tag{18.31}$$

With the help of (18.29), we also can write

$$\begin{aligned} K_L\big(x_f, t_f; k_i, t_i\big) = &\sum_{r=-\infty}^{+\infty} \int \frac{dp}{2\pi\hbar} \\ &\times \left[\exp\left\{\frac{-\mathrm{i}}{\hbar}\frac{p^2}{2m}\big(t_f - t_i\big) + \frac{\mathrm{i}}{\hbar}\big(2rL + x_f - x_i\big)p\right\}\right. \\ &\left. -\exp\left\{-\frac{\mathrm{i}}{\hbar}\frac{p^2}{2m}\big(t_f - t_i\big) + \frac{\mathrm{i}}{\hbar}\big(2rL - x_f - x_i\big)p\right\}\right] . \end{aligned}$$

At every turning point we pick up a phase $2Lp/\hbar$. We continue to write

$$\begin{aligned} K_L\big(x_f, t_f; x_i, t_i\big) = &\int \frac{dp}{2\pi\hbar} \exp\left[-\frac{\mathrm{i}}{\hbar}\frac{p^2}{2m}\big(t_f - t_i\big)\right] \exp\left[-\frac{\mathrm{i}}{\hbar}px_i\right] 2\mathrm{i} \\ &\times \underbrace{\frac{\mathrm{e}^{(\mathrm{i}/\hbar)x_f p} - \mathrm{e}^{-(\mathrm{i}/\hbar)x_f p}}{2\mathrm{i}}}_{\sin[(p/\hbar)x_f]} \left(\sum_{r=-\infty}^{+\infty} \exp\left[\frac{2\mathrm{i}rLp}{\hbar}\right]\right) . \end{aligned}$$

At this stage we use Poisson's formula

$$\begin{aligned} \sum_{r=-\infty}^{+\infty} \mathrm{e}^{(\mathrm{i}/\hbar)2Lpr} &= \sum_{n=-\infty}^{+\infty} \delta\left(p\frac{L}{\pi\hbar} - n\right) \\ &= \frac{\pi\hbar}{L} \sum_{n=-\infty}^{+\infty} \delta\left(p - n\frac{\pi\hbar}{L}\right) \end{aligned}$$

and obtain for our propagator

$$\begin{aligned} K_L\big(x_f, t_f; x_i, t_i\big) = &\frac{\mathrm{i}}{L} \sum_{n=-\infty}^{+\infty} \exp\left[-\frac{\mathrm{i}}{\hbar}E_n\big(t_f - t_i\big)\right] \\ &\times \exp\big[-\mathrm{i}k_n x_i\big] \sin\big(k_n x_f\big) \,. \end{aligned} \tag{18.32}$$

The appearance of the δ-function implies energy and momentum quantization:

$$E_n = \frac{1}{2m}\frac{\pi^2\hbar^2}{L^2}n^2 , \quad k_n = \frac{\pi n}{L} . \tag{18.33}$$

Since the term $n = 0$ vanishes in (18.32), we can combine positive and negative terms to get ($t_f > t_i$)

$$K_L(x_f, t_f; x_i, t_i) = \frac{\mathrm{i}}{L}\Bigg(\sum_{n=1}^{\infty} + \underbrace{\sum_{-\infty}^{n=-1}}_{\sum_{-1}^{-\infty}\exp[-(\mathrm{i}/\hbar)E_n(t_f-t_i)]\exp[-\mathrm{i}k_n x_i]\sin(k_n x_f)}\Bigg)$$

$$= \sum_{n=1}^{\infty} \mathrm{e}^{-(\mathrm{i}/\hbar)E_n(t_f-t_i)}\,\mathrm{e}^{\mathrm{i}k_n x_i}\left[-\sin(k_n x_f)\right]$$

$$\overset{\substack{E_n=E_{-n}\\ -k_n=k_{-n}}}{=} \frac{\mathrm{i}}{L}\sum_{n=1}^{\infty}\exp\left[-\frac{\mathrm{i}}{\hbar}E_n(t_f-t_i)\right]2\mathrm{i}\underbrace{\frac{\mathrm{e}^{-\mathrm{i}k_n x_i}-\mathrm{e}^{\mathrm{i}k_n x_i}}{2\mathrm{i}}}_{-\sin(k_n x_i)}\sin(k_n x_f) .$$

So our final result reads ($t_f > t_i$)

$$K_L(x_f, t_f; x_i, t_i) = \frac{2}{L}\sum_{n=1}^{\infty}\exp\left\{-\frac{\mathrm{i}}{\hbar}E_n(t_f-t_i)\right\}\sin k_n x_i \,\sin k_n x_f . \tag{18.34}$$

Later on we shall derive the same formula in a somewhat different manner.

As our next example we consider a particle whose one degree of freedom is constrained to move on a closed path, i.e., a curve that can be deformed continuously into a circle S^1. Let the ring be parametrized by s and its length be L. The coordinate on S^1 is φ with $0 \leqslant \varphi \leqslant 2\pi$ with $\varphi = 0$ and $\varphi = 2\pi$ identified. So we are dealing with a problem with periodic boundary conditions.

Now, as is well known, the universal covering space for the circle S^1 is the line $\mathbb{R}$, which, of course, is simply connected. Hence, when projecting the motion of our circulating particle onto $\mathbb{R}$, we are dealing once again with a free particle whose propagator we can write down immediately:

$$K_{\mathbb{R}}\left(x_f^{(n)}, t_f; x_i, t_i\right) = \left(\frac{m}{2\pi\mathrm{i}\hbar(t_f-t_i)}\right)^{1/2}\exp\left[\frac{\mathrm{i}}{\hbar}\frac{m}{2}\frac{(x_f^{(n)}-x_i)^2}{t_f-t_i}\right] .$$

The superscript (n) in $x_f^{(n)}$ reminds us that while the particle is going from x_i to $x_f^{(n)}$ on $\mathbb{R}$, it has covered the ring n times. The last formula can then be carried over to our ring with length L using the relation ($x_i = s_i$):

$$x_f^{(n)} - x_i = (s_f + nL) - s_i \equiv \bar{s} + nL , \quad \tau = t_f - t_i ,$$

$$K_L\big(s_f, t_f; s_i, t_i\big) = \left(\frac{m}{2\pi \mathrm{i}\hbar\tau}\right)^{1/2} \exp\left[\frac{\mathrm{i}}{\hbar}\frac{m}{2}\frac{(\bar{s}+nL)^2}{\tau}\right] .$$

With the aid of the identity

$$\frac{1}{\sqrt{\tau}}\exp\left[\frac{\mathrm{i}}{\tau}x^2\right] = \sqrt{\frac{\mathrm{i}}{4\pi}}\int dp \exp\left[-\mathrm{i}\frac{\tau}{4}p^2 + \mathrm{i}px\right]$$

one obtains

$$K_L(\bar{s},\tau) = \left(\frac{m}{2\pi \mathrm{i}\hbar}\right)^{1/2}\left(\frac{\mathrm{i}}{4\pi}\right)^{1/2}\int dp \exp\left[-\mathrm{i}\frac{\tau}{4}p^2 + \mathrm{i}p\sqrt{\frac{m}{2\hbar}}(\bar{s}+nL)\right]$$

or, employing the substitution $p \to \sqrt{(2/\hbar m)}p$:

$$\begin{aligned} K_L(\bar{s},\tau) &= \int \frac{dp}{2\pi\hbar}\exp\left[-\mathrm{i}\frac{\tau p^2}{2m\hbar} + \frac{\mathrm{i}}{\hbar}p(\bar{s}+nL)\right] \\ &= \int \frac{dp}{2\pi\hbar}\mathrm{e}^{(\mathrm{i}/\hbar)pLn}\exp\left[-\frac{\mathrm{i}}{\hbar}\frac{p^2}{2m}\tau + \frac{\mathrm{i}}{\hbar}p\bar{s}\right] . \end{aligned}$$

Let us recognize that we cannot distinguish between those particles which start at s_i and reach s_f directly and those which only arrive after numerous orbitings. In other words, the winding number is not observed and therefore has to be summed over. Again, as in the double-slit experiment, we cannot say via which of the possible paths the particle has reached s_f. Hence the propagator for our system should read more properly

$$\begin{aligned} K_L(\bar{s},\tau) &= \sum_{n=-\infty}^{+\infty}\left(\frac{m}{2\pi \mathrm{i}\hbar\tau}\right)^{1/2}\exp\left[\frac{\mathrm{i}}{\hbar}\frac{m}{2}\frac{(\bar{s}+nL)^2}{\tau}\right] \\ &= \int \frac{dp}{2\pi\hbar}\left(\sum_{n=-\infty}^{+\infty}\mathrm{e}^{(\mathrm{i}/\hbar)pLn}\right)\exp\left[-\frac{\mathrm{i}}{\hbar}\frac{p^2}{2m}\tau + \frac{\mathrm{i}}{\hbar}p\bar{s}\right] . \end{aligned} \tag{18.35}$$

This is still not the final answer.

Now we use Poisson's summation formula once again,

$$\sum_{n=-\infty}^{+\infty}\mathrm{e}^{\mathrm{i}2\pi nx} = \sum_{\mu=-\infty}^{+\infty}\delta(x-\mu)$$

and obtain in this manner ($x \equiv pL/2\pi\hbar$)

$$\sum_{n=-\infty}^{+\infty}\mathrm{e}^{(\mathrm{i}/\hbar)pLn} = \sum_{\mu=-\infty}^{+\infty}\delta\left(\frac{pL}{2\pi\hbar}-\mu\right) = \frac{2\pi\hbar}{L}\sum_{\mu=-\infty}^{+\infty}\delta\left(p-\frac{2\pi\hbar}{L}\mu\right) .$$

With this result our propagator reads

$$K_L(\bar{s}, \tau) = \frac{1}{L} \int dp \sum_{n=-\infty}^{+\infty} \delta\left(p - \frac{2\pi\hbar}{L} n\right) \exp\left[-\frac{\mathrm{i}}{\hbar} \frac{p^2}{2m}\tau + \frac{\mathrm{i}}{\hbar} p\bar{s}\right] . \tag{18.36}$$

Here we want to emphasize that p and s or $\bar{s}$ are canonically conjugate variables. Performing the p-integration in (11.36) we obtain

$$K_L(\bar{s}, \tau) = \frac{1}{L} \sum_{n=-\infty}^{+\infty} \exp\left[-\frac{\mathrm{i}}{\hbar} \frac{1}{2m} \left(\frac{2\pi\hbar}{L} n\right)^2 \tau + 2\pi \mathrm{i} n \frac{\bar{s}}{L}\right] . \tag{18.37}$$

For a circular motion with radius R, the natural parameter is the angle φ. This suggests rewriting (18.37) by setting

$$\bar{s} = R(\varphi_f - \varphi_i) = R\varphi , \quad L = 2\pi R , \quad \tau = t_f - t_i , \quad I = mR^2 .$$

On the left-hand side of (18.37) we then obtain $K_L(R\varphi, \tau)$, so that our propagator for a particle moving on a circular orbit becomes

$$K_L(R\varphi, \tau) = \frac{1}{2\pi R} \sum_{n=-\infty}^{+\infty} \exp\left[-\mathrm{i}\frac{\hbar}{2I} n^2 \tau + \mathrm{i}n\varphi\right] \tag{18.38}$$

or, introducing $K_L(R\varphi, \tau) = (1/R)K(\varphi, \tau)$:

$$K(\varphi_f, t_f; \varphi_i, t_i) = \frac{1}{2\pi} \sum_{n=-\infty}^{+\infty} \exp\left[-\mathrm{i}\frac{\hbar}{2I} n^2 \tau + \mathrm{i}n\varphi\right] . \tag{18.39}$$

The δ-function in (18.36) tells us that

$$p = \frac{2\pi\hbar}{L} n : \quad \frac{pL}{2\pi} = \hbar n , \quad n \in \mathbb{Z} .$$

For a circular motion this implies quantization of the orbital angular momentum and the energy:

$$p = mR\dot{\varphi} , \quad L = 2\pi R :$$

$$\frac{pL}{2\pi} = \frac{mR\dot{\varphi}}{2\pi} 2\pi R = mR^2 \dot{\varphi} = L_z = n\hbar \tag{18.40}$$

$$E_n = \frac{p^2}{2m} = \frac{\hbar^2}{2mR^2} n^2 = \frac{\hbar^2}{2I} n^2 . \tag{18.41}$$

Notice that we have never mentioned the name, "Hermitian operator." In fact we are witnessing the first sign of a topological quantization procedure, in particular of the orbital angular momentum.

Let us rewrite (18.39) a bit with the aid of Jacobi's θ_3-function:

$$\theta_3(z|t) = \sum_{n=-\infty}^{+\infty} \mathrm{e}^{\mathrm{i}\pi t n^2 + \mathrm{i}2nz} = \theta_3(-z|t) . \tag{18.42}$$

Setting $z = \varphi/2$ and $t = -\hbar\tau/2\pi I$, we obtain

$$K(\varphi, \tau) = \frac{1}{2\pi}\theta_3\left(\frac{\varphi}{2}\middle| -\frac{\hbar\tau}{2\pi I}\right) . \tag{18.43}$$

The θ_3-function satisfies the Poisson identity

$$\theta_3(z|t) = (-\mathrm{i}t)^{-1/2}\,\mathrm{e}^{z^2/\mathrm{i}\pi t}\,\theta_3\left(\frac{z}{t}\middle| -\frac{1}{t}\right) . \tag{18.44}$$

This allows us to write for the propagator

$$\begin{aligned} K\big(\varphi_f, t_f; \varphi_i, t_i\big) &= \left(\frac{I}{2\pi\mathrm{i}\hbar(t_f - t_i)}\right)^{1/2} \exp\left[\frac{\mathrm{i}}{\hbar}\frac{I}{2}\frac{(\varphi_f - \varphi_i)^2}{t_f - t_i}\right] \\ &\quad \times \theta_3\left(\frac{\pi I(\varphi_f - \varphi_i)}{\hbar(t_f - t_i)}\middle| \frac{2\pi I}{\hbar(t_f - t_i)}\right) . \end{aligned} \tag{18.45}$$

Let us go back to (18.35), which takes a slightly different form when written for a circular orbit:

$$\begin{aligned} K_L(\bar{s}, \tau) &= \sum_{n=-\infty}^{+\infty} \left(\frac{m}{2\pi\mathrm{i}\hbar\tau}\right)^{1/2} \exp\left[\frac{\mathrm{i}}{\hbar}\frac{m}{2}\frac{(R\varphi + n2\pi R)^2}{\tau}\right] \\ &= \sum_{n=-\infty}^{+\infty} \left(\frac{m}{2\pi\mathrm{i}\hbar\tau}\right)^{1/2} \exp\left[\frac{\mathrm{i}}{\hbar}\frac{m}{2}R^2\frac{(\varphi + 2\pi n)^2}{\tau}\right] \end{aligned} \tag{18.46}$$

or, upon introducing $K(\varphi, \tau)$ via $K_L(\bar{s}, \tau) = K(\varphi, \tau)/R$:

$$\begin{aligned} K(\varphi, \tau) &= \sum_{n=-\infty}^{+\infty} \left(\frac{mR^2}{2\pi\mathrm{i}\hbar\tau}\right)^{1/2} \exp\left[\frac{\mathrm{i}}{\hbar}\frac{mR^2}{2}\frac{(\varphi + 2\pi n)^2}{\tau}\right] \\ &= \sum_{n=-\infty}^{+\infty} \left(\frac{I}{2\pi\mathrm{i}\hbar\tau}\right)^{1/2} \exp\left[\frac{\mathrm{i}}{\hbar}\frac{I}{2}\frac{(\varphi + 2\pi n)^2}{\tau}\right] \end{aligned} \tag{18.47}$$

$$=: \sum_{n=-\infty}^{+\infty} K_n , \tag{18.48}$$

where

$$K_n = \left(\frac{I}{2\pi\mathrm{i}\hbar\tau}\right)^{1/2} \exp\left[\frac{\mathrm{i}}{\hbar}\frac{I}{2}\frac{(\varphi + 2\pi n)^2}{\tau}\right] . \tag{18.49}$$

We shall now assume (however, a proof can be given) that each term in (18.48) individually satisfies the Schrödinger equation. Then it follows that

$$K(\varphi, \tau) = \sum_{n=-\infty}^{+\infty} A_n K_n , \quad A_n = \mathrm{e}^{\mathrm{i}n\theta} , \quad 0 \leqslant \theta < 2\pi , \tag{18.50}$$

is also a legitimate candidate for a circular propagator. The justification for this can be found in the standard literature. Consequently, we find for the θ-propagator

$$K^\theta(\varphi,\tau) = \sum_{n=-\infty}^{+\infty} \left(\frac{I}{2\pi i\hbar\tau}\right)^{1/2} \exp\left[in\theta + \frac{i}{\hbar}\frac{I}{2}\frac{(\varphi+2\pi n)^2}{\tau}\right].$$

The term in the exponential is given by

$$\frac{i}{\hbar}\frac{I}{2}\frac{\varphi^2}{\tau} + i\pi\left(\frac{2\pi I}{\hbar\tau}\right)n^2 + 2in\left(\frac{\theta}{2}+\frac{\pi I\varphi}{\hbar\tau}\right).$$

Thus we obtain

$$K^\theta(\varphi,\tau) = \sum_{n=-\infty}^{+\infty} \left(\frac{I}{2\pi i\hbar\tau}\right)^{1/2} \exp\left[\frac{i}{\hbar}\frac{I}{2}\frac{\varphi^2}{\tau}\right] \times \exp\left[i\pi\left(\frac{2\pi I}{\hbar\tau}\right)n^2 + 2in\left(\frac{\theta}{2}+\frac{\pi I\varphi}{\hbar\tau}\right)\right],$$

$$K^\theta(\varphi,\tau) = \left(\frac{I}{2\pi i\hbar\tau}\right)^{1/2} \exp\left[\frac{i}{\hbar}\frac{I}{2}\frac{\varphi^2}{\tau}\right] \theta_3\left(\frac{\theta}{2}+\frac{\pi I\varphi}{\hbar\tau}\middle|\frac{2\pi I}{\hbar\tau}\right). \tag{18.51}$$

All values of θ are equally acceptable. To understand this "quantum ambiguity", let us write K^θ in a form that will allow us to identify the wave functions and the energy spectrum of the underlying problem:

$$K^\theta(\varphi_f,t_f;\varphi_i,t_i) = \frac{1}{2\pi}\exp\left[-i\left(\frac{\theta\varphi}{2\pi}+\frac{\theta^2\hbar\tau}{8I\pi^2}\right)\right] \times \theta_3\left(\frac{\varphi}{2}+\frac{\hbar\tau\theta}{4\pi I}\middle|-\frac{\hbar\tau}{2\pi I}\right). \tag{18.52}$$

Proof:

$$\theta_3\left(\frac{\varphi}{2}+\frac{\hbar\tau\theta}{4\pi I}\middle|-\frac{\hbar\tau}{2\pi I}\right)$$

$$= \sum_{n=-\infty}^{+\infty} \exp\Bigg[i\pi\overbrace{\left(-\frac{\hbar\tau}{2\hbar I}\right)}^{:=t}n^2 + i2n\overbrace{\left(\frac{\varphi}{2}+\frac{\hbar\tau\theta}{4\pi I}\right)}^{:=z}\Bigg] \tag{18.53}$$

$$= (-it)^{-1/2}\, e^{z^2/i\pi t}\, \theta_3\left(\frac{z}{t}\middle|-\frac{1}{t}\right)$$

$$= \left(\frac{i\hbar\tau}{2\pi I}\right)^{-1/2} \exp\left[\frac{2Ii}{\hbar\tau}\left(\frac{\varphi}{2}+\frac{\hbar\tau\theta}{4\pi I}\right)^2\right] \theta_3\left(-\frac{\pi I\varphi}{\hbar\tau}-\frac{\theta}{2}\middle|\frac{2\pi I}{\hbar\tau}\right)$$

$$\theta_3(z|t) = \theta_3(-z|t) := \left(\frac{2\pi I}{\mathrm{i}\hbar\tau}\right)^{1/2} \exp\left[\mathrm{i}\frac{2I}{\hbar\tau}\left(\frac{\varphi}{2} + \frac{\hbar\tau\theta}{4\pi I}\right)^2\right] \times \theta_3\left(\frac{\pi I\varphi}{\hbar\tau} + \frac{\theta}{2}\middle|\frac{2\pi I}{\hbar\tau}\right) .$$

If we multiply this result by the factor in front of θ_3 in (18.52), we indeed obtain (18.51). On the other hand, we can simply substitute (18.53) in (18.52). Then we can easily show that the Green's function can be written as

$$K^\theta(\varphi_f, t_f; \varphi_i, t_i) = \frac{1}{2\pi} \sum_{n=-\infty}^{+\infty} \exp\left[-\frac{\mathrm{i}}{\hbar}\frac{\hbar^2}{2I}\left(n - \frac{\theta}{2\pi}\right)^2 (t_f - t_i)\right] \times \mathrm{e}^{\mathrm{i}(n-\theta/2\pi)(\varphi_f-\varphi_i)} \tag{18.54}$$

$$= \sum_{n=-\infty}^{+\infty} \mathrm{e}^{-(\mathrm{i}/\hbar)E_n(t_f-t_i)} u_n(\varphi_f) u_n^*(\varphi_i) . \tag{18.55}$$

So we obtain for the eigenfunctions

$$\psi_n^\theta(\varphi, t) = u_n^\theta(\varphi)\, \mathrm{e}^{-(\mathrm{i}/\hbar)E_n^\theta t}$$

with

$$u_n^\theta(\varphi) = \frac{1}{\sqrt{2\pi}} \mathrm{e}^{\mathrm{i}(n-\theta/2\pi)\varphi} \tag{18.56}$$

and for the energy spectrum

$$E_n^\theta = \frac{\hbar^2}{2I}\left(n - \frac{\theta}{2\pi}\right)^2 , \quad n \in \mathbb{Z} . \tag{18.57}$$

The u_n^θ are no longer periodic:

$$u_n^\theta(2\pi) = \mathrm{e}^{-\mathrm{i}\theta}\, u_n^\theta(0) \tag{18.58}$$

or, more generally,

$$u_n^\theta(2\pi m) = \mathrm{e}^{-\mathrm{i}m\theta}\, u_n^0(0) . \tag{18.59}$$

Different values of θ correspond to different energy spectra, i.e., different physics. For the interpretation of the θ-angle, we refer to the literature, in particular, to the Aharonov-Bohm effect: $\theta = e\Phi/\hbar c$.

Let us now return to the beginning of this chapter and consider, rather than the simple Lagrangian for a free particle, a more general quadratic form:

$$L = a(t)x^2 + b(t)x\dot{x} + c(t)\dot{x}^2 + d(t)x + e(t)\dot{x} + f(t) . \tag{18.60}$$

Again we disturb the system relative to the classical path and write

$$x(t) = \bar{x}(t) + y(t)\,, \quad \dot{x}(t) = \dot{\bar{x}} + \dot{y}(t)$$

with

$$y(t_1) = 0 = y(t_2)\,.$$

Subsequently, we expand $L(x(t), \dot{x}(t); t)$ in a Taylor series around $\bar{x}(t)$, $\dot{\bar{x}}(t)$. This series terminates after the second term because of the special form (18.60). Then

$$\begin{aligned} L(x, \dot{x}; t) =& L(\bar{x}, \dot{\bar{x}}; t) + \left.\frac{\partial L}{\partial x}\right|_{\bar{x}} y + \left.\frac{\partial L}{\partial \dot{x}}\right|_{\dot{\bar{x}}} \dot{y} \\ &+ \frac{1}{2}\left.\left(\frac{\partial^2 L}{\partial x^2} y^2 + 2\frac{\partial^2 L}{\partial x \partial \dot{x}} y\dot{y} + \frac{\partial^2 L}{\partial \dot{x}^2}\dot{y}^2\right)\right|_{\bar{x},\dot{\bar{x}}}\,. \end{aligned}$$

From here we obtain for the action

$$\begin{aligned} S = \int_{t_1}^{t_2} dt\, L(x, \dot{x}; t) =& \int_{t_1}^{t_2} dt\, L(\bar{x}, \dot{\bar{x}}; t) + \int_{t_1}^{t_2} dt \left.\left(\frac{\partial L}{\partial x} y + \frac{\partial L}{\partial \dot{x}}\dot{y}\right)\right|_{\bar{x},\dot{\bar{x}}} \\ &+ \int_{t_1}^{t_2} dt \left(a(t) y^2 + b(t) y\dot{y} + c(t)\dot{y}^2\right)\,. \end{aligned}$$

Integration by parts and use of the Euler-Lagrange equation makes the second term on the right-hand side vanish. So we are left with

$$S = S_{\mathrm{cl}} + \int_{t_1}^{t_2} dt \left(a(t) y^2 + b(t) y\dot{y} + c(t)\dot{y}^2\right)$$

and the propagator can be written as

$$\begin{aligned} K(x_2, t_2; x_1, t_1) =& \mathrm{e}^{(\mathrm{i}/\hbar)S_{\mathrm{cl}}} \int_{y(t_1)=0}^{y(t_2)=0} [dy(t)] \\ &\times \exp\left\{\frac{\mathrm{i}}{\hbar}\int_{t_1}^{t_2} dt \left(a(t) y^2 + b(t) y\dot{y} + c(t)\dot{y}^2\right)\right\}\,. \end{aligned} \quad (18.61)$$

Since x_1 and x_2 do not appear in the path integral, the latter can only depend on t_1 and t_2:

$$K(x_2, t_2; x_1, t_1) = A(t_2, t_1)\, \mathrm{e}^{(\mathrm{i}/\hbar)S_{\mathrm{cl}}}\,. \quad (18.62)$$

If, furthermore, the coefficients a, b, and c are time-independent, then it follows [as in the case $c(t) = m/2$] that A is a function of the time difference: $A(t_2 - t_1)$. If $A(t_2, t_1)$ is known for a Lagrangian L, then $A(t_2, t_1)$ is also known for all Lagrangians of the type $L' = L + d(t)x + e(t)\dot{x} + f(t)$, because the linear terms did not appear in the calculation of $A(t_2, t_1)$. Here is an example: a particle in a constant external field: $L = m\dot{x}^2/2 + Fx$. The associated classical action was calculated in (3.34):

$$S_{\mathrm{cl}} = \frac{m}{2}\frac{(x_2 - x_1)^2}{t_2 - t_1} + \frac{F(t_2 - t_1)}{2}(x_1 + x_2) - \frac{F^2(t_2 - t_1)^3}{24m} \,. \tag{18.63}$$

Since the coefficients a and b are equal to zero, we immediately obtain from (18.61) with $c = m/2$:

$$K\big(x_2, t_2; x_1, t_1\big) = \mathrm{e}^{(\mathrm{i}/\hbar)S_{\mathrm{cl}}} \int_{y(t_1)=0}^{y(t_2)=0} [dy(t)] \exp\left[\frac{\mathrm{i}}{\hbar}\int_{t_1}^{t_2} dt \frac{m}{2}\dot{y}^2(t)\right] .$$

But we have already calculated the path integral in (18.17), so that the propagator of a particle in a constant external field is

$$\begin{aligned} K\big(x_2, t_2; x_1, t_1\big) = &\sqrt{\frac{m}{2\pi\mathrm{i}\hbar(t_2 - t_1)}} \exp\left\{\frac{\mathrm{i}}{\hbar}\left[\frac{m}{2}\frac{(x_2 - x_1)^2}{t_2 - t_1}\right.\right. \\ &\left.\left. + \frac{F}{2}(x_1 + x_2)(t_2 - t_1) - \frac{F^2(t_2 - t_1)^3}{24m}\right]\right\} . \end{aligned} \tag{18.64}$$

As final and most important example we consider the linear harmonic oscillator:

$$L = \frac{m}{2}\dot{x}^2 - \frac{m}{2}\omega^2 x^2 \,. \tag{18.65}$$

Since the Lagrangian is quadratic, for the propagator we again arrive at a form of the kind ($T = t_2 - t_1$)

$$K\big(x_2, T; x_1, 0\big) = A(T)\,\mathrm{e}^{(\mathrm{i}/\hbar)S_{\mathrm{cl}}} \,. \tag{18.66}$$

The classical action was calculated in (3.45) with the result

$$S_{\mathrm{cl}} = \frac{m\omega}{2\sin(\omega T)}\left[\big(x_2^2 + x_1^2\big)\cos(\omega T) - 2x_1x_2\right] , \quad \omega T \neq n\pi \,, \quad n \in \mathbb{Z} \,. \tag{18.67}$$

If, however, $\omega T = n\pi$, corresponding to a half, or complete, period, then we encounter difficulties. After one-half of a period, the particle is at the opposite position, and after a complete period, it is at the original point once again. At this stage, it is useful to recall the results worked out at the end of Chap. 4. There we discussed in detail the conjugate points (caustics) of the harmonic oscillator; cf. in particular, (4.58) and (4.72).

What we are mainly interested in right now is the factor $A(t)$, which we want to obtain with the aid of the group property (16.20), as we have already done for the case of a free particle. We recall that as a result of time-translation invariance, we have

$$\begin{aligned} \delta\big(x_2 - x_1\big) &= \int_{-\infty}^{+\infty} dx\, K\big(x_2, t; x, 0\big) K\big(x, 0; x_1, t\big) \\ &= \int_{-\infty}^{+\infty} dx\, K\big(x_2, t; x, 0\big) K^*\big(x_1, t; x, 0\big) \,. \end{aligned} \tag{18.68}$$

Writing

$$K(x_2, t; x, 0) = A(t)\,e^{(i/\hbar)S_{cl}(x_2,t;x,0)}$$
$$K^*(x_1, t; x, 0) = A^*(t)\,e^{-(i/\hbar)S_{cl}(x_1,t;x,0)}$$

we obtain

$$\delta(x_2 - x_1) = |A(t)|^2 \int_{-\infty}^{+\infty} dx\, e^{(i/\hbar)\alpha(x)(x_2-x_1)} \tag{18.69}$$

with

$$\begin{aligned}\alpha(x) &= \frac{\partial S_{cl}(x_1, t; x, 0)}{\partial x_1} = \frac{\partial}{\partial x_1}\left\{\frac{m\omega}{2\sin(\omega t)}\left[(x_1^2 + x^2)\cos(\omega t) - 2x_1 x\right]\right\} \\ &= \frac{m\omega}{\sin(\omega t)}\left(x_1 \cos(\omega t) - x\right) .\end{aligned}$$

Again, $\alpha(x)$ is a linear function of x, so that $d\alpha(x)/dx$ is independent of x:

$$\frac{d\alpha}{dx} = -\frac{m\omega}{\sin(\omega t)} = \frac{\partial^2 S_{cl}(x_1, t; x, 0)}{\partial x_1 \partial x} .$$

Changing integration variables in (18.69),

$$\delta(x_2 - x_1) = |A(t)|^2 \frac{2\pi\hbar}{|(d\alpha/dx)|} \int_{-\infty}^{+\infty} \frac{d\alpha}{2\pi\hbar}\, e^{-(i/\hbar)\alpha(x_2-x_1)} ,$$

we obtain

$$\begin{aligned}|A(t)|^2 &= \frac{1}{2\pi\hbar}\left|\frac{\partial^2 S_{cl}(x_1, t; x, 0)}{\partial x_1 \partial x}\right| \\ &= \frac{1}{2\pi\hbar}\frac{m\omega}{\sin(\omega t)} \equiv \frac{1}{2\pi\hbar}\frac{m}{t}\frac{1}{\sin(\omega t)/\omega t} .\end{aligned}$$

The phase can be determined from our knowledge of the free particle propagator. Using $\lim_{x\to 0}\sin(x)/x = 1$, we obtain from (18.14)

$$A(t) \underset{\omega\to 0}{\longrightarrow} \sqrt{\frac{m}{2\pi\hbar t}}\, e^{-i(\pi/4)} .$$

Thereby we have found the propagator for the linear harmonic oscillator ($T = t_2 - t_1$):

$$\begin{aligned}K(x_2, T; x_1, 0) =& \sqrt{\frac{m\omega}{2\pi i\hbar \sin(\omega T)}} \\ &\times \exp\left\{\frac{i}{\hbar}\frac{m}{2}\frac{\omega}{\sin(\omega T)}\left[(x_1^2 + x_1^2)\right.\right. \\ &\left.\left.\times \cos(\omega T) - 2x_1 x_2\right]\right\} .\end{aligned} \tag{18.70}$$

Formula (18.70) is, however, not absolutely correct. The problematic points are the caustics of $K(x_2, T; x_1, 0)$ at $T_n = n\pi/\omega$. Now we recall that below the first caustic, i.e., $T < \pi/\omega$, formula (18.70) is correct, as it agrees with the free particle propagator for small T. Then it holds that $(1/\sqrt{\mathrm{i}} = \mathrm{e}^{-\mathrm{i}\pi/4})$

$$K\big(x_2, T; x_1, 0\big) = \mathrm{e}^{-\mathrm{i}\pi/4}\sqrt{\frac{m\omega}{2\pi\hbar\,\sin(\omega T)}} \times \exp\left\{\frac{\mathrm{i}}{\hbar}\frac{m}{2}\frac{\omega}{\sin(\omega T)}\left[\big(x_2^2 + x_1^2\big)\cos(\omega T) - 2x_2x_1\right]\right\} .$$

If $T = (\pi/2\omega)$, this expression reduces to

$$K\left(x_2, \frac{1}{2}\frac{\pi}{\omega}; x_1, 0\right) = \mathrm{e}^{-\mathrm{i}\pi/4}\sqrt{\frac{m\omega}{2\pi\hbar}}\exp\left[-\frac{\mathrm{i}}{\hbar}m\omega x_2x_1\right] .$$

From the group property and the time-translation invariance, it follows further that

$$\begin{aligned} K\left(x_2, \frac{\pi}{\omega}; x_1, 0\right) &= \int_{-\infty}^{+\infty} dx\, K\left(x_2, \frac{1}{2}\frac{\pi}{\omega}; x, 0\right) K\left(x, \frac{1}{2}\frac{\pi}{\omega}; x_1, 0\right) \\ &= \mathrm{e}^{-\mathrm{i}\pi/2}\frac{m\omega}{\hbar}\int_{-\infty}^{+\infty}\frac{dx}{2\pi}\mathrm{e}^{-(\mathrm{i}/\hbar)m\omega(x_2+x_1)x} = \mathrm{e}^{-\mathrm{i}\pi/2}\delta\big(x_2 + x_1\big) . \end{aligned}$$

This expression is thus valid at the first caustic. In exactly the same manner we can proceed with the second caustic:

$$\begin{aligned} K\left(x_2, 2\frac{\pi}{\omega}; x_1, 0\right) &= \int_{-\infty}^{+\infty} dx\, K\left(x_2, \frac{\pi}{\omega}; x, 0\right) K\left(x, \frac{\pi}{\omega}; x_1, 0\right) \\ &= \mathrm{e}^{-\mathrm{i}\pi}\int_{-\infty}^{+\infty} dx\, \delta\big(x_2 + x\big)\delta\big(x + x_1\big) = \mathrm{e}^{-\mathrm{i}\pi}\,\delta\big(x_2 - x_1\big) . \end{aligned}$$

If we continue in this way, we obtain for the propagator at any caustic:

$$K\left(x_2, n\frac{\pi}{\omega}; x_1, 0\right) = \mathrm{e}^{-\mathrm{i}n\pi/2}\,\delta\big(x_2 - (-1)^n x_1\big) .$$

The propagator at a caustic serves as initial condition for the propagator in the following section, i.e., we have to require that

$$\lim_{T\to(n\pi/\omega)^+} K\big(x_2, T; x_1, 0\big) = \mathrm{e}^{-\mathrm{i}n\pi/2}\,\delta\left(x_2 - (-1)^n x_1\right) .$$

Now let us consider the expression

$$\lim_{T\to(n\pi/\omega)^+} \mathrm{e}^{-\mathrm{i}\pi/4}\sqrt{\frac{m\omega}{2\pi\hbar|\sin(\omega T)|}}\exp\left\{\frac{\mathrm{i}m\omega}{2\hbar\,\sin(\omega T)} \times \left[\big(x_2^2 + x_1^2\big)\cos(\omega T) - 2x_2x_1\right]\right\} .$$

If we now set $\omega T = n\pi + \omega\delta t$ ($0 < \delta t \ll 1$), and use $\sin(\omega T) = \sin(n\pi + \omega\delta t) = (-1)^n \sin(\omega\delta t) \simeq (-1)^n \omega\delta t$, we get

$$\lim_{\delta t\to 0} \mathrm{e}^{-\mathrm{i}\pi/4}\sqrt{\frac{m\omega}{2\pi\hbar\omega\delta t}}\exp\left\{\frac{\mathrm{i}m\omega}{2\hbar(-1)^n\omega\delta t}\left[\left(x_2^2+x_1^2\right)(-1)^n - 2x_2x_1\right]\right\}$$

$$= \lim_{\delta t\to 0}\sqrt{\frac{m}{2\pi\mathrm{i}\hbar\delta t}}\exp\left\{\frac{\mathrm{i}}{\hbar}\frac{m}{2}\frac{1}{\delta t}\left[x_2-(-1)^n x_1\right]^2\right\} = \delta\left(x_2-(-1)^n x_1\right) .$$

The last equal sign follows from a comparison with the propagator for the free particle:

$$\delta\left(x_2-x_1\right) = \lim_{\delta t\to 0}\sqrt{\frac{m}{2\pi\mathrm{i}\hbar\delta t}}\exp\left[\frac{\mathrm{i}}{\hbar}\frac{m}{2}\frac{(x_2-x_1)^2}{\delta t}\right] .$$

So in order to obtain the correct boundary conditions at the caustics, we need only include the additional phase factor ($T = t_2 - t_1$):

$$\exp\left\{-\mathrm{i}\frac{\pi}{2}\left[\frac{\omega}{\pi}T\right]\right\} ,$$

where $[x]$ denotes the largest integer smaller than x, e.g., $[3/2] = 1$. In our case, $\omega T/\pi < 1$ yields $[\omega T/\pi] = 0$; $\omega T/\pi < 2$ gives $[\omega T/\pi] = 1$, etc.

The correct propagator for the linear harmonic oscillator is therefore given by the following formula:

$$K\left(x_2,T;x_1,0\right) = \exp\left[-\mathrm{i}\frac{\pi}{2}\left(\frac{1}{2}+\left[\frac{\omega}{\pi}T\right]\right)\right]\sqrt{\frac{m\omega}{2\pi\hbar|\sin(\omega T)|}}$$
$$\times\exp\left\{\frac{\mathrm{i}}{\hbar}\frac{m}{2}\frac{\omega}{\sin(\omega T)}\left[\left(x_2^2+x_1^2\right)\cos(\omega T) - 2x_1x_2\right]\right\} . \qquad (18.71)$$

Formula (18.71) is called the Feynman–Soriau Formula and clearly takes care of the behavior of the propagator at the caustics. This formula was first written down in the above form in 1975. We shall hardly concern ourselves with the caustics and shall forthwith primarily use the form (18.70).

19. Direct Evaluation of Path Integrals

Until now we have always used a trick to calculate the path integral in

$$K\big(x_2, t_2; x_1, t_1\big) = \mathrm{e}^{(\mathrm{i}/\hbar)S_{\mathrm{cl}}[x(t)]} \int_{y(t_1)=0}^{y(t_2)=0} [dy(t)] \times \exp\left\{\frac{\mathrm{i}}{\hbar}\int_{t_1}^{t_2} dt\big(a(t)y^2 + b(t)y\dot{y} + c(t)\dot{y}^2\big)\right\} . \tag{19.1}$$

The path integral in (19.1) requires integration over all possible paths $y(t)$ from $(0, t_1)$ to $(0, t_2)$ with the associated action

$$S\big\{(0, t_2), (0, t_1)\big\} = \int_{t_1}^{t_2} dt\big(a(t)y^2 + b(t)y\dot{y}^2 + c(t)\dot{y}^2\big) . \tag{19.2}$$

In order to calculate K directly, we divide the time interval $T = t_2 - t_1$ in N steps of width ε: $T = N\varepsilon$, $t_2 \geqslant t_1$.

$$\tau_0 := t_1 \text{ (fixed)}, \; \tau_1 = t_1 + \varepsilon, \ldots, \; \tau_{N-1} = t_1 + (N-1)\varepsilon, \; \tau_N := t_2 \text{ (fixed)} .$$

Every time τ_n is assigned a point y_n. We now connect the individual points with a classical path $y(\tau)$. $y(\tau)$ is not necessarily the (on-shell trajectory) extremum of the classical action. It can be any path between τ_n and τ_{n-1} specified by the classical Lagrangian $L(y, \dot{y}, t)$. The action along one step of the path from (n) to $(n+1)$ is thus

$$S_{\mathrm{cl}}(n+1, n) = \int_{\tau_n}^{\tau_{n+1}} d\tau\, L(y, \dot{y}; \tau) \tag{19.3}$$

and the action along the entire path, accordingly,

$$S_{\mathrm{cl}}(N, 0) = \sum_{n=0}^{N-1} S_{\mathrm{cl}}(n+1, n) . \tag{19.4}$$

The contribution of each infinitesimal step of the path to the phase integral is then (Dirac, Feynman)

$$\phi_{n+1,n} = \text{const.}\, \mathrm{e}^{(\mathrm{i}/\hbar)S_{\mathrm{cl}}(n+1,n)} . \tag{19.5}$$

The sum over all path contributions is, at last,

$$\int_{y(t_1)=0}^{y(t_2)=0} [dy(t)]\, \mathrm{e}^{(\mathrm{i}/\hbar)S[y(t)]} = \text{const.} \int_{-\infty}^{+\infty} dy_1 \ldots \int_{-\infty}^{+\infty} dy_{N-1} \times \exp\left[\frac{\mathrm{i}}{\hbar}\sum_{n=0}^{N-1} S_{\mathrm{cl}}(n+1,n)\right] . \tag{19.6}$$

We do not integrate over $y_0 = y(\tau_0) = y(t_1) = 0$ and $y_N = y(\tau_N) = y(t_2) = 0$, since these are the fixed, chosen initial values of $y(\tau)$. In order to obtain all paths between t_1 and t_2 in the (y, τ)-plane, we have to form the limit $\varepsilon = (t_2 - t_1)/N \to 0$ or $N \to \infty$. In order for this limit to exist, the constant in (19.6), i.e., the integration measure, has to be chosen properly.

We know from experience with our examples from Chap. 18 that

$$\text{const.} = A^{-N} , \quad A(\delta t = \varepsilon) = \sqrt{\frac{2\pi \mathrm{i}\hbar\varepsilon}{m}} . \tag{19.7}$$

For this case, the limit exists and we can write

$$\int_{y(t_1)=0}^{y(t_2)=0} [dy(t)]\, \mathrm{e}^{(\mathrm{i}/\hbar)S[y(t)]} = \lim_{\varepsilon\to 0} \frac{1}{A} \int_{-\infty}^{+\infty} \frac{dy_1}{A} \cdots \int_{-\infty}^{+\infty} \frac{dy_{N-1}}{A} \times \exp\left[\frac{\mathrm{i}}{\hbar}\sum_{n=0}^{N=1} S_{\mathrm{cl}}(n+1,n)\right] . \tag{19.8}$$

Thus, our formula for calculating the propagator reads

$$K\big(x_2, t_2; x_1, t_1\big) = \mathrm{e}^{(\mathrm{i}/\hbar)S_{\mathrm{cl}}(2|1)} \lim_{\varepsilon\to 0} \frac{1}{A} \int_{-\infty}^{+\infty} \frac{dy_1}{A} \cdots \int_{-\infty}^{+\infty} \frac{dy_{N-1}}{A} \times \exp\left[\frac{\mathrm{i}}{\hbar}\sum_{n=0}^{N-1} S_{\mathrm{cl}}(n+1,n)\right] . \tag{19.9}$$

We want to now prove that this instruction for the calculation of a propagator indeed leads to this goal, at least for a free particle. The Lagrangian in the (y, τ)-plane is in this case

$$L = \frac{m}{2}\dot{y}^2(\tau) . \tag{19.10}$$

The classical action between $n := (y_n, \tau_n)$ and $n+1 := (y_{n+1}, \tau_{n+1})$ is, accordingly,

$$S_{\mathrm{cl}}(n+1, n) = \frac{m}{2}\frac{(y_{n+1} - y_n)^2}{\varepsilon} . \tag{19.11}$$

Then it follows that

$$\int_0^0 [dy(t)] \exp\left[\frac{\mathrm{i}}{\hbar}\int_{t_1}^{t_2} dt \frac{m}{2}\dot{y}^2(t)\right]$$
$$= \lim_{\varepsilon\to 0} \frac{1}{A}\int_{-\infty}^{+\infty} \cdots \int_{-\infty}^{+\infty} \frac{dy_1}{A} \cdots \frac{dy_{N-1}}{A}$$
$$\times \exp\left[\frac{\mathrm{i}}{\hbar}\sum_{n=0}^{N=1} \frac{m}{2}\frac{(y_{n+1}-y_n)^2}{\varepsilon}\right] . \tag{19.12}$$

To calculate the first integration, dy_1, we combine all terms which contain y_1:

$$\int_{-\infty}^{+\infty} \frac{1}{A^2} dy_1 \exp\left\{\frac{\mathrm{i}}{\hbar}\frac{m}{2\varepsilon}\underbrace{\left[(y_2-y_1)^2 + (y_1 - \overbrace{y_0}^{=0})^2\right]}_{2(y_1-y_2/2)^2+y_2^2/2}\right\}$$
$$= \frac{1}{A^2}\exp\left[\frac{\mathrm{i}}{\hbar}\frac{m}{2\varepsilon}\frac{y_2^2}{2}\right]\underbrace{\int_{-\infty}^{+\infty} dy_1 \exp\left[\frac{\mathrm{i}}{\hbar}\frac{m}{\varepsilon}\left(y_1 - \frac{1}{2}y_2\right)^2\right]}_{=\sqrt{\mathrm{i}\pi\hbar\varepsilon/m}} , \qquad A^2 = \frac{2\pi\mathrm{i}\hbar\varepsilon}{m}$$
$$= \sqrt{\frac{m}{2\pi\mathrm{i}\hbar(2\varepsilon)}}\exp\left[\frac{\mathrm{i}}{2\hbar}\frac{m}{2\varepsilon}y_2^2\right] . \tag{19.13}$$

One can show by induction that the n-th step of integration yields the following result:

$$\sqrt{\frac{m}{2\pi\mathrm{i}\hbar(n+1)\varepsilon}}\exp\left[\frac{\mathrm{i}}{2\hbar}\frac{m}{(n+1)\varepsilon}y_{n+1}^2\right] . \tag{19.14}$$

After $(N-1)$ integrations one obtains

$$\lim_{\varepsilon\to 0}\sqrt{\frac{m}{2\pi\mathrm{i}\hbar N\varepsilon}}\exp\left\{\frac{\mathrm{i}}{2\hbar}\frac{m}{N\varepsilon}\Big(\underbrace{y_N}_{=0} - \underbrace{y_0}_{=0}\Big)^2\right\} \underset{N\varepsilon=T}{=} \sqrt{\frac{m}{2\pi\mathrm{i}\hbar T}} .$$

So we again find for the propagator of a free particle

$$K\big(x_2,t_2;x_1,t_1\big) = \sqrt{\frac{m}{2\pi\mathrm{i}\hbar T}}\exp\left[\frac{\mathrm{i}}{\hbar}\frac{m}{2}\frac{(x_2-x_1)^2}{T}\right] . \tag{19.15}$$

For the linear harmonic oscillator we write in the same fashion

$$K\big(x_2,t_2;x_1,t_1\big) = \mathrm{e}^{(\mathrm{i}/\hbar)S_{\mathrm{cl}}}\int_0^0 [dy(t)]$$
$$\times \exp\left[\frac{\mathrm{i}}{\hbar}\int_{t_1}^{t_2} dt \frac{m}{2}\left(\dot{y}^2 - \omega^2 y^2\right)\right] . \tag{19.16}$$

The value of the path integral before the first caustic is repeated here:

$$\int_0^0 [dy(t)] \exp\left[\frac{\mathrm{i}}{\hbar}\int_{t_1}^{t_2} dt \frac{m}{2}(\dot{y}^2 - \omega^2 y^2)\right] = \sqrt{\frac{m\omega}{2\pi \mathrm{i}\hbar \sin[\omega(t_2 - t_1)]}} . \quad (19.17)$$

At this point we want to establish contact with the conventional approach to quantum mechanics, which is based on the Schrödinger equation.

We found a dynamical equation (integral equation) in (16.10) that describes the time development of the Schrödinger amplitude. As promised, we now wish to show that this equation is equivalent to the Schrödinger equation. To that end, we again consider the one-dimensional motion of a particle in a potential $V(x)$. The Lagrangian is then

$$L(x, \dot{x}) = \frac{m}{2}\dot{x}^2 - V(x) \quad (19.18)$$

with the propagator

$$K\big(x, t_2; y, t_1\big) = \int_{x(t_1)=y}^{x(t_2)=x} [dx(t)] \exp\left\{\frac{\mathrm{i}}{\hbar}\int_{t_1}^{t_2} dt \left[\frac{m}{2}\dot{x}^2 - V(x)\right]\right\} . \quad (19.19)$$

Writing $t_1 = t$ and $t_2 = t + \varepsilon$, our integral equation reads

$$\psi(x, t+\varepsilon) = \int_{-\infty}^{+\infty} dy\, K(x, t+\varepsilon; y, t)\psi(y, t) . \quad (19.20)$$

Since ε is assumed very small, we can use the following approximation for K in (19.20):

$$K(x, t+\varepsilon; y, t) \simeq \frac{1}{A}\exp\left\{\frac{\mathrm{i}}{\hbar}\left[\frac{m}{2}\frac{(x-y)^2}{\varepsilon^2} - V(y)\right]\varepsilon\right\} . \quad (19.21)$$

The constant A is still to be determined. Using (19.21) in (19.20), we obtain

$$\psi(x, t+\varepsilon) \simeq \frac{1}{A}\int_{-\infty}^{+\infty} dy \exp\left\{\frac{\mathrm{i}}{\hbar}\left[\frac{m}{2}\frac{(x-y)^2}{\varepsilon} - V(y)\varepsilon\right]\right\}\psi(y, t)$$

$$y = x + \xi; \; = \frac{1}{A}\int_{-\infty}^{+\infty} d\xi \exp\left\{\frac{\mathrm{i}}{\hbar}\left[\frac{m}{2}\frac{\xi^2}{\varepsilon} - \varepsilon V(x+\xi)\right]\right\}\psi(x+\xi, t) . \quad (19.22)$$

Because of the smallness of ε, ξ^2/ε is the dominant term in the exponential. For large ξ, the integrand oscillates very rapidly and makes no contributions. The main contribution to the integral comes from values $\xi \sim \sqrt{\varepsilon}$, or, better: $-\sqrt{\varepsilon\hbar/m} \leqslant \xi \leqslant \sqrt{\varepsilon\hbar/m}$. In the case of expansion of the integral up to linear terms in ε, the integrand must be expanded up to quadratic terms in ξ. Thus, with

$$\psi(x+\xi, t) \simeq \psi(x, t) + \xi\frac{\partial\psi(x, t)}{\partial x} + \frac{1}{2}\xi^2\frac{\partial^2\psi(x, t)}{\partial x^2}$$

we get in (19.22)

$$\psi(x, t+\varepsilon) \simeq \frac{1}{A} \int_{-\infty}^{+\infty} d\xi \exp\left\{\frac{\mathrm{i}}{\hbar} \frac{m}{2} \frac{\xi^2}{\varepsilon}\left(1 - \frac{\mathrm{i}}{\hbar}\varepsilon V(x) + \dots\right)\right\}$$
$$\times \left[\psi(x,t) + \underbrace{\xi \frac{\partial \psi}{\partial x}}_{\text{odd in } \xi} + \frac{1}{2}\xi^2 \frac{\partial^2}{\partial x^2}\psi(x,t) + \dots\right] .$$

Using the integrals

$$\int_{-\infty}^{+\infty} \exp\left[\frac{\mathrm{i}}{\hbar}\frac{m}{2}\frac{\xi^2}{\varepsilon}\right] d\xi = \left(\frac{2\pi \mathrm{i}\hbar\varepsilon}{m}\right)^{1/2} =: A' ,$$
$$\int_{-\infty}^{+\infty} \exp\left[\frac{\mathrm{i}}{\hbar}\frac{m}{2}\frac{\xi^2}{\varepsilon}\right] \xi d\xi = 0 ,$$
$$\int_{-\infty}^{+\infty} \exp\left[\frac{\mathrm{i}}{\hbar}\frac{m}{2}\frac{\xi^2}{2}\right] \xi^2 d\xi = \sqrt{2\pi}\left(\frac{\mathrm{i}\hbar\varepsilon}{m}\right)^{3/2}$$

it then follows that

$$\psi(x,t) + \varepsilon\frac{\partial\psi(x,t)}{\partial t} + \dots = \psi(x, t+\varepsilon) \cong \frac{1}{A}A'\psi(x,t)$$
$$+ \varepsilon\left[A'\frac{1}{A}\frac{\mathrm{i}\hbar}{m}\frac{1}{2}\frac{\partial^2\psi(x,t)}{\partial x^2} - A'\frac{1}{A}\frac{\mathrm{i}}{\hbar}V(x)\psi(x,t)\right] .$$

If we take the limit $\varepsilon \to 0$, then we can identify

$$A = A' = \left(\frac{2\pi \mathrm{i}\hbar\varepsilon}{m}\right)^{1/2} .$$

A comparison of the linear terms in ε yields, finally,

$$\frac{\partial}{\partial t}\psi(x,t) = \frac{\mathrm{i}\hbar}{2m}\frac{\partial^2\psi(x,t)}{\partial x^2} - \frac{\mathrm{i}}{\hbar}V(x)\psi(x,t)$$

or

$$\mathrm{i}\hbar\frac{\partial\psi(x,t)}{\partial t} = -\frac{\hbar^2}{2m}\frac{\partial^2}{\partial x^2}\psi(x,t) + V(x)\psi(x,t) . \tag{19.23}$$

This is the well-known Schrödinger equation of a particle in the potential $V(x)$.

The infinitesimal propagator obtained in this manner (19.21) can be factorized as follows:

$$K(x, t+\varepsilon; y, t) \sim \sqrt{\frac{m}{2\pi \mathrm{i}\hbar\varepsilon}}\exp\left[\frac{\mathrm{i}}{\hbar}\frac{m}{2}\frac{(x-y)^2}{\varepsilon}\right]\exp\left[-\frac{\mathrm{i}}{\hbar}V(x)\varepsilon\right]$$
$$= K_0(x, t+\varepsilon; y, t)\phi_C^I(x, t+\varepsilon; y, t) , \tag{19.24}$$

where K_0 is the infinitesimal free particle propagator and ϕ_C^I is the phase factor that corresponds to the interaction

$$\phi_C^I = \exp\left[-\frac{\mathrm{i}}{\hbar}\int_C dt\, V(x)\right] . \tag{19.25}$$

The methods just described are now to be applied to the direct calculation of the path integral (with potential):

$$K\big(x_f, t_f; x_i, t_i\big) = \int_{x(t_i)=x_i}^{x(t_f)=x_f} [dx(t)] \exp\left[\frac{\mathrm{i}}{\hbar}\int_{t_i}^{t_f} dt\, L(x, \dot{x}; t)\right] . \tag{19.26}$$

To this end, we again divide the time interval between t_i and t_f in N equal parts, $\varepsilon = (t_f - t_i)/N$:

$$t_0 = t_i \text{ (fixed)} , \quad t_1 = t_i + \varepsilon , \dots , \quad t_{N-1} = t_i + \varepsilon(N-1) , \quad t_N = t_f \text{ (fixed)} .$$

Now we recall the definition of the path integral:

$$\begin{aligned} K(f|i) &= \int_{x(t_i)=x_i}^{x(t_f)=x_f} [dx(t)] \exp\left\{\frac{\mathrm{i}}{\hbar} S[x(t)]\right\} \\ &= \int_{-\infty}^{+\infty} \dots \int_{-\infty}^{+\infty} dx_1 \dots dx_{N-1} K\big(\underbrace{x_N}_{=x_f}, \underbrace{t_N}_{t_f}; x_{N-1}, t_{N-1}\big) \dots \\ &\quad \times K\big(x_1, t_1; \underbrace{x_0}_{=x_i}, \underbrace{t_0}_{t_i}\big) . \end{aligned}$$

For $N \to \infty$, or $\varepsilon \to 0$, we again use infinitesimal propagators:

$$\begin{aligned} K(y, t+\varepsilon; x, t) \simeq \left(\frac{m}{2\pi \mathrm{i}\hbar\varepsilon}\right)^{1/2} &\exp\left[\frac{\mathrm{i}}{\hbar}\frac{m}{2}\frac{(x-y)^2}{\varepsilon}\right] \\ \times &\exp\left[-\frac{\mathrm{i}}{\hbar}\varepsilon V\left(\frac{x+y}{2}\right)\right] . \end{aligned} \tag{19.27}$$

Then we get for the propagator $K(f|i)$:

$$\begin{aligned} &\int_{x(t_i)=x_i}^{x(t_f)=x_f} [dx(t)] \exp\left\{\frac{\mathrm{i}}{\hbar}\int_{t_i}^{t_f} dt \left[\frac{m}{2}\dot{x}^2 - V(x)\right]\right\} \\ &= \lim_{N\to\infty} \left(\frac{m}{2\pi \mathrm{i}\hbar\varepsilon}\right)^{N/2} \int dx_1 \dots \int dx_{N-1} \\ &\quad \times \exp\left[\frac{\mathrm{i}}{\hbar}\sum_{k=1}^{N} \frac{m}{2}\frac{(x_k - x_{k-1})^2}{\varepsilon} - V\left(\frac{x_k + x_{k-1}}{2}\right)\varepsilon\right] . \end{aligned} \tag{19.28}$$

For the path integral, one often simply finds the expression

$$\begin{aligned} &\sim \int dx_1 \dots \int dx_{N-1} \\ &\quad \times \exp\left[\frac{\mathrm{i}}{\hbar}\sum_{k=1}^{N} \frac{m}{2}\frac{(x_k - x_{k-1})^2}{\varepsilon} - V\left(\frac{x_k + x_{k-1}}{2}\right)\varepsilon\right] . \end{aligned} \tag{19.29}$$

Here, the integration measure has been "forgotten". This can sometimes be justified by calculating, instead of a single path integral, the ratio of two path integrals, both of which describe a particle of mass m. Then the measure drops out and we have simply

$$\int_{x(t_i)=x_i}^{x(t_f)=x_f}[dx(t)]\exp\left\{\frac{\mathrm{i}}{\hbar}S[x(t)]\right\} \Big/ \int_{x(t_i)=x_i}^{x(t_f)=x_f}[dx(t)]\exp\left\{\frac{\mathrm{i}}{\hbar}S_0[x(t)]\right\}$$

$$= \lim_{N\to\infty}\frac{\int_{-\infty}^{+\infty}\dots\int_{-\infty}^{+\infty}dx_1\dots dx_{N-1}\exp\left\{\frac{\mathrm{i}}{\hbar}\sum_{k=1}^{N}\frac{m}{2}\frac{(x_k-x_{k-1})^2}{\varepsilon}-V\left(\frac{x_k+x_{k-1}}{2}\right)\varepsilon\right\}}{\int_{-\infty}^{+\infty}\dots\int_{-\infty}^{+\infty}dx_1\dots dx_{N-1}\exp\left\{\frac{\mathrm{i}}{\hbar}\sum_{k=1}^{N}\frac{m}{2}\frac{(x_k-x_{k-1})^2}{\varepsilon}-V_0\left(\frac{x_k+x_{k-1}}{2}\right)\varepsilon\right\}}\,. \tag{19.30}$$

Here, S_0 refers to the free propagator, so that we indeed are calculating the ratio of the interacting propagator to the free propagator.

If we want to calculate a path integral, we need not necessarily sum piecewise over straight lines in (x, t)-space. Other complete classes of paths can equally well be used. For example, if we wish to calculate the propagator

$$K(0, T; 0, 0)$$

we normally begin by dividing an arbitrary path $x(t)$ in segments which connect the intermediate points $x_1 = x(t_1)$ up to $x_{N-1} = x(t_{N-1})$. Now, however, we shall approximate the path $x(t)$ by using a "Fourier path", i.e., a path of the form

$$\tilde{x}(t) = \sum_{k=1}^{N-1} a_k \sin\left(\frac{\pi k}{T}t\right)\,. \tag{19.31}$$

If we then choose the coefficients a_k in such a manner that

$$x_j = \sum_{k=1}^{N-1} a_k \sin\left(\frac{\pi k}{N}j\right) \tag{19.32}$$

then the Fourier paths obviously go through the same intermediary points of (x, t)-space.

The approximated path is then once again completely characterized by the vertex coordinates $(x_1, \dots, x_{N-1})$, and we again can sum up the contributions of the various paths by integrating over the vertex coordinates. In practice, however, it is more convenient to integrate over the Fourier components $(a_1, \dots, a_{N-1})$. Since the relation (19.32) is one-to-one between the $(x_1, \dots, x_{N-1})$ and the $(a_1, \dots, a_{N-1})$, it is immediately evident that

$$\int\dots\int\{./.\}dx_1\dots dx_{N-1} = \int\dots\int\{./.\}\left|\left(\frac{\partial x_i}{\partial a_j}\right)\right|da_1\dots da_{N-1}\,.$$

Since the transformation (19.32) is linear, the Jacobi determinant is independent of $(a_1, \dots, a_{N-1})$! Its value is thus unimportant (drops out) when calculating the ratio again:

$$\int_{x(0)=0}^{x(T)=0}[dx(t)]\exp\left\{\frac{\mathrm{i}}{\hbar}S[x(t)]\right\}\Big/\int_{x(0)=0}^{x(T)=0}[dx(t)]\exp\left\{\frac{\mathrm{i}}{\hbar}S_0[x(t)]\right\}$$
$$=\lim_{N\to\infty}\frac{\int da_1\ldots da_{N-1}\exp\left\{\frac{\mathrm{i}}{\hbar}S\left[\sum_{k=1}^{N-1}a_k\sin\left(\frac{\pi k}{T}t\right)\right]\right\}}{\int da_1\ldots da_{N-1}\exp\left\{\frac{\mathrm{i}}{\hbar}S_0\left[\sum_{k=1}^{N-1}a_k\sin\left(\frac{\pi k}{T}t\right)\right]\right\}}\,. \tag{19.33}$$

In order to illustrate how this procedure works, we again return to the example of the harmonic oscillator, which we all treasure:

$$K\big(x_2,T;x_1,0\big)=\exp\left\{\frac{\mathrm{i}}{\hbar}S\big[x_{\mathrm{cl}}\big]\right\}\int_{x(0)=0}^{x(T)=0}[dx(t)]$$
$$\times\exp\left\{\frac{\mathrm{i}}{\hbar}\int_0^T dt\left[\frac{m}{2}\dot{x}^2-\frac{m}{2}\omega^2x^2\right]\right\}. \tag{19.34}$$

[The path integral is the same as in (19.17), where we used $y(t)$ instead of $x(t)$.] According to (19.33), we can rewrite (19.34) as

$$K\big(x_2,T;x_1,0\big)=\exp\left\{\frac{\mathrm{i}}{\hbar}S\big[x_{\mathrm{cl}}\big]\right\}K_0(0,T;0,0)$$
$$\times\lim_{N\to\infty}\frac{\int\ldots\int da_1\ldots da_{N-1}\exp\left\{\frac{\mathrm{i}}{\hbar}S[./.]\right\}}{\int\ldots\int da_1\ldots da_{N-1}\exp\left\{\frac{\mathrm{i}}{\hbar}S_0[./.]\right\}}\,. \tag{19.35}$$

The free propagator ($\omega=0$) is known to be given by

$$K_0(0,T;0,0)=\left(\frac{m}{2\pi\mathrm{i}\hbar T}\right)^{1/2}.$$

So now we have to calculate the remaining multiple integrals over the a_i. First we want to show that

$$S\left[\sum_{k=1}^{N-1}a_k\sin\left(\frac{\pi k}{T}t\right)\right]=\frac{mT}{4}\sum_{k=1}^{N-1}a_k^2\left(\frac{k^2\pi^2}{T^2}-\omega^2\right). \tag{19.36}$$

Proof:

$$x(t)\underset{(31)}{=}\sum_{k=1}^{N-1}a_k\sin\left(\frac{\pi k}{T}t\right)=\sum_{k=1}^{N-1}a_k\sin\big(\omega_kt\big)\,;\quad \omega_k=\frac{k\pi}{T}$$

$$\dot{x}=\sum_{k=1}^{N-1}a_k\omega_k\cos\big(\omega_kt\big)\,,\quad \dot{x}^2=\sum_{k=1}^{N-1}\sum_{j=1}^{N-1}a_ka_j\omega_k\omega_j\cos\big(\omega_kt\big)\cos\big(\omega_jt\big)\,.$$

We need

$$\int_0^T dt\,\dot{x}^2=\sum_{k,j}a_ka_j\omega_k\omega_j\int_0^T dt\cos\left(\frac{k\pi}{T}t\right)\cos\left(\frac{j\pi}{T}t\right)=\frac{T}{2}\sum_{k=1}^{N-1}a_k^2\omega_k^2\,.$$

Similarly, it follows that

$$\int_0^T dt\, x^2 = \frac{T}{2}\sum_{k=1}^{N-1} a_k^2 \, .$$

Therefore, we find indeed

$$\exp\left[\frac{\mathrm{i}}{\hbar}\int_0^T dt \frac{m}{2}\left(\dot{x}^2 - \omega^2 x^2\right)\right] = \exp\left[\frac{\mathrm{i}}{\hbar}\frac{mT}{4}\sum_{k=1}^{N-1}\left(\frac{k^2\pi^2}{T^2} - \omega^2\right) a_k^2\right] .$$

Now the integration over the a_i is simple to perform, since the exponential is not only quadratic but also diagonal in $(a_1, \ldots, a_{N-1})$.

$$\begin{aligned}&\int \cdots \int da_1 \ldots da_{N-1} \exp\left\{\frac{\mathrm{i}}{\hbar} S\left[\sum_{k=1}^{N-1} a_k \sin\left(\frac{k\pi}{T}t\right)\right]\right\} \\ &= \int \cdots \int da_1 \ldots da_{N-1} \exp\left[\frac{\mathrm{i}}{\hbar}\frac{mT}{4}\sum_{k=1}^{N-1}\left(\frac{k^2\pi^2}{T^2} - \omega^2\right) a_k^2\right] . \end{aligned} \tag{19.37}$$

Using

$$\int_{-\infty}^{+\infty} \mathrm{e}^{\mathrm{i}\alpha x^2}\, dx = \sqrt{\frac{\pi \mathrm{i}}{\alpha}} \quad :$$

$$\int_{-\infty}^{+\infty} da_k \exp\left[\frac{\mathrm{i}}{\hbar}\frac{mT}{4}\left(\omega_k^2 - \omega^2\right) a_k^2\right] = \sqrt{\frac{4\pi \mathrm{i}\hbar}{mT(\omega_k^2 - \omega^2)}}$$

we obtain in (19.37)

$$\left(\sqrt{\frac{4\pi\mathrm{i}\hbar}{mT}}\right)^{N-1} \prod_{k=1}^{N-1}\left(\frac{k^2\pi^2}{T^2} - \omega^2\right)^{-1/2} .$$

The free particle propagator is obtained simply by setting $\omega = 0$. The forefactor $(\sqrt{4\pi\mathrm{i}\hbar/mT})^{N-1}$ drops out when constructing the ratio in (19.35).

With

$$\frac{\prod_{k=1}^{N-1}(k^2\pi^2/T^2 - \omega^2)^{-1/2}}{\prod_{k=1}^{N-1}(k^2\pi^2/T^2)^{-1/2}} = \prod_{k=1}^{N-1}\left(1 - \frac{\omega^2 T^2}{k^2\pi^2}\right)^{-1/2}$$

we find

$$\begin{aligned} K\left(x_2, T; x_1, 0\right) &= \exp\left\{\frac{\mathrm{i}}{\hbar} S\left[x_{\mathrm{cl}}\right]\right\} \\ &\times \sqrt{\frac{m}{2\pi\mathrm{i}\hbar T}} \lim_{N\to\infty}\prod_{k=1}^{N-1}\left(1 - \frac{\omega^2 T^2}{k^2\pi^2}\right)^{1/2} . \end{aligned} \tag{19.38}$$

Euler's famous product formula for the sine function

$$\left(\frac{z}{\sin z}\right)^{1/2} = \prod_{k=1}^{\infty}\left(1 - \frac{z^2}{k^2\pi^2}\right)^{-1/2} , \qquad z \equiv \omega T$$

therefore yields for $N \to \infty$ or $\varepsilon \to 0$ in (19.38):

$$K\big(x_2, T; x_1, 0\big) = \sqrt{\frac{m\omega}{2\pi \mathrm{i}\hbar \sin(\omega T)}} \exp\left\{\frac{\mathrm{i}}{\hbar} S_{\mathrm{cl}}^{\mathrm{H.O.}}[x]\right\} .$$

We already know from our earlier considerations that the phase dependence is not yet correctly described. Let us recall, however,

$$\int_{-\infty}^{+\infty} dx\, \mathrm{e}^{\mathrm{i}\lambda x^2} = \sqrt{\frac{\pi}{|\lambda|}}\, \mathrm{e}^{(\mathrm{i}\pi/4)\operatorname{sign}\lambda} = \begin{cases} \sqrt{\frac{\pi}{|\lambda|}}\, \mathrm{e}^{\mathrm{i}\pi/4} , & \lambda > 0 , \\ \sqrt{\frac{\pi}{|\lambda|}}\, \mathrm{e}^{-\mathrm{i}\pi/4} , & \lambda < 0 , \end{cases}$$

and return to the action expressed in the Fourier path,

$$\exp\left[\frac{\mathrm{i}}{\hbar} S\right] = \exp\left[\frac{\mathrm{i}}{\hbar}\frac{mT}{4}\sum_{k=1}^{N-1} a_k^2 \left(\frac{\pi^2 k^2}{T^2} - \omega^2\right)\right] .$$

Then we see that the analogue of λ is negative if $k < \omega T/\pi$, and positive if $k > \omega T/\pi$. Once again, the additional phase factor $\exp\{-(\mathrm{i}\pi/2)[\omega T/\pi]\}$ has to be included, so that our multiple integral is given more precisely by

$$\begin{aligned} &\int \cdots \int da_1 \ldots da_{N-1} \exp\left[\frac{\mathrm{i}}{\hbar}\frac{mT}{4}\sum_{k=1}^{N-1} a_k^2 \left(\frac{\pi^2 k^2}{T^2} - \omega^2\right)\right] \\ &\quad = \exp\left\{-\mathrm{i}\frac{\pi}{2}\left[\frac{\omega T}{\pi}\right]\right\}\left(\mathrm{e}^{\mathrm{i}\pi/4}\sqrt{\frac{4\pi\hbar}{mT}}\right)^{N-1} \prod_{k=1}^{N-1}\left|\frac{\pi^2 k^2}{T^2} - \omega^2\right|^{-1/2} . \end{aligned} \qquad (19.39)$$

(For the free propagator this ambiguity does not appear.) The correct formula for the propagator of a particle in the potential of the harmonic oscillator well is therefore

$$\begin{aligned} K\big(x_2, T; x_1, 0\big) = &\exp\left[-\mathrm{i}\left(\frac{\pi}{4} + \frac{\pi}{2}\left[\frac{\omega T}{\pi}\right]\right)\right]\sqrt{\frac{m\omega}{2\pi\hbar|\sin(\omega T)|}} \\ &\times \exp\left\{\frac{\mathrm{i}}{\hbar} S_{\mathrm{cl}}^{\mathrm{H.O.}}[x]\right\} . \end{aligned} \qquad (19.40)$$

This is precisely the Feynman–Soriau formula (18.71).

20. Linear Oscillator with Time-Dependent Frequency

Here is another important example of a path integral calculation, namely the time-dependent oscillator whose Lagrangian is given by

$$L = \frac{m}{2}\dot{x}^2 - \frac{m}{2}W(t)x^2 \,. \tag{20.1}$$

Since L is quadratic, we again expand around a classical solution so that later on we will be dealing again with the calculation of the following path integral:

$$\int_{x(t_i)=0}^{x(t_f)=0} [dx(t)] \exp\left\{ \frac{i}{\hbar}\frac{m}{2} \int_{t_i}^{t_f} dt \left[\left(\frac{dx}{dt}\right)^2 - W(t)x^2 \right] \right\} \,. \tag{20.2}$$

Using $x(t_i) = 0 = x(t_f)$, we can integrate by parts and obtain

$$S[x(t)] = -\frac{m}{2} \int_{t_i}^{t_f} dt \left[x(t)\frac{d^2x}{dt^2} + W(t)x^2 \right] \,; \tag{20.3}$$

i.e.,

$$\int_{x(t_i)=0}^{x(t_f)=0} [dx(t)] \exp\left\{ -\frac{i}{\hbar}\frac{m}{2} \int_{t_i}^{t_f} dt\, x(t) \left[\frac{d^2}{dt^2} + W(t) \right] x(t) \right\} \,. \tag{20.4}$$

Here we are dealing with a generalized Gaussian integral. In order to calculate it, we should diagonalize the Hermitean operator,

$$\frac{d^2}{dt^2} + W(t) \,. \tag{20.5}$$

But at first we shall proceed somewhat differently. Using an appropriate transformation of variables, one can transform the action into that of a free particle.

Let $f(t)$ be the solution of

$$\left\{ \frac{d^2}{dt^2} + W(t) \right\} f(t) = 0 \,. \tag{20.6}$$

$f(t)$ is mostly arbitrary, up to the restriction that at the initial point t_i

$$f(t_i) \neq 0 \,. \tag{20.7}$$

Thus $f(t)$ is not an allowed path, as it violates the boundary condition. With this f we now construct the following linear transformation, where $x(t)$ is replaced by the path $y(t)$:

$$x(t) = f(t) \int_{t_i}^{t} ds \frac{\dot{y}(s)}{f(s)} . \tag{20.8}$$

Differentiation of (20.8) gives

$$\dot{x}(t) = \dot{f}(t) \int_{t_i}^{t} ds \frac{\dot{y}(s)}{f(s)} + \dot{y}(t) = \frac{\dot{f}(t)}{f(t)} x(t) + \dot{y}(t) , \tag{20.9}$$

so that the inverse transformation of (20.8) is given by

$$y(t) = x(t) - \int_{t_i}^{t} ds \frac{\dot{f}(s)}{f(s)} x(s) . \tag{20.10}$$

Note that $y(t)$ satisfies the boundary condition $y(t_i) = 0$, since $x(t_i) = 0$. If we differentiate (20.9) once again,

$$\ddot{x}(t) = \ddot{f}(t) \int_{t_i}^{t} ds \frac{\dot{y}(s)}{f(s)} + \dot{f}(t) \frac{\dot{y}(t)}{f(t)} + \ddot{y}(t) ,$$

we obtain

$$\left\{ \frac{d^2}{dt^2} + W(t) \right\} x(t)$$

$$= \underbrace{\{\ddot{f}(t) + W(t) f(t)\}}_{=0} \int_{t_i}^{t_f} ds \frac{\dot{y}(s)}{f(s)} + \frac{\dot{f}(t)\dot{y}(t)}{f(t)} + \ddot{y}(t) .$$

So far we can write:

$$S[x(t)] = -\frac{m}{2} \int_{t_i}^{t_f} dt\, x(t) \left[\frac{d^2}{dt^2} + W(t) \right] x(t)$$

$$= -\frac{m}{2} \int_{t_i}^{t_f} dt \left[f(t) \int_{t_i}^{t} ds \frac{\dot{y}(s)}{f(s)} \left(\frac{\dot{f}(t)\dot{y}(t)}{f(t)} + \ddot{y}(t) \right) \right] ,$$

$$F(t) := \int_{t_i}^{t} ds \frac{\dot{y}(s)}{f(s)}$$

$$= -\frac{m}{2} \int_{t_i}^{t_f} dt [F(t) \dot{f}(t) \dot{y}(t) + F(t) f(t) \ddot{y}(t)] .$$

An integration by parts on the second term yields

$$S[x(t)] = -\frac{m}{2} \left(\int_{t_i}^{t_f} dt \{ F \dot{f} \dot{y} - \dot{F} f \dot{y} - F \dot{f} \dot{y} \} + \underbrace{F f}_{=x} \dot{y} \Big]_{t_i}^{t_f} \right)$$

$$= \frac{m}{2} \int_{t_i}^{t_f} dt \frac{\dot{y}(t)}{f(t)} f(t) \dot{y}(t) - \frac{m}{2} \underbrace{\left[x(t) \dot{y}(t) \right]_{t_i}^{t_f}}_{=0} .$$

So we obtain

$$S[x(t)] = \int_{t_i}^{t_f} dt \frac{m}{2} \dot{y}^2 \,, \tag{20.11}$$

which is, as promised, the action of the free particle transformed to the path $y(t)$ of (20.10).

The only complication we have to deal with concerns the boundary value condition at the endpoint t_f. The boundary conditions for $x(t)$ are transformed into those for $y(t)$ according to [cf. (20.8) and (20.10)]:

$$y(t_i) = 0 \,, \quad \int_{t_i}^{t_f} ds \frac{\dot{y}(s)}{f(s)} = 0 \,. \tag{20.12}$$

The second boundary condition is nonlocal and thus not easy to use directly. We therefore use a trick based on the representation of the δ-function:

$$\delta\big(x(t_f)\big) = \frac{1}{2\pi} \int d\alpha \exp\big[-\mathrm{i}\alpha x(t_f)\big] \,.$$

This allows us to rewrite the path integral in the form

$$\begin{aligned}
&\int_{x(t_i)=0}^{x(t_f)=0} [dx(t)] \exp\left\{\frac{\mathrm{i}}{\hbar} S[x(t)]\right\} \\
&= \frac{1}{2\pi} \int_{x(t_i)=0}^{x(t_f)=\text{arb.}} [dx(t)] \int_{-\infty}^{+\infty} d\alpha\, \mathrm{e}^{-\mathrm{i}\alpha x(t_f)} \exp\left\{\frac{\mathrm{i}}{\hbar} S[x(t)]\right\} \\
&\underset{(8)}{=} \frac{1}{2\pi} \int_{y(t_i)=0}^{y(t_f)=\text{arb.}} [dy(t)] \left|\frac{\delta x}{\delta y}\right| \int_{-\infty}^{+\infty} d\alpha \\
&\quad \times \exp\left[-\mathrm{i}\alpha f(t_f) \int_{t_i}^{t_f} ds \frac{\dot{y}(s)}{f(s)}\right] \exp\left[\frac{\mathrm{i}}{\hbar}\frac{m}{2} \int_{t_i}^{t_f} dt\, \dot{y}(t)^2\right] .
\end{aligned} \tag{20.13}$$

The infinite dimensional generalization of the Jacobian is independent of $y(t)$, because the transformation (20.10) is linear in y. Let us write the exponents in (20.13) somewhat differently:

$$\begin{aligned}
&\frac{\mathrm{i}}{\hbar}\frac{m}{2} \int_{t_i}^{t_f} dt \left(\dot{y}^2 - \frac{2h}{m} \alpha f(t_f) \frac{\dot{y}(t)}{f(t)}\right) \\
&= \frac{\mathrm{i}}{\hbar}\frac{m}{2} \int_{t_i}^{t_f} dt \Bigg[\underbrace{\left(\dot{y} - \frac{\hbar\alpha}{m}\frac{f(t_f)}{f(t)}\right)^2}_{=\dot{\gamma}(t)} - \frac{\hbar^2\alpha^2}{m^2}\frac{f^2(t_f)}{f^2(t)}\Bigg] ,
\end{aligned}$$

$$\gamma(t) = y(t) - \frac{\hbar\alpha}{m} f(t_f) \int_{t_i}^{t_t} \frac{ds}{f(s)} \,.$$

We then obtain for (20.13)

$$\frac{1}{2\pi}\left|\frac{\delta x}{\delta y}\right| \int_{-\infty}^{+\infty} d\alpha \exp\left[-\frac{\mathrm{i}\hbar}{2m}\alpha^2 f^2(t_f)\int_{t_i}^{t_f}\frac{dt}{f^2(t)}\right]$$

$$\times \int_{\gamma(t_i)=0}^{\gamma(t_f)\text{arb.}} [d\gamma(t)]\exp\left[\frac{\mathrm{i}}{\hbar}\frac{m}{2}\int_{t_i}^{t_f} dt\,\dot{\gamma}^2(t)\right] . \tag{20.14}$$

Now comes the pleasant surprise: we can perform the α-integration (Gaussian integral). Furthermore, the path integral is easy to handle, since only the free particle propagator appears:

$$\int_{\gamma(t_i)=0}^{\gamma(t_f)\text{arb.}} [d\gamma(t)]\exp\left[\frac{\mathrm{i}}{\hbar}\frac{m}{2}\int_{t_i}^{t_f} dt\,\dot{\gamma}^2(t)\right] = \int_{-\infty}^{+\infty} dx\, K_0(x, t_f; 0, t_i)$$

$$= \sqrt{\frac{m}{2\pi\mathrm{i}\hbar(t_f - t_i)}}\int_{-\infty}^{+\infty} dx \exp\left[\frac{\mathrm{i}}{\hbar}\frac{m}{2}\frac{x^2}{t_f - t_i}\right] = 1 .$$

This result should not surprise anyone, since by virtue of its construction, the path integral represents the probability amplitude that the particle will be found anywhere at time t_f. The path integral being sought thus can be reduced to the simple expression

$$\int_{x(t_i)=0}^{x(t_f)=0} [dx(t)]\exp\left\{\frac{\mathrm{i}}{\hbar}\frac{m}{2}\int_{t_i}^{t_f} dt\left[\dot{x}^2 - W(t)x^2\right]\right\}$$

$$= \left|\frac{\delta x}{\delta y}\right|\sqrt{m\Big/2\pi\mathrm{i}\hbar f^2(t_f)\int_{t_i}^{t_f}\frac{dt}{f^2(t)}} .$$

We shall now calculate the Jacobian, using a rather naive procedure that is not always "clean", but produces the correct result. First we again replace the paths $x(t)$ and $y(t)$ by the points

$$(x_0, x_1, \ldots, x_N) , \quad (y_0, y_1, \ldots, y_N) ; \quad x_k = x(t_k) , \quad y_k = y(t_k) .$$

We then write the transformation (20.10) as

$$y_n = x_n - \frac{T}{N}\sum_{k=1}^{n}\frac{\dot{f}(t_k)}{f(t_k)}\frac{(x_k + x_{k-1})}{2}$$

$$= \sum_{k=1}^{N}\delta_{nk}x_k - \frac{1}{2}\sum_{k=1}^{n}\frac{\dot{f}(t_k)}{f(t_k)}\frac{T}{N}x_k - \frac{1}{2}\sum_{k=0}^{n-1}\frac{\dot{f}(t_{k+1})}{f(t_{k+1})}\frac{T}{N}x_k .$$

Now, the determinant is given by the diagonal elements, so that

$$J_N := \left|\frac{\partial y_i}{\partial x_j}\right| = \prod_{k=1}^{N}\left(1 - \frac{1}{2}\frac{\dot{f}(t_k)}{f(t_k)}\frac{T}{N}\right) .$$

If we now perform the limit $N \to \infty$, we obtain

$$
\begin{aligned}
\left|\frac{\delta y}{\delta x}\right| &= \lim_{N\to\infty} J_N = \lim_{N\to\infty} \exp\left\{\log\prod_{k=1}^{N}\left(1-\frac{1}{2}\frac{\dot f(t_k)}{f(t_k)}\frac{T}{N}\right)\right\} \\
&= \lim_{N\to\infty} \exp\left\{\sum_{k=1}^{N}\log\left(1-\frac{1}{2}\frac{\dot f(t_k)}{f(t_k)}\frac{T}{N}\right)\right\} \\
&= \lim_{N\to\infty} \exp\left[-\frac{1}{2}\left(\sum_{k=1}^{N}\frac{\dot f(t_k)}{f(t_k)}\frac{T}{N}\right)\right] = \exp\left[-\frac{1}{2}\int_{t_i}^{t_f}\frac{\dot f(t)}{f(t)}dt\right] \\
&= \exp\left[-\frac{1}{2}\log\left(\frac{f(t_f)}{f(t_i)}\right)\right] = \sqrt{\frac{f(t_i)}{f(t_f)}} \,.
\end{aligned}
$$

Therefore

$$
\left|\frac{\delta x}{\delta y}\right| = \left|\frac{\delta y}{\delta x}\right|^{-1} = \sqrt{\frac{f(t_f)}{f(t_i)}} \,.
$$

We thus obtain the remarkably simple result: the path integral which corresponds to the quadratic action is given by

$$
\begin{aligned}
&\int_{x(t_i)=0}^{x(t_f)=0} [dx(t)] \exp\left\{\frac{\mathrm{i}}{\hbar} S[x(t)]\right\} \\
&\quad = \sqrt{m \Big/ 2\pi \mathrm{i}\hbar f(t_i) f(t_f) \int_{t_i}^{t_f} \frac{dt}{f^2(t)}} \,,
\end{aligned} \tag{20.15}
$$

where f is an arbitrary function with the property:

$$
\left[\frac{d^2}{dt^2} + W(t)\right] f(t) = 0 \,, \quad f(t_i) \neq 0 \,. \tag{20.16}
$$

It is easy to see that the free particle propagator ($W(t) = 0$) and the harmonic oscillator ($W(t) = \omega^2$) are contained in (20.15). We only have to set $f(t) = 1$ or $f(t) = \sin\left[\omega(t \quad t_i)\right]$.

Now that in (20.15) we have access to a formula for the propagator of a quadratic Lagrangian, we consider it once again from another point of view. Since the action is quadratic, i.e.,

$$
S[x(t)] = -\frac{m}{2}\int_{t_i}^{t_f} dt\, x(t) \left[\frac{d^2}{dt^2} + W(t)\right] x(t) \,, \tag{20.17}
$$

we can diagonalize it. In order to do this, we consider the Hermitean operator

$$
-\frac{d^2}{dt^2} - W(t) \,, \tag{20.18}
$$

which acts on the space of paths $x(t)$, where the following boundary conditions are to be satisfied:

$$x(t_i) = 0 = x(t_f) \, . \tag{20.19}$$

The operator (20.18) possesses a complete system of normalized eigenfunctions $\phi_n(t)$:

$$\left[-\frac{d^2}{dt^2} - W(t)\right]\phi_n(t) = \lambda_n\phi_n(t) \, , \quad \phi_n(t_i) = 0 = \phi_n(t_f) \, ,$$
$$\int_{t_i}^{t_f} dt\, \phi_n(t)\phi_{n'}(t) = \delta_{nn'} \, . \tag{20.20}$$

A given path can now be approximated by

$$\tilde{x}_N(t) = \sum_{n=1}^{N} a_n\phi_n(t) \, , \quad a_n = \int_{t_i}^{t_f} dt\, \phi_n(t)\tilde{x}_N(t) \, . \tag{20.21}$$

The action containing the approximated path then reads:

$$S[\tilde{x}_N(t)] = \frac{m}{2}\int_{t_i}^{t_f} dt\, \tilde{x}_N(t)\left[-\frac{d^2}{dt^2} - W(t)\right]\tilde{x}_N(t)$$
$$\underset{(20,21)}{=} \frac{m}{2}\sum_n\sum_{n'}\lambda_n a_n a_{n'}\underbrace{\int_{t_i}^{t_f} dt\, \phi_n(t)\phi_{n'}(t)}_{=\delta_{nn'}} = \frac{m}{2}\sum_{n=1}^{N}\lambda_n a_n^2 \, . \tag{20.22}$$

Hence we obtain

$$\int da_1 \ldots \int da_N \exp\left\{\frac{\mathrm{i}}{\hbar}S[\tilde{x}(t)]\right\} = \left(\sqrt{\frac{2\pi\mathrm{i}\hbar}{m}}\right)^N \sqrt{\frac{1}{\lambda_1 \ldots \lambda_N}} \, . \tag{20.23}$$

In the limit $N \to \infty$ the approximated paths fill up the whole path space and the path integral (20.23) is essentially given by

$$\left(\prod_{n=1}^{\infty}\lambda_n\right)^{-1/2} = \left(\det\left[-\frac{d^2}{dt^2} - W(t)\right]\right)^{-1/2} \, , \tag{20.24}$$

i.e., by an infinite product of eigenvalues. Of course the determinant will become divergent, but we can "regularize" it by calculating the ratio of two determinants.

So up to now we have found that the path integral which corresponds to a quadratic Lagrangian is essentially given by the determinant of the associated differential operator:

$$\int_{x(t_i)=0}^{x(t_f)=0}[dx(t)]\exp\left\{\frac{\mathrm{i}}{\hbar}\frac{m}{2}\int_{t_i}^{t_f} dt\, x(t)\left[-\frac{d^2}{dt^2} - W(t)\right]x(t)\right\}$$
$$= \Delta\left(\det\left[-\frac{d^2}{dt^2} - W(t)\right]\right)^{-1/2} \, , \tag{20.25}$$

where the right-hand side is in reality to be interpreted as the limit

$$\lim_{N\to\infty} \Delta(N) \left(\prod_{n=1}^{N} \lambda_n\right)^{-1/2} = \lim_{N\to\infty} \Delta(N) \int \cdots \int \left(\sqrt{\frac{m}{2\pi \mathrm{i}\hbar}}\right)^N d^N a \\ \times \exp\left[\frac{\mathrm{i}}{\hbar}\frac{m}{2}\sum_{n=1}^{N} \lambda_n a_n^2\right] . \tag{20.26}$$

On the right-hand side one can see that the correct integration measure for the calculation of the determinant is given by

$$\prod_{k=1}^{N} \sqrt{\frac{m}{2\pi \mathrm{i}\hbar}} da_k . \tag{20.27}$$

Since we already know how to calculate the path integral (20.25), we want to try to derive a relation for the determinants; in order to avoid divergence problems, we consider the ratio

$$\frac{\det\left[-\partial_t^2 - W(t)\right]}{\det\left[-\partial_t^2 - V(t)\right]} = \frac{f_W(t_i) f_W(t_f) \int_{t_i}^{t_f} dt/ f_W^2(t)}{f_V(f_i) f_V(t_f) \int_{t_i}^{t_f} dt/ f_V^2(t)} . \tag{20.28}$$

We have previously assumed that $f_W(t_i) \neq 0$ and $f_V(t_i) \neq 0$. We now want to study the limit $f_{W,V}(t_i) = 0$. For this reason, let f_W^0 be the solution with boundary conditions:

$$\left[-\partial_t^2 - W(t)\right] f_W^0(t) = 0 , \quad f_W^0(t_i) = 0 , \quad \frac{d}{dt} f_W^0(t_i) = 1 . \tag{20.29}$$

Similarly, let f_W^1 be the solution with boundary conditions

$$f_W^1(t_i) = 1 , \quad \frac{d}{dt} f_W^1(t_i) = 0 . \tag{20.30}$$

In (20.28) we put

$$\begin{aligned} f_W &= f_W^0 + \varepsilon f_W^1 , \\ f_V &= f_V^0 + \varepsilon f_V^1 , \end{aligned} \tag{20.31}$$

and so obtain for the limit $\varepsilon \to 0$:

$$\lim_{\varepsilon\to 0} \frac{f_W(t_i)}{f_V(t_i)} = 1 , \quad \lim_{\varepsilon\to 0} \frac{f_W(t_f)}{f_V(t_f)} = \frac{f_W^0(t_f)}{f_V^0(t_f)} . \tag{20.32}$$

The limit of the integral in (20.28),

$$\int_{t_i}^{t_f} \frac{dt}{[f_W(t)]^2} ,$$

diverges, however, since f_W^0 vanishes at the lower limit t_i. But since the main contribution to the integral comes from the infinitesimal neighborhood around t_i (in the limit $\varepsilon \to 0$), the integral diverges as

$$\int_{t_i}^{t_f} \frac{dt}{(t-t_i)^2} \, .$$

So we find

$$\lim_{\varepsilon \to 0} \frac{\int_{t_i}^{t_f} [dt/f_W^2(t)]}{\int_{t_i}^{t_f} [dt/f_V^2(t)]} = 1 \, . \tag{20.33}$$

Thus, (20.28) reduces to the simple relation

$$\frac{\det\left[-\partial_t^2 - W(t)\right]}{\det\left[-\partial_t^2 - V(t)\right]} = \frac{f_W^0(t_f)}{f_V^0(t_f)} \, . \tag{20.34}$$

A simple example is supplied by

$$V(t) = 0 : \quad f_V^0 = t - t_i \, ,$$

which satisfies $f_V^0(t_i) = 0$ and $(d/dt) f_V^0(t_i) = 1$. Then the ratio of the determinants becomes simply

$$\frac{\det\left[-\partial_t^2 - W(t)\right]}{\det\left[-\partial_t^2\right]} = \frac{f_W^0(t_f)}{t_f - t_i} \, . \tag{20.35}$$

This can be used to calculate the propagator of a quadratic Lagrangian, relative to the free propagator K_0:

$$\begin{aligned} K\big(x_2, t_2; x_1, t_1\big) &= K\big(0, t_2; 0, t_1\big)\, \mathrm{e}^{(\mathrm{i}/\hbar) S[x_{\mathrm{cl}}]} \\ &= \left[\frac{\det\left(-\partial_t^2 - W(t)\right)}{\det\left(-\partial_t^2\right)}\right]^{-1/2} K_0\big(0, t_2, 0, t_1\big)\, \mathrm{e}^{(\mathrm{i}/\hbar) S[x_{\mathrm{cl}}]} \\ &= \sqrt{\frac{t_2 - t_1}{f_W^0(t_2)}} \sqrt{\frac{m}{2\pi \mathrm{i}\hbar (t_2 - t_1)}}\, \mathrm{e}^{(\mathrm{i}/\hbar) S[x_{\mathrm{cl}}]} \\ &= \sqrt{\frac{m}{2\pi \mathrm{i}\hbar f_W^0(t_2)}}\, \mathrm{e}^{(\mathrm{i}/\hbar) S[x_{\mathrm{cl}}]} \, . \end{aligned} \tag{20.36}$$

For a simple check of the formula we can take the linear harmonic oscillator again: $W(t) = \omega^2$, $f_W^0(t) = \sin[\omega(t - t_1)]/\omega$. This example shows that we are allowed to add to the argument of f_W^0 the initial time and hence rewrite (20.29) a bit, namely:

$$\left[\frac{d^2}{dt^2} + W(t)\right] f_W^0\big(t, t_1\big) = 0 \tag{20.37}$$

with

$$f_W^0(t,t_1)\Big|_{t=t_1}=0\,,\qquad \frac{d}{dt}f_W^0(t,t_1)\Big|_{t=t_1}=1\,, \tag{20.38}$$

and therefore

$$K(x_2,t_2;x_1,t_1)\underset{(36)}{=}\sqrt{\frac{m}{2\pi\mathrm{i}\hbar f_W^0(t_2,t_1)}}\exp\left\{\frac{\mathrm{i}}{\hbar}S[x_{\mathrm{cl}}]\right\}\,. \tag{20.39}$$

At this stage it is very useful to return to the end of Chap. 4. If we then compare (20.38) with (4.65–66), we can see that f_W^0 and $J=\partial x(p,t)/\partial p$ satisfy exactly the same (Jacobi) equation and have proportional boundary conditions. Since, given $L=m\dot{x}^2/2-mW(t)x^2/2$, we find that

$$\frac{d}{dt}\left(\frac{\partial^2 L}{\partial\dot{x}^2}\dot{f}_W^0\right)+\left[\frac{d}{dt}\left(\frac{\partial^2 L}{\partial x\partial\dot{x}}\right)-\frac{\partial^2 L}{\partial x^2}\right]f_W^0=0$$

reduces to

$$m\ddot{f}_W^0+mWf_W^0=0:\qquad \left[\frac{d^2}{dt^2}+W(t)\right]f_W^0(t,t_1)=0\,.$$

Moreover, since by (4.69)

$$\frac{1}{J}=-\frac{\partial^2 S}{\partial x_2\partial x_1}$$

it follows that

$$K(x_2,t_2;x_1,t_1)=\sqrt{\frac{\mathrm{i}}{2\pi\hbar}\frac{\partial^2 S_{\mathrm{cl}}}{\partial x_2\partial x_1}}\exp\left[\frac{\mathrm{i}}{\hbar}S_{\mathrm{cl}}\right]\,. \tag{20.40}$$

In N dimensions this formula goes into

$$K(x_2,t_2;x_1,t_1)=\sqrt{\det_N\left(\frac{\mathrm{i}}{2\pi\hbar}\frac{\partial^2 S_{\mathrm{cl}}}{\partial x_{2i}\partial x_{1j}}\right)}\exp\left[\frac{\mathrm{i}}{\hbar}S_{\mathrm{cl}}\right] \tag{20.41}$$

or

$$\begin{aligned}K(x_2,t_2;x_1,t_1)&=\left(\frac{1}{2\pi\mathrm{i}\hbar}\right)^{N/2}\sqrt{D}\exp\left[\frac{\mathrm{i}}{\hbar}S_{\mathrm{cl}}\right]\\&=\left(\frac{1}{2\pi\mathrm{i}\hbar}\right)^{N/2}\left(\det_N J_{ij}\right)^{1/2}\exp\left[\frac{\mathrm{i}}{\hbar}S_{\mathrm{cl}}\right]\,;\end{aligned} \tag{20.42}$$

$$D=\det_N\left(-\frac{\partial^2 S}{\partial x_{2i}\partial x_{1j}}\right)\,. \tag{20.43}$$

This propagator might be called the WKB propagator . D, as we know from Chap. 4, is the Van-Vleck determinant.

Note that the WKB propagator is exact for Lagrangians which terminate with quadratic terms. Here are three well-known examples to illustrate (20.42):

a) Free particle in $N = 1$ dimension:

$$
\begin{aligned}
S_{\text{cl}} &= \frac{m}{2}\frac{(x_2 - x_1)^2}{t_2 - t_1} \\
D &= (-1)^N \left|\frac{\partial^2 S}{\partial x_2 \partial x_1}\right| = -1\left(-\frac{m}{t_2 - t_1}\right) = \frac{m}{t_2 - t_1} \\
K\big(x_2, t_2; x_1, t_1\big) &= \sqrt{\frac{m}{2\pi \mathrm{i}\hbar(t_2 - t_1)}}\exp\left[\frac{\mathrm{i}}{\hbar}S_{\text{cl}}\right] .
\end{aligned}
$$

b) Particle in a constant field, $N = 1$:

$$
\begin{aligned}
S_{\text{cl}} &= \frac{m}{2}\frac{(x_2 - x_1)^2}{t_2 - t_1} + \frac{F}{2}\big(t_2 - t_1\big)\big(x_1 + x_2\big) - \frac{F^2}{24m}\big(t_2 - t_1\big)^3 \\
D &= (-1)^N \left|\frac{\partial^2 S}{\partial x_2 \partial x_1}\right| = \frac{m}{t_2 - t_1} \\
K\big(x_2, t_2; x_1, t_1\big) &= \sqrt{\frac{m}{2\pi \mathrm{i}\hbar(t_2 - t_1)}}\exp\left[\frac{\mathrm{i}}{\hbar}S_{\text{cl}}\right] .
\end{aligned}
$$

c) Linear harmonic oscillator, $N = 1$:

$$
\begin{aligned}
S_{\text{cl}} &= \frac{m\omega}{2\cos[\omega(t_2 - t_1)]}\left[\big(x_2^2 + x_1^2\big)\sin\big[\omega\big(t_2 - t_1\big)\big] - 2x_2x_1\right] \\
D\big(t_2 - t_1\big) &= (-1)^N \left|\frac{\partial^2 S}{\partial x_2 \partial x_1}\right| = \frac{m\omega}{\sin\big[\omega\big(t_2 - t_1\big)\big]} \\
2K\big(x_2, t_2; x_1, t_1\big) &= \sqrt{\frac{1}{2\pi \mathrm{i}\hbar}}\sqrt{D}\exp\left[\frac{\mathrm{i}}{\hbar}S_{\text{cl}}\right] \\
&= \sqrt{\frac{m\omega}{2\pi \mathrm{i}\hbar\,\sin[\omega(t_2 - t_1)]}}\exp\left[\frac{\mathrm{i}}{\hbar}S_{\text{cl}}\right] ,
\end{aligned}
\tag{20.44}
$$

which is exactly the same result as that of many of our former lengthy calculations of

$$
K\big(x_2, t_2; x_1, t_1\big) = \int_{x(t_1)=x_1}^{x(t_2)=x_2} [dx(t)] \exp\left\{\frac{\mathrm{i}}{\hbar}\int_{t_2}^{t_1} dt \left[\frac{m}{2}\dot{x}^2 - \frac{m}{2}\omega^2 x^2\right]\right\} .
$$

A more sophisticated problem arises when considering the harmonic oscillator with time-dependent frequency – the title of this chapter. For that reason, let us go back to (8.42):

$$
L(t) = \frac{m}{2}\dot{x}^2 - \frac{m}{2}\omega^2(t)x^2 \tag{20.45}
$$

with the equation of motion

$$\left[\frac{d^2}{dt^2} + \omega^2(t)\right] x(t) = 0 . \tag{20.46}$$

To solve this equation we use the ansatz

$$x(t) = f(t)\, \mathrm{e}^{\mathrm{i}g(t)} \tag{20.47}$$

and obtain the expressions (8.22, 23):

$$\ddot{f} - f\dot{g}^2 + \omega^2 f = 0 , \tag{20.48}$$

$$2\dot{f}\dot{g} + f\ddot{g} = 0 . \tag{20.49}$$

The last equation is equivalent to

$$f^2\dot{g} = C^2 \qquad \text{or} \qquad \dot{g} = \frac{C^2}{f^2} \tag{20.50}$$

with C^2 a constant which is independent of the terminal conditions $x(t_1) = x_1, x(t_2) = x_2$. Substituting (20.50) into (20.48) produces an equation which f has to satisfy:

$$\ddot{f} - C^4 \frac{1}{f^3} + \omega^2(t) f = 0 . \tag{20.51}$$

Now, the equation we are interested in is stated in (20.48):

$$\left(-\frac{d^2}{dt^2} + \dot{g}^2 - \omega^2(t)\right) f(t) = 0$$

or

$$\left(-\frac{d^2}{dt^2} - W(t)\right) f(t) = 0 , \tag{20.52}$$

with

$$W(t) := -\dot{g}^2(t) + \omega^2(t) . \tag{20.53}$$

Introducing the abbreviations $g(t) = g,\ g(t_1) = g_1,\ \dot{g}(t_1) = \dot{g}_1$, etc., we can easily prove that a function that meets the requirements (20.37) and (20.38) is given by

$$f_W^0(t, t_1) = \frac{\sin(g - g_1)}{\sqrt{\dot{g}\dot{g}_1}} \tag{20.54}$$

with

$$f_W^0(t_1, t_1) = 0 , \quad \frac{d}{dt} f_W^0(t, t_1)\Big|_{t=t_1} = 1 .$$

For this case, i.e., for the time-dependent harmonic oscillator, we therefore obtain, according to (20.39):

$$K(x_2, t_2; x_1, t_1) = \sqrt{\frac{m\sqrt{\dot{g}_2\dot{g}_1}}{2\pi i\hbar \sin(g_2 - g_1)}}\, e^{(i/\hbar)S_{cl}} . \tag{20.55}$$

Here we need the classical action of the present problem. This information is also needed if we want to obtain the propagator via (20.40), where two derivatives of S_{cl} are to be taken. Hence, let us first solve the classical problem of the harmonic oscillator with time-dependent frequency.

Here it is convenient to solve (20.46) with the ansatz

$$x(t) = f(t)[A \cos g(t) + B \sin g(t)] . \tag{20.56}$$

With the end point conditions $x(t_1) = x_1$, $x(t_2) = x_2$, we can easily compute the constant coefficients A and B using Cramer's rule and then obtain for the classical trajectory

$$x(t) = \frac{f(t)}{\sin(g_2 - g_1)}\left(\frac{x_1}{f_1}\sin(g_2 - g) - \frac{x_2}{f_2}\sin(g_1 - g)\right), \tag{20.57}$$
$$g(t_2) - g(t_1) \neq n\pi .$$

Now we turn to the object of interest, the classical action:

$$\begin{aligned} S_{cl} &= \int_{t_1}^{t_2} dt\, L(x, \dot{x}; t) = \frac{m}{2}\int_{t_1}^{t_2} dt \left[\overbrace{\left(\frac{dx}{dt}\right)^2}^{\text{integr. by parts}} - \omega^2(t)x^2\right] \\ &= \frac{m}{2}\left[x\dot{x}\right]_{t_1}^{t_2} - \frac{m}{2}\int_{t_1}^{t_2} \underbrace{dt\, x(t)\left[\frac{d^2}{dt^2} + \omega^2(t)\right]x(t)}_{=0} \\ &= \frac{m}{2}\left[x_2\dot{x}_2 - x_1\dot{x}_1\right] . \end{aligned} \tag{20.58}$$

Here we have to express the terminal velocities in terms of x_1 and x_2. Taking the time derivative of (20.57) gives, first of all,

$$\begin{aligned} \dot{x} = &\frac{1}{\sin(g_2 - g_1)}\left[\frac{\dot{f}x_1}{f_1}\sin(g_2 - g) - \frac{\dot{f}x_2}{f_2}\sin(g_1 - g)\right. \\ &\left. - \frac{f\dot{g}x_1}{f_1}\cos(g_2 - g) + \frac{f\dot{g}x_2}{f_2}\cos(g_1 - g)\right] . \end{aligned}$$

After a bit of algebra we obtain

$$\begin{aligned} \left[\dot{x}_2 x_2 - \dot{x}_1 x_1\right] = &\frac{\dot{f}_2 x_2^2}{f_2} - \frac{\dot{f}_1 x_1^2}{f_1} + \left(\dot{g}_2 x_2^2 + \dot{g}_1 x_1^2\right)\cot(g_2 - g_1) \\ &- \underbrace{\left(\frac{f_2\dot{g}_2}{f_1} + \frac{f_1\dot{g}_1}{f_2}\right)}_{=./.} x_1 x_2 \frac{1}{\sin(g_2 - g_1)} , \qquad \dot{g} = \frac{C^2}{f^2} \end{aligned}$$

$$./. = \frac{f_2^2 \dot{g}_2 + f_1^2 \dot{g}_1}{f_1 f_2} = \frac{2C^2}{C^2/\sqrt{\dot{g}_1 \dot{g}_2}} = 2\sqrt{\dot{g}_1 \dot{g}_2} \,.$$

Finally we obtain for the classical action

$$S_{\rm cl} = \frac{m}{2} \left[\frac{\dot{f}_2 x_2^2}{f_2} - \frac{\dot{f}_1 x_1^2}{f_1} + \left(\dot{g}_2 x_2^2 + \dot{g}_1 x_1^2\right) \cot(g_2 - g_1) \right.$$
$$\left. -2x_2 x_1 \sqrt{\dot{g}_1 \dot{g}_2}\, \frac{1}{\sin(g_2 - g_1)} \right] . \tag{20.59}$$

From here we get

$$\frac{\partial^2 S_{\rm cl}}{\partial x_2 \partial x_1} = -\frac{m\sqrt{\dot{g}_1 \dot{g}_2}}{\sin(g_2 - g_1)}$$

which implies the desired result,

$$K\left(x_2, t_2; x_1, t_1\right) = \sqrt{\frac{\mathrm{i}}{2\pi\hbar} \frac{\partial^2 S_{\rm cl}}{\partial x_2 \partial x_1}} \exp\left[\frac{\mathrm{i}}{\hbar} S_{\rm cl}\right]$$
$$= \sqrt{\frac{m\sqrt{\dot{g}_1 \dot{g}_2}}{2\pi \mathrm{i}\hbar \, \sin(g_2 - g_1)}} \exp\left[\frac{\mathrm{i}}{\hbar} S_{\rm cl}\right] , \tag{20.60}$$

where $S_{\rm cl}$ is given by (20.59).

In (8.43) we found an interesting invariant associated with the time-dependent oscillator, which we rewrite a bit ($\dot{x} = p/m$):

$$I(t) = \frac{1}{2m} \left[m^2 \frac{x^2}{\varrho^2} + (\varrho p - m x \dot{\varrho})^2 \right] , \quad \dot{I} \equiv \frac{\partial I}{\partial t} + \frac{1}{\mathrm{i}\hbar}[I, H] = 0 \,. \tag{20.61}$$

From the quantum mechanical point of view, $I(t)$ is an invariant operator, while $H(t)$ is not. p in (20.61) means $p = (\hbar/\mathrm{i})\partial/\partial x$. We also repeat the differential equation that $\varrho(t)$ has to satisfy:

$$\ddot{\varrho} + \omega^2(t)\varrho - \frac{1}{\varrho^3} = 0 \,. \tag{20.62}$$

But this equation is equivalent to (20.50, 51) if we absorb the constant C in $f : f/C \to f$, so that (20.50) becomes

$$\dot{g} = \frac{1}{f^2} \tag{20.63}$$

and (20.51) takes the form

$$\left(\ddot{f} + \omega^2(t) f - \frac{1}{f^3}\right) = 0 \,. \tag{20.64}$$

The existence of the Hermitean invariant operator $I(t)$ is of utmost importance in solving the time-dependent quantum mechanical problem given by

$$H(t) = \frac{p^2}{2m} + \frac{m}{2}\omega^2(t)x^2 = -\frac{\hbar^2}{2m}\frac{\partial^2}{\partial x^2} + \frac{m}{2}\omega^2(t)x^2 . \tag{20.65}$$

If we write the time-dependent Schrödinger equation for the problem as

$$\mathrm{i}\hbar\frac{\partial}{\partial t}\psi(x,t) = H(t)\psi(x,t) \tag{20.66}$$

then Lewis and Riesenfeld have shown that the general solution is given by the superposition

$$\psi(x,t) = \sum_n c_n\, \mathrm{e}^{\mathrm{i}\alpha_n(t)}\psi_n(x,t) \tag{20.67}$$

where $\psi_n(x,t)$ are the normalized eigenfunctions of the invariant operator I:

$$I\psi_n(x,t) = \lambda_n\psi_n(x,t) \tag{20.68}$$

with time-independent eigenvalues λ_n. The coefficients c_n in (20.67) are constants, while the time-dependent phases $\alpha_n(t)$ are to be obtained from the equation

$$\hbar\frac{d\alpha_n(t)}{dt} = \left\langle\psi_n\middle|\mathrm{i}\hbar\frac{\partial}{\partial t} - H\middle|\psi_n\right\rangle . \tag{20.69}$$

From (20.67) we have

$$c_n\,\mathrm{e}^{\mathrm{i}\alpha_n(t)} = \int dx\, \psi_n^*(x,t)\psi(x,t) \tag{20.70}$$

so that it holds that

$$\begin{aligned}
\psi(x_2,t_2) &= \sum_n \mathrm{e}^{-\mathrm{i}\alpha_n(t_1)}\left[\int dx_1\, \psi_n^*(x_1,t_1)\psi(x_1,t_1)\right]\mathrm{e}^{\mathrm{i}\alpha_n(t_2)}\psi_n(x_2,t_2) \\
&= \int dx_1 \sum_n \mathrm{e}^{\mathrm{i}(\alpha_n(t_2)-\alpha_n(t_1))}\psi_n(x_2,t_2)\psi_n^*(x_1,t_1)\psi(x_1,t_1) \qquad (20.71) \\
&= \int dx_1\, K(x_2,t_2;x_1,t_1)\psi(x_1,t_1) , \quad t_2 > t_1 . \qquad (20.72)
\end{aligned}$$

The last two equations allow us to identify the Feynman propagator:

$$K(x_2,t_2;x_1,t_1) = \sum_n \mathrm{e}^{\mathrm{i}(\alpha_n(t_2)-\alpha_n(t_1))}\psi_n(x_2,t_2)\psi_n^*(x_1,t_1) , \; t_2 > t_1 . \tag{20.73}$$

Here we see that the propagator admits an expansion in terms of the eigenfunctions of the invariant operator $I(t)$.

Now with the explicit knowledge of K at hand as given by (20.59) and (20.60), we should be able to write down the eigenfunctions and eigenvalues of $I(t)$. This can indeed be done and yields

$$\alpha_n(t) = -\left(n + \tfrac{1}{2}\right) g(t)$$

$$\psi_n(x,t) = \left(\frac{1}{2^n n!}\left(\frac{m\dot{g}}{\hbar}\right)^{1/2}\right)^{1/2} \exp\left[\frac{\mathrm{i}}{\hbar}\frac{m}{2}\left(\frac{\dot{f}}{f} + \mathrm{i}\dot{g}\right)x^2\right] H_n\left(\left(\frac{m\dot{g}}{\hbar}\right)^{1/2} x\right)$$

with

$$I\psi_n(x,t) = \hbar\left(n + \tfrac{1}{2}\right)\psi_n(x,t)\,, \quad n = 0,\, 1,\, 2,\, \ldots$$

while $\alpha_n(t)$ satisfies (20.69).

21. Propagators for Particles in an External Magnetic Field

In order to describe the propagation of a scalar particle in an external potential, we begin again with the path integral

$$K(\boldsymbol{r}', t'; \boldsymbol{r}, 0) = \int_{\boldsymbol{r}(0)}^{\boldsymbol{r}'(t')} [d\boldsymbol{r}(t)] \exp\left\{\frac{\mathrm{i}}{\hbar} S[\boldsymbol{r}(t)]\right\} \tag{21.1}$$

with

$$S[\boldsymbol{r}(t)] = \int_0^{t'} dt\, L(\boldsymbol{r}, \dot{\boldsymbol{r}}) \,.$$

Classical electrodynamics tells us that in the presence of an external classical field $(\boldsymbol{A}, \phi)$, $\boldsymbol{B} = \nabla \times \boldsymbol{A}$, the Lagrangian is modified as follows:

$$L = \frac{m}{2}\left(\frac{d\boldsymbol{r}}{dt}\right)^2 + \frac{e}{c}\frac{d\boldsymbol{r}}{dt} \cdot \boldsymbol{A} - e\phi(\boldsymbol{r}) \,. \tag{21.2}$$

We now pursue the calculation of the particle propagator as it was done in (19.18)ff. We take a wave function at time t and let it be propagated toward $t + \varepsilon$:

$$\begin{aligned}\psi(\boldsymbol{r}', t+\varepsilon) = \int_{-\infty}^{+\infty} d^3r \frac{1}{A^3(\varepsilon)} \exp\left\{\frac{\mathrm{i}}{\hbar}\varepsilon\left[\frac{m}{2}\frac{(\boldsymbol{r}'-\boldsymbol{r})^2}{\varepsilon^2} + \frac{e}{c}\frac{(\boldsymbol{r}'-\boldsymbol{r})}{\varepsilon}\right.\right.\\ \left.\left.\times \boldsymbol{A}\left(\frac{\boldsymbol{r}'+\boldsymbol{r}}{2}\right) - e\phi\left(\frac{\boldsymbol{r}'+\boldsymbol{r}}{2}\right)\right]\right\} \psi(\boldsymbol{r}, t) \,.\end{aligned}$$

As usual, we have $A = (2\pi \mathrm{i}\hbar\varepsilon/m)^{1/2}$. Now we define the variable $\boldsymbol{\varrho}$: $\boldsymbol{\varrho} = \boldsymbol{r}' - \boldsymbol{r}$ and obtain

$$\begin{aligned}\psi(\boldsymbol{r}', t+\varepsilon) = \int_{-\infty}^{+\infty} \frac{d^3\varrho}{A^3(\varepsilon)} \exp\left\{\frac{\mathrm{i}}{\hbar}\frac{m}{2\varepsilon}\boldsymbol{\varrho}^2 + \frac{\mathrm{i}e}{\hbar c}\boldsymbol{\varrho} \cdot \boldsymbol{A}\left(\frac{2\boldsymbol{r}'-\boldsymbol{\varrho}}{2}\right)\right.\\ \left. - \frac{\mathrm{i}}{\hbar}\varepsilon e\phi\left(\frac{2\boldsymbol{r}'-\boldsymbol{\varrho}}{2}\right)\right\} \psi(\boldsymbol{r}' - \boldsymbol{\varrho}, t) \,.\end{aligned} \tag{21.3}$$

In analogy to the procedure on the way to the Schrödinger equation of Chap. 19, we now expand up to terms linear in ε or quadratic in $\boldsymbol{\varrho}$:

$$
\begin{aligned}
\psi(\boldsymbol{r}',t) + \varepsilon\frac{\partial}{\partial t}\psi(\boldsymbol{r}',t) + \ldots &= \int \frac{d^3\varrho}{A^3(\varepsilon)} \exp\left[\frac{\mathrm{i}}{\hbar}\frac{m}{2\varepsilon}\varrho^2\right] \\
&\times \exp\left\{\frac{\mathrm{i}e}{\hbar c}\left[\boldsymbol{\varrho}\cdot\boldsymbol{A}(\boldsymbol{r}') - \frac{1}{2}\underbrace{\boldsymbol{\varrho}\cdot\left((\boldsymbol{\varrho}\cdot\nabla')\boldsymbol{A}(\boldsymbol{r}')\right)}_{=\varrho_n\varrho_m\partial'_m A_n(\boldsymbol{r}')}\right]\right\} \\
&\times \left(1 - \frac{\mathrm{i}e}{\hbar}\varepsilon\phi(\boldsymbol{r}') + \ldots\right)\left(\psi(\boldsymbol{r}',t) - (\boldsymbol{\varrho}\cdot\nabla')\psi(\boldsymbol{r}',t)\right. \\
&\left. + \frac{1}{2}(\boldsymbol{\varrho}\cdot\nabla')^2\psi(\boldsymbol{r}',t) + \ldots\right) .
\end{aligned}
$$

In the next step we also expand the exponential on the right-hand side, which contains the vector potential:

$$
\begin{aligned}
\psi(\boldsymbol{r}',t) + \varepsilon\frac{\partial}{\partial t}\psi(\boldsymbol{r}',t) + \ldots &= \int \frac{d^3\varrho}{A^3(\varepsilon)} \exp\left\{\frac{\mathrm{i}}{\hbar}\frac{m}{2\varepsilon}\varrho^2\right\} \\
&\times \left(1 + \frac{\mathrm{i}\varepsilon}{\hbar c}\left[\boldsymbol{\varrho}\cdot\boldsymbol{A}(\boldsymbol{r}') - \frac{1}{2}\boldsymbol{\varrho}\cdot(\boldsymbol{\varrho}\cdot\nabla')\boldsymbol{A}(\boldsymbol{r}') + \ldots\right]\right. \\
&\left. - \frac{1}{2}\frac{e^2}{\hbar^2c^2}\left[\left(\boldsymbol{\varrho}\cdot\boldsymbol{A}(\boldsymbol{r}')\right)^2 + \ldots\right]\right)\left(1 - \frac{\mathrm{i}}{\hbar}\varepsilon e\phi(\boldsymbol{r}') + \ldots\right) \\
&\times \left(\psi(\boldsymbol{r}',t) - (\boldsymbol{\varrho}\cdot\nabla')\psi(\boldsymbol{r}',t) + \frac{1}{2}(\boldsymbol{\varrho}\cdot\nabla')^2\psi(\boldsymbol{r}',t) + \ldots\right) \\
&= \int \frac{d^3\varrho}{A^3(\varepsilon)} \exp\left\{\frac{\mathrm{i}}{\hbar}\frac{m}{2\varepsilon}\varrho^2\right\}\left(\psi - \frac{\mathrm{i}}{\hbar}\varepsilon e\phi\psi - (\boldsymbol{\varrho}\cdot\nabla')\psi + \frac{1}{2}(\boldsymbol{\varrho}\cdot\nabla')^2\psi\right. \\
&+ \frac{\mathrm{i}}{\hbar}\frac{e}{c}\left[\boldsymbol{\varrho}\cdot\boldsymbol{A}(\boldsymbol{r}') - \frac{1}{2}\boldsymbol{\varrho}\cdot(\boldsymbol{\varrho}\cdot\nabla')\boldsymbol{A}(\boldsymbol{r}')\right]\psi \\
&\left. - \frac{\mathrm{i}}{\hbar}\frac{e}{c}\left(\boldsymbol{\varrho}\cdot\boldsymbol{A}(\boldsymbol{r}')\right)(\boldsymbol{\varrho}\cdot\nabla')\psi - \frac{1}{2}\frac{e^2}{\hbar^2c^2}\left(\boldsymbol{\varrho}\cdot\boldsymbol{A}(\boldsymbol{r}')\right)^2\psi + \ldots\right) .
\end{aligned}
$$

At this point we make use of the integrals

$$
\begin{aligned}
&\int \frac{d^3\varrho}{A^3}\exp\left[\frac{\mathrm{i}}{\hbar}\frac{m}{2\varepsilon}\varrho^2\right] = 1 , \\
&\int \frac{d^3\varrho}{A^3}\exp\left[\frac{\mathrm{i}}{\hbar}\frac{m}{2\varepsilon}\varrho^2\right]\boldsymbol{\varrho} = 0 , \\
&\int \frac{d^3\varrho}{A^3}\exp\left[\frac{\mathrm{i}}{\hbar}\frac{m}{2\varepsilon}\varrho^2\right]\varrho^2 = \frac{\mathrm{i}\hbar\varepsilon}{m} .
\end{aligned}
$$

So we finally end up with

$$
\begin{aligned}
\psi(\boldsymbol{r}',t) + \varepsilon\frac{\partial}{\partial t}\psi(\boldsymbol{r}',t) + \ldots &= \psi(\boldsymbol{r}',t) - \frac{\mathrm{i}}{\hbar}\varepsilon e\phi(\boldsymbol{r}')\psi(\boldsymbol{r}',t) + \frac{1}{2}\frac{\mathrm{i}\hbar\varepsilon}{m}\nabla'^2\psi(\boldsymbol{r}',t) \\
&- \frac{\mathrm{i}\hbar\varepsilon}{2m}\frac{\mathrm{i}e}{\hbar c}\left(\nabla'\cdot\boldsymbol{A}(\boldsymbol{r}')\right)\psi(\boldsymbol{r}',t) - \frac{\mathrm{i}\hbar\varepsilon}{m}\frac{\mathrm{i}e}{\hbar c}\left(\boldsymbol{A}(\boldsymbol{r}')\cdot\nabla'\right)\psi(\boldsymbol{r}',t) \\
&- \frac{\mathrm{i}\hbar\varepsilon}{m}\frac{1}{2}\frac{e^2}{\hbar^2c^2}\boldsymbol{A}^2(\boldsymbol{r}')\psi(\boldsymbol{r}',t) .
\end{aligned}
$$

A comparison of the terms linear in ε yields ($\boldsymbol{r}' \to \boldsymbol{r}$):

$$\begin{aligned}\frac{\partial}{\partial t}\psi(\boldsymbol{r},t) =& \frac{1}{\mathrm{i}\hbar}e\phi(\boldsymbol{r})\psi(\boldsymbol{r},t) + \frac{1}{2m}\mathrm{i}\hbar\nabla^2\psi(\boldsymbol{r},t) + \frac{1}{2m}\frac{e}{c}\big(\nabla\cdot\boldsymbol{A}(\boldsymbol{r})\big)\psi(\boldsymbol{r},t)\\ &+ \frac{1}{m}\frac{e}{c}\big(\boldsymbol{A}(\boldsymbol{r})\cdot\nabla\big)\psi(\boldsymbol{r},t) + \frac{1}{\mathrm{i}\hbar}\frac{e^2}{2mc^2}\boldsymbol{A}^2(\boldsymbol{r})\psi(\boldsymbol{r},t)\end{aligned}$$

or

$$\begin{aligned}\mathrm{i}\hbar\frac{\partial}{\partial t}\psi(\boldsymbol{r},t) &= -\frac{\hbar^2}{2m}\nabla^2\psi(\boldsymbol{r},t) + \frac{\mathrm{i}e\hbar}{2mc}(\nabla\cdot\boldsymbol{A})\psi(\boldsymbol{r},t) + \frac{\mathrm{i}e\hbar}{mc}(\boldsymbol{A}\cdot\nabla)\psi(\boldsymbol{r},t)\\ &\quad+ \frac{e^2}{2mc^2}\boldsymbol{A}^2\psi(\boldsymbol{r},t) + e\phi(\boldsymbol{r})\psi(\boldsymbol{r},t)\\ &= \left(-\frac{\hbar^2}{2m}\nabla^2 + \frac{\mathrm{i}e\hbar}{2mc}(\nabla\cdot\boldsymbol{A}) + \frac{\mathrm{i}e\hbar}{mc}(\boldsymbol{A}\cdot\nabla)\right.\\ &\quad\left.+ \frac{e^2}{2mc^2}\boldsymbol{A}^2(\boldsymbol{r}) + e\phi(\boldsymbol{r})\right)\psi(\boldsymbol{r},t)\,.\end{aligned}$$

Recalling the $\boldsymbol{r}$-representation of the momentum operator, we obtain the well-known equation

$$\mathrm{i}\hbar\frac{\partial}{\partial t}\psi(\boldsymbol{r},t) = \left[\frac{1}{2m}\left(\boldsymbol{p} - \frac{e}{c}\boldsymbol{A}(\boldsymbol{r})\right)^2 + e\phi(\boldsymbol{r})\right]\psi(\boldsymbol{r},t) \tag{21.4}$$

since

$$\begin{aligned}\frac{1}{2m}\left(\boldsymbol{p} - \frac{e}{c}\boldsymbol{A}\right)^2\psi &= \frac{1}{2m}\left(\boldsymbol{p}^2 - \boldsymbol{p}\cdot\frac{e}{c}\boldsymbol{A} - \frac{e}{c}\boldsymbol{A}\cdot\boldsymbol{p} + \frac{e^2}{c^2}\boldsymbol{A}^2\right)\psi\\ &= \frac{1}{2m}\Big(-\hbar^2\nabla^2\psi + \mathrm{i}\hbar\frac{e}{c}\underbrace{\nabla\cdot(\boldsymbol{A}\psi)}_{\psi(\nabla\cdot\boldsymbol{A})+(\boldsymbol{A}\cdot\nabla)\psi} + \mathrm{i}\hbar\frac{e}{c}(\boldsymbol{A}\cdot\nabla)\psi + \frac{e^2}{c^2}\boldsymbol{A}^2\psi\Big)\\ &= -\frac{\hbar^2}{2m}\nabla^2\psi + \frac{\mathrm{i}e\hbar}{2mc}\big(\psi(\nabla\cdot\boldsymbol{A}) + (\boldsymbol{A}\cdot\nabla)\psi + (\boldsymbol{A}\cdot\nabla)\psi\big) + \frac{e^2}{2mc^2}\boldsymbol{A}^2\psi\\ &= -\frac{\hbar^2}{2m}\nabla^2\psi + \frac{\mathrm{i}e\hbar}{2mc}\psi(\nabla\cdot\boldsymbol{A}) + \frac{\mathrm{i}e\hbar}{mc}(\boldsymbol{A}\cdot\nabla)\psi + \frac{e^2}{2mc^2}\boldsymbol{A}^2\psi\,.\end{aligned}$$

The above equation (21.4) is the Schrödinger equation of a particle of mass m and charge e in presence of an external electromagnetic field. In this way we have demonstrated that the wave function propagated by the path integral (kernel) follows a development in time which is fixed by the Schrödinger equation.

The complete expression for the path integral is then

$$\begin{aligned}K^A(\boldsymbol{r}',t';\boldsymbol{r}) &= \int_{\boldsymbol{r}(0)}^{\boldsymbol{r}'(t')}[d\boldsymbol{r}(t)]\\ &\quad\times\exp\left\{\frac{\mathrm{i}}{\hbar}\int_0^{t'}dt\left[\frac{m}{2}\dot{\boldsymbol{r}}^2 - e\phi(\boldsymbol{r}) + \frac{e}{c}\dot{\boldsymbol{r}}\cdot\boldsymbol{A}\right]\right\}\,.\end{aligned} \tag{21.5}$$

A gauge transformation with respect to A,

$$A \to A' = A + \nabla\chi(r) \tag{21.6}$$

where $\chi(r)$ is a scalar function, leads to an additive term in the exponential of (21.5):

$$\frac{ie}{\hbar c}\int_0^{t'} dt\,\dot{r}\cdot\nabla\chi = \frac{ie}{\hbar c}\int_r^{r'}\nabla\chi\cdot dr = \frac{ie}{\hbar c}\left[\chi(r') - \chi(r)\right].$$

This value is the same for all paths, i.e., is path-independent. So we have found that the gauge transformation (21.6) induces a transformation on the propagator:

$$\begin{aligned} K' &:= K^{A+\nabla\chi}(r', t'; r) \\ &= \exp\left[\frac{\mathrm{i}}{\hbar}\frac{e}{c}\chi(r')\right] K^{A}(r', t'; r)\exp\left[-\frac{\mathrm{i}}{\hbar}\frac{e}{c}\chi(r)\right]. \end{aligned} \tag{21.7}$$

This corresponds to a change of the phase of the wave function

$$\psi^{A+\nabla\chi}(r) =: \psi'(r) = \exp\left[\frac{\mathrm{i}}{\hbar}\frac{e}{c}\chi(r)\right]\psi^{A}(r) =: \exp\left[\frac{\mathrm{i}}{\hbar}\frac{e}{c}\chi(r)\right]\psi(r). \tag{21.8}$$

If at the same time the wave function is subjected to a phase transformation while A is changed according to (21.6),

$$\begin{aligned} A &\to A' = A + \nabla\chi, \\ \psi &\to \psi' = \exp\left[\frac{\mathrm{i}}{\hbar}\frac{e}{c}\chi\right]\psi, \end{aligned} \tag{21.9}$$

then one can say that the quantum mechanics is gauge invariant. One can, of course, prove this fact directly, using the Schrödinger equation (21.4).

In order to confirm (21.8), we recall that the Schrödinger wave function satisfies the following integral equation:

$$\psi(r_2, t_2) = \int d^3r_1 K(r_2, t_2; r_1, t_1)\psi(r_1, t_1). \tag{21.10}$$

In order to find the gauge-transformed wave function ψ', we first note that it has to satisfy an integral equation similar to (21.10):

$$\begin{aligned} \psi'(r_2, t_2) &= \int d^3r_1 K'(r_2, t_2; r_1, t_1)\psi'(r_1, t_1) \\ &\underset{(7)}{=} \int d^3r_1 \exp\left[\frac{\mathrm{i}}{\hbar}\frac{e}{c}\chi(r_2)\right] K(r_2, t_2; r_1, t_1) \\ &\qquad \times \exp\left[-\frac{\mathrm{i}}{\hbar}\frac{e}{c}\chi(r_1)\right]\psi'(r_1, t_1). \end{aligned}$$

The first exponential under the integral sign is independent of r_1 and therefore may be moved to the left-hand side:

$$\exp\left[-\frac{\mathrm{i}}{\hbar}\frac{e}{c}\chi\left(\boldsymbol{r}_2\right)\right]\psi'\left(\boldsymbol{r}_2,t_2\right)$$
$$=\int d^3\boldsymbol{r}_1\,K\left(\boldsymbol{r}_2t_2;\boldsymbol{r}_1,t_1\right)\exp\left[-\frac{\mathrm{i}}{\hbar}\frac{e}{c}\chi\left(\boldsymbol{r}_1\right)\right]\psi'\left(\boldsymbol{r}_1,t_1\right).$$

If we compare this expression with the integral equation (21.10), we immediately obtain

$$\exp\left[-\frac{\mathrm{i}}{\hbar}\frac{e}{c}\chi(\boldsymbol{r})\right]\psi'(\boldsymbol{r},t)=\psi(\boldsymbol{r},t)$$

or

$$\psi(\boldsymbol{r},t)\to\psi'(\boldsymbol{r},t)=\exp\left[\frac{\mathrm{i}}{\hbar}\frac{e}{c}\chi(\boldsymbol{r})\right]\psi(\boldsymbol{r},t)$$

which is exactly equation (21.9).

22. Simple Applications of Propagator Functions

Let us first summarize what we know until now about the Feynman propagator, thinking first, for simplicity, of a one-dimensional system, described by the following Lagrangian:

$$L(x, \dot{x}) = \frac{m}{2}\dot{x}^2 - V(x) \ . \tag{22.1}$$

Then we know that

$$\begin{aligned} 1) \quad K\left(x_f, t_f; x_i, t_i\right) = \int_{x(t_i)=x_i}^{x(t_f)=x_f} & [dx(t)] \\ & \times \exp\left\{\frac{\mathrm{i}}{\hbar}\int_{t_i}^{t_f} dt\left[\frac{m}{2}\dot{x}^2 - V(x)\right]\right\} \ . \end{aligned} \tag{22.2}$$

$$2) \quad \mathrm{i}\hbar\frac{\partial}{\partial t_f}K\left(x_f, t_f; x_i, t_i\right) = \left[-\frac{\hbar^2}{2m}\frac{\partial^2}{\partial x_f^2} + V\left(x_f\right)\right]K\left(x_f, t_f; x_i, t_i\right) \ . \tag{22.3}$$

$$K\left(x_f, t_i; x_i, t_i\right) = \delta\left(x_f - x_i\right) \ .$$

$$3) \quad K\left(x_f, t_f; x_i, t_i\right) = \sum_{n=0}^{\infty}\phi_n\left(x_f\right)\phi_n^*\left(x_i\right)\mathrm{e}^{-(\mathrm{i}/\hbar)E_n(t_f - t_i)} \ . \tag{22.4}$$

We have already seen in some examples (particle in a square well, or constrained to move on a ring) that the representation (22.4) exists. More generally, (22.4) can be shown as follows: we know that the propagator for fixed x_i, t_i solves the Schrödinger equation. This Schrödinger function can be decomposed as follows:

$$\begin{aligned} \psi_{(x_i,t_i)}\left(x_f, t_f\right) = K\left(x_f, t_f; x_i, t_i\right) &= \sum_n \left\langle x_f\right| \mathrm{e}^{-(\mathrm{i}/\hbar)Ht_f}\left|\left(|n\rangle\langle n|\right)\right| x_i, t_i\big\rangle \\ &= \sum_n a_n\left(x_i, t_i\right)\psi_n\left(x_f\right)\mathrm{e}^{-(\mathrm{i}/\hbar)E_n t_f} \ . \end{aligned}$$

For $t_f = t_i$ we have the condition

$$\delta\left(x_f - x_i\right) = \sum_n a_n\left(x_i, t_i\right)\psi_n\left(x_f\right)\mathrm{e}^{-(\mathrm{i}/\hbar)E_n t_i} \ .$$

Since the left-hand side is time-independent, we are forced to choose

$$a_n(x_i, t_i) = \psi_n^*(x_i)\, \mathrm{e}^{+(\mathrm{i}/\hbar)E_n t_i} \,.$$

Check:

$$\delta(x_f - x_i) = \sum_n \psi_n^*(x_i)\psi_n(x_f) = \sum_n \langle x_f|n\rangle\langle n|x_i\rangle$$

$$= \langle x_f| \underbrace{\left(\sum_n |n\rangle\langle n|\right)}_{=1} |x_i\rangle = \delta(x_f - x_i) \,.$$

With (22.4) the propagator can be calculated if one knows the wave function and the energy spectrum. For example, it holds for the free particle that

$$K(x_2, t_2; x_1, t_1) = \int dp\, \psi_p^*(x_1)\psi_p(x_2) \exp\left[-\frac{\mathrm{i}}{\hbar}\frac{p^2}{2m}(t_2 - t_1)\right]$$

$$\psi_p(x) = \frac{1}{\sqrt{2\pi\hbar}}\mathrm{e}^{(\mathrm{i}/\hbar)xp} : \quad = \int \frac{dp}{2\pi\hbar} \exp^{(\mathrm{i}/\hbar)p(x_2-x_1)} \exp\left[-\frac{\mathrm{i}}{\hbar}\frac{p^2}{2m}(t_2 - t_1)\right]$$

$$\underset{(16.29)}{=} \left(\frac{m}{2\pi\mathrm{i}\hbar(t_2 - t_1)}\right)^{1/2} \exp\left[\frac{\mathrm{i}}{\hbar}\frac{m}{2}\frac{(x_2 - x_1)^2}{t_2 - t_1}\right] \,.$$

Now let us consider the time-development operator

$$\mathrm{e}^{-(\mathrm{i}/\hbar)HT} \,. \tag{22.5}$$

In coordinate representation, it is given by

$$\langle x_2|\mathrm{e}^{-(\mathrm{i}/\hbar)HT}|x_1\rangle$$

where

$$x|x_{\frac{1}{2}}\rangle = x_{\frac{1}{2}}|x_{\frac{1}{2}}\rangle \,.$$

Let $|n\rangle$ be a complete system of states of the Hamiltonian H:

$$H|n\rangle = E_n|n\rangle \,, \quad \sum_n |n\rangle\langle n| = \mathbb{1} \,.$$

Then we expand any state, e.g., the eigenstate vector $|x_1\rangle$, according to

$$|x_1\rangle = \sum_n |n\rangle\langle n|x_1\rangle = \sum_n \psi_n^*(x_1)|n\rangle \,.$$

Furthermore, we use

$$\mathrm{e}^{-(\mathrm{i}/\hbar)HT}|n\rangle = \mathrm{e}^{-(\mathrm{i}/\hbar)E_n T}|n\rangle$$

so that

$$\begin{aligned}\langle x_2|e^{-(i/\hbar)HT}|x_1\rangle &= \sum_n \langle x_2|(|n\rangle\langle n|)\,e^{-(i/\hbar)HT}|x_1\rangle \\ &= \sum_n \psi_n^*(x_1)\psi_n(x_2)\,e^{-(i/\hbar)E_nT} .\end{aligned} \tag{22.6}$$

Then we arrive at the following important result:

4) The propagator can be written as matrix element of the time-development operator $\exp[-(i/\hbar)HT]$:

$$\begin{aligned}K(x_2,T;x_1,0) &= \langle x_2|e^{-(i/\hbar)HT}|x_1\rangle \underset{T=t_2-t_1}{=} \langle x_2|e^{-(i/\hbar)H(t_2-t_1)}|x_1\rangle \\ &= \left(\langle x_2|e^{-(i/\hbar)Ht_2}\right)\left(e^{(i/\hbar)Ht_1}|x_1\rangle\right)\end{aligned}$$

which implies

$$\langle x_2,t_2| = \langle x_2|e^{-(i/\hbar)Ht_2} , \quad |x_1,t_1\rangle = e^{(i/\hbar)Ht_1}|x_1\rangle . \tag{22.7}$$

We now want to use the explicit form of the propagator in order to find the energy spectrum of a particle in a potential. To this end we consider the trace of the time-development operator:

$$\begin{aligned}G(T) &= \mathrm{Tr}\left\{e^{-(i/\hbar)HT}\right\} = \int dx_0\,\langle x_0|e^{-(i/\hbar)HT}|x_0\rangle \\ &\underset{4)}{=} \int dx_0\,K(x_0,T;x_0,0) \underset{3)}{=} \int dx_0 \sum_{n=0}^{\infty} |\phi_n(x_0)|^2\,e^{-(i/\hbar)E_nT} \\ &= \sum_{n=0}^{\infty} \langle n|\underbrace{\left(\int dx_0|x_0\rangle\langle x_0|\right)}_{=1}|n\rangle e^{-(i/\hbar)E_nT} = \sum_{n=0}^{\infty} e^{-(i/\hbar)E_nT} ,\end{aligned}$$

i.e., we have found

$$G(T) = \mathrm{Tr}\left\{e^{-(i/\hbar)HT}\right\} = \sum_{n=0}^{\infty} e^{-(i/\hbar)E_nT} . \tag{22.8}$$

With the Fourier transform ($\mathrm{Im}\,E > 0$),

$$G(E) = \frac{i}{\hbar}\int_0^{\infty} dT\,e^{(i/\hbar)ET}G(T)$$

we immediately obtain

$$G(E) = \sum_{n=0}^{\infty} \frac{i}{\hbar}\int_0^{\infty} dT\,e^{(i/\hbar)(E-E_n)T} = \sum_{n=0}^{\infty} \frac{1}{E_n - E} . \tag{22.9}$$

Since we have the propagator for the harmonic oscillator at hand, we ought to be able to calculate the energy spectrum by forming the trace:

$$
\begin{aligned}
G(T) &= \int dx_0\, K\big(x_0, T; x_0, 0\big) = \sqrt{\frac{m\omega}{2\pi \mathrm{i}\hbar \sin(\omega T)}} \int dx_0 \\
&\quad \times \exp\left[\frac{\mathrm{i}}{\hbar}\frac{m\omega}{\sin(\omega T)} x_0^2 \big(\cos(\omega T) - 1\big)\right] \\
&= \frac{1}{\sqrt{2(\cos(\omega T) - 1)}} = \frac{1}{2\mathrm{i}\,\sin(\omega T/2)} = \frac{\mathrm{e}^{-\mathrm{i}\omega T/2}}{1 - \mathrm{e}^{-\mathrm{i}\omega T}} \\
&= \mathrm{e}^{-\mathrm{i}\omega T/2} \sum_{n=0}^{\infty} \mathrm{e}^{-\mathrm{i}n\omega T} = \sum_{n=0}^{\infty} \mathrm{e}^{-\mathrm{i}(n+1/2)\omega T} .
\end{aligned}
\tag{22.10}
$$

Comparison with (22.8) yields

$$
E_n^{\mathrm{H.O.}} = \left(n + \tfrac{1}{2}\right) \hbar\omega , \quad n = 0,\, 1,\, 2 \ldots .
\tag{22.11}
$$

From the above, something can also be learned about the free propagator:

$$
K(x, T; 0, 0) = \left(\frac{m}{2\pi \mathrm{i}\hbar T}\right)^{1/2} \exp\left[\frac{\mathrm{i}}{\hbar}\frac{m}{2}\frac{x^2}{T}\right] .
$$

The Fourier transform is given by

$$
G(x; E) = \frac{\mathrm{i}}{\hbar} \int dT \left(\frac{m}{2\pi \mathrm{i}\hbar T}\right)^{1/2} \exp\left[\frac{\mathrm{i}}{\hbar}\left(ET + \frac{mx^2}{2T}\right)\right] .
$$

With the aid of the identity

$$
\int_0^{\infty} du \exp\left[-\frac{a}{u^2} - bu^2\right] = \sqrt{\frac{\pi}{4b}}\, \mathrm{e}^{-2\sqrt{ab}} , \quad T =: u^2
$$

we can easily show that

$$
G(x; E) = \frac{\mathrm{i}}{\hbar} \sqrt{\frac{m}{2E}} \exp\left[-\frac{\mathrm{i}}{\hbar}\sqrt{2mE}x\right] ,
$$

which gives the well-known branch point in the E-plane at $E = 0$.

In the case of the harmonic oscillator, the following statements can be made. First of all, we recall that

$$
K(x', t; x'', 0) = \sum_{n=0}^{\infty} \psi_n(x')\psi_n^*(x'')\, \mathrm{e}^{-\mathrm{i}(n+1/2)\omega t} .
\tag{22.12}
$$

Then it can be shown that K is periodic in t with the period $(2\pi/\omega)$:

$$
K(x', t + m2\pi/\omega; x'') = (-1)^m K(x', t; x'') .
\tag{22.13}
$$

Proof:

$$K(x', t + m2\pi/\omega; x'') = \sum_n \psi_n(x')\psi_n^*(x'')\,\mathrm{e}^{-\mathrm{i}(n+1/2)\omega(t+m2\pi/\omega)} .$$

Now we write

$$\begin{aligned}\mathrm{e}^{-\mathrm{i}(n+1/2)\omega m2\pi/\omega} &= \mathrm{e}^{-\mathrm{i}(n+1/2)2m\pi} \\ &= \underbrace{\left(\mathrm{e}^{-\mathrm{i}m\pi}\right)}_{=(-1)^m}{}^{(2n+1)} = (-1)^m(-1)^{m2n} = (-1)^m .\end{aligned}$$

Therefore, we indeed obtain

$$K(x', t + m2\pi/\omega; x'') = (-1)^m K(x', t; x'') .$$

Furthermore, it holds that

$$K\big(x', t + (2m+1)\pi/\omega; x''\big) = (-\mathrm{i})^{2m+1} K(x', t; -x'') \tag{22.14}$$

since

$$K\big(x', t + (2m+1)\pi/\omega; x''\big) = \sum_n \psi_n(x')\psi_n^*(x'')\,\mathrm{e}^{-\mathrm{i}(n+1/2)[\omega t+(2m+1)\pi]} .$$

Because

$$\begin{aligned}\mathrm{e}^{-\mathrm{i}(n+1/2)(2m+1)\pi} &= \mathrm{e}^{-\mathrm{i}(n+1/2)2m\pi}\,\mathrm{e}^{-\mathrm{i}(n+1/2)\pi} \\ &= (-1)^m\,\mathrm{e}^{-\mathrm{i}n\pi}\,\mathrm{e}^{-\mathrm{i}\pi/2} = (-1)^m(-1)^n(-\mathrm{i}) \\ &= (-\mathrm{i})^{2m}(-\mathrm{i})(-1)^n = (-\mathrm{i})^{2m+1}(-1)^n\end{aligned}$$

we can continue to write

$$(-\mathrm{i})^{2m+1}\sum_n \psi_n(x')(-1)^n\psi_n^*(x'')\,\mathrm{e}^{-\mathrm{i}(n+1/2)\omega t} .$$

Now we know from the $\psi_n(x)$ ($\propto$ Hermite polynomials) that they are even for even n and odd for odd n:

$$\psi_n(-x) = (-1)^n\psi_n(x) \ :\ \psi_n^*(-x) = (-1)^n\psi_n^*(x) .$$

Then we immediately get

$$\begin{aligned}K\big(x', t + (2m+1)\pi/\omega; x''\big) &= (-\mathrm{i})^{2m+1}\sum_{n=0}^{\infty}\psi_n(x')\psi_n^*(-x'')\,\mathrm{e}^{-\mathrm{i}(n+1/2)\omega t} \\ &= (-\mathrm{i})^{2m+1}K(x', t; -x'') . \quad \text{q.e.d.}\end{aligned}$$

The physical meaning of (22.13) and (22.14) becomes clear when looking at

$$\begin{aligned}\langle x' t|\psi\rangle = \psi(x', t) &= \int \langle x', t|x'', 0\rangle dx'' \langle x'', 0|\psi\rangle \\ &= \int dx''\, K(x', t; x'', 0)\psi(x'', 0) ,\end{aligned}$$

which yields

$$\begin{aligned}\psi\big(x, t_0 + m2\pi/\omega\big) &= \int dx''\, K\big(x, t_0 + m2\pi/\omega; x''\big)\psi(x'', 0) \\ &\underset{(13)}{=} (-1)^m \int dx''\, K\big(x, t_0; x''\big)\psi(x'', 0) \\ &= (-1)^m \psi\big(x, t_0\big) . \end{aligned} \tag{22.15}$$

On the other hand, (22.14) gives us

$$\begin{aligned}\psi\big(x, t_0 + (2m+1)\pi/\omega\big) &= \int dx''\, K\big(x, t_0 + (2m+1)\pi/\omega; x''\big)\psi(x'', 0) \\ &= (-\mathrm{i})^{2m+1} \int dx''\, K^{\mathrm{H.O.}}\big(x, t_0; -x''\big)\psi(x'', 0) \\ &= (-\mathrm{i})^{2m+1} \int dx''\, K^{\mathrm{H.O.}}\big(-x, t_0; x''\big)\psi(x'', 0) \\ &= (-\mathrm{i})^{2m+1}\psi\big(-x, t_0\big) \qquad (22.16) \\ &= \frac{1}{\mathrm{i}}(-1)^m \psi\big(-x, t_0\big) .\end{aligned}$$

We now present a highly interesting application of the just-derived formulae – valid only for the oscillator potential.

Let $\psi(x, t_0)$ be the wave function of a particle centered around $x_0 = x(t_0)$, and let this particle move with an average momentum p_0, so that we can write a wave packet of the form

$$\psi\big(x, t_0\big) = \mathrm{e}^{(\mathrm{i}/\hbar)p_0 x}\, f\big(x - x(t_0)\big) . \tag{22.17}$$

f is real and takes its maximum when the argument is zero. After a time interval of π/ω, i.e., corresponding to one-half of the classical period, (22.16) tells us that, with $m = 0$,

$$\psi\big(x, t_0 + \pi/\omega\big) = -\mathrm{i}\psi\big(-x, t_0\big) = -\mathrm{i}\,\mathrm{e}^{-(\mathrm{i}/\hbar)p_0 x} f\big(-x - x_0\big) , \tag{22.18}$$

so that the wave packet is now centered around $x = -x_0$, *unchanged* in form, and where its initial average momentum is now turned around: $-p_0$. After one period, $2\pi/\omega$, (22.15) tells us, with $m = 1$,

$$\psi\big(x, t_0 + 2\pi/\omega\big) = (-1)\psi\big(x, t_0\big) = -\mathrm{e}^{(\mathrm{i}/\hbar)p_0 x} f\big(x - x_0\big) , \tag{22.19}$$

so that the wave packet has again reached its initial state, unchanged in shape and with its initial average momentum, p_0. This motion is repeated arbitrarily often, whereby the wave packet moves like a classical particle. One should note that this conclusion requires no special form of the wave packet; it applies to every wave packet (cf. special case of the Gaussian wave packet – ground state wave function of the harmonic oscillator – studied by Schrödinger himself). We want to retrace Schrödinger's calculations, considering, at time $t = 0$, the wave function

$$\psi(x,0) = \left(\frac{\alpha}{\pi}\right)^{1/4} \exp\left[-\frac{\alpha}{2}(x-x_0)^2 + \mathrm{i}k_0 x\right] , \quad \alpha := \frac{m\omega}{\hbar} .$$

At a later time, $t > 0$, ψ develops as

$$\begin{aligned}
\psi(x,t) &= \int dx'\, K^{\mathrm{H.O.}}(x,t;x',0)\psi(x',0) \qquad (22.20)\\
&= \left(\frac{\alpha}{\hbar}\right)^{1/4} \left(\frac{m\omega}{2\pi\mathrm{i}\hbar\sin(\omega t)}\right)^{1/2} \int dx' \\
&\quad\times \exp\left\{\frac{im\omega}{2\hbar}\left[(x^2+x'^2)\cot(\omega t) - \frac{2xx'}{\sin(\omega t)}\right] - \frac{\alpha}{2}(x'-x_0)^2 + ik_0x'\right\}\\
&= \left(\frac{\alpha}{\pi}\right)^{1/4} \left(\frac{m\omega}{2\pi\mathrm{i}\hbar\sin(\omega t)}\right)^{1/2} \exp\left\{\frac{im\omega}{2\hbar}x^2\cot(\omega t) - \frac{\alpha}{2}x_0^2\right\}\\
&\quad\times \int dx' \exp\left\{\left[\frac{im\omega}{2\hbar}\cot(\omega t) - \frac{\alpha}{2}\right]x'^2\right.\\
&\quad\left.-\left[\frac{im\omega}{\hbar\sin(\omega t)}x - \alpha x_0 - ik_0\right]x'\right\}\\
&= \left(\frac{\alpha}{\pi}\right)^{1/4} \left(\frac{m\omega}{2\pi\mathrm{i}\hbar\sin(\omega t)}\right)^{1/2}\\
&\quad\times \exp\left\{\frac{im\omega}{2\hbar}x^2\cot(\omega t) - \frac{\alpha}{2}x^2 - \frac{1}{2}\frac{\left(\frac{im\omega x}{\hbar\sin(\omega t)} - \mathrm{i}k_0 - \alpha x_0\right)^2}{\frac{im\omega}{\hbar}\cot(\omega t) - \alpha}\right\}\\
&\quad\times \int dx' \exp\left\{\left[\frac{im\omega}{2\hbar}\cot(\omega t) - \frac{\alpha}{2}\right]\left(x' - \frac{\frac{im\omega x}{\hbar\sin(\omega t)} - \mathrm{i}k_0 - \alpha x_0}{\frac{im\omega}{\hbar}\cot(\omega t) - \alpha}\right)^2\right\}\\
&= \left(\frac{\alpha}{\pi}\right)^{1/4} \left(\frac{m\omega}{2\pi\mathrm{i}\hbar\sin(\omega t)}\right)^{1/2} \left(\frac{\pi}{\frac{\alpha}{2} - \frac{im\omega}{2\hbar}\cot(\omega t)}\right)^{1/2}\\
&\quad\times \exp\left[\frac{im\omega}{2\hbar}\cot(\omega t)x^2 - \frac{\alpha}{2}x_0^2 + \frac{1}{2}\frac{\alpha + \frac{im\omega}{\hbar}\cot(\omega t)}{\alpha^2 + \left(\frac{m\omega}{\hbar}\cot(\omega t)\right)^2}\right.\\
&\quad\left.\times\left(\frac{im\omega x}{\hbar\sin(\omega t)} - \mathrm{i}k_0 - \alpha x_0\right)^2\right] . \qquad (22.21)
\end{aligned}$$

If we set $\alpha = m\omega/\hbar$, $k_0 = 0$, $x_0 = a$, then we get for this special case of an initial Gaussian probability distribution centered around $x_0 = a$:

$$\begin{aligned}\psi(x,t) &= \left(\frac{\alpha}{\pi}\right)^{1/4}\left(\frac{\alpha}{2\pi \mathrm{i}\sin(\omega t)}\right)^{1/2}\left(\frac{2\pi}{\alpha(1-\mathrm{i}\cot(\omega t))}\right)^{1/2} \\ &\quad \times \exp\left[\frac{\mathrm{i}\alpha}{2}x^2\cot(\omega t) - \frac{\alpha}{2}a^2 + \frac{\alpha}{2}\frac{1+\mathrm{i}\cot(\omega t)}{1+\cot^2(\omega t)}\left(\frac{\mathrm{i}x}{\sin(\omega t)} - a\right)^2\right] \\ &= \left(\frac{\alpha}{\pi}\right)^{1/4} \exp\left[-\frac{\mathrm{i}\omega t}{2} - \frac{\alpha}{2}x^2 + \alpha a x\,\mathrm{e}^{-\mathrm{i}\omega t} - \frac{\alpha}{4}a^2\left(1+\mathrm{e}^{-2\mathrm{i}\omega t}\right)\right] \\ &= \left(\frac{m\omega}{\pi\hbar}\right)^{1/4} \exp\left\{-\frac{\mathrm{i}\omega t}{2} - \frac{m\omega}{2\hbar}\left[x^2 - 2ax\,\mathrm{e}^{-\mathrm{i}\omega t}\right.\right. \\ &\quad \left.\left. + \frac{1}{2}a^2\left(1+\mathrm{e}^{-2\mathrm{i}\omega t}\right)\right]\right\} . \end{aligned} \tag{22.22}$$

In particular, we obtain for $t = n2\pi/\omega$:

$$\begin{aligned}\psi\left(x, n\frac{2\pi}{\omega}\right) &= \left(\frac{m\omega}{\pi\hbar}\right)^{1/4} \exp\left[-\mathrm{i}n\pi - \frac{m\omega}{2\hbar}(x^2 - 2ax + a^2)\right] \\ &= (-1)^n \psi(x,0)\end{aligned}$$

and for $t = (n+1/2)2\pi/\omega$:

$$\begin{aligned}\psi\left(x, \left(n+\frac{1}{2}\right)\frac{2\pi}{\omega}\right) &= \left(\frac{m\omega}{\pi\hbar}\right)^{1/4} \exp\left[-\mathrm{i}\pi\left(n+\frac{1}{2}\right) - \frac{m\omega}{2\hbar}(x^2 + 2ax + a^2)\right] \\ &= \frac{1}{\mathrm{i}}(-1)^n \psi(-x,0) .\end{aligned}$$

In general we obtain:

$$\psi\left(x, \frac{\pi}{\omega}n\right) = \left(\frac{1}{\mathrm{i}}\right)^n \psi((-1)^n x, 0) .$$

From this we learn that

$$K\left(x, n\frac{\pi}{\omega}; x', 0\right) = \left(\frac{1}{\mathrm{i}}\right)^n \delta\left(x - (-1)^n x'\right) ,$$

which is known to us from p. 215.

We now split $\psi(x,t)$ into its modulus and the phase, in order to study the shape of $\psi(x,t)$:

$$
\begin{aligned}
\psi(x,t) &= \left(\frac{m\omega}{\pi\hbar}\right)^{1/4} \exp\Bigg\{-\frac{\mathrm{i}\omega t}{2} - \frac{m\omega}{2\hbar}\Bigg[x^2 - 2ax\big(\cos(\omega t) - \mathrm{i}\,\sin(\omega t)\big) \\
&\quad + \frac{1}{2}a^2 \underbrace{\Big(\underbrace{1 + \underbrace{\cos(2\omega t)}_{\cos^2(\omega t)-\sin^2(\omega t)}}_{2\cos^2(\omega t)} - \mathrm{i}\underbrace{\sin(2\omega t)}_{2\sin(\omega t)\cos(\omega t)}\Big)}_{2\cos^2(\omega t) - 2\mathrm{i}\,\sin(\omega t)\cos(\omega t)}\Bigg]\Bigg\} \\
&= \left(\frac{m\omega}{\pi\hbar}\right)^{1/4} \exp\left[-\frac{\mathrm{i}\omega t}{2} - \mathrm{i}\frac{m\omega}{\hbar}\sin(\omega t)\left(ax - \frac{a^2}{2}\cos(\omega t)\right)\right] \\
&\quad \times \exp\left[-\frac{m\omega}{2\hbar}\big(x^2 - 2ax\,\cos(\omega t) + a^2\,\cos^2(\omega t)\big)\right] \\
&= \left(\frac{m\omega}{\pi\hbar}\right)^{1/4} \exp\left[-\frac{\mathrm{i}\omega t}{2} - \mathrm{i}\frac{m\omega}{\hbar}\sin(\omega t)\left(ax - \frac{a^2}{2}\cos(\omega t)\right)\right] \\
&\quad \times \exp\left[-\frac{m\omega}{2\hbar}\big(x - a\,\cos(\omega t)\big)^2\right] . \qquad (22.23)
\end{aligned}
$$

Apart from the complicated phase factor, $\psi(x,t)$ has the same form as (22.20) with $k_0 = 0$, where it now holds that

$$x_0 = a\,\cos(\omega t) .$$

The corresponding probability distribution reads, therefore, simply

$$P(x,t) = |\psi(x,t)|^2 = \left(\frac{m\omega}{\pi\hbar}\right)^{1/2} \exp\left[-\frac{m\omega}{\hbar}\big(x - a\,\cos(\omega t)\big)^2\right] . \qquad (22.24)$$

This is still (for $t > 0$) a Gaussian distribution, only this time centered around $x_0 = a\,\cos(\omega t)$. We are dealing here with a highly interesting result: the wave packet oscillates back and forth, following the same path as a classical particle. For $a = 0$, (22.24) implies a stationary probability distribution:

$$a = 0: \quad P(x,t) = \left(\frac{m\omega}{\pi\hbar}\right)^{1/2} \exp\left[-\frac{m\omega}{\hbar}x^2\right] . \qquad (22.25)$$

This corresponds to a particle that is sitting on the bottom of the harmonic oscillator potential. The associated ground state is found from (22.23) with $a = 0$:

$$
\begin{aligned}
\psi(x,t) &= \left(\frac{m\omega}{\pi\hbar}\right)^{1/4} \exp\left[-\frac{m\omega}{2\hbar}x^2\right] \exp\left[-\mathrm{i}\frac{\omega}{2}t\right] \qquad (22.26) \\
&= \psi(x,0)\,\mathrm{e}^{-\mathrm{i}\omega t/2} = \psi(x,0)\,\mathrm{e}^{\ (\mathrm{i}/\hbar)E_0} , \quad E_0 = \frac{\hbar\omega}{2} .
\end{aligned}
$$

We now want to determine the lowest eigenfunctions from the propagator of the harmonic oscillator. Let us recall

$$K(x',t;x'',0) = \left(\frac{m\omega}{2\pi i\hbar\,\sin(\omega t)}\right)^{1/2} \times \exp\left\{\frac{im\omega}{2\hbar\,\sin(\omega t)}\left[\left(x'^2+x''^2\right)\cos(\omega t)-2x'x''\right]\right\}$$

and use here

$$2i\,\sin(\omega t) = e^{i\omega t}\left(1-e^{-2i\omega t}\right),$$
$$2\,\cos(\omega t) = e^{i\omega t}\left(1+e^{-2i\omega t}\right).$$

This yields, in K:

$$K(x',t;x'',0) = \left(\frac{m\omega}{\pi\hbar}\right)^{1/2} e^{-i\omega t/2}\left(1-e^{-2i\omega t}\right)^{-1/2} \times \exp\left\{\frac{m\omega}{\hbar}\left[-\left(x'^2+x''^2\right)\frac{1+e^{-2i\omega t}}{2(1-e^{-2i\omega t})} +2x'x''\frac{e^{-i\omega t}}{1-e^{-2i\omega t}}\right]\right\}.$$

Now we make use of the series expansion

$$(1-x)^{-1/2} = 1+\tfrac{1}{2}x+\ \dots\,, \qquad (1-x)^{-1} = 1+x+\ \dots \qquad |x|<1$$

and write

$$K(x',t;x'',0) = \left(\frac{m\omega}{\pi\hbar}\right)^{1/2} e^{-i\omega t/2}\left(1+\frac{1}{2}e^{-2i\omega t}+\ \dots\right) \times \exp\left\{\frac{m\omega}{\hbar}\left[-\frac{1}{2}\left(x'^2+x''^2\right)\left(1+e^{-2i\omega t}\right)\left(1+e^{-2i\omega t}+\ \dots\right) + 2x'x''\,e^{-i\omega t}\left(1+e^{-2i\omega t}+\ \dots\right)\right]\right\}.$$

Expanding up to quadratic terms yields

$$= \left(\frac{m\omega}{\pi\hbar}\right)^{1/2} e^{-i\omega t/2}\left(1+\frac{1}{2}e^{-2i\omega t}+\ \dots\right) \times \exp\left[-\frac{m\omega}{2\hbar}\left(x'^2+x''^2\right)\right]\exp\left[-\frac{m\omega}{\hbar}\left(x'^2+x''^2\right)e^{-2i\omega t} +\frac{2m\omega}{\hbar}x'x''\,e^{-i\omega t}\right].$$

At this point we also expand $\exp[\dots e^{-in\omega t}]$ and so obtain

$$= \left(\frac{m\omega}{\pi\hbar}\right)^{1/2} e^{-i\omega t/2}\left(1+\frac{1}{2}e^{-2i\omega t}\right)\exp\left[-\frac{m\omega}{2\hbar}\left(x'^2+x''^2\right)\right] \times\left(1-\frac{m\omega}{\hbar}\left(x'^2+x''^2\right)e^{-2i\omega t}+\frac{2m\omega}{\hbar}x'x''\,e^{-i\omega t} +\frac{2m^2\omega^2}{\hbar^2}x'^2x''^2\,e^{-2i\omega t}+\ \dots\right). \tag{22.27}$$

Using $E_n = (n + 1/2)\hbar\omega$ and (22.12),

$$K(x', t; x'', 0) = \sum_{n=0}^{\infty} \psi_n(x')\psi_n^*(x'')\,\mathrm{e}^{-(\mathrm{i}/\hbar)E_n t}$$

and comparing the first terms with (22.27), we obtain:

$$\begin{aligned}
n = 0 : E_0 = \frac{1}{2}\hbar\omega;\quad & \left(\frac{m\omega}{\pi\hbar}\right)^{1/2} \exp\left[-\frac{m\omega}{2\hbar}(x'^2 + x''^2)\right] \mathrm{e}^{-\mathrm{i}\omega t/2} \\
&\equiv \psi_0(x')\psi_0^*(x'')\,\mathrm{e}^{-(\mathrm{i}/\hbar)E_0 t} \\
&\Rightarrow \psi_0(x) = \left(\frac{m\omega}{\pi\hbar}\right)^{1/4} \exp\left[-\frac{m\omega}{2\hbar}x^2\right] . \\
n = 1 : E_1 = \frac{3}{2}\hbar\omega;\quad & \left(\frac{m\omega}{\pi\hbar}\right)^{1/2} \frac{2m\omega}{\hbar}x'x'' \exp\left[-\frac{m\omega}{2\hbar}(x'^2 + x''^2)\right] \mathrm{e}^{-3\mathrm{i}\omega t/2} \\
&\equiv \psi_1(x')\psi_1^*(x'')\,\mathrm{e}^{-(\mathrm{i}/\hbar)E_1 t} \\
&\Rightarrow \psi_1(x) = \left(\frac{m\omega}{\pi\hbar}\right)^{1/4} \sqrt{\frac{2m\omega}{\hbar}}\,x \exp\left[-\frac{m\omega}{2\hbar}x^2\right] \\
n = 2 : E_2 = \frac{5}{2}\hbar\omega;\quad & \left(\frac{m\omega}{\pi\hbar}\right)^{1/2} \exp\left[-\frac{m\omega}{2\hbar}(x'^2 + x''^2)\right] \\
&\times \left(\frac{1}{2} - \frac{m\omega}{\hbar}(x'^2 + x''^2) + \frac{2m^2\omega^2}{\hbar^2}x'^2x''^2\right) \mathrm{e}^{-5\mathrm{i}\omega t/2} \\
&= \left(\frac{m\omega}{\pi\hbar}\right)^{1/2} \exp\left[-\frac{m\omega}{2\hbar}(x'^2 + x''^2)\right] \frac{1}{2}\left(\frac{2m\omega}{\hbar}x'^2 - 1\right) \\
&\times \left(\frac{2m\omega}{\hbar}x''^2 - 1\right) \mathrm{e}^{-5\mathrm{i}\omega t/2} \\
&\equiv \psi_2(x')\psi_2^*(x'')\,\mathrm{e}^{-(\mathrm{i}/\hbar)E_2 t} \\
&\Rightarrow \psi_2(x) = \frac{1}{\sqrt{2}}\left(\frac{m\omega}{\pi\hbar}\right)^{1/4} \left(\frac{2m\omega}{\hbar}x^2 - 1\right) \exp\left[-\frac{m\omega}{2\hbar}x^2\right] .
\end{aligned}$$

From the quantum mechanics of the harmonic oscillator, one gets for the eigenfunctions

$$\psi_n(x) = \frac{1}{\sqrt{2^n n!}}\left(\frac{m\omega}{\pi\hbar}\right)^{1/4} H_n\left(\sqrt{\frac{m\omega}{\hbar}}\,x\right) \mathrm{e}^{-m\omega x^2/2\hbar} \tag{22.28}$$

with the Hermite polynomials

$$H_0(\xi) = 1 , \quad H_1(\xi) = 2\xi , \quad H_2(\xi) = 4\xi^2 - 2 , \ \ldots .$$

Knowing the propagator functions e.g., for the free particle or for the particle in the harmonic oscillator potential, we are now in a position to quickly give the density matrix – in configuration space, for example. This can simply be achieved by going over to the propagator with “imaginary time”, i.e., by the substitution

$$t \to \frac{1}{\mathrm{i}}\beta\hbar\,, \quad \beta = \frac{1}{kT}\,.$$

$$L = \frac{m}{2}\dot{x}^2 : \quad K(x', t; x'', 0) = \sqrt{\frac{m}{2\pi \mathrm{i}\hbar t}} \exp\left[\frac{\mathrm{i}}{\hbar}\frac{m}{2}\frac{(x'-x'')^2}{t}\right]$$

$$\Rightarrow \varrho(x', x''; \beta) = \sqrt{\frac{m}{2\pi\hbar^2\beta}} \exp\left[-\frac{m}{2\hbar^2\beta}(x'-x'')^2\right]\,. \tag{22.29}$$

For the important case, $L = m\dot{x}^2/2 - m\omega^2 x^2/2$, we get from

$$K(x', t; x'', 0) = \sqrt{\frac{m\omega}{2\pi\mathrm{i}\hbar\,\sin(\omega t)}} \exp\left\{\frac{\mathrm{i}}{\hbar}\frac{m\omega}{2\sin(\omega t)}\left[(x'^2 + x''^2)\right.\right.$$
$$\left.\left.\times \cos(\omega t) - 2x'x''\right]\right\}$$

$$\varrho(x', x''; \beta) = \sqrt{\frac{m\omega}{2\pi\hbar\,\sinh(\hbar\omega\beta)}} \exp\left\{-\frac{m\omega}{2\hbar\,\sinh(\hbar\omega\beta)}\left[(x'^2 + x''^2)\right.\right.$$
$$\left.\left.\times \cosh(\hbar\omega\beta) - 2x'x''\right]\right\}\,. \tag{22.30}$$

Here, as a reminder, the most important properties of the density operator.

Let $|a'\rangle$ be a complete orthonormal basis. In this basis the operator A can be represented as

$$A = \sum_{a'} a'|a'\rangle\langle a'|\,. \tag{22.31}$$

If we now take the number $w_{a'}$ for the numbers a', where $w_{a'}$ is the probability of finding the system in the state $|a'\rangle$, then a new operator, the density operator, can be written as:

$$\varrho = \sum_{a'} w_{a'}|a'\rangle\langle a'| \tag{22.32}$$

with

$$w_{a'} \geqslant 0 \quad \text{and} \quad \sum_{a'} w_{a'} = 1\,. \tag{22.33}$$

From (22.32) it is obvious that ϱ is Hermitean: $\varrho = \varrho^\dagger$. Then the expectation value of an operator O can be expressed as

$$\langle O\rangle = \mathrm{Tr}(\varrho O) = \sum_{a,a',a''} \langle a|a'\rangle w_{a'}\langle a'|a''\rangle\langle a''|O|a\rangle \tag{22.34}$$
$$= \sum_{a'} w_{a'}\langle a'|O|a'\rangle = \sum_{a'} w_{a'}\langle O\rangle_{a'}\,,$$

where $\langle a'|O|a'\rangle$ is the expectation value of O in the state $|a'\rangle$. For $O = \mathbb{1}$ we get

$$\mathrm{Tr}(\varrho) = 1 \ . \tag{22.35}$$

Pure states $|a'\rangle$ are those for which $w_{a'} = \delta_{a'a''}$, i.e.,

$$\varrho = |a'\rangle\langle a'| : \quad \varrho^2 = |a'\rangle\langle a'|\,|a'\rangle\langle a'| = |a'\rangle\langle a'| = \varrho \ ,$$

i.e.,

$$\varrho^2 = \varrho \ . \tag{22.36}$$

The expectation value is then simply written as

$$\begin{aligned}\langle O\rangle = \mathrm{Tr}(\varrho O) &= \sum_{a,a'',a'''} \overbrace{\langle a|a'\rangle}^{\delta_{aa'}} \underbrace{\langle a'|a''\rangle}_{\delta_{a'a''}} \langle a''|O|a'''\rangle \overbrace{\langle a'''|a\rangle}^{\delta_{a'''a}} \\ &= \langle a'|O|a'\rangle \ . \end{aligned} \tag{22.37}$$

Let us again recall that the operator ϱ is suitable for describing a system whose probability of being found in the state $|a'\rangle$ is equal to $w_{a'}$. These can be both pure and mixed states, e.g., the orientation of the spin of the silver atoms in the Stern-Gerlach experiment prior to entrance into the inhomogenous magnetic field or an unpolarized beam of photons.

In x-representation we write the density operator as

$$\varrho(x,x') = \langle x'|\varrho|x\rangle = \sum_{a'} \langle x'|a'\rangle w_{a'} \langle a'|x\rangle = \sum_{a'} w_{a'} a'(x') a'^*(x) \ .$$

The expectation value is likewise

$$\begin{aligned}\langle O\rangle &= \int dx \langle x|\varrho|x'\rangle dx' \langle x'|O|x\rangle \\ &= \int dx\, dx'\, \varrho(x,x') O(x,x') = \mathrm{Tr}(\varrho O) \ . \end{aligned} \tag{22.38}$$

If the states change in time, (22.32) becomes

$$\varrho(t) = \sum_{a'} w_{a'} |a'(t)\rangle\langle a'(t)| \ . \tag{22.39}$$

Now H, the Hamilton operator, generates the development in time, so that with

$$H|n\rangle = E_n|n\rangle \ , \quad \sum_n |n\rangle\langle n| = 1\!\!1$$

and

$$|a'\rangle = \sum_n |n\rangle\langle n|a'\rangle$$

as well as (for Schrödinger state kets)

$$|a',t\rangle = \sum_n |n\rangle\langle n \underbrace{|a',t\rangle}_{=\mathrm{e}^{-(\mathrm{i}/\hbar)Ht}|a'\rangle}$$

it follows that

$$|a',t\rangle = \sum_n |n\rangle\,\mathrm{e}^{-(\mathrm{i}/\hbar)E_n t}\,\langle n|a'\rangle\,.$$

If we substitute $|a'(t)\rangle = \mathrm{e}^{-(\mathrm{i}/\hbar)Ht}|a'\rangle$ in (22.39), we then get

$$\begin{aligned}\varrho(t) &= \sum_{a'} w_{a'}\,\mathrm{e}^{-(\mathrm{i}/\hbar)Ht}|a'\rangle\langle a'|\,\mathrm{e}^{(\mathrm{i}/\hbar)Ht}\\ &= \mathrm{e}^{-(\mathrm{i}/\hbar)Ht}\left(\sum_{a'} w_{a'}|a'\rangle\langle a'|\right)\mathrm{e}^{(\mathrm{i}/\hbar)Ht}\\ &= \mathrm{e}^{-(\mathrm{i}/\hbar)Ht}\,\varrho(0)\,\mathrm{e}^{(\mathrm{i}/\hbar)Ht}\,.\end{aligned} \tag{22.40}$$

The time derivative yields

$$\frac{\partial\varrho}{\partial t} = -\frac{\mathrm{i}}{\hbar}H\varrho(t) + \varrho(t)\frac{\mathrm{i}}{\hbar}H$$

or

$$\frac{\partial\varrho}{\partial t} = -\frac{\mathrm{i}}{\hbar}[H,\varrho(t)]\,.$$

By way of illustration, let us consider a canonical ensemble from statistical mechanics. Let $|n\rangle$ and E_n be eigenstate and eigenvalue of the Hamilton operator H. Then the probability of finding the system in state $|n\rangle$ with the energy E_n is given by

$$w_n = \frac{\mathrm{e}^{-\beta E_n}}{\sum_m \mathrm{e}^{-\beta E_m}}\,, \qquad H|n\rangle = E_n|n\rangle\,. \tag{22.41}$$

Then the density operator becomes

$$\varrho = \sum_n w_n|n\rangle\langle n| = \sum_n \frac{\mathrm{e}^{-\beta E_n}}{\sum_m \mathrm{e}^{-\beta E_m}}|n\rangle\langle n| = \frac{\mathrm{e}^{-\beta H}}{\mathrm{Tr}(\mathrm{e}^{-\beta H})}\,.$$

Thus we have

$$\varrho = \frac{\mathrm{e}^{-\beta H}}{\mathrm{Tr}(\mathrm{e}^{-\beta H})} =: \frac{1}{Q}\mathrm{e}^{-\beta H}\,, \tag{22.42}$$

with the partition function

$$Q := \mathrm{Tr}\big(\mathrm{e}^{-\beta H}\big) = \sum_n \mathrm{e}^{-E_n/kT} := \mathrm{e}^{-F/kT} = \mathrm{e}^{-\beta F} \tag{22.43}$$

and the free energy

$$F = -kT \ln Q = -kT \ln \sum_n e^{-E_n/kT} .$$

Incidentally, we have for the entropy

$$S = -k \sum_n W_n \ln W_n \tag{22.44}$$

with

$$W_n = \frac{1}{Q} e^{-E_n/kT} .$$

We now consider the density operator as a function of β:

$$\varrho(\beta) = \frac{e^{-\beta H}}{\mathrm{Tr}(e^{-\beta H})}$$

or, with the non-normalized ϱ:

$$\varrho_u(\beta) = e^{-\beta H} , \quad Q = \mathrm{Tr}(\varrho_u) .$$

In the following, we drop the index u and obtain in the energy representation

$$\varrho_{nm}(\beta) = \delta_{nm}\, e^{-\beta E_n} ,$$

which implies

$$\frac{\partial \varrho_{nm}}{\partial \beta} = \delta_{nm}(-E_n)\, e^{-\beta E_n} = -E_n \varrho_{nm}(\beta)$$

or

$$-\frac{\partial \varrho(\beta)}{\partial \beta} = H\varrho(\beta) , \quad \varrho(O) = 1 . \tag{22.45}$$

In configuration space we thus obtain

$$-\frac{\partial \varrho(x, x'; \beta)}{\partial \beta} = H_x \varrho(x, x'; \beta) ; \quad \varrho(x, x'; 0) = \delta(x - x') . \tag{22.46}$$

For a free particle with $H_x = p_x^2/2m$ we get the differential equation

$$-\frac{\partial \varrho(x, x'; \beta)}{\partial \beta} = -\frac{\hbar^2}{2m} \frac{\partial^2}{\partial x^2} \varrho(x, x'; \beta) ; \quad \varrho(x, x', 0) = \delta(x - x') . \tag{22.47}$$

Note that the substitution $\beta \to (\mathrm{i}/\hbar)t$ brings us back to the Schrödinger equation. This analogy makes it easy to write down the solution of the differential equation (22.47):

$$\varrho(x, x'; \beta) = \sqrt{\frac{m}{2\pi\hbar^2\beta}} \exp\left[-\frac{m}{2\hbar^2\beta}(x - x')^2\right] . \tag{22.48}$$

For the harmonic oscillator with $H_x = p_x^2/2m + m\omega^2 x^2/2$, we obtain likewise

$$-\frac{\partial\varrho}{\partial\beta} = -\frac{\hbar^2}{2m}\frac{\partial^2}{\partial x^2}\varrho + \frac{m\omega^2}{2}x^2\varrho \tag{22.49}$$

with the solution

$$\begin{aligned} \varrho(x, x'; \beta) = &\sqrt{\frac{m\omega}{2\pi\hbar\,\sinh(\hbar\omega/kT)}} \exp\left\{-\frac{m\omega}{2\hbar\,\sinh(\hbar\omega/kT)}\left[\left(x^2 + x'^2\right)\right.\right. \\ &\left.\left.\times \cosh\left(\frac{\hbar\omega}{kT}\right) - 2xx'\right]\right\} . \end{aligned} \tag{22.50}$$

For a free particle, the above result (22.48) originates from the calculation of the path integral ($U = \hbar\beta$)

$$\varrho(x, x'; U) = \int_{x(0)=x'}^{x(U)=x} [dx(u)] \exp\left[-\frac{1}{\hbar}\int_0^U du \frac{m}{2}\dot{x}^2(u)\right] . \tag{22.51}$$

For a particle in the potential V, it holds analogously that

$$\begin{aligned} \varrho(x, x'; U) = &\int_{x(0)=x'}^{x(U)=x} [dx(u)] \\ &\times \exp\left\{-\frac{1}{\hbar}\int_0^U du\left[\frac{m}{2}\dot{x}^2(u) + V\big(x(u)\big)\right]\right\} . \end{aligned} \tag{22.52}$$

The trace is also interesting:

$$\begin{aligned} e^{-\beta F} = Q &= \int dx\, \varrho(x, x; U) \\ &= \int dx \int_{x(0)=x}^{x(U)=x} [dx(u)] \exp\left\{-\frac{1}{\hbar}\int_0^U du\left[\frac{m}{2}\dot{x}^2(u) + V\big(x(u)\big)\right]\right\} \\ &= \int_{\text{all closed paths}} [dx(u)] \\ &\times \exp\left\{-\frac{1}{\hbar}\int_0^U du\left[\frac{m}{2}\dot{x}^2(u) + V\big(x(u)\big)\right]\right\} . \end{aligned} \tag{22.53}$$

This kind of path integral representation of the partion function is frequently used in statistical mechanics.

23. The WKB Approximation

In this chapter we shall develop an important semiclassical method which has come back into favor again, particularly in the last few years, since it permits a continuation into field theory. Here, too, one is interested in nonperturbative methods.

As a starting point we consider the propagation of a particle in a constant field:

$$H = \frac{p^2}{2m} - Fx \,. \tag{23.1}$$

The Heisenberg equations of motion then read

$$\dot{x} = \frac{p}{m} \,, \qquad \dot{p} = F \tag{23.2}$$

with the solutions $(x \equiv x(0),\ p \equiv p(0))$

$$p(t) = p + Ft$$
$$x(t) = x + \frac{p}{m}t + \frac{1}{2}\frac{F}{m}t^2 \,.$$

Note that the first equation is simpler, so that we prefer to work in the p-representation:

$$\mathrm{i}\hbar\frac{\partial}{\partial t}\langle p,t|p',o\rangle = \left\langle p,t\left|\frac{p^2}{2m} - Fx\right|p',0\right\rangle = \left(\frac{p^2}{2m} - F\mathrm{i}\hbar\frac{\partial}{\partial p}\right)\langle p,t|p'\rangle \,.$$

Using

$$\langle p,t|p'\rangle = \langle p|\mathrm{e}^{-(\mathrm{i}/\hbar)Ht}|p'\rangle = \int \underbrace{\langle p|E\rangle}_{\psi(p)} dE\, \mathrm{e}^{-(\mathrm{i}/\hbar)Et} \underbrace{\langle E|p'\rangle}_{\psi^*(p')} \tag{23.3}$$

we easily obtain

$$\left(\frac{p^2}{2m} - F\mathrm{i}\hbar\frac{\partial}{\partial p}\right)\psi(p) = E\psi(p) \,. \tag{23.4}$$

This can be rewritten as

$$\frac{\partial}{\partial p}\log\psi(p) = \frac{1}{F\mathrm{i}\hbar}\left(\frac{p^2}{2m} - E\right) \,,$$

which is solved by

$$\psi(p) = C\exp\left[-\frac{\mathrm{i}}{\hbar F}\left(\frac{p^3}{6m} - Ep\right)\right] = \langle p|E\rangle\,.$$

The constant is determined by the δ-normalization in E:

$$\begin{aligned}\delta(E-E') &= \int \langle E|p\rangle dp\langle p|E'\rangle = |C|^2\int dp\exp\left[-\frac{\mathrm{i}}{\hbar F}(E-E')p\right]\\ &= |C|^2 2\pi\hbar F\delta(E-E')\end{aligned}$$

so that

$$C = \frac{1}{\sqrt{2\pi\hbar F}}$$

and

$$\psi(p) = \langle p|E\rangle = \frac{1}{\sqrt{2\pi\hbar F}}\exp\left[\frac{\mathrm{i}}{\hbar F}\left(Ep - \frac{p^3}{6m}\right)\right]\,. \tag{23.5}$$

If we substitute this result into (23.3), we obtain

$$\begin{aligned}\langle p,t|p'\rangle &= \exp\left[-\frac{\mathrm{i}}{\hbar 6mF}(p^3-p'^3)\right]\frac{1}{2\pi\hbar F}\int dE\exp\left[\frac{\mathrm{i}}{\hbar F}(p-p'-Ft)E\right]\\ &= \delta(p-p'-Ft)\exp\left[-\frac{\mathrm{i}}{\hbar}\frac{1}{6mF}(p^3-p'^3)\right]\,.\end{aligned} \tag{23.6}$$

In the limit $F \to 0$, we reproduce a well-known result:

$$\begin{aligned}\langle p,t|p',0\rangle &\underset{F\to 0}{\longrightarrow} \delta(p-p'-Ft)\exp\left[-\frac{\mathrm{i}}{\hbar}\frac{1}{6mF}(p^3-(p-Ft)^3)\right]\\ &= \delta(p-p')\exp\left[-\frac{\mathrm{i}}{\hbar}\frac{p^2}{2m}t\right]\,,\end{aligned}$$

i.e., for $F = 0$, there is only one value of p (or E), namely p' $(E(p') = p'^2/2m > 0)$; whereas for $F \neq 0$, the spectrum is continuous – Ft is an arbitrary number. The only value for $p'(E(p'))$ mentioned above comes from the fact that for $F \to 0$, the amplitude $\langle p|E\rangle$ oscillates so rapidly that no contribution exists – except for the case in which the phase becomes stationary at a certain point p':

$$\frac{\partial}{\partial p}\left(Ep - \frac{p^3}{6m}\right)\bigg|_{p=p'} = 0 = E - \frac{p'^2}{2m}\,.$$

In order to calculate the configuration space wave function $\psi(x) = \langle x|E\rangle$, we write

$$\begin{aligned}\psi(x) = \langle x|E\rangle &= \int \langle x|p\rangle dp\langle p|E\rangle\\ &= \int \frac{dp}{\sqrt{2\pi\hbar}}\mathrm{e}^{(\mathrm{i}/\hbar)xp}\frac{1}{\sqrt{2\pi\hbar}}\frac{1}{\sqrt{F}}\exp\left[\frac{\mathrm{i}}{\hbar F}\left(Ep-\frac{p^3}{6m}\right)\right]\\ &= \int\frac{dp}{2\pi\hbar\sqrt{F}}\exp\left\{\frac{\mathrm{i}}{\hbar}\left[xp+\frac{1}{F}\left(Ep-\frac{p^3}{6m}\right)\right]\right\}\,.\end{aligned} \tag{23.7}$$

Now we introduce a new integration variable,

$$u = -(2m\hbar F)^{-1/3} p \,,$$

and write

$$\alpha := \left(\frac{2m}{\hbar^2} F\right)^{1/3} , \quad q = \left(x + \frac{E}{F}\right)\alpha \,.$$

Then our wave function takes the form

$$\psi(x) = \frac{\alpha}{\pi\sqrt{F}} \int_0^\infty du \, \cos\left(\frac{u^3}{3} - qu\right) \,.$$

The integral in this expression can also be written with the definition of the Airy function,

$$\mathrm{Ai}(q) = \frac{1}{\sqrt{\pi}} \int_0^\infty du \, \cos\left(\frac{u^3}{3} + uq\right) \,,$$

as

$$\psi(x) = \frac{\alpha}{\sqrt{\pi F}} \mathrm{Ai}(-q) \,. \tag{23.8}$$

We now return to our solution $\langle x|E\rangle$ in the form (23.7) and consider the semiclassical (WKB) approximation, where x and E are to be taken so that the phase in the integrand of (23.7) is very large relative to $\hbar$. Then we are dealing with rapid oscillations which become washed out – except for the stationary points. These stationary values are determined by

$$\frac{\partial}{\partial p}\left[px + \frac{1}{F}\left(Ep - \frac{p^3}{6m}\right)\right]\bigg|_{p=p_0} = 0 \,, \tag{23.9}$$

i.e.,

$$Fx + E - \frac{p_0^2}{2m} = 0$$

or

$$E = \frac{p_0^2}{2m} - Fx \,.$$

Here we meet the classical energy-momentum relation again:

$$p_0 = \pm\sqrt{(E + Fx)2m} \,. \tag{23.10}$$

Fx can take positive and negative values.

The value of the integral in (23.7) can, under certain conditions which have been given above, be dominated by the points p_0 of (23.10). The classical x-regions are

$$\begin{aligned} \text{allowed:} \quad & x > -E/F\,, \quad p_0 \text{ real}\,, \\ \text{forbidden:} \quad & x < -E/F\,, \quad p_0 \text{ imaginary}\,. \end{aligned}$$

Let us first consider the classically allowed region $x > -E/F$ and write

$$\exp\left\{\frac{\mathrm{i}}{\hbar}\left[px + \frac{1}{F}\left(Ep - \frac{p^3}{6m}\right)\right]\right\} =: \mathrm{e}^{\mathrm{i}\varphi(p)}$$

and then expand $\varphi(p)$ around the stationary value p_0:

$$\varphi(p) = \varphi(p_0) + \underbrace{(p - p_0)\left.\frac{\partial\varphi}{\partial p}\right|_{p_0}}_{=0} + \frac{1}{2}(p - p_0)^2\left.\frac{\partial^2\varphi}{\partial p^2}\right|_{p_0} + \ldots\,.$$

Now we have

$$\varphi(p_0) = \frac{1}{\hbar}p_0\underbrace{\left(x + \frac{E}{F}\right)}_{(9):\ p_0^2/2mF} - \frac{1}{\hbar F}\frac{p_0^3}{6m} = \frac{1}{\hbar F}\frac{p_0^3}{3m}$$

and

$$\left.\frac{\partial^2\varphi}{\partial p^2}\right|_{p_0} = -\frac{p_0}{m\hbar F}\,.$$

The condition that allows us to neglect the third derivative in the above expansion for $\varphi(p)$ will be given later. So far we have found the asymptotic behavior of $\psi(x)$ (there are two values of p_0!):

$$\psi_\pm(x) \sim \frac{1}{2\pi\hbar\sqrt{F}}\mathrm{e}^{\mathrm{i}\varphi(p_0)}\int dp \exp\left[-\mathrm{i}\frac{(p - p_0)^2}{2}\frac{p_0}{m\hbar F}\right]\,.$$

Writing $q = p - p_0$ and using

$$\int_{-\infty}^{+\infty} dq\, \mathrm{e}^{-\mathrm{i}aq^2} = \sqrt{\frac{\pi}{|a|}}\mathrm{e}^{-(\mathrm{i}\pi/4)\operatorname{sign} a}\,,$$

where $a = p_0/2m\hbar F$, we finally obtain

$$\begin{aligned} \psi(x) = \psi_+(x) + \psi_-(x) &= \frac{1}{2\pi\hbar\sqrt{F}}\sqrt{\frac{2\pi m\hbar F}{p_0}} \\ &\times \left(\mathrm{e}^{-\mathrm{i}\pi/4}\exp\left[\frac{\mathrm{i}}{\hbar F}\frac{p_0^3}{3m}\right] + \mathrm{e}^{\mathrm{i}\pi/4}\exp\left[\left(-\frac{\mathrm{i}}{\hbar F}\right)\frac{p_0^3}{3m}\right]\right)\,. \qquad (23.11) \\ p_0 &:= |p_0| \end{aligned}$$

With this we get the asymptotic formula,

$$\psi(x) \sim \sqrt{\frac{2m}{\pi \hbar p_0}} \cos\left(\frac{p_0^3}{3\hbar F m} - \frac{\pi}{4}\right)$$

with

$$p_0 = +\sqrt{2m(E + Fx)}$$

and

$$\frac{p_0^3}{3\hbar F m} = \frac{1}{\hbar} \int_{x_0 = -E/F}^{x} dx' \, p_0(x') \,,$$

which can be proved as follows:

$$p_0^2 = 2m(E + Fx) \quad : \quad \frac{\partial}{\partial x} p_0^2 = 2mF \,.$$

Furthermore,

$$\begin{aligned} \frac{\partial}{\partial x} \frac{p_0^3}{3\hbar m F} &= \frac{1}{3\hbar m F} \frac{\partial}{\partial x} \left(p_0^2\right)^{3/2} = \frac{1}{3\hbar m F} \frac{\partial p_0^2}{\partial x} \frac{\partial}{\partial p_0^2} \left(p_0^2\right)^{3/2} \\ &= \frac{1}{3\hbar m F} 2mF \frac{3}{2} p_0 = \frac{p_0}{\hbar} \,, \end{aligned}$$

integration yields

$$\frac{p_0^3}{3\hbar m F} = \frac{1}{\hbar} \int_{x_0}^{x} dx' \, p_0(x') \,, \quad x_0 = -\frac{E}{F} \,.$$

So we obtain

$$\begin{aligned} \psi(x) &\sim \sqrt{\frac{2m}{\pi \hbar p_0}} \cos\left(\frac{1}{\hbar} \int_{x_0}^{x} dx' \, p_0(x') - \frac{\pi}{4}\right) \\ &= \sqrt{\frac{m}{2\pi \hbar p_0}} \left(\exp\left[\frac{\mathrm{i}}{\hbar} \int_{x_0}^{x} dx' \, p_0(x') - \frac{\mathrm{i}\pi}{4}\right] \right. \\ &\qquad \left. + \exp\left[-\frac{\mathrm{i}}{\hbar} \int_{x_0}^{x} dx' \, p_0(x') + \frac{\mathrm{i}\pi}{4}\right]\right) \\ &=: \psi_+(x) + \psi_-(x) \,. \end{aligned} \tag{23.12}$$

What we have found is the superposition of a wave that is moving toward the left and another wave that is moving toward the right. The total phase change (one "bounce") is $\pi/2$. All our considerations apply to stationary states, so that we have inward bound and outward bound particles at all times.

If p_0 were constant, then we would obtain in (23.12) $\exp[\pm(\mathrm{i}/\hbar)xp]$, i.e., free particles. The form we have obtained for $\psi(x)$ takes the slowly changing momentum (in configuration space) into account, $p_0(x)$. If $p_0(x)$ does not vary very much, then we know that the derivatives are small, and this provides us with the condition under which we can neglect the third derivatives in the final expansion of $\varphi(p)$:

$$1)\quad \frac{\hbar}{p_0}\frac{\partial}{\partial x}p_0(x) = \hbar\frac{\partial}{\partial x}\ln p_0(x) \ll p_0 \qquad \left(\frac{h}{p_0} = \lambda(x)\right)$$

$$2)\quad \frac{\hbar}{p_0}\frac{\partial}{\partial x}\frac{p_0^2}{2} = \frac{\hbar}{p_0}mF \ll p_0^2\,.$$

From 2) it follows that $p_0^3 \gg \hbar mF$ or $p_0^3/\hbar mF \equiv 3\varphi(p_0) \gg 1$. Using the classical energy momentum relation, we also can write

$$[2m(E+Fx)]^{3/2} \gg \hbar mF$$

or

$$\left(x+\frac{E}{F}\right)^{3/2}(2mF)^{3/2} \gg \hbar mF\,,$$

i.e.,

$$\left(x+\frac{E}{F}\right)^{3/2} = (x-x_0)^{3/2} \gg \frac{\hbar}{\sqrt{mF}}\,.$$

So if x is far enough to the right of the classical turning point $x_0 = -E/F$, the results which have been derived so far are valid:

$$(x-x_0) \gg \left(\frac{\hbar^2}{mF}\right)^{1/3}\,.$$

Under these conditions we found

$$\begin{aligned}\psi(x) \sim &\exp\left[\frac{\mathrm{i}}{\hbar}\int_{x_0}^{x}dx'\,p_0(x') - \mathrm{i}\frac{\pi}{4}\right]\\ &+\exp\left[-\frac{\mathrm{i}}{\hbar}\int_{x_0}^{x}dx'\,p_0(x') + \mathrm{i}\frac{\pi}{4}\right] = \psi_+ + \psi_-\,.\end{aligned}$$

This is our former superposition of stationary states. If we measure p, then the same probability exists for measuring $+$ or $-$. The coordinate measurement is given by

$$|\psi(x)|^2 = \frac{2m}{\pi\hbar p_0}\cos^2\left[\int_{x_0}^{x}dx'\,p_0(x') - \frac{\pi}{4}\right]\,.$$

We obtain an interference pattern between the incoming and outgoing particles. With increasing x, p_0 increases in order to satisfy $p_0 = \sqrt{2m(E+Fx)}$. Accordingly, the "wave lengths" decrease and the modes move closer together. In the classical limit of very large x, the modes get so close to each other that an interference pattern is no longer visible, and we thus again are dealing with the classical case.

In the following we wish to establish the formal identity between the WKB method and the path integral method, where the path integral is dominated by the classical trajectory of the particle, which we assume to be moving in a one-dimensional potential $V(x)$.

We now do not want to assume that $V(x)$ is harmonic ("weak coupling" limit). Rather than expanding $V(x)$ around its minimum, $V(x) = V(0) + m\omega^2 x^2/2 + \dots$, we shall expand the entire action $S[x(t)]$ around the extremum path, i.e., around the classical trajectory. This solves Newton's equation of motion:

$$m\frac{d^2\bar{x}(t)}{dt^2} = -V'\big(\bar{x}(t)\big) \tag{23.13}$$

with the initial condition

$$\bar{x}\big(t_1\big) = \bar{x}_1 \,, \quad \bar{x}\big(t_2\big) = \bar{x}_2 \,.$$

Note that it is only for the classical path that it makes sense to speak of the conserved energy of the particle (on-shell):

$$\frac{dE}{dt} = \frac{d}{dt}\left[\frac{m}{2}\dot{\bar{x}}^2 + V(\bar{x})\right] = \dot{\bar{x}}[m\ddot{\bar{x}} + V'(\bar{x})] = 0 \,.$$

For any path other than $\bar{x}(t)$, the energy is not constant. Besides, the energy is a function of the terminal configurations (x_2, t_2), (x_1, t_1):

$$t_2 - t_1 = T = \int_{x_1}^{x_2} dx \left(\frac{m}{2[E - V(x)]}\right)^{1/2} \,. \tag{23.14}$$

So let us begin with the expansion of $S[x(t)]$ around the classical solution:

$$x(t) = \bar{x}(t) + y(t) \,, \quad y\big(t_1\big) = 0 = y\big(t_2\big)$$
$$V\big(x(t)\big) = V\big(\bar{x}(t)\big) + V'\big(\bar{x}(t)\big)y(t) + \tfrac{1}{2}V''\big(\bar{x}(t)\big)y^2(t) + \dots \,.$$

When substituted in $S[x(t)]$, this yields, using (23.13):

$$\begin{aligned} S[x(t)] &= \int_{t_1}^{t_2} dt\left[\frac{m}{2}(\dot{\bar{x}} + \dot{y})^2 - V\big(x(t)\big)\right] \\ &= \int_{t_1}^{t_2} dt\left[\frac{m}{2}\dot{\bar{x}}^2 - V\big(\bar{x}(t)\big)\right] + \int_{t_1}^{t_2} dt\left[\frac{m}{2}\dot{y}^2 - \frac{1}{2}V''(\bar{x})y^2(t)\right] \,, \end{aligned}$$

$$S[x(t)] = S[\bar{x}(t)] + \frac{m}{2}\int_{t_1}^{t_2} dt\, y(t)\left[-\frac{d^2}{dt^2} - \frac{1}{m}V''\big(\bar{x}(t)\big)\right]y(t) \,. \tag{23.15}$$

The propagator now can be written as

$$\begin{aligned} K\big(x_2, t_2; x_1, t_1\big) &= \int [dx(t)]\, \mathrm{e}^{(\mathrm{i}/\hbar)S[x(t)]} \\ &= \mathrm{e}^{(\mathrm{i}/\hbar)S[\bar{x}]} \int_{y(t_1)=0}^{y(t_2)=0} [dy(t)] \exp\left\{\frac{\mathrm{i}}{\hbar}\int_{t_1}^{t_2} dt\left[\frac{m}{2}\dot{y}^2 - \frac{1}{2}V''(\bar{x})y^2\right]\right\} \,. \end{aligned} \tag{23.16}$$

The path integral was calculated in (20.15). There we found

$$\int_0^0 [dy(t)]./. = \sqrt{\frac{m}{2\pi i\hbar f(t_1) f(t_2) \int_{t_1}^{t_2} dt/f^2(t)}}\,, \tag{23.17}$$

where

$$\left[m\frac{d^2}{dt^2} + V''\big(\bar{x}(t)\big)\right] f(t) = 0\,. \tag{23.18}$$

We can easily relate $f(t)$ to the classical path. In order to do so, we differentiate Newton's equation (23.13) once more and get

$$m\frac{d^3\bar{x}(t)}{dt^3} = -V''\big(\bar{x}(t)\big)\frac{d\bar{x}(t)}{dt}\,,$$

i.e., we can set $f(t) = d\bar{x}(t)/dt$. This yields

$$\int_0^0 [dy(t)]./. = \left(\frac{2\pi i\hbar}{m}\dot{\bar{x}}(t_1)\dot{\bar{x}}(t_2)\int_{t_1}^{t_2}\frac{dt}{\dot{\bar{x}}(t)^2}\right)^{-1/2} \tag{23.19}$$

$$= \left(2\pi i k(x_2)k(x_1)\int_{x_1}^{x_2}\frac{dx}{k(x)^3}\right)^{-1/2}\,, \tag{23.20}$$

where

$$k(x) = \frac{1}{\hbar}\big[2m\big(E_{\rm cl} - V(x)\big)\big]^{1/2} \tag{23.21}$$

and $E_{\rm cl} = E_{\rm cl}(x_2, t_2; x_1, t_1)$ is defined by

$$t_2 - t_1 = \int_{x_1}^{x_2}\frac{dx}{v(x)} = \int_{x_1}^{x_2} dx \left(\frac{m}{2[E - V(x)]}\right)^{1/2}\,. \tag{23.22}$$

Until now we have for the propagator

$$K\big(x_2, t_2; x_1, t_1\big) = e^{(i/\hbar)S[\bar{x}]}\left(2\pi i k(x_2)k(x_1)\int_{x_1}^{x_2}\frac{dx}{k(x)^3}\right)^{-1/2}\,. \tag{23.23}$$

For a potential of the form $V(x) = V_0 + mgx + m\omega^2x^2/2$, the above formula (23.23) for the propagator is exact, since $V''(x)$ is a constant. If the potential does not have this specific structure, then we obtain

$$S[\bar{x}] = \int_{t_1}^{t_2} dt\left[\frac{m}{2}\dot{\bar{x}}^2 - V(\bar{x})\right] \underset{E=T+V}{=} \int_{t_1}^{t_2} dt\big[m\dot{\bar{x}}^2 - E_{\rm cl}\big]$$

$$= \int_{t_1}^{t_2} dx\big\{2m\big[E_{\rm cl} - V(\bar{x})\big]\big\}^{1/2} - E_{\rm cl}\big(t_2 - t_1\big)\,. \tag{23.24}$$

Therefore, the semiclassical approximation for the propagator is

$$K\big(x_2,t_2;x_1,t_1\big) = \left(2\pi\mathrm{i}k\big(x_2\big)k\big(x_1\big)\int_{x_1}^{x_2}\frac{dx}{k(x)^3}\right)^{-1/2} \times \exp\left[\mathrm{i}\int_{x_1}^{x_2}dx\,k(x) - \frac{\mathrm{i}}{\hbar}E_{\mathrm{cl}}\big(t_2-t_1\big)\right] . \tag{23.25}$$

An alternative representation of K exists in the form of

$$\begin{aligned} K\big(x_2,t_2;x_1,t_1\big) &= \langle x_2|\mathrm{e}^{-(\mathrm{i}/\hbar)H(t_2-t_1)}|x_1\rangle \\ &= \sum_n \langle x_2|n\rangle\langle n|x_1\rangle\,\mathrm{e}^{-(\mathrm{i}/\hbar)E_n(t_2-t_1)} \\ &= \sum_n \psi_n\big(x_2\big)\psi_n^*\big(x_1\big)\,\mathrm{e}^{-(\mathrm{i}/\hbar)E_n(t_2-t_1)} . \end{aligned} \tag{23.26}$$

Here we can use the explicit form of the WKB wave function (23.12) to our advantage ($p_0(x) = \hbar k(x)$):

$$\psi_n(x) = \frac{1}{\hbar}\sqrt{\frac{m}{2\pi k_n(x)}}\exp\left[\pm\frac{\mathrm{i}}{\hbar}\int_a^x dx'\,k_n(x')\right] , \tag{23.27}$$

where a is a parameter and $k_n(x)$ is given by

$$k_n(x) = \frac{1}{\hbar}\big\{2m\big[E_n - V(x)\big]\big\}^{1/2} .$$

Then our propagator becomes

$$\begin{aligned} K\big(x_2,t_2;x_1,t_1\big) &= \int dE\,\psi\big(x_2\big)\psi^*\big(x_1\big)\,\mathrm{e}^{-(\mathrm{i}/\hbar)E(t_2-t_1)} \\ &= \sum_{\varepsilon=\pm1}\frac{m}{2\pi\hbar^2}\int\frac{dE}{[k(x_1)k(x_2)]^{1/2}}\,\mathrm{e}^{-(\mathrm{i}/\hbar)E(t_2-t_1)} \\ &\quad\times\exp\left[\mathrm{i}\varepsilon\int_{x_1}^{x_2}dx\,k(x)\right] . \end{aligned} \tag{23.28}$$

$\varepsilon = \pm1$ takes care of the twofold degeneracy of the energy eigenstates. We shall compute the integral in (23.28) using the saddle point method ("stationary phase approximation"). To this end we consider the following one-dimensional integral:

$$\int_{-\infty}^{+\infty} dx\,\mathrm{e}^{\mathrm{i}f(x)}g(x) ,$$

and look for points x_0, where the phase becomes stationary, i.e., $f'(x_0) = 0$. We then expand around this point:

$$\begin{aligned} f(x) &\simeq f\big(x_0\big) + \tfrac{1}{2}f''\big(x_0\big)\big(x-x_0\big)^2 , \\ g(x) &\simeq g\big(x_0\big) + g'\big(x_0\big)\big(x-x_0\big) . \end{aligned}$$

The above integral is hereby reduced again to a Gaussian integral:

$$\int_{-\infty}^{+\infty} dx\, \mathrm{e}^{\mathrm{i}f(x)} g(x) \simeq g(x_0)\, \mathrm{e}^{\mathrm{i}f(x_0)} \sqrt{\frac{2\pi\mathrm{i}}{f''(x_0)}} . \tag{23.29}$$

In our case it holds that

$$f(E) = \frac{\varepsilon}{\hbar} \int_{x_1}^{x_2} dx \{2m[E - V(x)]\}^{1/2} - \frac{E}{\hbar}(t_2 - t_1) ,$$

so that

$$\frac{\partial f}{\partial E} = -\frac{(t_2 - t_1)}{\hbar} + \frac{\varepsilon}{\hbar} \int_{x_1}^{x_2} dx \left(\frac{m}{2[E - V(x)]}\right)^{1/2} = 0 . \tag{23.30}$$

Assuming $x_2 > x_1$, so that only $\varepsilon = +1$ makes a contribution, we can determine E from (23.30):

$$t_2 - t_1 = \int_{x_1}^{x_2} dx \left(\frac{m}{2[E - V(x)]}\right)^{1/2} .$$

So we find $E = E_{\mathrm{cl}}$, i.e., the classical energy dominates in the energy integral in (23.28). Since, furthermore,

$$\frac{\partial^2 f}{\partial E^2} = -\frac{m^2}{\hbar} \int_{x_1}^{x_2} \frac{dx}{\{2m[E - V(x)]\}^{3/2}} = -\frac{m^2}{\hbar^4} \int_{x_1}^{x_2} \frac{dx}{k(x)^3}$$

we find

$$\begin{aligned} K(x_2, t_2; x_1, t_1) &= \frac{m}{2\pi\hbar^2} \frac{1}{(k(x_1)k(x_2))^{1/2}} \left(\frac{2\pi\mathrm{i}}{-(m^2/\hbar^4) \int_{x_1}^{x_2} dx/k^3(x)}\right)^{1/2} \\ &\quad \times \mathrm{e}^{-(\mathrm{i}/\hbar)E_{\mathrm{cl}}(t_2 - t_1)} \exp\left[\mathrm{i} \int_{x_1}^{x_2} dx\, k(x)\right] \\ &= \left(2\pi\mathrm{i} k(x_2) k(x_1) \int_{x_1}^{x_2} \frac{dx}{k(x)^3}\right)^{-1/2} \\ &\quad \times \exp\left[\mathrm{i} \int_{x_1}^{x_2} dx\, k(x) - \frac{\mathrm{i}}{\hbar} E_{\mathrm{cl}}(t_2 - t_1)\right] , \end{aligned}$$

which is identical to (23.25).

24. Computing the Trace

So far we have been interested in the general expression for the WKB-propagation function. Now we turn our attention to the trace of that propagator, since we want to exhibit the energy eigenvalues of a given potential. From earlier discussions we know that the energy levels of a given Hamiltionian are provided by the po'es of the Green's function:

$$G(E) = \mathrm{Tr}\frac{1}{H-E} = \sum_n \frac{1}{E_n - E} = \frac{\mathrm{i}}{\hbar}\int_0^\infty dT \mathrm{e}^{\frac{\mathrm{i}}{\hbar}ET} G(T) \tag{24.1}$$

where $G(T) = \mathrm{Tr}\left[\mathrm{e}^{-\frac{\mathrm{i}}{\hbar}HT}\right] = \int dx_0 \langle x_0 \mid \mathrm{e}^{-\frac{\mathrm{i}}{\hbar}HT} \mid x_0\rangle .$ (24.2)

The transition amplitude $\langle x_0 \mid \mathrm{e}^{-\frac{\mathrm{i}}{\hbar}HT} \mid x_0\rangle$ will be written as Feynman path integral

$$\langle x_0 \mid \mathrm{e}^{-\frac{\mathrm{i}}{\hbar}HT} \mid x_0\rangle = \int_{x(0)=x_0}^{x(T)=x_0} [dx(t)] \mathrm{e}^{\frac{\mathrm{i}}{\hbar}S[x(t)]}, \tag{24.3}$$

where $S[x(t)]$ is the action functional along the path $x(t)$:

$$S[x(t)] = \int_0^T dt \left[\frac{m}{2}\dot{x}^2 - V(x(t))\right] \tag{24.4}$$

and the integral in (24.3) goes over all closed paths that begin at x_0 at $t = 0$ and end at x_0 at $t = T$. Proceeding as before, we obtain for equation (24.2)

$$\begin{aligned} G(T) &= \int_{-\infty}^{+\infty} dx_0 \int_{x(0)=x_0}^{x(T)=x_0} [dx(t)] \exp\left\{\frac{\mathrm{i}}{\hbar}S[x(t)]\right\} \\ &= \int_{\substack{x(0)=x(T) \\ \text{arbitrary}}} [dx(t)] \exp\left\{\frac{\mathrm{i}}{\hbar}S[x(t)]\right\} \end{aligned} \tag{24.5}$$

$$\begin{aligned} &= \int_{-\infty}^{+\infty} dx_0 \exp\left\{\frac{\mathrm{i}}{\hbar}S[\overline{x}(t)]\right\} \int_{y(0)=0}^{y(T)=0} [dy(t)] \\ &\quad \exp\left\{\frac{\mathrm{i}}{\hbar}\frac{m}{2}\int_0^T dt y(t)\left[-\frac{d^2}{dt^2} - \frac{1}{m}V''(\overline{x}(t))\right] y(t)\right\}. \end{aligned} \tag{24.6}$$

The path integral appearing in (24.6) was already calculated in (23.17), so that we may write

$$\int_0^0 [dy(t)] \,\%\, = \sqrt{\frac{m}{2\pi i\hbar f(0) f(T) \int_0^T \frac{dt}{f^2(t)}}} \tag{24.7}$$

and we can set $f(t) = \dot{\overline{x}}(t)$. This brings us to

$$G(T) = \int_{-\infty}^{+\infty} dx_0 \exp\left\{\frac{i}{\hbar} S[\overline{x}(t)]\right\} \sqrt{\frac{m}{2\pi i\hbar \dot{\overline{x}}(0)\dot{\overline{x}}(T) \int_0^T \frac{dt}{(\dot{\overline{x}}(t))^2}}} \,. \tag{24.8}$$

Next, we have to perform the x_0 integration, again by stationary phase approximation. Hence we need those classical paths which produce stationary phases in the variable x_0:

$$\begin{aligned} 0 &= \frac{\partial}{\partial x_0} S[x_0, T; x_0, 0] \\ &= \frac{\partial}{\partial x_f} S[x_f, T; x_i, 0] \,|_{x_i=x_f=x_0} + \frac{\partial}{\partial x_i} S[x_f, T; x_i, 0] \,|_{x_i=x_f=x_0} \\ &= p(t=T) - p(t=0) = p_f - p_i. \end{aligned} \tag{24.9}$$

The latter equality follows from (1.34), (1.36). Equation (24.9) shows that the path not only has the same initial and final position, $x(0) = x(T) = x_0$, but the same momentum as well, i.e., the path is periodic. Hence the stationary phase approximation selects periodic classical paths only. These orbits can be parametrized by their fundamental period T. For every "primitive" periodic orbit with fundamental period T, orbits with periods $nT, n = 1, 2, 3 \ldots$ exist which are realized if the primitive orbit is traversed n times. By definition, a primitive orbit is not the iteration of another one with a smaller period. For a given T, $G(T)$ will receive contributions from all periodic orbits whose fundamental period is an integral fraction of T:

$$x_T(t), x_{\frac{T}{2}}(t), \ldots, x_{\frac{T}{n}}(t).$$

A primitive orbit of period T/n, traversed n times, gives rise to a periodic orbit of the desired period T. As a result, $G(T)$ takes the form

$$G(T) = \int_{-\infty}^{+\infty} dx_0 \sum_{n=1}^{\infty} \exp\left\{\frac{i}{\hbar} nS\left[\frac{T}{n}\right]\right\} \frac{1}{\left|\dot{x}_{\frac{T}{n}}(0)\right|} \sqrt{\frac{m}{2\pi i\hbar}} \left[\int_0^T \frac{dt}{\left(\dot{x}_{\frac{T}{n}}\right)^2}\right]^{-\frac{1}{2}} . \tag{24.10}$$

The action of the periodic orbits does not depend upon the choice of the starting point; this is also true for the expression $\int_0^T dt\, (\dot{x}^2)^{-1}$. The only contribution to the x_0-integral comes from

$$\int_{-\infty}^{+\infty} dx_0 \frac{1}{\left|\dot{x}_{\frac{T}{n}}(0)\right|} = 2\int_{x_1}^{x_2} dx_0 \frac{1}{\frac{dx_0}{dt}} = \oint dt = \frac{T}{n}. \tag{24.11}$$

Next, we have to compute the integral

$$\int_0^T \frac{dt}{\left(\dot{x}_{\frac{T}{n}}\right)^2} = 2n \int_{x_1}^{x_2} \frac{dx}{\dot{x}^3} = 2n \int_{x_1}^{x_2} dx \left(\frac{2\left[E\left(\frac{T}{n}\right) - V(x)\right]}{m} \right)^{-\frac{3}{2}}, \tag{24.12}$$

where the classical orbit may have completed n traverses of a fundamental periodic path in time T, and x_1, x_2 are the turning points. In the last equality in (24.12) we have used $\frac{m}{2}\dot{x}^2 = E - V(x)$. We have also dropped the bar in $\overline{x}$, indicating that everything corresponds to the classical orbit.

Let us pause for a moment and note that the integral (24.12) shows up by a different reasoning. So far we have classified every orbit by its fundamental period $T : x = x_T(t)$. Furthermore, every orbit is associated with a certain energy $E = E(T)$. Now we know from classical mechanics that $\frac{\partial S}{\partial t_i} = E$ and $\frac{\partial S}{\partial t_f} = -E$. Putting $t_f = T$ we obtain

$$\frac{\partial S}{\partial T} = -E, \qquad E = E(T), \tag{24.13}$$

where $S(T)$ stands for the classical action of the periodic orbit $x_T(t)$, and E is the classical energy.

Introducing a Legendre transform with respect to the variable T we get

$$W(E) = S(T) + ET = S(T) - \frac{\partial S}{\partial T}T, \qquad T = T(E) \tag{24.14}$$

and

$$\begin{aligned} \frac{dW(E)}{dE} &= \frac{\partial S}{\partial T}\frac{dT}{dE} + T + E\frac{dT}{dE} \\ &= -E\frac{dT}{dE} + T + E\frac{dT}{dE} = T. \end{aligned} \tag{24.15}$$

Returning to the action S we obtain

$$S(T) = W(E) - TE = W(E) - \frac{dW}{dE}E$$

$$\text{where} \quad \frac{dS}{dT} = -E, \quad E = E(T). \tag{24.16}$$

If we now recall the relation $E = \frac{m}{2}\dot{x}^2 + V(x)$, we can express equation (24.14) in the form

$$\begin{aligned} W(E) = S(T) + ET &= \int_0^T dt \left(\frac{m}{2}\dot{x}^2 - V + E\right) = \int_0^T dtm\dot{x}^2 \\ &= 2n \int_{x_1}^{x_2} dx\sqrt{2m(E - V)} \end{aligned} \tag{24.17}$$

which holds true for an orbit with n traverses.

Differentiating (24.17) twice with respect to E we obtain

$$\frac{d^2W}{dE^2} = \frac{dT}{dE} = -\frac{2n}{m}\int_{x_1}^{x_2} dx \left(\frac{2}{m}(E-V)\right)^{-\frac{3}{2}}$$

which implies the desired relation $\displaystyle\int_0^T \frac{dt}{\left(\dot{x}_{\frac{T}{n}}\right)^2} = -m\frac{dT}{dE}.$ (24.18)

Collecting our results so far we get

$$\begin{aligned} G(T) &= \sum_{n=1}^{\infty} \exp\left\{\frac{\mathrm{i}}{\hbar} nS\left[\frac{T}{n}\right]\right\} \frac{T}{n}\sqrt{\frac{m}{2\pi \mathrm{i}\hbar}}\left(-m\frac{dT}{dE}\right)^{-\frac{1}{2}} \\ &= \sum_{n=1}^{\infty} \exp\left\{\frac{\mathrm{i}}{\hbar} nS\left[\frac{T}{n}\right]\right\} \frac{T}{n}\sqrt{\frac{\mathrm{i}}{2\pi\hbar}}\left|\frac{dT}{dE}\right|^{\frac{1}{2}}_{E\equiv E_{\mathrm{cl}}=E\left(\frac{T}{n}\right)}. \end{aligned} \tag{24.19}$$

With this expression for $G(T)$ we return to equation (24.1):

$$\begin{aligned} G(E) &= \frac{\mathrm{i}}{\hbar}\int_0^{\infty} dT \exp\left\{\frac{\mathrm{i}}{\hbar} ET\right\} G(T) \\ &= \frac{\mathrm{i}}{\hbar}\sqrt{\frac{\mathrm{i}}{2\pi\hbar}} \sum_{n=1}^{\infty}\int_0^{\infty} dT \exp\left\{\frac{\mathrm{i}}{\hbar}\left(ET + nS\left[\frac{T}{n}\right]\right)\right\} \frac{T}{n}\left|\frac{dE_{\mathrm{cl}}}{dT}\right|^{\frac{1}{2}}. \end{aligned} \tag{24.20}$$

Here it is convenient to replace the integration variable T by $\tau = T/n$, wich is the fundamental period of the periodic orbit in question. We also need to rewrite

$$\frac{dE_{\mathrm{cl}}\left[\frac{T}{n}\right]}{dT} = \frac{dE_{\mathrm{cl}}(\tau)}{d\tau}\frac{d\tau}{dT} = \frac{dE_{\mathrm{cl}}}{d\tau}\frac{1}{n}$$

so that equation (24.20) becomes

$$G(E) = \frac{\mathrm{i}}{\hbar}\sqrt{\frac{\mathrm{i}}{2\pi\hbar}} \sum_{n=1}^{\infty}\int d\tau \exp\left\{\frac{\mathrm{i}}{\hbar} n\left(E\tau + S(\tau)\right)\right\} \tau\sqrt{n}\sqrt{\left|\frac{dE_{\mathrm{cl}}}{d\tau}\right|}. \tag{24.21}$$

Again, we compute the τ-integration by use of the stationary phase approximation. For the phase in (24.21) to be stationary we require

$$0 = \frac{\partial}{\partial\tau}\left(E\tau + S(\tau)\right) = E - \frac{\partial S}{\partial\tau} = E - E_{\mathrm{cl}}.$$

Thus, for a given value of E we obtain the main contribution from the periodic orbit which has exactly the classical energy E_{cl}. This enables us to rewrite (24.21) as [cf. (21.29)]

$$G(E) = \frac{\mathrm{i}}{\hbar}\sqrt{\frac{\mathrm{i}}{2\pi\hbar}} \sum_{n=1}^{\infty}\sqrt{\frac{2\pi \mathrm{i}\hbar}{n\frac{d^2S_{\mathrm{cl}}}{d\tau^2}}} \exp\left\{\frac{\mathrm{i}}{\hbar} n\left(E\tau(E) + S(\tau)\right)\right\} \tau(E)\sqrt{n}\sqrt{\left|\frac{dE_{\mathrm{cl}}}{d\tau}\right|}.$$

Using the fact that $\frac{d^2 S_{\rm cl}}{d\tau^2} = -\frac{dE}{d\tau}$ and $W(E) = S(\tau) + E\tau(E)$, we end up with

$$\begin{aligned} G(E) &= \frac{\mathrm{i}}{\hbar}\sqrt{\frac{\mathrm{i}}{2\pi\hbar}}\sum_{n=1}^{\infty}\sqrt{\frac{2\pi\hbar}{\mathrm{i}}}\,\tau(E)\exp\left\{\frac{\mathrm{i}}{\hbar}nW(E)\right\} \\ &= \frac{\mathrm{i}}{\hbar}\tau(E)\sum_{n=1}^{\infty}\exp\left\{\frac{\mathrm{i}}{\hbar}nW(E)\right\} = \frac{\mathrm{i}}{\hbar}\tau(E)\frac{\mathrm{e}^{\frac{\mathrm{i}}{\hbar}W(E)}}{1-\mathrm{e}^{\frac{\mathrm{i}}{\hbar}W(E)}}. \end{aligned} \tag{24.22}$$

This form of $G(E)$ allows us to identify the poles which are located at the energies E_n satisfying

$$W(E_n) = 2\pi n\hbar \tag{24.23}$$

for n integer. Using

$$W(E) = S + E\tau = \int_0^T dT\left(\frac{m}{2}\dot{x}^2 - V + E\right) = m\int_0^T dt\dot{x}^2 = 2\int_{x_1}^{x_2} pdx$$

equation (24.23) assumes the form

$$\oint pdq = 2\pi\hbar n. \tag{24.24}$$

So we have rediscovered the old Bohr–Sommerfeld quantization rule.

We can, however, do better than that. What we are aiming at is an improvement of formula (24.23), i.e., we want to generate the correct half-integer quantum number $(n+1/2)$. The reason that we have not been able to produce the quantum mechanical correction is due to the fact that we have not taken care of the phase ambiguity of the path integral. However, the same problem occured with the harmonic oscillator. Expression (24.18) is completely analogous to the result contained in (16.70). In both cases we have to carefuly study the behavior of the propagator at the caustics. In the present case, the additional phase factor stems from the singularities of the integrand in $\left(\int_0^T \frac{dt}{f^2(t)}\right)^{-\frac{1}{2}}$. These singular points correspond to the zeros of $f(t) = \dot{x}(t)$, i.e., the turning points x_1 and x_2. Each time we run through a turning point, we pick up an additional phase factor $\mathrm{e}^{\mathrm{i}\frac{\pi}{2}}$, and since there are $2n$ turning points in the orbit $x_{T/n}$, we gain $\mathrm{e}^{\mathrm{i}\pi n} = (-1)^n$ altogether. It is this factor which, when included in (24.22), will produce the correct pole structure:

$$\begin{aligned} G(E) &= \frac{\mathrm{i}}{\hbar}\tau(E)\sum_{n=1}^{\infty}(-1)^n\exp\left\{\frac{\mathrm{i}}{\hbar}nW(E)\right\} \\ &= -\frac{\mathrm{i}}{\hbar}\tau(E)\frac{\mathrm{e}^{\frac{\mathrm{i}}{\hbar}W(E)}}{1+\mathrm{e}^{\frac{\mathrm{i}}{\hbar}W(E)}}. \end{aligned} \tag{24.25}$$

Therefore the poles are correctly shifted to

$$W(E_n) = 2\pi\left(n+\tfrac{1}{2}\right)\hbar. \tag{24.26}$$

25. Partition Function for the Harmonic Oscillator

We start by making the following changes from Minkowski real time $t = x_0$ to Euclidean "time" $\tau = t_E$:

$$\tau = it = \beta \, . \tag{25.1}$$

Here we put $\hbar = 1$, $T =$ temperature of the system $= \beta^{-1}$, $k = 1$. Then we write for the partition function of our one-dimensional quantum mechanical system:

$$Z = \int_{\substack{x(0)=x(\beta)\\ \text{arbitr.}}} [dx(\tau)] \exp\left\{ - \int_0^\beta d\tau \left[\frac{m}{2}\dot{x}^2(\tau) + V\big(x(\tau)\big) \right] \right\} . \tag{25.2}$$

The exponential is obtained from

$$\begin{aligned} iS &= i \int_0^t dt' \left[\frac{m}{2}\dot{x}^2(t') - V\big(x(t')\big) \right] \\ &= - \int_0^\beta d\tau \left[\frac{m}{2}\dot{x}^2(\tau) + V\big(x(\tau)\big) \right] . \end{aligned}$$

In the functional integration we require $x(\tau)$ to be periodic with period $\beta : x(\tau) = x(\tau + \beta)$ or, if we put $\tau = 0$,

$$x(0) = x(\beta) \, .$$

In particular, we obtain for the harmonic oscillator ($m = 1$), where $V(x) = \omega^2 x^2/2$,

$$Z = \int_\beta [dx(\tau)] \exp\left\{ -\frac{1}{2} \int_0^\beta d\tau \left[\dot{x}^2(\tau) + \omega^2 x^2(\tau) \right] \right\} ,$$

or, performing an integration by parts on the first term in the exponential,

$$Z = \int_\beta [dx(\tau)] \exp\left\{ -\frac{1}{2} \int_0^\beta d\tau \, x(\tau) \left[-\frac{d^2}{d\tau^2} + \omega^2 \right]_\beta x(\tau) \right\} . \tag{25.3}$$

The subscript β in $[-d^2/d\tau^2 + \omega^2]_\beta$ indicates that the differential operator is restricted to the function space defined by $x(0) = x(\beta)$.

$$\Omega_\tau := \left[-\frac{d^2}{d\tau^2} + \omega^2\right]_\beta \tag{25.4}$$

Ω_τ is a positive definite elliptic operator acting on $x_n(\tau)$ defined on a compact circular τ-manifold. Ω_τ has a complete set of orthonormal (real) eigenfunctions $x_n(\tau)$ and associated eigenvalues λ_n:

$$\left[-\frac{d^2}{d\tau^2} + \omega^2\right]_\beta x_n(\tau) = \lambda_n x_n(\tau) \tag{25.5}$$

with

$$x_n(0) = x_n(\beta)\ , \qquad \int_0^\beta d\tau\, x_m(\tau)x_n(\tau) = \delta_{mn}\ . \tag{25.6}$$

Explicitly,

$$x_n(\tau) = \begin{cases} \sqrt{\frac{2}{\beta}}\sin\left(\frac{2\pi}{\beta}n\tau\right)\ , \\ \sqrt{\frac{2}{\beta}}\cos\left(\frac{2\pi}{\beta}n\tau\right)\ , \end{cases} \qquad \lambda_n = \left(\frac{2\pi n}{\beta}\right)^2 + \omega^2\ , \quad n \in \mathbb{Z}\ . \tag{25.7}$$

Check:

$$\frac{d^2}{d\tau^2}\sqrt{\frac{2}{\beta}}\sin\left(\frac{2\pi}{\beta}n\tau\right) = -\left(\frac{2\pi}{\beta}n\right)^2\sqrt{\frac{2}{\beta}}\sin\left(\frac{2\pi}{\beta}n\tau\right)\ ,$$

$$\left[-\frac{d^2}{d\tau^2} + \omega^2\right]\sqrt{\frac{2}{\beta}}\sin\left(\frac{2\pi}{\beta}n\tau\right) = \left[\left(\frac{2\pi}{\beta}n\right)^2 + \omega^2\right]\sqrt{\frac{2}{\beta}}\sin\left(\frac{2\pi}{\beta}\tau n\right)$$
$$= \lambda_n\sqrt{\frac{2}{\beta}}\sin\left(\frac{2\pi}{\beta}n\tau\right)\ .$$

Periodicity is also obvious:

$$x_n(\tau + \beta) = x_n(\tau)\ ,$$

and orthonormality follows from

$$\frac{2}{\beta}\int_0^\beta d\tau\, \cos\left(\frac{2\pi}{\beta}m\tau\right)\cos\left(\frac{2\pi}{\beta}n\tau\right) = \delta_{mn}\ .$$

Now, as we have done with the real-time oscillator, we approximate the path $x(\tau)$ by a finite number of basis functions

$$\tilde{x}_N(\tau) = \sum_{n=1}^{N} a_n x_n(\tau)\ , \quad a_n = \int_0^\beta d\tau\, x_n(\tau)\tilde{x}_N(\tau)\ , \tag{25.8}$$

and write for the Euclidean action

$$S_{\mathrm{E}}[x(\tau)] = \frac{1}{2}\int_0^{\beta} d\tau\, x(\tau)\Omega_\tau x(\tau)$$

or

$$S_{\mathrm{E}}\big[\tilde{x}_N(\tau)\big] = \frac{1}{2}\int_0^{\beta} d\tau\, \tilde{x}_N(\tau)\Omega_\tau \tilde{x}_N(\tau)\,.$$

Substituting (25.8) and using $\Omega_\tau x_n(\tau) = \lambda_n x_n(\tau)$, we find

$$\begin{aligned} S_{\mathrm{E}}[\tilde{x}(\tau)] &= \frac{1}{2}\sum_{m,n}^{N} \lambda_n a_m a_n \underbrace{\int_0^{\beta} d\tau\, x_m(\tau) x_n(\tau)}_{\delta_{mn}} \\ &= \frac{1}{2}\sum_{n=1}^{N} \lambda_n a_n^2\,. \end{aligned}$$

From here on we are on familiar ground:

$$\begin{aligned} &\int_{-\infty}^{+\infty} da_1 \ldots \int_{-\infty}^{\infty} da_N \exp\left\{-S_{\mathrm{E}}\big[\tilde{x}_N(\tau)\big]\right\} \\ &\quad = (2\pi)^{N/2}\sqrt{\frac{1}{\lambda_1 \ldots \lambda_N}} = (2\pi)^{N/2}\left(\prod_{n=1}^{N}\lambda_n\right)^{-1/2}\,. \end{aligned} \tag{25.9}$$

If we change the integration measure,

$$\prod_{n=1}^{N} da_n \to \prod_{n=1}^{N} da_n \sqrt{\frac{1}{2\pi}}$$

and finally let N go to infinity, $N \to \infty$, we obtain

$$Z = \int_{\beta} [dx(\tau)]\, \mathrm{e}^{-S_{\mathrm{E}}[x(\tau)]} \tag{25.10}$$

or

$$Z = \left(\det \Omega_\tau\right)^{-1/2}, \quad \Omega_\tau = \left[-\frac{d^2}{d\tau^2} + \omega^2\right]_{\beta}\,. \tag{25.11}$$

There are various ways of computing this determinant. This is a good setting for introducing the so-called ζ-function evaluation of determinants. So let us consider an operator A with positive real discrete eigenvalues $\lambda_1, \ldots, \lambda_n, \ldots$ and let the eigenfunctions be $f_n(x) : A f_n(x) = \lambda_n f_n(x)$. Then we form the expression

$$\zeta_A(s) = \sum_n \frac{1}{\lambda_n^s}\,. \tag{25.12}$$

This is the ζ-function associated to A. For the one-dimensional oscillator $H \equiv A$, ζ is, except for the zero-point energy, Riemann's ζ-function. In (25.12) the sum goes over all the eigenvalues, and s is a variable, real or complex, chosen such that the series (25.12) converges.

Writing $\lambda_n^{-s} = \mathrm{e}^{-s \ln \lambda_n}$, we have

$$\zeta_A(s) = \sum_{n=0}^{\infty} \mathrm{e}^{-s \ln \lambda_n} ,$$

or, upon formal differentiation with respect to s,

$$\zeta_A'(s) = \frac{d}{ds} \sum_{n=0}^{\infty} \mathrm{e}^{-s \ln \lambda_n} = - \sum_{n=0}^{\infty} \ln \lambda_n \, \mathrm{e}^{-s \ln \lambda_n}$$

and, putting $s = 0$,

$$\zeta_A'(s)\big|_{s=0} = - \sum_{n=0}^{\infty} \ln \lambda_n ,$$

we finally obtain

$$\det A = \prod_n \lambda_n = \exp \left[\sum_{n=0}^{\infty} \ln \lambda_n \right] = \mathrm{e}^{-\zeta_A'(0)} , \tag{25.13}$$

or

$$\ln \det A = -\zeta_A'(0) , \tag{25.14}$$

where $\zeta'(0)$ has to be determined by analytic continuation from the domain where the defining series (25.12) actually converges. Now let us return to (25.11) and apply (25.13):

$$\ln Z = -\frac{1}{2} \ln \det \Omega = \frac{1}{2} \zeta_\Omega'(0) \tag{25.15}$$

where

$$\zeta_\Omega(s) = \sum_{n \in \mathbb{Z}} \lambda_n^{-s}$$

with

$$\lambda_n = \left(\frac{2\pi n}{\beta} \right)^2 + \omega^2 , \quad n \in \mathbb{Z} . \tag{25.16}$$

Expression (25.16) is also known as the "Matsubara rule" for Bose particles. In order to evaluate (25.15), we have to discuss the thermal ζ-function,

$$\zeta_\Omega(s) = \sum_{n=-\infty}^{+\infty}\left[\left(\frac{2\pi}{\beta}\right)^2 n^2 + \omega^2\right]^{-s} = \left(\frac{2\pi}{\beta}\right)^{-2s}\sum_{n=-\infty}^{+\infty}\left[n^2 + \left(\frac{\omega\beta}{2\pi}\right)^2\right]^{-s}$$
$$=: \left(\frac{\beta}{2\pi}\right)^{2s} D\left(s, \frac{\omega\beta}{2\pi}\right)$$

or

$$\zeta_\Omega(s) = \left(\frac{\beta}{2\pi}\right)^{2s} D(s, \nu) \tag{25.17}$$

where

$$D(s,\nu) := \sum_{n=-\infty}^{+\infty}\left(n^2+\nu^2\right)^{-s}, \quad \nu := \frac{\omega\beta}{2\pi}. \tag{25.18}$$

The series (25.18) converges for $\mathrm{Re}\, s > 1/2$, and its analytic continuation defines a meromorphic function of s, analytic at $s = 0$. Further properties are:

$$D(0,\nu) = 0, \tag{25.19}$$

which implies $\zeta_\Omega(0) = 0$, and since

$$\det(\mu\Omega) = \mu^{\zeta_\Omega(0)}\det\Omega = \det\Omega, \tag{25.20}$$

there is no dependence on the overall scale of Ω. In fact the first equal sign in (25.20) follows from

$$\det(\mu\Omega) = \mathrm{e}^{-\zeta'_{\mu\Omega}(0)} = \exp\left[-\frac{d}{ds}\left(\zeta_{\mu\Omega}(s)\right)\Big|_{s=0}\right].$$

Here we need the scaling property

$$\zeta_{\mu\Omega}(s) = \sum_n (\mu\lambda_n)^{-s} = \mu^{-s}\sum_n \lambda_n^{-s} = \mu^{-s}\zeta_\Omega(s),$$
$$= \mathrm{e}^{-s\ln\mu}\,\zeta_\Omega(s)$$

which yields

$$\frac{d}{ds}\zeta_{\mu\Omega}(s) = -\ln\mu\,\mathrm{e}^{-s\ln\mu}\,\zeta_\Omega(s) + \mathrm{e}^{-s\ln\mu}\,\zeta'_\Omega(s)$$

and, putting $s = 0$,

$$-\frac{d}{ds}\zeta_{\mu\Omega}(s)\Big|_{s=0} = \ln(\mu)\zeta_\Omega(0) - \zeta'_\Omega(0),$$

we obtain indeed

$$\mathrm{e}^{-\zeta'_{\mu\Omega}(0)} = \mu^{\zeta_\Omega(0)}\,\mathrm{e}^{-\zeta'_\Omega(0)} = \mathrm{e}^{-\zeta'_\Omega(0)}\,.$$

In addition to (25.18), we list the following properties:

$$\frac{\partial}{\partial s}D(s,\nu)|_{s=0} = -2\ln\bigl(2\sinh(\pi\nu)\bigr)\,,$$
$$D(1,\nu) = \frac{\pi}{\nu}\coth(\pi\nu)\,,$$
$$D\left(-\tfrac{1}{2},0\right) = -\tfrac{1}{6}\,,\quad D\left(-\tfrac{3}{2},0\right) = \tfrac{1}{60}\,.$$

$D(s,\nu)$ has poles at

$$s = \tfrac{1}{2}, -\tfrac{1}{2}, -\tfrac{3}{2}, -\tfrac{5}{2}, \ldots\,.$$

We are interested in

$$\zeta'_\Omega(s)|_{s=0} = \frac{d}{ds}\left(\frac{\beta}{2\pi}\right)^{2s}\Big|_{s=0}\overbrace{D(0,\nu)}^{=0} + \left(\frac{\beta}{2\pi}\right)^{2s}\Big|_{s=0}\frac{d}{ds}D(s,\nu)|_{s=0}$$
$$= -2\ln[2\sinh(\pi\nu)] = -2\ln\left[2\sinh\left(\frac{\omega\beta}{2}\right)\right]$$
$$= -2\left[\ln 2 + \ln\sinh\left(\frac{\omega\beta}{2}\right)\right]\,.$$

Setting $x := \omega\beta/2$, we can continue to write

$$\zeta'_\Omega(s)|_{s=0} = -2\left[\ln 2 + \ln\left(\frac{\mathrm{e}^x - \mathrm{e}^{-x}}{2}\right)\right]$$
$$= -2\bigl[\ln 2 - \ln 2 + \ln\bigl(\mathrm{e}^x(1-\mathrm{e}^{-2x})\bigr)\bigr]$$
$$= -2\bigl[x + \ln\bigl(1-\mathrm{e}^{-2x}\bigr)\bigr]\,.$$

So we finally obtain

$$\zeta'_\Omega(s)|_{s=0} = -\omega\beta - 2\ln\bigl(1-\mathrm{e}^{-\omega\beta}\bigr)$$

or, according to (25.15),

$$\ln Z = \tfrac{1}{2}\zeta'_\Omega(0) = -\tfrac{1}{2}\omega\beta - \ln\bigl(1-\mathrm{e}^{-\omega\beta}\bigr)\,. \qquad (25.21)$$

Note that we have clearly isolated the zero-point energy. Another useful form is

$$\ln Z = \frac{1}{2}\zeta'_\Omega(0) = -\ln\left[2\sinh\left(\frac{\omega\beta}{2}\right)\right]$$

which implies

$$Z = \frac{1}{2\sinh(\omega\beta/2)}\,. \qquad (25.22)$$

Admittedly, the above procedure is a little far-fetched. But it is exactly the way that is used to great advantage in field theory. This, however, is a totally different story and can be looked up in the authors' contributions on ζ-function regularization in quantum field theory as published in Springer's Lecture Notes in Physics, Nos. 220 and 244.

26. Introduction to Homotopy Theory

Consider two manifolds X and Y together with a set of continuous maps $f, g, \ldots$

$$f:\ X \to Y\,,\quad x \to f(x) = y\,;\quad x \in X\,,\quad y \in Y\,.$$

Then two maps are defined to be homotopic if they can be continuously distorted into one another. That is, f is homotopic to g, $f \sim g$, if there exists an intermediate family of continuous maps $H(x, t)$, $0 \leqslant t \leqslant 1$,

$$H:\ X \times I \to Y\,,\quad I = [0, 1] \tag{26.1}$$

such that

$$H(x, 0) = f(x)\,,\quad H(x, 1) = g(x)\,. \tag{26.2}$$

H is then called a homotopy between f and g.

Next we define a product of paths:

$$f * g:\ [0, 1] \to Y$$

or

$$x \to (f * g)(x) := \begin{cases} f(2x)\,, & x \in \left[0,\, \frac{1}{2}\right]\,, \\ g(2x - 1)\,, & x \in \left[\frac{1}{2},\, 1\right]\,. \end{cases}$$

The inverse of f is

$$f^{-1}:\ [0, 1] \to Y$$

or

$$x \to f^{-1}(x) := f(1 - x)\,,\quad \forall x \in [0, 1]\,.$$

The homotopy relation, usually indicated by $\sim$, is an equivalence relation on the set of continuous maps $X \to Y$:

(a) reflexive: $f \sim f$

(b) symmetric: $f_1 \sim f_2 \Rightarrow f_2 \sim f_1$

(c) transitive: $f_1 \sim f_2 \wedge f_2 \sim f_3 \Rightarrow f_1 \sim f_3$.

Let us verify (b) and (c) with the aid of equations (26.1) and (26.2).

(b′) Define $G(x,t) := F(x, 1-t)$, $t \in [0,1]$, then

$$\begin{array}{lcl} F: f_1 \sim f_2 & & G: f_2 \sim f_1 \\ F(x,0) = f_1(x) & \Rightarrow & G(x,0) = f_2(x) \\ F(x,1) = f_2(x) & & G(x,1) = f_1(x) \end{array}$$

$$\text{since} \quad \begin{aligned} G(x,0) &= F(x,1) = f_2(x) \\ G(x,1) &= F(x,0) = f_1(x)\ . \end{aligned}$$

(c′) $F: f_1 \sim f_2 \quad \wedge \quad G: f_2 \sim f_1 \quad \Rightarrow \quad H: f_1 \sim f_3$

where H is defined by $H(x,t) = \begin{cases} F(x,2t)\ , & 0 \leqslant t \leqslant \frac{1}{2} \\ G(x, 2t-1)\ , & \frac{1}{2} \leqslant t \leqslant 1\ . \end{cases}$

$$\begin{array}{lll} F(x,0) = f_1(x) & G(x,0) = f_2(x) & H(x,0) = f_1(x) \\ F(x,1) = f_2(x) & G(x,1) = f_3(x) & H(x,1) = f_3(x) \end{array}$$

$$\text{since} \quad \begin{aligned} H(x,0) &= F(x,0) = f_1(x) \\ H\left(x,\tfrac{1}{2}\right) &= F(x,1) = f_2(x) \\ H\left(x,\tfrac{1}{2}\right) &= G(x,0) = f_2(x) \\ H(x,1) &= G(x,1) = f_3(x)\ . \end{aligned}$$

So homotopically equivalent maps form an equivalence class. Hence the equivalence relation $\sim$ decomposes the set of maps $\{f: X \to Y\}$ into equivalence classes called homotopy classes H_{F_i}:

$$H_{F_i} = \left\{f:\ X \to Y | f \sim F_i\right\}\ ,$$

where F_i denotes a representative of the equivalence class H_{F_i}.

Next we want to show that homotopy classes can assume a group structure under multiplication. This can best be seen by taking a specific example for X, namely, the closed line interval $I[0,1]$ with end points identified. This manifold is topologically equivalent to a circle S^1. We want to restrict ourselves to continuous maps f which satisfy $f(0) = f(1) = y_0$, a fixed point in Y; then the equivalence classes $f, g, \ldots$ form a group under the multiplication defined on the previous page. Notice that the multiplication is independent of our choice of representatives because

$$(f_1 \sim f_2) \wedge (g_1 \sim g_2) \Rightarrow f_1 g_1 \sim f_2 g_2\ .$$

In particular,

$$(f \sim F_i) \wedge (g \sim F_j) \Rightarrow fg \sim F_i F_j$$

or

$$f \in H_{F_i}\ , \quad g \in H_{F_j} \Rightarrow fg \in H_{F_i F_j}\ .$$

Also note that the identity element $e = ff^{-1}$ is the class of mappings homotopic to the constant mapping C:

$$x \to C(x) = y_0 , \quad \forall x \in [0, 1] .$$

So we have discovered that our homotopy classes form a group under multiplication. Our particular example with $X = S^1$ is called the first homotopy group of Y:

$$\Pi_1(Y)$$

with group properties

(1) $H_{F_i} H_{F_j} = H_{F_i F_j}$

(2) neutral element: C (all mappings which are homotopic to the constant mapping)

(3) inverse element: $H_{F_i}^{-1} = H_{F_i^{-1}}$.

Group manifolds are topological spaces, and it is possible to calculate the first homotopy group for any Lie group G. Let us demonstrate this explicitly for $G =$ U(1), so that we have

$$X = S^1 , \quad Y = \mathrm{U}(1) ,$$

i.e., $f : S^1 \to \mathrm{U}^1$ or $S^1 \to S^1$. What, then, is $\Pi_1(U_1)$? Let us parametrize

$$X = S^1 = \{(\cos\theta, \sin\theta) | \theta \in [0, 2\pi]\} , \quad Y = \mathrm{U}(1) = \left\{ \mathrm{e}^{\mathrm{i}\alpha} | \alpha \in [0, 2\pi] \right\} .$$

We shall now decompose these mappings into homotopy classes.

$$f : \; [0, 2\pi] \to \mathrm{U}(1) , \quad \theta \to f(\theta) = \mathrm{e}^{\mathrm{i}\alpha(\theta)} \tag{26.3}$$

with

$$f(0) = f(2\pi) = y_0 = 1 \in \mathrm{U}(1)$$

so that

$$f(0) = \mathrm{e}^{\mathrm{i}\alpha(0)} = \mathrm{e}^{\mathrm{i}\alpha(2\pi)} = f(2\pi) ,$$

which yields

$$\alpha(2\pi) = \alpha(0) + 2\pi n , \quad n \in \mathbb{Z} . \tag{26.4}$$

Hence, $\theta \to f(\theta) = \mathrm{e}^{\mathrm{i}\alpha(\theta)}$ becomes decomposed into homotopy classes which are classified by an integer $n \in \mathbb{Z}$. We now give some examples.

$$\begin{array}{lll} 1) & f_1(\theta) = \mathrm{e}^{\mathrm{i}\,\sin\,\theta}\,, & \alpha_1(\theta) = \sin\,\theta \\ & f_2(\theta) = 1\,, & \alpha_2(\theta) = 0 \\ & & \alpha_{1,2}(0) = \alpha_{1,2}(\pi)\,; \quad n_{1,2} = 0\,. \end{array}$$

f_1 and f_2 are homotopic since they can be continuously deformed into each other via a sequence of mappings, namely a homotopy given by

$$F(\theta, t) := \mathrm{e}^{\mathrm{i}t\,\sin\,\theta}$$

so that

$$F(\theta, 0) = 1 = f_2(\theta)$$

and

$$F(\theta, 1) = \mathrm{e}^{\mathrm{i}\,\sin\,\theta} = f_1(\theta)\,.$$

So, all mappings with $\alpha(0) = \alpha(2\pi) = 0$ (which includes the identity mapping) belong to the same homotopy class.

$$\begin{array}{lll} 2) & f_1(\theta) = \mathrm{e}^{\mathrm{i}\theta}\,, & \alpha_1(\theta) = \theta\,, \\ & f_2(\theta) = 1\,, & \alpha_2(\theta) = 0\,, \\ & \alpha_1(2\pi) = \alpha_1(0) + 1 \cdot 2\pi\,, & n_1 = 1\,, \\ & \alpha_2(2\pi) = \alpha_2(0)\,, & n_2 = 0\,. \end{array}$$

This example yields two mappings, f_1 and f_2, which are not homotopic; i.e., there is no way to continuously deform f_1 into f_2, although f_1 of the present example is not so very different from f_1 of the first example. $f_1(\theta) = \mathrm{e}^{\mathrm{i}\theta}$ identifies a new homotopy class which does not include the identity.

$$\begin{array}{llll} 3) & f_1(\theta) = \mathrm{e}^{\mathrm{i}n\theta}\,, & \alpha_1 = n\theta\,, & \\ & f_2(\theta) = \mathrm{e}^{\mathrm{i}m\theta}\,, & \alpha_2 = m\theta\,, & m, n \in \mathbb{Z} \\ & \alpha_1(2\pi) = \alpha_1(0) + 2\pi n\,, & n_1 = n\,, & \\ & \alpha_2(2\pi) = \alpha_2(0) + 2\pi m\,, & n_2 = m\,. & \end{array}$$

f_1 and f_2 are not homotopic for $m \neq n$, and they are homotopic if, and only if, $m = n$.

It is clear that any mapping $f : \theta \to f(\theta)$, because it has to satisfy (26.4), is homotopic to one, and only one, of the mappings

$$f_n(\theta) = \mathrm{e}^{\mathrm{i}n\theta}\,, \quad n \in \mathbb{Z}\,, \tag{26.5}$$

so that

$$\Pi_1\big(\mathrm{U}(1)\big) = \mathbb{Z}\,. \tag{26.6}$$

The group property can be illustrated as follows:

Inverse element of $\{f|f \sim f_n\}$ is given by $\{f|f \sim f_{-n}\}$.

Neutral element:

$$\{f|f \sim f_0\}\,, \quad f_0(\theta) = 1 = y_0\,.$$

Composition (multiplication):

$$f_1 \sim f_n,\, f_2 \sim f_m \Rightarrow f_1 f_2 \sim f_n f_m = f_{n+m}$$

since

$$f_n(\theta) f_m(\theta) = \mathrm{e}^{\mathrm{i}n\theta}\, \mathrm{e}^{\mathrm{i}m\theta} = \mathrm{e}^{\mathrm{i}(n+m)\theta} = f_{n+m}(\theta)\,.$$

Obviously, when θ varies from 0 to 2π, so that $n\theta = 0, \ldots 2\pi n$, $f_n(\theta) = \mathrm{e}^{\mathrm{i}n\theta}$ goes around the U(1) circle n times. Hence, one round-trip in $X = S^1$ corresponds to n round-trips in $Y = \mathrm{U}(1)$, or n points in $X = S^1$ have the same image point in U(1).

For obvious reasons, n is called the winding number, and any mapping homotopic to $f_n(\theta)$ also winds around the group space U(1) n times.

After having explained what the first homotopy group is all about, we give the definition of the nth homotopy group $\Pi_n(Y)$: the nth homotopy group $\Pi_n(Y)$ is the set of all equivalence classes of maps of the unit sphere S^n into a topological space Y. As has already been stated above, two mappings are said to be homotopic if they can be continuously deformed into each other, i.e., the mappings

$$f:\ S^n \to Y$$

are decomposed into equivalence classes which are the group elements of

$$\Pi_n(Y)\,.$$

Unfortunately, all of this is hard to visualize for the higher homotopy groups. Nevertheless, here are some results which can be understood, to some extent, by analogy with Π_1:

$$\begin{aligned}
&\Pi_1\big(\mathrm{U}(1)\big) = \Pi_1\big(S^1\big) = \mathbb{Z}\\
&\Pi_1\big(\mathrm{SU}(2)\big) = \Pi_1\big(S^3\big) = 0\\
&\Pi_1\big(\mathrm{U}(2)\big) = \mathbb{Z}\\
&\Pi_n\big(S^m\big) = 0\,,\ n < m\\
&\Pi_n\big(S^n\big) = \mathbb{Z}\\
&\Pi_n\big(S^1\big) = 0\,,\ n > 1\\
&\Pi_3\big(\mathrm{SU}(2)\big) = \Pi_3\big(S^3\big) = \mathbb{Z}\\
&\Pi_3\big(\mathrm{SO}(4)\big) = \mathbb{Z} \times \mathbb{Z}\\
&\Pi_1\big(\mathrm{Sp}(2N)\big) = \mathbb{Z}\,.
\end{aligned}$$

27. Classical Chern–Simons Mechanics

We are interested in a completely integrable Hamiltonian system $(\mathcal{M}_{2N}, \omega, H)$. Local coordinates on the $2N$-dimensional phase space $\mathcal{M}_{2N}$ are denoted by $\eta^a = (p, q)$, $a = 1, 2, \ldots 2N$ and the symplectic 2-form ω is given by

$$\omega = \tfrac{1}{2}\omega_{ab} d\eta^a \wedge d\eta^b$$

with $d\omega = 0$ and $\det(\omega_{ab}) \neq 0$. For simplicity we assume that ω_{ab} is η^a-independent. In canonical form we have $\eta^a = (p^i, q^i)$, i $= 1, 2, \ldots, N$. If $N = 1$, e.g., $\eta^1 = p$, $\eta^2 = q$, $\omega_{pq} = -\omega_{qp} = 1$ and $\omega = dp \wedge dq$. By definition, the system has N conserved quantitites $J_i(\eta^a)$, which are in involution:

$$\{H, J_i\} \equiv \frac{\partial H}{\partial \eta^a}\omega^{ab}\frac{\partial J_i}{\partial \eta^b} = 0 \,, \tag{27.1}$$

$$\{J_i, J_k\} \equiv \frac{\partial J_i}{\partial \eta^a}\omega^{ab}\frac{\partial J_k}{\partial \eta^b} = 0 \,. \tag{27.2}$$

In canonical form:

$$\omega_{ab} = \operatorname{diag}\left[\begin{pmatrix} 0 & 1 \\ -1 & 0 \end{pmatrix}, \ldots, \begin{pmatrix} 0 & 1 \\ -1 & 0 \end{pmatrix}\right] . \tag{27.3}$$

The Poisson brackets (27.1, 2) are defined in terms of the matrix ω^{ab}, the inverse of ω_{ab}:

$$\omega_{ab}\omega^{bc} = \delta_a^c \,. \tag{27.4}$$

Hamilton's equations of motion read:

$$\dot{\eta}^a = \omega^{ab}\frac{\partial H}{\partial \eta^b} \equiv \omega^{ab}\partial_b H \tag{27.5}$$

and the total time derivative of some function of the canonical variables $\eta^a(t)$, $A = A(\eta^a(t))$ is given by

$$\dot{A}(t) = \partial_a A \dot{\eta}^a = \partial_a A \omega^{ab} \partial_b H \tag{27.6}$$

or

$$\frac{d}{dt}A(\eta(t)) = \{A, H\} , \tag{27.7}$$

where

$$\{A, B\} = \partial_a A \omega^{ab} \partial_b B . \tag{27.8}$$

To make contact with the usual form of Hamilton's equations, let us write (27.5) in components:

$$\dot{p} = \omega^{pq} \partial_q H = -\frac{\partial H}{\partial q} ,$$
$$\dot{q} = \omega^{qp} \partial_p H = \frac{\partial H}{\partial p} ,$$

which results in

$$\omega^{qp} = -\omega^{pq} = 1 , \tag{27.9}$$

and, together with (27.4), i.e.,

$$\omega_{pq}\omega^{qp} = 1 = \omega_{qp}\omega^{pq} \tag{27.10}$$

yields, as it should,

$$\omega_{qp} = -1 = -\omega_{pq} . \tag{27.11}$$

The Lagrangian that produces the correct equations of motion is given by

$$L = \tfrac{1}{2}\eta^a \omega_{ab} \dot{\eta}^b - H(\eta^a) . \tag{27.12}$$

To show this, let us look at the variation of the action,

$$S = \int_{t_1}^{t_2} dt\, L , \tag{27.13}$$

under $\eta \to \eta + \delta\eta$. According to the action principle, it holds that

$$\delta S = \delta \int_{t_1}^{t_2} dt\, L = G[t_2] - G[t_1] . \tag{27.14}$$

For our particular Lagrangian (27.12), we obtain:

$$\delta S = \int_{t_1}^{t_2} dt \left[\tfrac{1}{2}\delta\eta^a \omega_{ab} \dot{\eta}_b + \tfrac{1}{2}\eta^a \omega_{ab} \delta\dot{\eta}_b - \delta H(\eta^a)\right] . \tag{27.15}$$

Writing

$$\delta H(\eta) = \delta\eta^a \partial_a H(\eta)$$

and

$$
\begin{aligned}
\eta^a \omega_{ab} \delta \dot{\eta}_b &= \frac{d}{dt}\left(\eta^a \omega_{ab} \delta \eta^b\right) - \dot{\eta}^a \omega_{ab} \delta \eta^b \\
&= \frac{d}{dt}\left(\eta^a \omega_{ab} \delta \eta^b\right) + \delta \eta^a \omega_{ab} \dot{\eta}^b
\end{aligned}
\tag{27.16}
$$

we find

$$
\begin{aligned}
\delta S &= \int_{t_1}^{t_2} dt \left[\delta \eta^a \omega_{ab} \dot{\eta}^b - \delta \eta^a \partial_a H(\eta) + \frac{1}{2} \frac{d}{dt} \left(\eta^a \omega_{ab} \delta \eta^b\right) \right] \\
&= \int_{t_1}^{t_2} dt \frac{d}{dt} \left[\frac{1}{2} \eta^a \omega_{ab} \delta \eta^b \right] + \int_{t_1}^{t_2} dt \left[\delta \eta^a \left(\omega_{ab} \dot{\eta}^b - \partial_a H(\eta) \right) \right] \\
&= G_2 - G_1 \,.
\end{aligned}
\tag{27.17}
$$

From (27.17) we conclude

$$
\begin{aligned}
&\omega_{ab} \dot{\eta}^b = \partial_a H(\eta) \,, \\
&\Rightarrow \omega^{ca} \omega_{ab} \dot{\eta}^b = \omega^{ca} \partial_a H(\eta) \,, \\
&\Rightarrow \dot{\eta}^c = \omega^{ca} \partial_a H(\eta) \quad \text{or} \\
&\dot{\eta}^a = \omega^{ab} \partial_b H\big(\eta(t)\big) \,,
\end{aligned}
\tag{27.18}
$$

which brings us back to (27.5). The surface term in (27.17) is given by

$$
\left[G\right]_1^2 = \left[\tfrac{1}{2} \eta^a \omega_{ab} \delta \eta^b\right]_1^2 \,. \tag{27.19}
$$

At this point we leave the on-shell theory; i.e., we are still given a Hamiltonian H but we are not assuming that Hamilton's equations (27.5) are satisfied; i.e., we are concerned with "off-shell" mechanics.

Next we study the response of the action (27.13) under an infinitesimal variation of the form

$$
\delta \eta^a - \sum_{i=1}^{N} \varepsilon_i \omega^{ab} \partial_b J_i\left(\eta^c\right) \,, \tag{27.20}
$$

where the "charges" $J_i(\eta^a)$ are the generators of the infinitesimal canonical transformation (27.20). The ε_i are constant parameters – as in a global gauge transformation. Accordingly, a path on $\mathcal{M}_{2N}$, $\eta^a(t)$, transforms as

$$
\delta \eta^a(t) = \varepsilon_i \omega^{ab} \partial_b J_i\big(\eta^c(t)\big) \,. \tag{27.21}
$$

Here and in the following, a summation over $i = 1, 2, \ldots, N$ is understood. Variation of the action yields:

$$
\begin{aligned}
\delta S &= \int_{t_1}^{t_2} dt \frac{d}{dt}\left[\frac{1}{2}\eta^a \omega_{ab}\delta\eta^b\right] + \int_{t_1}^{t_2} dt \left[\delta\eta^a \omega_{ab}\dot{\eta}^b - \delta\eta^a \partial_a H(\eta)\right] \\
&= \int_{t_1}^{t_2} dt \frac{d}{dt}\left[\frac{1}{2}\eta^a \omega_{ab}\varepsilon_i \omega^{bc}\partial_c J_i\right] + \varepsilon_i \int_{t_1}^{t_2} dt \Big[\underbrace{\omega^{ac}\omega_{ab}}_{-\delta^c_b}\dot{\eta}^b \partial_c J_i - \underbrace{\partial_b J_i \omega^{ab}\partial_a H}_{=\{H,J_i\}=0}\Big] \\
&= \frac{1}{2}\varepsilon_i \left[\eta^a \partial_a J_i\right]_{t_1}^{t_2} - \varepsilon_i \int_{t_1}^{t_2} dt \frac{d}{dt} J_i\big(\eta(t)\big) = \varepsilon_i \left[\frac{1}{2}\eta^a \partial_a J_i - J_i\right]_{t_1}^{t_2} .
\end{aligned}
\tag{27.22}
$$

This is a pure surface term.

Next we gauge our $U(1)^N$-symmetry; i.e., we make (27.20) a local symmetry transformation by allowing ε to depend on time. As in ordinary gauge field theory, this requires the introduction of a $U(1)^N$-gauge field $A_i(t)$ which couples to the "matter field" $\eta^a(t)$:

$$
L_0 = \frac{1}{2}\eta^a \omega_{ab}\dot{\eta}^b - H(\eta) - A_i(t) J_i\big(\eta(t)\big) , \tag{27.23}
$$

$$
\begin{aligned}
S_0\left[\eta^a, A_i\right] &= \int_{t_1}^{t_2} dt\, L_0 \\
&= \int_{t_1}^{t_2} dt \left[\frac{1}{2}\eta^a \omega_{ab}\dot{\eta}^b - H(\eta) - A_i(t) J_i\big(\eta(t)\big)\right] .
\end{aligned}
\tag{27.24}
$$

Variation of S_0 with respect to η^a and A_i yields:

$$
\begin{aligned}
\delta S_0 &= \int_{t_1}^{t_2} dt \Big[\frac{1}{2}\delta\eta^a \omega_{ab}\dot{\eta}^b + \frac{1}{2}\underbrace{\eta^a \omega_{ab}\frac{d}{dt}\delta\eta^b}_{(d/dt)\left(\eta^a\omega_{ab}\delta\eta^b\right)+\delta\eta^a\omega_{ab}\dot{\eta}^b} \\
&\quad - \overbrace{\delta H}^{\delta\eta^a\partial_a H} - A_i \overbrace{\delta J_i}^{\delta\eta^a\partial_a J_i} - \delta A_i J_i\Big] \\
&= \int_{t_1}^{t_2} dt \frac{d}{dt}\left[\frac{1}{2}\eta^a \omega_{ab}\delta\eta^b\right] + \int_{t_1}^{t_2} dt \Big[\delta\eta^a \Big(\omega_{ab}\dot{\eta}^b \\
&\quad - \partial_a\big(H + A_i J_i\big)\Big) - \delta A_i J_i\Big] .
\end{aligned}
\tag{27.25}
$$

The equations of motion are given by

$$
\begin{aligned}
\delta\eta^a : \quad & \omega_{ab}\dot{\eta}^b = \partial_a\big(H + A_i J_i\big) \quad \text{or} \\
& \dot{\eta}^a = \omega^{ab}\partial_b\big(H(\eta) + A_i J_i(\eta)\big) ,
\end{aligned}
\tag{27.26}
$$

$$
\delta A_i : \quad J_i = 0 . \tag{27.27}
$$

Now we study the response of S_0 under the local gauge transformation – and consider off-shell dynamics again:

$$\delta\eta^a(t) = \varepsilon_i(t)\omega^{ab}\partial_b J_i\big(\eta^c(t)\big)\ , \tag{27.28}$$

$$\delta A_i(t) = \dot{\varepsilon}_i\ . \tag{27.29}$$

In (27.25) we need the expressions

$$\tfrac{1}{2}\eta^a\omega_{ab}\delta\eta^b = \tfrac{1}{2}\eta^a\omega_{ab}\varepsilon_i(t)\omega^{bc}\partial_c J_i = \tfrac{1}{2}\varepsilon_i(t)\eta^a\partial_a J_i(\eta) \tag{27.30}$$

$$\begin{aligned}\delta\eta^a\left[\omega_{ab}\dot{\eta}^b - \partial_a\big(H + A_j J_j\big)\right] &= \varepsilon_i(t)\omega^{ab}\partial_b J_i\left(\omega_{ac}\dot{\eta}^c - \partial_a H - A_j\partial_a J_j\right)\\ &= -\varepsilon_i(t)\dot{\eta}^b\partial_b J_i\big(\eta(t)\big) - \varepsilon_i(t)\partial_a H\omega^{ab}\partial_b J_i - \varepsilon_i(t)A_j(t)\partial_a J_j\omega^{ab}\partial_b J_i\\ &= -\varepsilon_i(t)\frac{d}{dt}J_i\big(\eta(t)\big) - \varepsilon_i(t)\underbrace{\{H, J_i\}}_{=0} - \varepsilon_i(t)A_j(t)\underbrace{\{J_j, J_i\}}_{=0}\ ,\end{aligned} \tag{27.31}$$

$$\delta A_i J_i = \dot{\varepsilon}_i J_i\ . \tag{27.32}$$

Using (27.30–32) in (27.25) we obtain:

$$\begin{aligned}\delta S_0 &= \int_{t_1}^{t_2} dt\frac{d}{dt}\left[\frac{1}{2}\varepsilon_i(t)\eta^a\partial_a J_i\right] - \int_{t_1}^{t_2} dt\ \varepsilon_i(t)\frac{d}{dt}J_i\\ &\quad - \int_{t_1}^{t_2} dt\quad \underbrace{\dot{\varepsilon}_i J_i\big(\eta(t)\big)}_{=(d/dt)\left(\varepsilon_i J_i\right) - \varepsilon_i(dJ_i/dt)}\\ &= \int_{t_1}^{t_2} dt\frac{d}{dt}\left[\frac{1}{2}\varepsilon_i(t)\eta^a\partial_a J_i - \varepsilon_i(t)J_i\right]\\ &= \left[\varepsilon_i(t)\left(\frac{1}{2}\eta^a\partial_a J_i - J_i\right)\right]_1^2\ .\end{aligned} \tag{27.33}$$

[The surface term vanishes identically if $J(\eta)$ is quadratic in η: $J(\eta) = Q_{ab}\eta^a\eta^b/2$, Q symmetric. Then $\eta^a\partial_a J/2 = \eta^a Q_{cb}\eta^b/2 = J$.]

Thus, S_0 will be invariant under (27.28, 29) if the surface terms vanish, which is certainly true for closed trajectories and "small" gauge transformations (cf. below): $\varepsilon_i(t_2) = \varepsilon_i(t_1)$. Then

$$\begin{aligned}\delta S_0 &= \left[\varepsilon_i(t)\left(\frac{1}{2}\eta^a\partial_a J_i\big(\eta(t)\big) - J_i\big(\eta(t)\big)\right)\right]_{t_1}^{t_2}\\ &= \varepsilon_i\big(t_1\big)\underbrace{\left[\frac{1}{2}\eta^a\partial_a J_i\big(\eta(t)\big) - J_i\big(\eta(t)\big)\right]_{t_1}^{t_2}}_{=0 \text{ for closed trajectory.}}\end{aligned} \tag{27.34}$$

Usually one adds to (27.24) a gauge invariant kinetic term like $F_{\mu\nu}F^{\mu\nu}$. However, in $0+1$ dimensions, such a term does not exist; hence, the only term which can be added to S containing the gauge field A_i alone is the Chern–Simons action:

$$S_{CS}\big[A_i\big] = k_i\int_{t_1}^{t_2} dt\ A_i(t)\ . \tag{27.35}$$

So far the k_i are arbitrary real constants. The variation of S_{CS} is given by

$$\begin{aligned}\delta S_{\text{CS}} &= k_i \int_{t_1}^{t_2} dt\, \delta A_i(t) = k_i \int_{t_1}^{t_2} dt\, \dot{\varepsilon}_i \\ &= k_i \big(\varepsilon_i(t_2) - \varepsilon_i(t_1)\big) . \end{aligned} \tag{27.36}$$

Evidently S_{CS} is invariant under "small" gauge transformations with $\varepsilon_i(t_2) - \varepsilon_i(t_1) = 0$, but it is not invariant under "large" gauge transformations with $\varepsilon_i(t_2) - \varepsilon_i(t_1) \neq 0$.

Now, the complete action of interest reads:

$$\begin{aligned} S\left[\eta^a, A_i\right] &= S_0\left[\eta^a, A_i\right] + S_{\text{CS}}\left[A_i\right] \\ &= \int_{t_1}^{t_2} dt \left[\frac{1}{2}\eta^a \omega_{ab} \dot{\eta}^b - H(\eta) - A_i\big(J_i(\eta) - k_i\big)\right] . \end{aligned} \tag{27.37}$$

To derive the classical equations of motion belonging to (27.37), we can follow the same steps as before [cf. (27.26, 27)]:

$$\delta\eta^a : \quad \dot{\eta}^a(t) = \omega^{ab}\partial_b\big(H + A_i J_i\big)\big(\eta(t)\big) , \tag{27.38}$$

$$\delta A_i : \quad J_i\big(\eta(t)\big) = k_i . \tag{27.39}$$

In the sequel we will show that, using appropriate boundary conditions, we can always gauge A_i to zero. As a result, we get back the usual equation of motion (27.18), $\dot{\eta}^a = \omega^{ab}\partial_b H$, but supplemented by the "Gauss law constraints" (27.39). Hence, only those trajectories $\eta^a(t)$ are admitted for which the J_i equal the constant coefficients k_i occuring in the Chern–Simons term (27.35). The level surfaces of $J_i(\eta^a)$ induce a foliation of phase space by N-tori and, since $\{H, J_i\} = 0$, a classical trajectory which starts on a given torus $T_N(k_i)$ will always stay on this particular torus. Thus we are dealing with different classical Chern–Simons theories, "living" on different tori $T_N(k_i)$, for different values of the parameters k_i.

Eventually we want to quantize the model (27.37) by way of a path integral. Therefore we need to know all closed classical trajectories of period T which serve as "background fields" for the one-loop approximation.

Let us begin by introducing action-angle variables (I_i, θ_i) as coordinates on phase space. So we perform a canonical transformation on $\mathcal{M}_{2N} : \eta^a \to (I_i, \theta_i)$. The actions I_i, $i = 1, 2, \ldots N$ fix certain tori on $\mathcal{M}_{2N}$. Since we are studying integrable systems, the solutions of Hamilton's equations simply read:

$$I_i(t) = I_{i0} \tag{27.40}$$

$$\vartheta_i = \theta_{i0} + \omega_i\big(I_0\big)t , \tag{27.41}$$

with the frequencies

$$\omega_i\big(I_0\big) = \left.\frac{\partial H(I)}{\partial I_i}\right|_{I=I_0} . \tag{27.42}$$

For closed trajectories it holds that

$$\omega_i\big(I_0\big) = \frac{2\pi}{T} p_i \,, \quad p_i \in \mathbb{Z} \,, \quad i = 1, 2, \ldots, N \,. \tag{27.43}$$

Thus, all closed classical solutions are described in action-angle variables by

$$I_i(t) = I_{i0} \,, \tag{27.44}$$

$$\theta_i(t) = \theta_{i0} + \frac{2\pi}{T} p_i t \,. \tag{27.45}$$

At this stage we return to the previously introduced gauge transformations. Let $\eta^a(t) \equiv (I_i(t), \theta_i(t))$, $t \in [0, T]$, be an arbitrary closed path on $\mathcal{M}_{2N}$ with period T, $\eta^a(0) = \eta^a(T)$, or

$$\theta_i(T) - \theta_i(0) = 2\pi p_i \,, \quad p_i \in \mathbb{Z} \,. \tag{27.46}$$

Since we will be interested in closed paths contributing to $\mathrm{Tr}(e^{-i\mathcal{H}T})$, our "field theory" defined by

$$S_0\left[\eta^a, A_i\right] = \int_0^T dt \left[\frac{1}{2}\eta^a \omega_{ab} \dot{\eta}^b - H - A_i J_i\right]$$

$$S_{CS}\left[A_i\right] = k_i \int_0^T dt\, A_i(t) \tag{27.47}$$

can be visualized as a theory of maps from the circle S^1 to the symplectic manifold $\mathcal{M}_{2N}$. In terms of action-angle variables, S_0 becomes

$$S_0\left[I_i, \theta_i, A_i\right] = \int_0^T dt\left[I_i(t)\dot{\theta}_i(t) - H\big(I(t)\big) - A_i(t) I_i(t)\right] \tag{27.48}$$

and the gauge transformations (27.28, 29) are replaced by

$$\begin{aligned} \Delta I_i(t) &= 0 \,, \\ \Delta\theta_i(t) &= \varepsilon_i(t) \,, \\ \Delta A_i(t) &= \dot{\varepsilon}_i(t) \,, \end{aligned} \tag{27.49}$$

where the use of Δ instead of δ indicates that we also allow for finite gauge transformations. In particular, there exists the possibility of topologically nontrivial ("large") transformations which cannot be continuously deformed to the identity. They are introduced in the following way: under a gauge transformation (27.49), we obtain for (27.46):

$$\theta_i'(T) - \theta_i'(0) = \underbrace{\theta_i(T) - \theta_i(0)}_{=2\pi p_i} + \varepsilon_i(T) - \varepsilon_i(0) \,. \tag{27.50}$$

If the gauge-transformed trajectory $(I'(t), \theta_i'(t))$, $t \in [0, T]$ is to be closed again, $\varepsilon_i(T) - \varepsilon_i(0)$ must be a multiple of 2π:

$$\varepsilon_i(T) - \varepsilon_i(0) = 2\pi \mathcal{N}_i \ , \qquad \mathcal{N}_i \in \mathbb{Z} \ . \tag{27.51}$$

Consequently, relation (27.46) is changed to

$$\theta_i'(T) - \theta_i'(0) = 2\pi\big(p_i + \mathcal{N}_i\big) \equiv 2\pi p_i' \ , \quad \text{i.e.,} \quad p_i' = p_i + \mathcal{N}_i \ . \tag{27.52}$$

Previously we considered ε_i infinitesimal, so that only $\mathcal{N}_i = 0$ was possible. These transformations are called "small" or topologically trivial transformations because they can be obtained by iterating infinitesimal ones. Now we also allow for "large" gauge transformations, i.e., topologically nontrivial ones: $\mathcal{N}_i \neq 0$. They change the revolution number p_i, i.e., the number of revolutions which the angle variables perform between $t = 0$ and $t = T$. Obviously, it is impossible to gauge transform two closed paths into each other by a small gauge transformation if they have different p_i's, i.e., belong to disjoint homotopy classes in the sense of $\Pi_1(S^1) = Z$.

The condition (27.51) implies a partition of the gauge fields $A_i(t)$ in different topological classes. Now, using small gauge transformations, $\mathcal{N}_i = 0$,

$$\delta A_i = \dot{\varepsilon}_i \ , \quad \varepsilon_i(T) - \varepsilon_i(0) = 0 \ , \tag{27.53}$$

every $A_i(t)$ can be transformed into a time-independent $\mathrm{U}(1)^N$-gauge field:

$$\tilde{A}_i = \frac{1}{T}\int_0^T dt\, A_i(t) \ . \tag{27.54}$$

This quantity is invariant under small gauge transformations but changes under large ones:

$$\Delta\tilde{A}_i = \frac{1}{T}\int_0^T dt\, \dot{\varepsilon}_i(t) = \frac{1}{T}\big[\varepsilon_i(T) - \varepsilon_i(0)\big] = \frac{2\pi}{T}\mathcal{N}_i \ . \tag{27.55}$$

Note that if $\tilde{A}_i$ is not an integer multiple of $2\pi/T$, A_i or $\tilde{A}_i$, respectively, cannot be gauged to zero. Since we wanted to work with the usual Hamiltonian equation of motion (27.18) by going to the $A_i = 0$-gauge, we impose the following restriction on the allowed gauge fields:

$$\tilde{A}_i = \frac{2\pi}{T} z_i \ , \quad z_i \in \mathbb{Z} \ . \tag{27.56}$$

$\{z_i\}$ enumerates the topological classes of the gauge fields. Thus, according to (27.54), the space of allowed gauge fields A_i (over which the path integral will be performed) is subject to the condition

$$\int_0^T dt\, A_i(t) = 2\pi z_i \ . \tag{27.57}$$

Using

$$\tilde{A}_i' = \tilde{A}_i + \frac{2\pi}{T}\mathcal{N}_i$$

or

$$\frac{2\pi}{T}z_i' = \frac{2\pi}{T}z_i + \frac{2\pi}{T}\mathcal{N}_i$$

we obtain

$$z_i' = z_i + \mathcal{N}_i \ , \tag{27.58}$$

i.e., large gauge transformations change the topological class of the gauge field.

Now let us go back to the action (27.48) which is invariant under infinitesimal gauge transformations. But it is also invariant under large gauge transformations:

$$\Delta S_0 = \int_0^T dt\big[I_i(t)\,\underbrace{\Delta\dot{\theta}_i}_{\dot{\varepsilon}_i(t)} - \underbrace{\Delta A_i(t)}_{\dot{\varepsilon}_i(t)}\,I_i(t)\big] \ . \tag{27.59}$$

So, the two terms in (27.59) cancel. That is precisely the reason for having introduced A_i. The Chern–Simons term,

$$S_{\text{CS}}\big[A_i\big] = k_i\int_0^T dt\ A_i(t) \ , \tag{27.60}$$

on the other hand, is invariant only under small gauge transformations. Under large ones it changes according to

$$\begin{aligned} S_{\text{CS}}\big[A_i'\big] &= k_i\int_0^T dt\ A_i' = k_i\int_0^T dt\ A_i + 2\pi\,\mathcal{N}_i k_i \\ &= S_{\text{CS}}\big[A_i\big] + 2\pi k_i\mathcal{N}_i \ , \\ \Rightarrow \Delta S_{\text{CS}} &= 2\pi k_i\mathcal{N}_i \ . \end{aligned} \tag{27.61}$$

Next we look at closed paths on $\mathcal{M}_{2N}$ which are solutions of the classical equations of motion derived from S_0 (27.48):

$$\dot{\theta}_i(t) = \frac{\partial H(I(t))}{\partial I_i(t)}\bigg|_{I_0} + A_i(t) = \omega_i\big(I_0\big) + A_i(t) \tag{27.62}$$

$$\dot{I}_i(t) = 0 \ , \tag{27.63}$$

with the solutions

$$\theta_i(t) = \theta_{i0} + \omega_i\big(I_0\big)t + \int_0^t dt'\ A_i(t') \ , \tag{27.64}$$

$$I_i(t) = I_{i0} \ . \tag{27.65}$$

A gauge transformation yields for (27.62, 63):

$$\dot{\theta}'_i(t) = \omega_i\big(I_0\big) + A_i(t) + \dot{\varepsilon}_i(t) \ , \tag{27.66}$$

$$\dot{I}'_i(t) = 0 \tag{27.67}$$

with the solutions

$$\begin{aligned} \theta'_i &= \theta_{i0} + \omega_i\big(I_0\big)t + \int_0^t dt'\, A_i(t') + \varepsilon_i(t) - \varepsilon_i(0) \\ &= \theta_i(t) + \varepsilon_i(t) - \varepsilon_i(0) \ , \\ I'_i(t) &= I_{0i} \ . \end{aligned} \tag{27.68}$$

If A_i satisfies (27.57), then (27.68) implies

$$\theta'_i(T) - \theta'_i(0) = 2\pi\big(p_i + z_i + \mathcal{N}_i\big) \ . \tag{27.69}$$

The p_i's are the numbers of revolutions of a classical solution if $A_i \equiv 0$. Adding a gauge field of topological class $\{z_i\}$ changes p_i to $p_i + z_i$. Finally, if the trajectory is gauge-transformed, the revolution numbers are changed from $p_i + z_i$ to $p_i + z_i + \mathcal{N}_i$.

Again, the number $p_i \in \mathbb{Z}$ tells us how often the trajectory (during the time T) winds around the i-th homology cycle γ_i of the torus defined by $\{I_i\}$. From the definition of the action variables we have:

$$I_i = \frac{1}{2\pi}\oint_{\gamma_i} p_i\, dq_i = \frac{1}{2\pi}\oint_{\gamma_i} \frac{1}{2}\eta^a \omega_{ab}\dot{\eta}^b \ . \tag{27.70}$$

The coordinate θ_i increases along the cycle γ_j from 0 to $2\pi\delta_{ij}$. Hence, for a path η^a winding around γ_i p_i-times, we get:

$$\int_0^T dt \frac{1}{2}\eta^a \omega_{ab}\dot{\eta}^b = 2\pi I_i p_i \ . \tag{27.71}$$

Therefore, we obtain for S_0, evaluated along a solution of Hamilton's equations:

$$\begin{aligned} S_0\big[\eta_{\rm cl} \equiv \big(I_{\rm cl}, \theta_{\rm cl}\big), A_i\big] = &\underbrace{\int_0^T dt \frac{1}{2}\eta^a \omega_{ab}\dot{\eta}^b}_{=2\pi\sum I_i p_i} \\ &- \underbrace{H(I)}_{=E}\underbrace{\int_0^T dt}_{=T} - \sum_{i=1}^N I_i \underbrace{\int_0^T dt\, A_i(t)}_{=2\pi z_i} \end{aligned}$$

or

$$S_0\big[\eta_{\rm cl}, A_i\big] = 2\pi\sum_{i=1}^N I_i\big(p_i - z_i\big) - ET \ . \tag{27.72}$$

The constants I_i and E are determined by the initial conditions: $I_i = I_i(\eta_{\text{cl}}(0))$, $E = H(\eta_{\text{cl}}(0))$. Again, we observe that S_0 is gauge invariant:

$$\begin{aligned} p_i &\to p_i' = p_i + \mathcal{N}_i\,, \quad z_i \to z_i' = z_i + \mathcal{N}_i \\ &\Rightarrow \left(p_i - z_i\right)' = p_i - z_i\,. \end{aligned}$$

If we had not introduced the $-\sum A_i I_i$-term, the action would have changed by $2\pi \sum I_i \mathcal{N}_i$.

Here, then, is a summary of how to gauge-transform Hamilton's equations of motion:

general coordinates η:

$\dot\eta^a = \omega^{ab}\partial_b\left[H + \sum_i A_i J_i\right](\eta)$	$\dot\eta'^a = \omega^{ab}\partial_b\left[H + \sum_i\left(A_i + \dot\varepsilon_i\right)J_i\right](\eta')$

action angle-coordinates:

$\dot\theta_i = \omega_i(I) + A_i(t)$	$\dot\theta_i' = \omega_i(I) + A_i(t) + \dot\varepsilon_i(t)$
$\dot I_i = 0$	$\dot I_i' = 0$

solutions:

$\theta_i(t) = \theta_i(0) + \omega_i(I)t + \int_0^t dt\, A_i(t)$	$\theta_i'(t) = \theta_i(t) + \varepsilon_i(t) - \varepsilon_i(0)$
$I_i(t) = I_i = \text{const.}$	$I_i'(t) = I_i = \text{const.}$

angles at $t = T$:

$\theta_i(T) - \theta_i(0) = \omega_i T + \int_0^T dt\, A_i$	$\theta_i'(T) - \theta_i'(0) =$
	$\theta_i(T) - \theta_i(0) + \varepsilon_i(T) - \varepsilon_i(0)$
$= 2\pi\left(p_i + z_i\right)$	$= 2\pi\left(p_i + z_i + \mathcal{N}_i\right)$

Littlejohn's winding number:

$\mathcal{W} = \sum_i \left(p_i + z_i\right) w_i$	$\mathcal{W}' = \sum_i \left(p_i + z_i + \mathcal{N}_i\right) w_i$

Hence

$$
\begin{aligned}
\Delta z_i &= \mathcal{N}_i && \text{(gauge field)},\\
\Delta p_i &= \mathcal{N}_i && \text{(classical trajectory)},\\
\Delta \mathcal{W} &= \textstyle\sum_i w_i \mathcal{N}_i && \text{(Littlejohn's winding number)}.
\end{aligned}
$$

The winding numbers $\mathcal{W}$ and w_i, respectively, will be studied in the following sections.

28. Semiclassical Quantization

We want to investigate the semiclassical or one-loop approximation of our Chern–Simons model:

$$S = S_0 + S_{CS} , \tag{28.1}$$

where

$$S_0[\eta, A] = \int_0^T dt \left[\frac{1}{2}\eta^a \omega_{ab} \dot{\eta}^b - H(\eta) - \sum_i A_i(t) J_i\big(\eta(t)\big) \right]$$

$$S_{CS}[A] = \int_0^T dt \sum_i k_i A_i(t) \tag{28.2}$$

and k_i is fixed. We shall see that consistency requires k_i to assume (half-) integer values only. In the following, all fields are defined on $[0, T]$ and are assumed to be periodic.

Now the question arises as to whether the gauge invariance present at the classical level is maintained at the quantum level. To answer this question, we define the "effective action" $\Gamma[A]$ by integrating out the "matter field" η^a:

$$e^{i\Gamma[A]} = \int \mathcal{D}\eta^a \exp\left\{ \frac{i}{\hbar} \int_0^T dt \left[\frac{1}{2}\eta^a \omega_{ab} \dot{\eta}^b - H(\eta) - A_i\big(J_i(\eta) - k_i\big) \right] \right\} . \tag{28.3}$$

This path integral must be evaluated for periodic boundary conditions $q(0) = q(T)$ in configuration space. In the sequel we are only interested in the one-loop approximation $\Gamma^{(1)}[A] : \Gamma[A] = \{1 + O(\hbar)\}\Gamma^{(1)}[A]$.

To do the calculation we separate a general path $\eta^a(t)$ into a classical path $\eta^a_{\text{cl}}(t)$ and a quantum fluctuation $\chi^a(t)$:

$$\eta^a(t) = \eta^a_{\text{cl}}(t) + \chi^a(t) . \tag{28.4}$$

Here, $\eta^a_{\text{cl}}(t)$ as well as the fluctuating field $\chi^a(t)$ are periodic with period T. Furthermore, $\eta^a_{\text{cl}}(t)$ satisfies the modified Hamiltonian equation

$$\dot{\eta}^a_{\text{cl}}(t) = \omega^{ab} \partial_b \mathcal{H}\left(\eta^a_{\text{cl}}(t)\right) , \tag{28.5}$$

where

$$\mathcal{H}(\eta^a) := H(\eta^a) + \sum_i A_i J_i(\eta^a) . \tag{28.6}$$

Since the equation of motion for A_i ("Gauss' law") requires $J_i(\eta^a_{\text{cl}}(t)) = k_i$, we use only those η^a_{cl} as backgrounds so that J_i equals k_i. Therefore, we consider only those classical trajectories which lie on the torus

$$T_N(k_i) := \{\eta \in \mathcal{M}_{2N} | J_i(\eta) = k_i\} . \tag{28.7}$$

If we integrate (28.3) over the gauge field, we obtain the δ-function $\delta[J_i(\eta) - k_i]$; i.e., quantum paths lie on $T_N(k_i)$ also. Consequently, $J_i(\eta^a) = J_i(\eta^a_{\text{cl}}) = k_i$, and we have to make sure that the fluctuation field χ^a is tangent to $T_N(k_i)$; i.e., we impose on the allowed χ's the restriction $\chi^a \partial_a J_i(\eta_{\text{cl}}) = 0$.

At last we come to the substitution of (28.4) into (28.1):

$$\begin{aligned} S_0[\eta, A] &= \int_0^T dt \left[\frac{1}{2}\eta^a \omega_{ab} \dot{\eta}^b - \mathcal{H}(\eta)\right] \\ &= \int_0^T dt \Big[\frac{1}{2}\left(\eta^a_{\text{cl}} + \chi^a\right) \omega_{ab} \left(\dot{\eta}^b_{\text{cl}} + \dot{\chi}^b\right) - \underbrace{\mathcal{H}(\eta_{\text{cl}} + \chi)}_{=\mathcal{H}(\eta_{\text{cl}})+\chi^a\partial_a\mathcal{H}(\eta_{\text{cl}})+\frac{1}{2}\chi^a\chi^b\partial_a\partial_b\mathcal{H}(\eta_{\text{cl}})+O(\chi^3)} \Big] \\ &= \int_0^T dt \Big[\frac{1}{2}\left(\eta^a_{\text{cl}}\omega_{ab}\dot{\eta}^b_{\text{cl}} + \eta^a_{\text{cl}}\omega_{ab}\dot{\chi}^b + \chi^a\omega_{ab}\dot{\eta}^b_{\text{cl}} + \chi^a\omega_{ab}\dot{\chi}^b\right) \\ &\quad - \mathcal{H}(\eta_{\text{cl}}) - \chi^a\partial_a\mathcal{H} - \frac{1}{2}\chi^a\chi^b\partial_a\partial_b\mathcal{H}\Big] . \end{aligned}$$

The second and third term in the integrand can be rewritten as follows:

$$\begin{aligned} \eta^a_{\text{cl}}\omega_{ab}\dot{\chi}^b &= \frac{d}{dt}\left(\eta^a_{\text{cl}}\omega_{ab}\chi^b\right) - \dot{\eta}^a_{\text{cl}}\omega_{ab}\chi^b \\ &= \frac{d}{dt}\left(\eta^a_{\text{cl}}\omega_{ab}\chi^b\right) - \partial_c\mathcal{H}\,\underbrace{\omega^{ac}\omega_{ab}}_{-\delta^c_b}\,\chi^b \\ &= \frac{d}{dt}\left(\eta^a_{\text{cl}}\omega_{ab}\chi^b\right) + \partial_a\mathcal{H}\,\chi^a \end{aligned}$$

$$\chi^a\omega_{ab}\dot{\eta}^a_{\text{cl}} = \chi^a\omega_{ab}\omega^{bc}\partial_c\mathcal{H} = \chi^a\partial_a\mathcal{H}$$

$$\begin{aligned} \Rightarrow S_0[\eta, A] = \int_0^T dt \Big[&\frac{1}{2}\eta^a_{\text{cl}}\omega_{ab}\dot{\eta}^b_{\text{cl}} - \mathcal{H}(\eta_{\text{cl}}) + \frac{1}{2}\frac{d}{dt}\left(\eta^a_{\text{cl}}\omega_{ab}\chi^b\right) \\ &+ \frac{1}{2}\chi^a\omega_{ab}\dot{\chi}^b - \frac{1}{2}\chi^a\chi^b\partial_a\partial_b\mathcal{H}(\eta_{\text{cl}})\Big] . \end{aligned}$$

So we obtain

$$S_0\left[\eta^a, A_i\right] = S_0\left[\eta^a_{\rm cl}, A_i\right] + \frac{1}{2}\int_0^T dt\, \chi^a\left[\omega_{ab}\partial_t - \partial_a\partial_b\mathcal{H}\left(\eta_{\rm cl}(t)\right)\right]\chi^b + \frac{1}{2}\underbrace{\left[\eta^a_{\rm cl}\omega_{ab}\chi^b\right]_0^T}_{=0 \text{ for periodic orbits}} + \mathrm{O}(\chi^3)$$

or

$$S_0\left[\eta^a, A_i\right] = S_0\left[\eta^a_{\rm cl}, A_i\right] + S_{\rm fl}\left[\chi^a, \eta^a_{\rm cl}, A_i\right] + \mathrm{O}(\chi^3)\,, \tag{28.8}$$

with the action of the fluctuation χ_a:

$$S_{\rm fl}\left[\chi^a, \eta^a_{\rm cl}, A_i\right] = \frac{1}{2}\int_0^T dt\, \chi^a\left[\omega_{ab}\partial_t - \partial_a\partial_b\mathcal{H}\left(\eta_{\rm cl}(t)\right)\right]\chi^b \tag{28.9}$$

$$\equiv \frac{1}{2}\int_0^T dt\, \chi^a\omega_{ab}\left[\partial_t - \tilde{M}(t)\right]^b_c\chi^c\,. \tag{28.10}$$

In (28.10) we introduced the matrix-valued field

$$\begin{aligned}\tilde{M}(t)^a{}_b &= \omega^{ac}\partial_c\partial_b\mathcal{H}\left(\eta_{\rm cl}(t)\right)\\ &= \omega^{ac}\partial_c\partial_b\left(H + A_iJ_i\right)\left(\eta_{\rm cl}(t)\right)\,.\end{aligned} \tag{28.11}$$

The local gauge transformations

$$\begin{aligned}\delta\eta^a(t) &= \varepsilon_i(t)\omega^{ab}\partial_bJ_i\left(\eta(t)\right)\,,\\ \delta A_i(t) &= \dot{\varepsilon}_i(t)\end{aligned} \tag{28.12}$$

decompose under $\eta^a = \eta^a_{\rm cl} + \chi^a$ according to

$$\begin{aligned}\delta\left(\eta^a_{\rm cl}(t) + \chi^a(t)\right) &= \varepsilon_i(t)\omega^{ab}\partial_b\underbrace{J_i\left(\eta_{\rm cl} + \chi\right)(t)}_{=J_i(\eta_{\rm cl}(t)) + \chi^c\partial_cJ_i(\eta_{\rm cl}) + \mathrm{O}(\chi^2)}\\ &= \varepsilon_i(t)\omega^{ab}\partial_bJ_i\left(\eta_{\rm cl}\right)\\ &\quad + \varepsilon_i(t)\left[\omega^{ab}\partial_b\partial_cJ_i\left(\eta_{\rm cl}(t)\right)\right]\chi^c(t)\end{aligned}$$

so that

$$\delta\eta^a_{\rm cl}(t) = \varepsilon_i(t)\omega^{ab}\partial_bJ_i\left(\eta_{\rm cl}(t)\right)\,, \tag{28.13}$$

$$\delta\chi^a(t) = \varepsilon_i(t)M_i(t)^a{}_b\chi^b(t)\,, \tag{28.14}$$

$$\delta A_i(t) = \dot{\varepsilon}_i(t)\,, \tag{28.15}$$

where

$$M_i(t)^a{}_b := \omega^{ac}\partial_c\partial_bJ_i\left(\eta_{\rm cl}(t)\right)\,. \tag{28.16}$$

$\eta^a_{\text{cl}}(t)$ [like $\eta^a(t)$] transforms as a coordinate (or rather, a path) on $\mathcal{M}_{2N}$, whereas $\chi^a(t)$ transforms (homogeneously!) like a tangent space vector. The transformations (28.13–15) induce the following transformation on $\tilde{M}$:

$$\delta\tilde{M} = \partial_t\big(\varepsilon_i M_i\big) + \big[\varepsilon_i M_i, \tilde{M}\big] . \tag{28.17}$$

(When we use an index-free notation, it is understood that the upper index of M is left, the lower one is right.)

Proof:

$$\begin{aligned}
\delta\tilde{M}^a{}_b &= \omega^{ac}\partial_c\partial_b\delta\underbrace{\big(H + A_i J_i\big)}_{=\mathcal{H}}\big(\eta_{\text{cl}}\big)\\
&= \omega^{ac}\partial_c\partial_b\Big[\partial_e\mathcal{H}\big(\eta_{\text{cl}}\big)\underbrace{\delta\eta^e_{\text{cl}}}_{\varepsilon_i\omega^{ed}\partial_d J_i} + \underbrace{\delta A_i}_{\dot\varepsilon_i} J_i\big(\eta_{\text{cl}}\big)\Big]\\
&= \varepsilon_i(t)\omega^{ac}\partial_c\partial_b\partial_e\mathcal{H}\big(\eta_{\text{cl}}\big)\omega^{ed}\partial_d J_i\big(\eta_{\text{cl}}\big) + \dot\varepsilon_i\omega^{ac}\partial_c\partial_b J_i\big(\eta_{\text{cl}}\big)\\
&=: \varepsilon_i(t)\ ./.\ + \dot\varepsilon M^a_{i\,b} .
\end{aligned}$$

Since $\{H, J_i\} = 0$, we can write:

$$\begin{aligned}
\partial_b| \quad & \partial_e\mathcal{H}\omega^{ed}\partial_d J_i = 0\\
\partial_c| \quad & \partial_b\partial_e\mathcal{H}\omega^{ed}\partial_d J_i = -\partial_e\mathcal{H}\omega^{ed}\partial_d\partial_b J_i\\
& \partial_c\partial_b\partial_e\mathcal{H}\omega^{ed}\partial_d J_i = -\partial_b\partial_e\mathcal{H}\omega^{ed}\partial_d\partial_c J_i - \partial_c\partial_e\mathcal{H}\omega^{ed}\partial_d\partial_b J_i\\
& \qquad\qquad - \partial_e\mathcal{H}\omega^{ed}\partial_d\partial_b\partial_c J_i
\end{aligned}$$

$$\begin{aligned}
./. = &-\underbrace{\underbrace{\underbrace{\partial_b\partial_e\mathcal{H}\omega^{ed}}_{-\omega^{de}\partial_b\partial_e\mathcal{H}}}_{-\tilde{M}^d{}_b}\ \underbrace{\partial_d\omega^{ac}\partial_c J_i}_{M^a_{i\,d}}}_{(M_i\tilde{M})^a{}_b} - \underbrace{\underbrace{\omega^{ac}\partial_c\partial_e\mathcal{H}}_{\tilde{M}^a{}_e}\ \underbrace{\omega^{ed}\partial_d\partial_b J_i}_{M^e_{i\,b}}}_{(\tilde{M}M_i)^a{}_b}\\
&-\underbrace{\underbrace{\underbrace{\partial_e\mathcal{H}\omega^{ed}}_{-\omega^{de}\partial_e\mathcal{H}}}_{-\dot\eta^d_{\text{cl}}}\ \underbrace{\partial_d\partial_b\omega^{ac}\partial_c J_i}_{(\partial/\partial\eta^d_{\text{cl}})M^a_{i\,b}}}_{(d/dt)M^a_{i\,b}(\eta_{\text{cl}}(t))}
\end{aligned}$$

$$\Rightarrow \delta\tilde{M} = \varepsilon_i\big[M_i, \tilde{M}\big] + \varepsilon_i\dot{M}_i + \dot\varepsilon_i M_i = \frac{d}{dt}\big(\varepsilon_i M_i\big) + \varepsilon_i\big[M_i, \tilde{M}\big] .$$

Now we can write down – in one-loop approximation:

$$\begin{aligned} e^{i\Gamma^{(1)}[A]} &= \sum_{\text{cct}} e^{iS[\eta_{\text{cl}}, A]} \int_{\chi^a(0)=\chi^a(T)} \mathcal{D}\chi \, e^{iS_{\text{fl}}[\chi, \eta_{\text{cl}}, A]} \\ &= \sum_{\text{cct}} e^{iS[\eta_{\text{cl}}, A]} \, e^{i\hat{\Gamma}[\eta_{\text{cl}}, A]} , \end{aligned} \tag{28.18}$$

where

$$S[\eta_{\text{cl}}, A] \equiv S_0[\eta_{\text{cl}}, A] + S_{\text{CS}}[A] \tag{28.19}$$

and

$$e^{i\hat{\Gamma}[\eta_{\text{cl}}, A]} = \int_{\chi^a(0)=\chi^a(T)} \mathcal{D}\chi \, e^{iS_{\text{fl}}[\chi, \eta_{\text{cl}}, A]} . \tag{28.20}$$

In (28.18), "$\sum_{\text{cct}}$" denotes the sum over all classical trajectories generated by $\mathcal{H}$ (or equivalently, by H) which have period T and for which $J_i(\eta_{\text{cl}}(0))$ takes the prescribed values k_i on the torus $T_N(k_i)$.

Finally, we determine the allowed values of k_i by requiring gauge invariance of $\exp\{i\Gamma^{(1)}[A]\}$. So let us consider a large gauge transformation of the topological class $\{\mathcal{N}_i\}$,

$$\varepsilon_i(T) - \varepsilon_i(0) = 2\pi \mathcal{N}_i , \tag{28.21}$$

with the gauge-transformed field

$$A_i \to A_i' = A_i + \dot{\varepsilon}_i . \tag{28.22}$$

Then, denoting the "summation variable" in [28.18] as η_{cl}'

$$e^{i\Gamma^{(1)}[A']} = \sum_{\eta_{\text{cl}}'} e^{iS[\eta_{\text{cl}}', A']} \, e^{i\hat{\Gamma}[\eta_{\text{cl}}', A']} . \tag{28.23}$$

Here we assumed that "$\sum_{\eta_{\text{cl}}'}$" is a gauge invariant measure, i.e., that we can replace the "η_{cl}"-integration by a "η_{cl}'"-integration or summation, where η_{cl}' is the gauge transform of η_{cl}. From (27.59) and (27.61) we know that S_0 is strictly gauge invariant but S_{CS} changes by the amount of $2\pi k_i \mathcal{N}_i$:

$$S_0[A', \eta_{\text{cl}}'] = S_0[A, \eta_{\text{cl}}] , \tag{28.24}$$

$$S_{\text{CS}}[A'] = S_{\text{CS}}[A] + 2\pi k_i \mathcal{N}_i . \tag{28.25}$$

Consequently, (28.23) can be written as

$$e^{i\Gamma^{(1)}[A']} = \sum_{\eta_{\text{cl}}} e^{i(2\pi k_i \mathcal{N}_i + \Delta\hat{\Gamma})} \, e^{iS[\eta_{\text{cl}}, A]} \, e^{i\hat{\Gamma}[\eta_{\text{cl}}, A]} . \tag{28.26}$$

Here we have introduced the change in the quantum action

$$\Delta\hat{\Gamma} := \hat{\Gamma}\left[\eta'_{\text{cl}}, A'\right] - \hat{\Gamma}\left[\eta_{\text{cl}}, A\right]. \tag{28.27}$$

A priori we do not know $\Delta\hat{\Gamma}$. It has to be calculated from (28.20), which is, for S_{fl} quadratic, formally given by

$$e^{i\hat{\Gamma}[\eta_{\text{cl}},A]} = \det{}^{-1/2}\left(\partial_t - \tilde{M}(t)\right). \tag{28.28}$$

It will be one of our goals in the sequel to study the behavior of this quantity under gauge transformations. We will find out that $\Delta\hat{\Gamma}$ depends only on the $\mathcal{N}_i$'s but not on the details of η_{cl} or A:

$$\Delta\hat{\Gamma} = -2\pi \sum_i \frac{\mu_i}{4} \mathcal{N}_i\,. \tag{28.29}$$

The constants μ_i are even integers which are related to certain winding numbers on the group manifold Sp(2N).

Altogether we obtain from (28.26) and (28.29):

$$e^{i\Gamma^{(1)}[A']} = \sum_{\substack{\eta_{\text{cl}} \\ J_i=k_i}} \exp\left[2\pi i \sum_i \left(k_i - \frac{\mu_i}{4}\right)\mathcal{N}_i\right] e^{iS[\eta_{\text{cl}},A]}\, e^{i\hat{\Gamma}[\eta_{\text{cl}},A]}\,. \tag{28.30}$$

Gauge invariance, $e^{i\Gamma^{(1)}[A']} = e^{i\Gamma^{(1)}[A]}$, obviously requires that $k_i - \mu_i/4$ be an integer:

$$k_i = n_i + \frac{\mu_i}{4}\,, \qquad n_i \in \mathbb{Z}\,. \tag{28.31}$$

Since k_i equals the action J_i of the classical "background" trajectory around which we expanded, we obtain the requirement:

$$J_i\left(\eta_{\text{cl}}\right) = n_i + \frac{\mu_i}{4}\,, \qquad n_i \in \mathbb{Z}\,. \tag{28.32}$$

Only if J_i is quantized in this manner is gauge invariance left intact at the quantum level. Equation (28.32) is the well-known semiclassical EBK quantization condition. The topological numbers μ_i coincide with the Maslov indices. In the present treatment they arise from a quantum mechanical anomaly: despite the fact that the classical action S_{fl} is gauge invariant, the associated action $\hat{\Gamma}$ (28.20) is gauge invariant only under small gauge transformations with $\mathcal{N}_i = 0$, but changes under large ones. In the next section we are going to evaluate μ_i for a specific example.

29. The "Maslov Anomaly" for the Harmonic Oscillator

Specializing the discussion of the previous section to the harmonic oscillator we have for $N = 1$, $\eta^a = (p, x)$, $a = 1, 2$, $\eta^1 \equiv p$, $\eta^2 \equiv x$

$$H(p, x) = \tfrac{1}{2}\eta^a\eta^a = \tfrac{1}{2}\left(p^2 + x^2\right) . \tag{29.1}$$

The only conserved quantity is $J = H$. In the action we need the combination

$$\frac{1}{2}\eta^a \omega_{ab}\dot{\eta}^b - \mathcal{H}(\eta) = \frac{1}{2}\eta^a\left[\omega_{ab}\frac{d}{dt} - \big(1 + A(t)\big)\,\mathbb{1}_{ab}\right]\eta^b \tag{29.2}$$

and

$$\tilde{M}^a{}_b = \omega^{ac}\partial_c\partial_b(H + AJ) = \big(1 + A(t)\big)\omega^{ac}\,\mathbb{1}_{cb}$$

or, compactly written:

$$\tilde{M} = \big(1 + A(t)\big)\Omega , \tag{29.3}$$

where we have introduced the notation $\Omega \equiv \omega^{-1} = (\omega^{ab})$. $A(t)$ is the gauge field for a single U(1) group associated with energy conservation. The fluctuation part of the action is now given by

$$S_{\mathrm{fl}} = \frac{1}{2}\int_0^T dt\, \chi^a\omega_{ab}\left[\frac{d}{dt} - \tilde{M}\right]^b_{\ c}\chi^c , \tag{29.4}$$

and the path integral for $\hat{\Gamma}$ becomes

$$\mathrm{e}^{\mathrm{i}\hat{\Gamma}[A]} = \int_{\chi(0)=\chi(T)} \mathcal{D}\chi \exp\left[\frac{\mathrm{i}}{2}\int_0^{T=2\pi} dt\, \chi^a\omega_{ab}\left[\frac{\partial}{\partial t} - \underbrace{\big(1 + A(t)\big)\Omega}_{=\tilde{M}}\right]^b_{\ c}\chi^c\right] \tag{29.5}$$

$$= \det{}^{-1/2}\left[\omega\left(\frac{\partial}{\partial t} - \tilde{M}\right)\right] , \qquad \det\omega = 1$$

$$= \det{}^{-1/2}\left(\frac{\partial}{\partial t} - \tilde{M}\right) . \tag{29.6}$$

For a two-dimensional space and canonical coordinates, ω and its inverse $\omega^{-1} = \Omega$ can be expressed by the Pauli matrix σ_2 : $\omega = \mathrm{i}\sigma_2$, $\Omega = -\mathrm{i}\sigma_2$. Furthermore, in

(29.5) we put $T = 2\pi$, since the classical trajectories of (29.1) are all 2π-periodic. Altogether we get:

$$e^{i\hat{\Gamma}[A]} = \det{}^{-1/2}\left[\frac{\partial}{\partial t} + i\big(1 + A(t)\big)\sigma_2\right] . \tag{29.7}$$

At this stage we have to investigate the eigenvalue problem of the antihermitian operator:

$$\begin{aligned} D := & \frac{\partial}{\partial t} + i\big(1 + A(t)\big)\sigma_2 \\ \equiv & \frac{\partial}{\partial t} + iB(t)\sigma_2 , \end{aligned} \tag{29.8}$$

on the space of periodic functions:

$$DF_m^\eta(t) = i\lambda_m^\eta F_m^\eta(t) , \quad \lambda_m^\eta \in \mathbb{R} , \tag{29.9}$$

with

$$F_m^\eta(t + 2\pi) = F_m^\eta(t) .$$

Here is the set of complete, orthonormal functions on $[0, 2\pi]$ that satisfy (29.9):

$$F_m^\eta(t) = \frac{1}{\sqrt{2}}\begin{pmatrix} f_m^\eta(t) \\ i\eta f_m^\eta(t) \end{pmatrix} , \tag{29.10}$$

where

$$f_m^\eta(t) = \frac{1}{\sqrt{2\pi}}\exp\left[i\lambda_m^\eta t - i\eta\int_0^t dt'\, B(t')\right] . \tag{29.11}$$

The F's are simultaneous eigenfunctions of D and σ_2:

$$\sigma_2 F_m^\eta = \eta F_m^\eta , \quad \eta = \pm 1 . \tag{29.12}$$

Periodicity $f_m^\eta(0) = f_m^\eta(2\pi)$ implies that

$$1 = \exp\left[i\lambda_m^\eta 2\pi - i\eta\int_0^{2\pi} dt\, B(t)\right]$$

or

$$\lambda_m^\eta 2\pi - \eta\int_0^{2\pi} dt\, B(t) = 2\pi m , \quad m \in \mathbb{Z} .$$

Hence the eigenvalues are given by

$$\lambda_m^\eta = m + \eta\frac{1}{2\pi}\int_0^{2\pi} dt\, B(t) , \quad m \in \mathbb{Z} , \quad \eta = \pm 1 . \tag{29.13}$$

Let us check some of the above-mentioned properties:

$$\sigma_2 F_m^\eta = \frac{1}{\sqrt{2}}\begin{pmatrix}0 & -\mathrm{i}\\ \mathrm{i} & 0\end{pmatrix}\begin{pmatrix}f_m^\eta\\ \mathrm{i}\eta f_m^\eta\end{pmatrix} = \frac{1}{\sqrt{2}}\begin{pmatrix}\eta f_m^\eta\\ \mathrm{i} f_m^\eta\end{pmatrix} \overset{\eta^2=1}{=} \eta\frac{1}{\sqrt{2}}\begin{pmatrix}f_m^\eta\\ \mathrm{i}\eta f_m^\eta\end{pmatrix} = \eta F_m^\eta$$

$$\begin{aligned} DF_m^\eta &= \left(\partial_t + \mathrm{i}\sigma_2 B\right)F_m^\eta = \left(\partial_t + \mathrm{i}\eta B\right)\frac{1}{\sqrt{2}}\begin{pmatrix}f_m^\eta\\ \mathrm{i}\eta f_m^\eta\end{pmatrix}\\ &= \frac{1}{\sqrt{2}}\begin{pmatrix}\left[\partial_t + \mathrm{i}\eta B\right]f_m^\eta\\ \mathrm{i}\eta\left[\partial_t + \mathrm{i}\eta B\right]f_m^\eta\end{pmatrix} = \mathrm{i}\lambda_m^\eta\frac{1}{\sqrt{2}}\begin{pmatrix}f_m^\eta\\ \mathrm{i}\eta f_m^\eta\end{pmatrix} = \mathrm{i}\lambda_m^\eta F_m^\eta\,, \end{aligned}$$

since f_m^η solves (29.9),

$$\left[\frac{\partial}{\partial t} + \mathrm{i}\eta B(t)\right] f_m^\eta = \mathrm{i}\lambda_m^\eta f_m^\eta\,. \tag{29.14}$$

The orthonormality follows from

$$\begin{aligned} \int_0^{2\pi} dt\, F_m^{\eta\dagger}(t)F_{m'}^{\eta'}(t) &= \frac{1}{2}\int_0^{2\pi} dt\left(f_m^{\eta^*}, -\mathrm{i}\eta f_m^{\eta^*}\right)\begin{pmatrix}f_{m'}^{\eta'}\\ \mathrm{i}\eta' f_{m'}^{\eta'}\end{pmatrix}\\ &= \frac{1}{2}\int_0^{2\pi} dt\left[f_m^{\eta^*} f_{m'}^{\eta'} + \eta\eta' f_m^{\eta^*} f_{m'}^{\eta'}\right] \overset{\eta=\eta'}{=} \frac{1}{2}\int_0^{2\pi} dt\left[f_m^{\eta^*} f_{m'}^{\eta} + f_m^{\eta^*} f_{m'}^{\eta}\right]\\ &\qquad = \frac{1}{2}\left(\delta_{mm'} + \delta_{mm'}\right) = \delta_{mm'}\,;\\ &\overset{\eta=-\eta'}{=} \frac{1}{2}\int_0^{2\pi} dt\left[f_m^{\eta^*} f_{m'}^{-\eta} - f_m^{\eta^*} f_{m'}^{-\eta}\right] = 0\,. \end{aligned}$$

Hence we have proved

$$\int_0^{2\pi} dt\, F_m^{\eta\dagger}(t)F_{m'}^{\eta'}(t) = \delta_{mm'}\delta_{\eta\eta'}\,. \tag{29.15}$$

Note that the spectrum (29.13) is symmetric around zero, and that under a gauge transformation $A' = A + \dot{\varepsilon}$, $\varepsilon(2\pi) - \varepsilon(0) = 2\pi\mathcal{N}$, it is mapped onto itself:

$$\int_0^{2\pi} dt\, A'(t) = \int_0^{2\pi} dt\, A(t) + \int_0^{2\pi} dt\, \dot{\varepsilon}(t) = \int_0^{2\pi} dt\, A(t) + 2\pi\mathcal{N}\,.$$

Upon using the definition $B(t) = 1 + A(t)$, we can continue to write

$$\int_0^{2\pi} dt\, B'(t) = \int_0^{2\pi} dt\, B(t) + 2\pi\mathcal{N}$$

or

$$\eta\frac{1}{2\pi}\int_0^{2\pi} dt\, B'(t) = \eta\frac{1}{2\pi}\int_0^{2\pi} dt\, B(t) + \eta\mathcal{N}\,. \tag{29.16}$$

If we add $m \in \mathbb{Z}$ on either side of (29.16) and employ (29.13), we obtain:

$$\lambda_m^\eta[A'] = \lambda_{m+\eta\mathcal{N}}^\eta[A] \tag{29.17}$$

with $m + \eta\mathcal{N} = m \pm \mathcal{N}$ again an integer.

If the gauge field is restricted according to (27.57),

$$\int_0^{2\pi} dt\, A(t) \equiv 2\pi\tilde{A} = 2\pi z\,, \quad z \in \mathbb{Z}\,,$$

the eigenvalue (29.13) becomes:

$$\begin{aligned}\lambda_m^\eta &= m + \eta\frac{1}{2\pi}\int_0^{2\pi} dt\big(1 + A(t)\big) = m + \eta + \eta\frac{1}{2\pi}\int_0^{2\pi} dt\, A(t)\\ &= m + \eta(1 + z)\,,\end{aligned} \tag{29.18}$$

which contains two zero modes:

$$\begin{aligned}\eta &= 1\,, \quad m = -(1 + z)\,,\\ \eta &= -1\,, \quad m = +(1 + z)\,.\end{aligned} \tag{29.19}$$

In this case the determinant in (29.7) vanishes and $\hat{\Gamma}$ is not defined for these fields. As a way out, we assume that for the computation of $\hat{\Gamma}[A]$, the condition $\tilde{A} = z \in \mathbb{Z}$ is *not* imposed on the gauge field. Instead we use $A(t)$'s for which $\tilde{A}$ is arbitrary. After having computed $\Delta\hat{\Gamma}$ for such fields, we let $\tilde{A} \to z \in \mathbb{Z}$ at the very end.

We now turn to the computation of $\Delta\hat{\Gamma}$ by a spectral flow argument. Before we do so, we read off a reality constraint for the functions F_m^η which follows from the explicit expressions given in (29.10) and (29.11):

$$\begin{aligned}\left(f_m^\eta\right)^* = f_{-m}^{-\eta} : \quad \left(F_m^\eta\right)^* &= \frac{1}{\sqrt{2}}\begin{pmatrix}\left(f_m^\eta\right)^*\\ -i\eta\left(f_m^\eta\right)^*\end{pmatrix}\\ &= \frac{1}{\sqrt{2}}\begin{pmatrix}f_{-m}^{-\eta}\\ -i\eta f_{-m}^{-\eta}\end{pmatrix} = F_{-m}^{-\eta}\,.\end{aligned} \tag{29.20}$$

This equation puts constraints on the coefficients in the expansion of the path integration variable $\chi^a(t)$ in terms of the complete set F_m^η:

$$\begin{aligned}\chi(t) &= \sum_{m=-\infty}^{+\infty}\sum_{\eta=\pm 1} C_m^\eta F_m^\eta(t) = \sum_{m,\eta} C_m^{\eta^*} F_m^{\eta^*}(t)\\ &= \sum_{m,\eta} C_m^{\eta^*} F_{-m}^{-\eta}(t) = \sum_{m,\eta} C_{-m}^{-\eta^*} F_m^\eta(t)\,.\end{aligned}$$

Therefore

$$\sum_{m,\eta}\left(C_m^\eta - C_{-m}^{-\eta^*}\right) F_m^\eta = 0\,,$$

meaning that for real χ, the expansion coefficients satisfy

$$\left(C_m^\eta\right)^* = C_{-m}^{-\eta} \ . \tag{29.21}$$

Let us return to the representation of the determinant as a Gaussian integral as given in (29.5) and (29.6):

$$\mathrm{e}^{\mathrm{i}\hat{\Gamma}[A]} = \int \mathcal{D}\chi \exp\left[\frac{\mathrm{i}}{2}\int_0^{2\pi} dt\ \chi^{\mathrm{T}}\omega D\chi\right] , \tag{29.22}$$

where we have used a two-component matrix notation in the exponential. The operator ωD is given by

$$\begin{aligned}\omega D &= \omega\left[\frac{\partial}{\partial t} + \mathrm{i}\sigma_2\big(1 + A(t)\big)\right] , \quad \mathrm{i}\sigma_2 = -\omega^{-1} \\ &= \omega\frac{\partial}{\partial t} - \big(1 + A(t)\big)\end{aligned}$$

and the integral in the exponential of (29.22) is

$$\int_0^{2\pi} dt\ \chi^{\mathrm{T}}\omega D\chi = \sum_{m,\eta}\sum_{m',\eta'} C_m^{\eta *} C_{m'}^{\eta'} \int_0^{2\pi} dt\ F_m^{\eta\dagger}\, \omega\, \underbrace{DF_{m'}^{\eta'}} , \quad \omega = \mathrm{i}\sigma_2$$

$$\begin{aligned}\int_0^{2\pi} ./. &= \mathrm{i}^2\lambda_{m'}^{\eta'}\int_0^{2\pi} dt\ F_m^{\eta\dagger}\underbrace{\sigma_2 F_{m'}^{\eta'}}_{=\eta' F_{m'}^{\eta'}} = -\eta'\lambda_{m'}^{\eta'}\underbrace{\int_0^{2\pi} dt\ F_m^{\eta\dagger} F_{m'}^{\eta'}}_{=\delta_{\eta\eta'}\delta_{mm'}} \\ &= -\eta\lambda_m^\eta\delta_{\eta\eta'}\delta_{mm'} \ .\end{aligned}$$

This leads to

$$\int_0^{2\pi} dt\ \chi^{\mathrm{T}}\omega D\chi = -\sum_{m=-\infty}^{+\infty}\sum_{\eta=\pm 1} \eta\lambda_m^\eta |C_m^\eta|^2 \ . \tag{29.23}$$

In order to reexpress the path integral measure in (29.22), we have to find a set of independent C's. Now, the C's are subject to constraints as stated in (29.21). Hence, a set of independent C's is given by

$$\begin{aligned}&\left\{C_m^\eta | m = 0 , \quad \eta = 1 \quad \text{and} \quad m > 0 , \quad \eta = \pm 1\right\} , \\ &\left\{C_m^\eta | m = 0 , \quad \eta = -1 \quad \text{and} \quad m > 0 , \quad \eta = \pm 1\right\} .\end{aligned}$$

We choose the first alternative if $\int_0^{2\pi} dt\ B(t) > 0$ and the second alternative if $\int_0^{2\pi} dt\ B(t) < 0$. Then, in view of

$$\lambda_m^\eta = m + \eta\frac{1}{2\pi}\int_0^{2\pi} dt\ B(t) , \tag{29.24}$$

we take those pairs (η, m), which have $\lambda_m^\eta > 0$. Thus the path integral becomes

$$e^{i\hat{\Gamma}[A]} = \int \prod_{\{m,\eta|\lambda_m^\eta>0\}} dC_m^\eta \, dC_m^{\eta*} \exp\left[-\frac{i}{2}\sum_{m,\eta} \eta\lambda_m^\eta |C_m^\eta|^2\right]\Bigg|_{C_{-m}^{-\eta}=C_m^{\eta*}} .$$

For a fixed pair (η, m) we have in the exponential:

$$\begin{aligned}-\frac{i}{2}\left\{\eta\lambda_m^\eta|C_m^\eta|^2 + (-\eta)\lambda_{-m}^{-\eta}|C_m^{\eta*}|^2\right\} &= -\frac{i}{2}\eta\left\{\lambda_m^\eta - \lambda_{-m}^{-\eta}\right\}|C_m^\eta|^2 \\ &= -i\eta\lambda_m^\eta|C_m^\eta|^2 ,\end{aligned}$$

where we have used the explicit formula (29.24) for λ_m^η to show that $\lambda_{-m}^{-\eta} = -\lambda_m^\eta$.

So far our result reads

$$e^{i\hat{\Gamma}[A]} = \prod_{\{m,\eta|\lambda_m^\eta>0\}} \int dC_m^\eta \, dC_m^{\eta*} \exp\left[-i\eta\lambda_m^\eta|C_m^\eta|^2\right] . \tag{29.25}$$

What we actually are looking for is not $\hat{\Gamma}$ itself, but the difference:

$$\Delta\hat{\Gamma} = \hat{\Gamma}[A'] - \hat{\Gamma}[A] = \hat{\Gamma}[A + \dot{\varepsilon}] - \hat{\Gamma}[A] , \tag{29.26}$$

where $\varepsilon(t)$ is a gauge transformation of winding number $\mathcal{N}$:

$$\varepsilon(2\pi) - \varepsilon(0) = 2\pi\mathcal{N} . \tag{29.27}$$

We are going to evaluate the difference (29.26) by spectral flow arguments. For this reason we introduce the following 1-parameter family of gauge potentials $A_s(t)$ interpolating between $A(t)$ and $A'(t)$ as s runs from minus to plus infinity:

$$A_s(t) \equiv A(t) + g(s)\dot{\varepsilon}(t) , \quad s \in (-\infty, +\infty) . \tag{29.28}$$

Here, $g(s)$ is an arbitrary smooth function with $g(s = -\infty) = 0$ and $g(s = +\infty) = 1$. Hence, $A_{-\infty}(t) = A(t)$ and $A_{+\infty}(t) = A'(t)$. We can derive $\Delta\hat{\Gamma}$ from the flow of the eigenvalues $\lambda_m^\eta \equiv \lambda_m^\eta(s)$ as the parameter s is varied. The spectrum $\{\lambda_m^\eta\}$ changes as follows:

$$\begin{aligned}\lambda_m^\eta(s) &= m + \eta\frac{1}{2\pi}\int_0^{2\pi} dt\big(1 + A_s(t)\big) = m + \frac{1}{2\pi}\eta\int_0^{2\pi} dt\big(1 + A(t) + g(s)\dot{\varepsilon}(t)\big) \\ &= m + \eta\frac{1}{2\pi}\int_0^{2\pi} dt\, B(t) + \frac{1}{2\pi}\eta g(s)2\pi\mathcal{N} \\ &= \big(m + \eta g(s)\mathcal{N}\big) + \eta\frac{1}{2\pi}\int_0^{2\pi} dt\, B(t) = \lambda_{m+\eta g(s)\mathcal{N}}^\eta(0) .\end{aligned} \tag{29.29}$$

We observe that as s runs from $-\infty$ to $+\infty$, the m index of the eigenvalues with $\eta = +1(-1)$ is shifted to $m + \mathcal{N}(m - \mathcal{N})$:

$$\lambda_m^{+1} \longrightarrow \lambda_{m+\mathcal{N}}^{+1} ,$$
$$\lambda_m^{-1} \longrightarrow \lambda_{m-\mathcal{N}}^{-1} .$$

What is important for the determination of $\Delta\hat{\Gamma}$ are the eigenvalues crossing zero for some value of s.

Now (29.29) tells us that for a gauge transformation with $\mathcal{N} > 0$ there are $\mathcal{N}$ eigenvalues with $\eta = +1$ which are negative for $s \to -\infty$ and which become positive for $s \to +\infty$. There are also $\mathcal{N}$ eigenvalues with $\eta = -1$ which cross zero in the opposite direction; i.e., they are positive for $s \to -\infty$ and become negative for $s \to +\infty$.

For a gauge transformation with $\mathcal{N} < 0$, the pattern is reversed: there are $|\mathcal{N}|$ zero-crossings of eigenvalues with $\eta = +1$, which go from positive to negative values, and $|\mathcal{N}|$ zero-crossings of eigenvalues with $\eta = -1$, which go from negative to positive ones.

For the interpolating gauge field $A_s(t)$, the path integral (29.25) is modified according to

$$e^{i\Gamma[A_s]} = \prod_{\{m,\eta|\lambda_m^\eta(s)>0\}} \int dC_m^\eta \, dC_m^{\eta*} \exp\left[-i\eta\lambda_m^\eta(s)|C_m^\eta|^2\right] . \tag{29.30}$$

Using the formula

$$\int dz\, dz^* \, e^{-ia|z|^2} = \frac{2\pi}{|a|} e^{-i\pi\,\mathrm{sign}(a)/2}$$

we obtain

$$e^{i\hat{\Gamma}[A_s]} = \prod_{\{m,\eta|\lambda_m^\eta(s)>0\}} \frac{2\pi}{\lambda_m^\eta(s)} e^{-i\pi\eta/2} , \tag{29.31}$$

since sign $(\eta\lambda_m^\eta(s)) = \eta$ for $\lambda_m^\eta(s) > 0$. We need

$$\begin{aligned} e^{i\Delta\hat{\Gamma}[A]} &= e^{i\hat{\Gamma}[A'] - i\hat{\Gamma}[A]} = e^{i\hat{\Gamma}[A_{s=+\infty}] - i\hat{\Gamma}[A_{s=-\infty}]} \\ &= \frac{e^{i\hat{\Gamma}[A_{+\infty}]}}{e^{i\hat{\Gamma}[A_{-\infty}]}} = \frac{\prod_{\lambda_m^\eta(+\infty)>0} 2\pi/\lambda_m^\eta(+\infty)}{\prod_{\lambda_m^\eta(-\infty)>0} 2\pi/\lambda_m^\eta(-\infty)} \\ &\quad \times \frac{\prod_{\lambda_m^\eta(+\infty)>0} e^{-i\pi\eta/2}}{\prod_{\lambda_m^\eta(-\infty)>0} e^{-i\pi\eta/2}} . \end{aligned} \tag{29.32}$$

The first factor in (29.32) is 1, since A and A' are related by a gauge transformation, and we found in (29.17) that the spectrum is gauge invariant. So we obtain:

$$e^{i\hat{\Gamma}[A]} = \frac{\prod_{\lambda_m^\eta(+\infty)>0} e^{-i\pi\eta/2}}{\prod_{\lambda_m^\eta(-\infty)>0} e^{-i\pi\eta/2}} =: \; e^{-i\pi\nu/2} . \tag{29.33}$$

A nonzero $\Delta\hat{\Gamma}$ can occur only if the number of factors of $\exp[-\mathrm{i}\pi\eta/2]$ is different for $s = -\infty$ and $s = +\infty$. This number is determined by the eigenvalues crossing zero. Writing

$$\Delta\hat{\Gamma} = -\frac{\pi}{2}\nu \equiv -\frac{\pi}{2}(\nu_1 - \nu_2) \pmod{2\pi} \tag{29.34}$$

we have in an obvious notation:

$$\begin{aligned} \nu_1 &= \#\left\{\eta = +1\,,\ \nearrow\right\} - \#\left\{\eta = +1\,,\ \searrow\right\}, \\ \nu_2 &= \#\left\{\eta = -1\,,\ \nearrow\right\} - \#\left\{\eta = -1\,,\ \searrow\right\}, \end{aligned} \tag{29.35}$$

where $\#\{\eta = +1,\ \nearrow\}$ denotes the number of eigenvalues with $\eta = +1$ crossing zero from below, etc. For a gauge transformation with $\mathcal{N} > 0$, we know from the explicit construction of the spectrum

$$\nu = (\mathcal{N} - 0) - (0 - \mathcal{N}) = 2\mathcal{N} \tag{29.36}$$

and for $\mathcal{N} < 0$:

$$\nu = (0 - |\mathcal{N}|) - (|\mathcal{N}| - 0) = -2|\mathcal{N}| = 2\mathcal{N}\ .$$

Hence, $\nu_1 = \mathcal{N}$ and $\nu_2 = -\mathcal{N}$, so that $\nu = 2\mathcal{N}$, $\mathcal{N} \in \mathbb{Z}$.

Our final result is therefore given by

$$\Delta\hat{\Gamma} = -\frac{\pi}{2} \cdot 2 \cdot \mathcal{N} \pmod{2\pi}\ . \tag{29.37}$$

In (25.29) we defined the Maslov index via

$$\Delta\hat{\Gamma} = -2\pi\frac{\mu}{4}\mathcal{N} = -\frac{\pi}{2}\mu\mathcal{N}\ . \tag{29.38}$$

This at last identifies the Maslov index for the linear harmonic oscillator: $\mu = 2$. The correct energy spectrum follows from (25.32): $E = n + 1/2$, $n = 0, \pm 1, \pm 2, \ldots$. Note our argument implies only $n \in \mathbb{Z}$; the actual range of n has to follow from other considerations. In the present case it is the positivity of $H \equiv J$ which implies $n \in \mathbb{N}$.

Since

$$\mathrm{e}^{\mathrm{i}\Delta\hat{\Gamma}} = \mathrm{e}^{-\mathrm{i}\pi\mathcal{N}} = \begin{Bmatrix} +1\,, & \mathcal{N}\ \text{even} \\ -1\,, & \mathcal{N}\ \text{odd} \end{Bmatrix} = (-1)^{\mathcal{N}} \tag{29.39}$$

we observe that in (29.7),

$$\mathrm{e}^{\mathrm{i}\hat{\Gamma}[B]} = \det{}^{-1/2}\left[\frac{\partial}{\partial t} + \mathrm{i}\sigma_2 B(t)\right], \tag{29.40}$$

the effect of a large gauge transformation $A' = A + \mathcal{N}$ or $B' = B + \mathcal{N}$ is at most a sign change of the square root of the determinant.

30. Maslov Anomaly and the Morse Index Theorem

Our starting point is again the phase space integral

$$e^{i\hat{\Gamma}[\tilde{M}]} = \int \mathcal{D}\chi^a \, e^{iS_{\text{fl}}[\chi,\tilde{M}]} \tag{30.1}$$

with periodic boundary conditions $\chi(0) = \chi(T)$ and

$$S_{\text{fl}}[\chi, \tilde{M}] = \frac{1}{2}\int_0^T dt \, \bar{\chi}_a(t) \left[\frac{\partial}{\partial t} - \tilde{M}(t)\right]^a_{\ b} \chi^b(t) \, . \tag{30.2}$$

Here we have indicated that S_{fl} and $\hat{\Gamma}$ depend on η^a_{cl} and A_i only through $\tilde{M}^a{}_b$:

$$\tilde{M}(t)^a{}_b = \omega^{ac}\partial_c\partial_b \mathcal{H}\big(\eta_{\text{cl}}(t)\big) = \omega^{ac}\partial_c\partial_b\big(H + A_i J_i\big)\big(\eta_{\text{cl}}(t)\big) \, . \tag{30.3}$$

We also have used the "dual" $\bar{\chi}_a \equiv \chi^b\omega_{ba}$ in (30.2). We decompose $\chi^a = (\pi_i, x_i)$, $a = 1, 2, \ldots 2N$; $i = 1, 2 \ldots, N$. Now, the Morse Index theorem works in configuration space. Therefore we have to convert the phase space path integral (30.1) to a configuration space integral by integrating out the momentum components π_i. So let us first write:

$$S_{\text{fl}} = \frac{1}{2}\int_0^T dt \left[\chi^a\omega_{ab}\dot{\chi}^b - \chi^a\partial_a\partial_b\mathcal{H}\big(\eta_{\text{cl}}(t)\big)\chi^b\right] \tag{30.4}$$

and define

$$Q_{ab}(t) := \partial_a\partial_b\mathcal{H}\big(\eta_{\text{cl}}(t)\big) =: \begin{pmatrix} Q^{\pi\pi}_{ij}(t) & Q^{\pi x}_{ij}(t) \\ Q^{x\pi}_{ij}(t) & Q^{xx}_{ij}(t) \end{pmatrix} \, . \tag{30.5}$$

Note that the Q's are time-dependent, and Q_{ab} is symmetrical. This leads to

$$S_{\text{fl}} = \int_0^T dt \left[\pi_i\dot{x}_i - \frac{1}{2}\left(\pi_i Q^{\pi\pi}_{ij}\pi_j + 2\pi_i Q^{\pi x}_{ij}x_j + x_i Q^{xx}_{ij}x_j\right)\right] \tag{30.6}$$

$$=: \int_0^T dt \, L_{\text{fl}} \tag{30.7}$$

with

$$L_{\text{fl}} := \pi_i\left(\dot{x}_i - Q^{\pi x}_{ij}x_j\right) - \tfrac{1}{2}\pi_i Q^{\pi\pi}_{ij}\pi_j - \tfrac{1}{2}x_i Q^{xx}_{ij}x_j \, . \tag{30.8}$$

Equation (30.6) is still in first-order form. Now we eliminate the momenta by means of their classical equations of motion to get the second-order form. Upon using

$$\frac{\partial L_{\mathrm{fl}}}{\partial \pi_i} = 0$$

we obtain

$$\dot{x}_i - Q^{\pi x}_{ij} x_j = Q^{\pi\pi}_{ij} \pi_j \;:\; \pi = \left(Q^{\pi\pi}\right)^{-1}\left(\dot{x} - Q^{\pi x} x\right).$$

Inserting this back into (30.7), we find:

$$\begin{aligned} L_{\mathrm{fl}} &= \pi_i Q^{\pi\pi}_{ij} \pi_j - \tfrac{1}{2}\pi_i Q^{\pi\pi}_{ij} \pi_j - \tfrac{1}{2} x_i Q_{ij} x_j \\ &= \tfrac{1}{2}\pi Q^{\pi\pi}\pi - \tfrac{1}{2} x Q^{xx} x \end{aligned}$$

or

$$L_{\mathrm{fl}}(x, \dot{x}) = \tfrac{1}{2}\left(\dot{x} - Q^{\pi x} x\right)_i \left(Q^{\pi\pi}\right)^{-1}_{ij} \left(\dot{x} - Q^{\pi x} x\right)_j - \tfrac{1}{2} x_i Q^{xx}_{ij} x_j .$$

If we substitute this expression into (30.7) and perform suitable integrations by parts, we may rewrite the new action as

$$\begin{aligned} S_{\mathrm{fl}} &= \frac{1}{2}\int_0^T dt\, x_i(t) \left[C^{(2)}_{ij}(t) \frac{d^2}{dt^2} + C^{(1)}_{ij}(t) \frac{d}{dt} + C^{(0)}_{ij}(t) \right] x_j(t) \\ &\equiv \frac{1}{2}\int_0^T dt\, x_i(t) \Delta_{ij} x_j(t) , \end{aligned} \tag{30.9}$$

where the hermitian operator has the form

$$\Delta_{ij} = C^{(2)}_{ij}(t) \frac{d^2}{dt^2} + C^{(1)}_{ij}(t) \frac{d}{dt} + C^{(0)}_{ij}(t) . \tag{30.10}$$

The C's could be expressed in terms of the Q's, but this relation is not important here. What is important is that the classical equation of motion belonging to L_{fl}, i.e.,

$$\Delta_{ij} \psi_j(t) = 0 \tag{30.11}$$

is equivalent to the Jacobi equation,

$$\left(\frac{\partial}{\partial t} - \tilde{M}\right)^a_{\ b} \psi^b(t) = 0 . \tag{30.12}$$

Let us recall that the zero modes ψ^a of the fluctuation operator $(\partial_t - \tilde{M})$ are called Jacobi fields. They follow from a solution $\eta^a_{\mathrm{cl}}(t)$ of Hamilton's equation $\dot{\eta}^a_{\mathrm{cl}}(t) = \omega^{ab}\partial_b \mathcal{H}(\eta^c_{\mathrm{cl}}(t))$ when we linearize according to $\eta^a(t) = \eta^a_{\mathrm{cl}}(t) + \psi^a(t)$. The Jacobi field in configuration space, ψ_j, is obtained from ψ^a by eliminating the momentum components.

Let us return to the path integral

$$e^{i\hat{\Gamma}[\tilde{M}]} = \int \mathcal{D}x_i \mathcal{D}\pi_i \exp\left[i\int_0^T dt\, L_{\text{fl}}\right] . \tag{30.13}$$

When we insert the first-order form (30.8) in (30.13) and integrate over the momenta π_i we obtain the following path integral over configuration space:

$$e^{i\hat{\Gamma}[\tilde{M}]} = \int d^N x^{(0)} \int \mathcal{D}' x_i(t) \exp\left[\frac{i}{2}\int_0^T dt\, x_i(t)\Delta_{ij}x_j(t)\right] . \tag{30.14}$$

Here we have indicated explicitly the integration over the terminal points of the path; the integration $\mathcal{D}'x_i(t)$ is over paths with the boundary condition $x_i(0) = x_i^{(0)} = x_i(T)$.

In order for the Morse theory to be applicable, we reduce the path integral over loops (in configuration space) based at $x_i^{(0)} = 0$. This is done by expanding the quantum path $x_i(t)$ around the Jacobic field:

$$x_i(t) = \psi_i(t) + y_i(t) , \quad \Delta_{ij}\psi_j = 0 . \tag{30.15}$$

We require the Jacobi field ψ_i to fulfill the condition $\psi_i(0) = x_i^{(0)} = \psi_i(T)$, so that y_i has to vanish at the end points: $y_i(0) = 0 = y_i(T)$. Then we obtain

$$\begin{aligned}\int_0^T dt\, x\Delta x &= \int_0^T dt(\psi + y)\Delta(\psi + y)\\ &= \int_0^T dt[\psi\Delta\psi + \underbrace{\psi\Delta}_{\substack{=0\\ \Delta=\Delta^\dagger}}\, y + y\, \underbrace{\Delta\psi}_{=0} + y\Delta y] .\end{aligned}$$

Therefore we get

$$\begin{aligned}e^{i\hat{\Gamma}[\tilde{M}]} &= \int d^N x^{(0)} \exp\left[\frac{i}{2}\int_0^T dt\, \psi\Delta\psi\right]\\ &\times \int_{y(0)=0=y(T)} \mathcal{D}y(t) \exp\left[\frac{i}{2}\int_0^T dt\, y\Delta y\right] .\end{aligned} \tag{30.16}$$

The first factor involves the classical action of the Jacobi field. Since this factor is gauge invariant, the remainder $\hat{\Gamma}_0$ has the same gauge variation (under large gauge transformations) as $\hat{\Gamma}$, $\Delta\hat{\Gamma}_0 = \Delta\hat{\Gamma}$, where

$$e^{i\hat{\Gamma}_0[\tilde{M}]} = \int_{y(0)=0=y(T)} \mathcal{D}y(t) \exp\left[\frac{i}{2}\int_0^T dt\, y_i\Delta_{ij}y_j\right] . \tag{30.17}$$

In the usual way, by expanding y_i in terms of a complete set of eigenfunctions of Δ,

$$\Delta y_n = \lambda_n y_n \ , \quad \int_0^T dt\ y_n y_m = \delta_{nm} \ , \quad \sum_n y_n(t) y_m(t') = \delta(t-t')$$

$$y(t) = \sum_n a_n y_n(t) \ : \quad \int_0^T dt\ y \Delta y = \sum_n \lambda_n a_n^2 \ ,$$

we obtain

$$\begin{aligned} e^{i\hat{\Gamma}_0[\tilde{M}]} &= \prod_n \int_{-\infty}^{+\infty} da_n e^{i\lambda_n a_n^2/2} \\ &= \prod_n \left(\frac{2\pi}{|\lambda_n|}\right)^{1/2} e^{i\pi\,\mathrm{sign}(\lambda_n)/2} \ . \end{aligned} \tag{30.18}$$

Under a large gauge transformation $\tilde{M} \to \tilde{M}'$, the product

$$\prod_n |\lambda_n| = |\det(\Delta)| \tag{30.19}$$

can be regularized gauge invariantly. The only change comes from the exponential in (30.18) with the signs of the eigenvalues. Let us choose a path $\tilde{M}_s(t)$, $s \in (-\infty, +\infty)$ which interpolates between $\tilde{M}$ and the gauge-transformed $\tilde{M}'$. As we vary s, some of the eigenvalues $\lambda_n = \lambda_n(s)$ of $\Delta = \Delta[M_s]$ will cross zero and might give rise to a change $\Delta\hat{\Gamma}_0$ of $\hat{\Gamma}_0$. With the notation used before, we have:

$$\begin{aligned} \Delta\hat{\Gamma} = \Delta\hat{\Gamma}_0 &= \hat{\Gamma}_0[\tilde{M}'] - \hat{\Gamma}_0[\tilde{M}] \\ &= -\frac{\pi}{2}(\#\{\searrow\} - \#\{\nearrow\}) \pmod{2\pi} \ . \end{aligned} \tag{30.20}$$

Now the Morse Index theorem (cf. Chap. 4) tells us that the index of $\delta^2 S$ is equal to the number of conjugate points to $q_{\mathrm{cl}}(0)$ with $q_{\mathrm{cl}}(t)$ restricted to $0 \leqslant t \leqslant T$. When this is applied to

$$\Delta \equiv \Delta[\tilde{M}(t)] \equiv \Delta\left[\omega^{ab}\partial_a\partial_b\mathcal{H}\left(\eta^a_{\mathrm{cl}}(t)\right)\right] \ , \quad \eta^a_{\mathrm{cl}}(t) = \left(q^i_{\mathrm{cl}}(t)\ ,\ p^i_{\mathrm{cl}}(t)\right)$$

we find that the number of negative eigenvalues of $\Delta[q^i_{\mathrm{cl}}(t)]$ equals the number of points conjugate to q^i_{cl} $(t=0)$ along the trajectory $q^i_{\mathrm{cl}}(t)$. Note that $q^i_{\mathrm{cl}}(0) = q^i_{\mathrm{cl}}(T)$.

Under a large gauge transformation, a trajectory is mapped onto a new one with a different number of revolutions around the torus $\{J_i = \text{const.}\}$ and a different winding number. Also, the number of conjugate points in configuration space changes:

$$\begin{aligned} \#\{\searrow\} - \#\{\nearrow\} &\equiv \text{index}\ [\Delta\,(q_{\mathrm{cl}}(t))] - \text{index}\ \left[\Delta\left(q'_{\mathrm{cl}}(t)\right)\right] \\ &= \#\ \{\text{conj. pts. along } q_{\mathrm{cl}}(t)\} - \#\ \left\{\text{conj. pts. along } q'_{\mathrm{cl}}(t)\right\} \ . \end{aligned} \tag{30.21}$$

Combining this with (30.20) we obtain:

$$\Delta\hat{\Gamma} = -\frac{\pi}{2}\left[\#\{\text{conj. pts. along } q_{\text{cl}}(t)\} - \#\left\{\text{conj. pts. along } q'_{\text{cl}}(t)\right\}\right] . \tag{30.22}$$

This shows that $\Delta\hat{\Gamma}$ can be obtained from purely classical data, namely by examining how often the final point $q^i_{\text{cl}}(T)$ is conjugate around the loop to the initial point $q^i_{\text{cl}}(0)$. This is easily done for the harmonic oscillator where $q_{\text{cl}}(2\pi)$ is conjugate to $q_{\text{cl}}(0)$ of "order 2", since the first point conjugate to $q_{\text{cl}}(0)$ appears after half-period already: $q_{\text{cl}}(\pi)$. On the other hand, the effect of a gauge transformation with winding number $\mathcal{N}$ is to increase the number of revolutions from p to $p + \mathcal{N}$. Hence the square bracket in (30.22) equals $2\mathcal{N}$, which yields

$$\Delta\hat{\Gamma} = -\pi\mathcal{N} , \tag{30.23}$$

as found in (26.37). One can generalize the result from a one-torus to an N-torus. One obtains [s. M. Reuter, Phys. Rev. D **42**, 2763 (1990)]

$$\Delta\hat{\Gamma} = -\pi\sum_{i=1}^{N}\mathcal{N}_i w_i = -\pi\Delta W , \tag{30.24}$$

where the integers w_i are certain winding numbers related to the topology of Sp(2N). The quantity $\sum_{i=1}^{N} w_i\mathcal{N}_i = \Delta W$ has been called Littlejohn's winding number in one of the previous sections. For its definition and a detailed discussion of its properties we have to refer to the original publications in Phys. Rep. **138**, 193 (1986) and Phys. Rev. A **36**, 2953 (1987). Comparing (30.24) and (28.29) we see that $\mu_i = 2w_i$ which is the most important result we can obtain in this approach.

Let us have a final look at the gauge variance of $\mathrm{e}^{\mathrm{i}\hat{\Gamma}[A]}$. Formally we may write:

$$\mathrm{e}^{\mathrm{i}\hat{\Gamma}[A]} = \left[\det\left(\frac{\partial}{\partial t} - \tilde{M}\right)\right]^{-1/2} \tag{30.25}$$

or

$$\mathrm{e}^{-2\mathrm{i}\hat{\Gamma}[A]} = \det\left(\frac{\partial}{\partial t} - \tilde{M}\right) . \tag{30.26}$$

Note that $\det(\partial_t - \tilde{M})$ is real since $\omega(\partial_t - \tilde{M})$ is hermitian and $\det[\omega(\partial_t - \tilde{M})] = \det(\partial_t - \tilde{M})$, $\det\omega = 1$. Our main result was that under a (large) gauge transformation

$$\Delta\hat{\Gamma} = -\pi\sum_{i=1}^{N}\mathcal{N}_i w_i = \pi \cdot (\text{integer}) . \tag{30.27}$$

Therefore

$$\mathrm{e}^{\mathrm{i}\Delta\hat{\Gamma}} = \mathrm{e}^{\mathrm{i}\pi(\text{integer})} = \pm 1 \tag{30.28}$$

but

$$\mathrm{e}^{2\mathrm{i}\Delta\hat{\Gamma}} = +1 . \tag{30.29}$$

Comparing (30.29) with (30.25, 26), we see that $\det(\partial_t - \tilde{M})$ is gauge invariant under large gauge transformations, but its formal square root $[\det(\partial_t - \tilde{M})]^{1/2}$ is not, because it is not a priori clear which sign the square root should have. Defined in this way, $[\det(\partial_t - \tilde{M})]^{1/2}$ is certainly invariant under infinitesimal gauge transformations – since the sign cannot change abruptly. But nothing guarantees that $[\det(\partial_t - \tilde{M})]^{1/2}$ is invariant under topologically nontrivial gauge transformations. In other words, if one continuously varies the gauge field from $A_i(t)$ to $A_i + 2\pi \mathcal{N}_i/T$, one will possibly end up with a square root of opposite sign.

The situation outlined here is very similar to Witten's global SU(2) anomaly. For the partition function $Z_{\text{Dirac}} = \det \not{D}$ of a massless Dirac fermion, there is no problem defining it in a gauge invariant way. However, the partition function of a Weyl fermion reads: $Z_{\text{Weyl}} = \sqrt{\det \not{D}}$, and the sign ambiguity $\pm\sqrt{\det \not{D}}$ leads to the global SU(2) anomaly. The correspondence is therefore:

$$Z_{\text{Dirac}} \equiv \det \not{D} \longleftrightarrow \frac{1}{\det(\partial_t - \tilde{M})} = \mathrm{e}^{2\mathrm{i}\hat{\Gamma}} , \quad \text{no anomaly} .$$

$$Z_{\text{Weyl}} \equiv \sqrt{\det \not{D}} \longleftrightarrow \frac{1}{\sqrt{\det(\partial_t - \tilde{M})}} = \mathrm{e}^{\mathrm{i}\hat{\Gamma}} , \quad \text{global anomaly} .$$

From this general discussion we learn that the effect of a large gauge transformation is at most a sign change:

$$\mathrm{e}^{\mathrm{i}\Delta\hat{\Gamma}} = \pm 1 = \mathrm{e}^{\mathrm{i}\pi(\text{integer})} . \tag{30.30}$$

31. Berry's Phase

Let a physical system be described by a Hamiltonian with two sets of variables $\boldsymbol{r}$ and $\boldsymbol{R}(t)$: $H(\boldsymbol{r}, \boldsymbol{R}(t))$. The dynamical degrees of freedom $\boldsymbol{r}$ (not necessarily space variables) are also called fast variables. The external time dependence is given by the slowly varying parameters $\boldsymbol{R}(t) = \{X(t), Y(t), \ldots, Z(t)\}$; consequently, the $\boldsymbol{R}(t)$ are called slow variables.

We will be interested in solving the Schrödinger equation for the state vector $|\psi(t)\rangle$:

$$\mathrm{i}\hbar\frac{\partial}{\partial t}|\psi(t)\rangle = H\big(\boldsymbol{R}(t)\big)|\psi(t)\rangle . \tag{31.1}$$

If the $\boldsymbol{R}(t)$ were independent of the external time parameter t, then a time-independent set of energy eigenstates $|n, \boldsymbol{R}\rangle$ with energy eigenvalues $E_n(\boldsymbol{R})$ would exist satisfying

$$H(\boldsymbol{R})|n, \boldsymbol{R}\rangle = E_n(\boldsymbol{R})|n, \boldsymbol{R}\rangle . \tag{31.2}$$

When $\boldsymbol{R}(t)$ changes in time, we can still take $|n, \boldsymbol{R}(t)\rangle$ as a basis; however the eigenvalue equation (31.2) is then only valid at an instantaneous moment t:

$$H\big(\boldsymbol{R}(t)\big)|n, \boldsymbol{R}(t)\rangle = E_n\big(\boldsymbol{R}(t)\big)|n, \boldsymbol{R}(t)\rangle . \tag{31.3}$$

This eigenvalue equation implies no relation between the (so far arbitrary) phases of the eigenstates $|n, \boldsymbol{R}(t)\rangle$ at different $\boldsymbol{R}(t)$. The states $|n, \boldsymbol{R}(t)\rangle$ are normalized according to

$$\langle n, \boldsymbol{R}(t)|m, \boldsymbol{R}(t)\rangle = \delta_{nm} . \tag{31.4}$$

The solution of the Schrödinger equation (31.1) can be expanded in terms of the complete basis set $|n, \boldsymbol{R}(t)\rangle$:

$$|\psi(t)\rangle = \sum_n a_n(t)\exp\left[-\frac{\mathrm{i}}{\hbar}\int_0^t dt'\, E_n\big(\boldsymbol{R}(t')\big)\right]|n, \boldsymbol{R}(t)\rangle . \tag{31.5}$$

Substituting this ansatz in (31.1) we obtain

$$\begin{aligned}
&\mathrm{i}\hbar\sum_n \exp\left[-\frac{\mathrm{i}}{\hbar}\int_0^t dt'\, E_n\big(\boldsymbol{R}(t')\big)\right]\left(\dot{a}_n + \frac{1}{\mathrm{i}\hbar}a_n E_n + a_n\frac{\partial}{\partial t}\right)|n, \boldsymbol{R}(t)\rangle \\
&\quad = \sum_n a_n(t)\exp\left[-\frac{\mathrm{i}}{\hbar}\int_0^t dt'\, E_n\big(\boldsymbol{R}(t')\big)\right]\underbrace{H\big(\boldsymbol{R}(t)\big)|n, \boldsymbol{R}(t)\rangle}_{E_n\left(\boldsymbol{R}(t)\right)|n, \boldsymbol{R}(t)\rangle} .
\end{aligned}$$

So we get

$$\sum_n \exp\left[-\frac{\mathrm{i}}{\hbar}\int_0^t dt'\, E_n\big(\boldsymbol{R}(t')\big)\right]\left(\dot{a}_n + a_n\frac{\partial}{\partial t}\right)|n, \boldsymbol{R}(t)\rangle = 0\,.$$

Taking the inner product of this equation with

$$\langle m, \boldsymbol{R}(t)|\exp\left[\frac{\mathrm{i}}{\hbar}\int_0^t dt'\, E_m\big(\boldsymbol{R}(t')\big)\right]$$

leads to

$$\dot{a}_m(t) = -\sum_n a_n(t)\exp\left\{\frac{\mathrm{i}}{\hbar}\int_0^t dt'\left[E_m\big(\boldsymbol{R}(t')\big) - E_n\big(\boldsymbol{R}(t')\big)\right]\right\} \times \langle m, \boldsymbol{R}(t)\left|\frac{\partial}{\partial t}\right|n, \boldsymbol{R}(t)\rangle\,. \tag{31.6}$$

In order to get rid of the time derivative of the base set we go back to the eigenvalue equation (31.3) and take the time derivative on both sides:

$$\frac{\partial H}{\partial t}|n, \boldsymbol{R}(t)\rangle + H\frac{\partial}{\partial t}|n, \boldsymbol{R}(t)\rangle = \frac{\partial E_n}{\partial t}|n, \boldsymbol{R}(t)\rangle + E_n\frac{\partial}{\partial t}|n, \boldsymbol{R}(t)\rangle\,.$$

Multiplying this equation from the left by $\langle m, \boldsymbol{R}(t)|$ we find

$$\langle m, \boldsymbol{R}(t)\left|\frac{\partial H}{\partial t}\right|n, \boldsymbol{R}(t)\rangle + \underbrace{\langle m, \boldsymbol{R}(t)|H}_{E_m\langle m,\boldsymbol{R}|}\left|\frac{\partial}{\partial t}\right|n, \boldsymbol{R}(t)\rangle$$

$$= \frac{\partial E_n}{\partial t}\underbrace{\langle m, \boldsymbol{R}(t)|n, \boldsymbol{R}(t)\rangle}_{=0,\, m\neq n} + E_n\langle m, \boldsymbol{R}(t)\left|\frac{\partial}{\partial t}\right|n, \boldsymbol{R}(t)\rangle\,.$$

This equation can be rewritten in the form

$$\langle m, \boldsymbol{R}(t)\left|\frac{\partial}{\partial t}\right|n, \boldsymbol{R}(t)\rangle\Big(E_m\big(\boldsymbol{R}(t)\big) - E_n\big(\boldsymbol{R}(t)\big)\Big) = -\langle m, \boldsymbol{R}(t)\left|\frac{\partial H}{\partial t}\right|n, \boldsymbol{R}(t)\rangle \quad m \neq n \tag{31.7}$$

or

$$\langle m, \boldsymbol{R}(t)\left|\frac{\partial}{\partial t}\right|n, \boldsymbol{R}(t)\rangle = \frac{\langle m, \boldsymbol{R}(t)|\partial H/\partial t|n, \boldsymbol{R}(t)\rangle}{E_n\big(\boldsymbol{R}(t)\big) - E_m\big(\boldsymbol{R}(t)\big)}\,, \quad m \neq n\,,$$

which is the desired expression in (31.6). Hence we end up with

$$\dot{a}_m(t) = -a_m(t)\langle m, \boldsymbol{R}(t)\left|\frac{\partial}{\partial t}\right|m, \boldsymbol{R}(t)\rangle - \sum_{n\neq m} a_n(t)\exp\left\{\frac{\mathrm{i}}{\hbar}\int_0^t dt'\left[E_m(t') - E_n(t')\right]\right\} \times \frac{\langle m, \boldsymbol{R}(t)|\partial H/\partial t|n, \boldsymbol{R}(t)\rangle}{E_n(t) - E_m(t)}\,. \tag{31.8}$$

At this stage we want to make the adiabatic approximation for the coefficients $a_m(t)$, which is equivalent to requiring

$$\langle m, \boldsymbol{R}(t) \left| \frac{\partial}{\partial t} \right| n, \boldsymbol{R}(t) \rangle = 0 \,, \quad m \neq n \,. \tag{31.9}$$

In other words, we want the base state vector $|n, \boldsymbol{R}(t)\rangle$ to undergo a parallel transport in parameter space. Equation (31.8) is then reduced to

$$\dot{a}_m(t) = -a_m(t) \langle m, \boldsymbol{R}(t) \left| \frac{\partial}{\partial t} \right| m, \boldsymbol{R}(t) \rangle \,. \tag{31.10}$$

The physical meaning of the above approximation is that the rate of change of the basis states is small compared to the Bohr period $\omega_{nm} = (E_n - E_m)/\hbar$ for the transition $m \to n$; the perturbation should be so slow – in fact, infinitely slow (adiabatic) – that no transitions between the energy eigenstates become possible. Therefore, neglecting the second term in (31.8) expresses the fact that $H(\boldsymbol{R}(t))$ does not change rapidly enough to induce transitions between the states. The system, once in the eigenstate $|m, \boldsymbol{R}(0)\rangle$, remains in this state also at a later time t, $|m, \boldsymbol{R}(t)\rangle$.

Now, let the system be in an eigenstate $|m, \boldsymbol{R}(0)\rangle$ at $t = 0$, so that according to (31.5), we obtain $a_m(0) = 1$. If we then integrate (31.10) with this initial condition we obtain

$$a_m(t) = \exp\left[-\int_0^t dt' \langle m, \boldsymbol{R}(t') \left| \frac{\partial}{\partial t'} \right| m, \boldsymbol{R}(t') \rangle \right] \underbrace{a_m(0)}_{=1} \,. \tag{31.11}$$

Using the normalization condition (31.4),

$$\underbrace{\left(\frac{\partial}{\partial t} \langle m, \boldsymbol{R}(t)| \right) |m, \boldsymbol{R}(t)\rangle}_{=\langle m, \boldsymbol{R}(t)|\partial/\partial t|m, \boldsymbol{R}(t)\rangle^*} + \langle m, \boldsymbol{R}(t) \left| \frac{\partial}{\partial t} \right| m, \boldsymbol{R}(t) \rangle = 0 \tag{31.12}$$

or

$$2 \,\mathrm{Re} \langle m, \boldsymbol{R}(t) \left| \frac{\partial}{\partial t} \right| m, \boldsymbol{R}(t) \rangle = 0 \,,$$

we see that the integrand in the exponential of (31.11) is purely imaginary:

$$a_m(t) = \mathrm{e}^{\mathrm{i}\gamma_m(t)} \,, \quad \gamma_m(t) \in \mathbb{R} \,, \tag{31.13}$$

where

$$\gamma_m(t) = \mathrm{i} \int_0^t dt' \langle m, \boldsymbol{R}(t') \left| \frac{\partial}{\partial t'} \right| m, \boldsymbol{R}(t') \rangle \,. \tag{31.14}$$

Finally, then, the adiabatic approximation yields, according to (31.5),

$$|\psi(t)\rangle = e^{i\gamma_m(t)} \exp\left[-\frac{i}{\hbar}\int_0^t dt'\, E_m\big(\boldsymbol{R}(t')\big)\right] |m, \boldsymbol{R}(t)\rangle \,. \tag{31.15}$$

Here, the question naturally arises as to whether the extra phase $\gamma_m(t)$ in (31.15) has any physical significance. The naive answer would be no; let us look at the state

$$e^{i\gamma_m(t)} |m, \boldsymbol{R}(t)\rangle =: \widetilde{|m, \boldsymbol{R}(t)\rangle} \,, \tag{31.16}$$

which appears in (31.15). Taking the time derivative of (31.16) yields

$$\frac{\partial}{\partial t}\widetilde{|m, \boldsymbol{R}(t)\rangle} = e^{i\gamma_m(t)} \left(i\dot{\gamma}_m + \frac{\partial}{\partial t}\right) |m, \boldsymbol{R}(t)\rangle \,,$$

or, multiplying from the left with $\widetilde{\langle m, \boldsymbol{R}(t)|}$ and using $\langle m, \boldsymbol{R}(t)|m, \boldsymbol{R}(t)\rangle = 1$, we find

$$\begin{aligned}\widetilde{\langle m, \boldsymbol{R}(t)}\left|\frac{\partial}{\partial t}\right|\widetilde{m, \boldsymbol{R}(t)\rangle} &= i\dot{\gamma}_m(t) + \underbrace{\langle m, \boldsymbol{R}(t)|\partial/\partial t|m, \boldsymbol{R}(t)\rangle}_{=:\, i\alpha_m(t)} \\ &= i\big(\dot{\gamma}_m(t) + \alpha_m(t)\big) \,. \end{aligned} \tag{31.17}$$

If we now choose γ_m to be

$$\gamma_m(t) = -\int_0^t dt'\, \alpha_m(t') \,, \tag{31.18}$$

equation (31.17) turns into

$$\widetilde{\langle m, \boldsymbol{R}(t)}\left|\frac{\partial}{\partial t}\right|\widetilde{m, \boldsymbol{R}(t)\rangle} = 0 \,.$$

If the phase γ_m has been chosen in this way, (31.18), dropping the tilde we simply obtain

$$\langle m, \boldsymbol{R}(t)\left|\frac{\partial}{\partial t}\right|m, \boldsymbol{R}(t)\rangle = 0$$

and hence γ_m would be absent from (31.15).

This was the state of affairs until Berry (1984) looked at the problem again, considering the case in which the $\boldsymbol{R}(t)$ change by moving along a closed path, or circuit C, in parameter space, returning to their original values at $t = T : \boldsymbol{R}(T) = \boldsymbol{R}(0)$. Since the states $|n, \boldsymbol{R}(t)\rangle$ only depend on t via the external parameters $\boldsymbol{R}(t)$, we may write (31.14) as

$$\begin{aligned}\gamma_m(C) &= i\int_0^T dt \frac{d\boldsymbol{R}(t)}{dt} \cdot \langle m, \boldsymbol{R}(t)|\nabla_{\boldsymbol{R}}|m, \boldsymbol{R}(t)\rangle \\ &= i\oint_C d\boldsymbol{R} \cdot \langle m, \boldsymbol{R}|\nabla_{\boldsymbol{R}}|m, \boldsymbol{R}\rangle \,, \end{aligned} \tag{31.19}$$

where the integral in the $\boldsymbol{R}$-space (slow parameter space) is along the circuit C parametrized by t.

Introducing a "vector potential" in $\boldsymbol{R}$-space (this need not be the ordinary three-dimensional space),

$$\boldsymbol{A}(\boldsymbol{R}) := \mathrm{i}\langle m, \boldsymbol{R}|\nabla_{\boldsymbol{R}}|m, \boldsymbol{R}\rangle \;, \tag{31.20}$$

the Berry phase becomes

$$\gamma_m(C) = \oint d\boldsymbol{R} \cdot \boldsymbol{A}(\boldsymbol{R}) \;. \tag{31.21}$$

If we now make an $\boldsymbol{R}$-dependent phase change of $|m, \boldsymbol{R}\rangle$ via

$$|m, \boldsymbol{R}\rangle \to \mathrm{e}^{\mathrm{i}\chi(\boldsymbol{R})}\,|m, \boldsymbol{R}\rangle \;, \tag{31.22}$$

$\boldsymbol{A}(\boldsymbol{R})$ changes as follows: first we need in (31.20)

$$\begin{aligned}&\nabla_{\boldsymbol{R}}\big[\mathrm{e}^{\mathrm{i}\chi(\boldsymbol{R})}|m, \boldsymbol{R}\rangle\big]\\ &\quad = \mathrm{e}^{\mathrm{i}\chi}\mathrm{i}\big(\nabla_{\boldsymbol{R}}\chi(\boldsymbol{R})\big)|m, \boldsymbol{R}\rangle + \mathrm{e}^{\mathrm{i}\chi}\nabla_{\boldsymbol{R}}|m, \boldsymbol{R}\rangle \;.\end{aligned}$$

Multiplying this equation from the left by $\langle m, \boldsymbol{R}|\mathrm{e}^{-\mathrm{i}\chi}$ we obtain, using (31.4) once more,

$$\mathrm{i}\nabla_{\boldsymbol{R}}\chi(\boldsymbol{R}) + \langle m, \boldsymbol{R}|\nabla_{\boldsymbol{R}}|m, \boldsymbol{R}\rangle \;.$$

This then says that $\boldsymbol{A}(\boldsymbol{R})$ changes by a gradient:

$$\boldsymbol{A}(\boldsymbol{R}) \to \boldsymbol{A}(\boldsymbol{R}) - \nabla_{\boldsymbol{R}}\chi(\boldsymbol{R}) \;, \tag{31.23}$$

i.e., the parameter vector potential transforms exactly the way in which an electromagnetic vector potential $\boldsymbol{A}$ changes when the states undergo a local phase transformation analogous to (31.22). Nevertheless, Berry's phase is gauge invariant because by Stoke's Law, the line integral in (31.21) may be converted to an integral of the curl of $\boldsymbol{A}$:

$$\gamma_m(C) = \oint_{C=\partial S} d\boldsymbol{R} \cdot \boldsymbol{A}(\boldsymbol{R}) = \int_S d\boldsymbol{S} \cdot (\nabla \times \boldsymbol{A}) \tag{31.24}$$

$$= \int_S d\boldsymbol{S} \cdot \boldsymbol{V} \,, \quad \boldsymbol{V} = \nabla \times \boldsymbol{A} \,, \tag{31.25}$$

where S is a surface in parameter space bounded by C, and use has been made of the "$\nabla\times$" notation as if that space were three dimensional. Nevertheless, let us assume in the sequel that we are indeed working in three dimensions.

The analogy with the electromagnetic potentials and fields does not mean that the effects are necessarily of electromagnetic origin. Hence, in order not to make any reference to electrodynamics, we call the vector potential $\boldsymbol{A}$ a connection and the field $\boldsymbol{V}$ a curvature. The Berry phase arises from the nontrivial topological properties of the space spanned by the parameters $\boldsymbol{R}$.

If $\boldsymbol{A}$ is a pure gauge $\boldsymbol{A} = \nabla\phi$, then Berry's phase will vanish, provided ϕ is nonsingular. However, we anticipate a Berry phase for a magnetic flux (Aharonov–Bohm) or a magnetic monopole configuration.

Let us continue to rewrite $\gamma_m(C)$ in (31.24), so as to get rid of the gradient of the state vector and obtain for $\gamma_m(C)$ an expression which is manifestly independent of the phase of $|m, \boldsymbol{R}\rangle$. For this reason we write, limiting ourselves to a three-dimensional parameter space:

$$\begin{aligned}(\nabla_{\boldsymbol{R}} \times \boldsymbol{A})_i &= \mathrm{i}\big(\nabla_{\boldsymbol{R}} \times \langle m, \boldsymbol{R}|\nabla_{\boldsymbol{R}}|m, \boldsymbol{R}\rangle\big)_i \\ \partial_i \equiv \frac{\partial}{\partial R_i}\,; &= \mathrm{i}\varepsilon_{ijk}\partial_j\big(\langle m, \boldsymbol{R}|\partial_k|m, \boldsymbol{R}\rangle\big) \\ &= \mathrm{i}\varepsilon_{ijk}\big(\partial_j\langle m, \boldsymbol{R}|\big)\big(\partial_k|m, \boldsymbol{R}\rangle\big) + \mathrm{i}\varepsilon_{ijk}\langle m, \boldsymbol{R}|\big(\partial_j\partial_k|m, \boldsymbol{R}\rangle\big) \\ &= \mathrm{i}\varepsilon_{ijk}\big(\partial_j\langle m, \boldsymbol{R}|\big)\big(\partial_k|m, \boldsymbol{R}\rangle\big)\end{aligned}$$

or

$$\begin{aligned}\nabla_{\boldsymbol{R}} \times \mathrm{i}\langle m, \boldsymbol{R}|\nabla_{\boldsymbol{R}}|m, \boldsymbol{R}\rangle &\equiv \nabla_{\boldsymbol{R}} \times \boldsymbol{A}(\boldsymbol{R}) \\ &= \mathrm{i}\big(\nabla_{\boldsymbol{R}}\langle m, \boldsymbol{R}|\big) \times \big(\nabla_{\boldsymbol{R}}|m, \boldsymbol{R}\rangle\big)\,.\end{aligned}$$

Hence we can rewrite (31.24) in the form

$$\gamma_m(C) = \mathrm{i}\int_S d\boldsymbol{S} \cdot \big(\nabla_{\boldsymbol{R}}\langle m, \boldsymbol{R}|\big) \times \big(\nabla_{\boldsymbol{R}}|m, \boldsymbol{R}\rangle\big)\,,$$

or, upon using the completeness of the energy eigenstates,

$$\sum_n |n, \boldsymbol{R}\rangle\langle n, \boldsymbol{R}| = \mathbb{1}$$

$$\gamma_m(C) = \sum_n \mathrm{i}\int_S d\boldsymbol{S} \cdot \big(\nabla_{\boldsymbol{R}}\langle m, \boldsymbol{R}|\big)|n, \boldsymbol{R}\rangle \times \langle n, \boldsymbol{R}|\nabla_{\boldsymbol{R}}|m, \boldsymbol{R}\rangle\,. \tag{31.26}$$

To discuss this expression further, let us begin by looking at the diagonal elements: first we repeat the equation which follows from the normalization (31.4):

$$\big(\nabla_{\boldsymbol{R}}\langle m, \boldsymbol{R}|\big)|m, \boldsymbol{R}\rangle + \langle m, \boldsymbol{R}|\nabla_{\boldsymbol{R}}|m, \boldsymbol{R}\rangle = 0\,.$$

This equation is employed in the integrand of (31.26):

$$\begin{aligned}&\big(\nabla_{\boldsymbol{R}}\langle m, \boldsymbol{R}|\big)|m, \boldsymbol{R}\rangle \times \langle m, \boldsymbol{R}|\nabla_{\boldsymbol{R}}|m, \boldsymbol{R}\rangle \\ &\quad = -\langle m, \boldsymbol{R}|\nabla_{\boldsymbol{R}}|m, \boldsymbol{R}\rangle \times \langle m, \boldsymbol{R}|\nabla_{\boldsymbol{R}}|m, \boldsymbol{R}\rangle \\ &\quad = \langle m, \boldsymbol{R}|\nabla_{\boldsymbol{R}}|m, \boldsymbol{R}\rangle \times \big(\nabla_{\boldsymbol{R}}\langle m, \boldsymbol{R}|\big)|m, \boldsymbol{R}\rangle \\ &\quad = -\big(\nabla_{\boldsymbol{R}}\langle m, \boldsymbol{R}|\big)|m, \boldsymbol{R}\rangle \times \langle m, \boldsymbol{R}|\nabla_{\boldsymbol{R}}|m, \boldsymbol{R}\rangle = 0\,.\end{aligned} \tag{31.27}$$

Therefore the sum in (31.26) can be taken over $n \neq m$. So we have to look at the off-diagonal elements in (31.26). For this reason we start out with the eigenvalue equation (31.3):

$$H(\boldsymbol{R})|m, \boldsymbol{R}\rangle = E_m(\boldsymbol{R})|m, \boldsymbol{R}\rangle \ . \tag{31.28}$$

Taking the gradient on both sides we obtain

$$\big(\nabla_{\boldsymbol{R}} H\big)|m, \boldsymbol{R}\rangle + H\nabla_{\boldsymbol{R}}(|m, \boldsymbol{R}\rangle) = \big(\nabla_{\boldsymbol{R}} E_m(\boldsymbol{R})\big)|m, \boldsymbol{R}\rangle + E_m(\boldsymbol{R})\big(\nabla_{\boldsymbol{R}}|m, \boldsymbol{R}\rangle\big) \ .$$

Now we multiply from the left by $\langle n, \boldsymbol{R}|$ to get

$$\begin{aligned}\langle n, \boldsymbol{R}|\big(\nabla_{\boldsymbol{R}} H\big)|m, \boldsymbol{R}\rangle &+ \underbrace{\langle n, \boldsymbol{R}|H}_{E_n(\boldsymbol{R})\langle n,\boldsymbol{R}|}\ \nabla_{\boldsymbol{R}}|m, \boldsymbol{R}\rangle \\ &= \nabla_{\boldsymbol{R}} E_m(\boldsymbol{R})\underbrace{\langle n, \boldsymbol{R}|m, \boldsymbol{R}\rangle}_{=0,\ n\neq m} + E_m(\boldsymbol{R})\langle n, \boldsymbol{R}|\nabla_{\boldsymbol{R}}|m, \boldsymbol{R}\rangle\end{aligned}$$

or

$$\langle n, \boldsymbol{R}|\nabla_{\boldsymbol{R}}|m, \boldsymbol{R}\rangle = \frac{\langle n, \boldsymbol{R}|(\nabla_{\boldsymbol{R}} H)|m, \boldsymbol{R}\rangle}{E_m(\boldsymbol{R}) - E_n(\boldsymbol{R})} \ , \qquad n \neq m \ .$$

Likewise,

$$\big(\nabla_{\boldsymbol{R}}\langle m, \boldsymbol{R}|\big)|n, \boldsymbol{R}\rangle = \frac{\langle m, \boldsymbol{R}|(\nabla_{\boldsymbol{R}} H)|n, \boldsymbol{R}\rangle}{E_m(\boldsymbol{R}) - E_n(\boldsymbol{R})} \ , \qquad n \neq m \ .$$

These last two results are needed in (31.26):

$$\begin{aligned}&\big(\nabla_{\boldsymbol{R}}\langle m, \boldsymbol{R}|\big)|n, \boldsymbol{R}\rangle \times \langle n, \boldsymbol{R}|\nabla_{\boldsymbol{R}}|m, \boldsymbol{R}\rangle \\ &= \frac{\langle m, \boldsymbol{R}|(\nabla_{\boldsymbol{R}} H)|n, \boldsymbol{R}\rangle \times \langle n, \boldsymbol{R}|(\nabla_{\boldsymbol{R}} H)|m, \boldsymbol{R}\rangle}{(E_m(\boldsymbol{R}) - E_n(\boldsymbol{R}))^2} \ .\end{aligned} \tag{31.29}$$

Therefore γ_m can be expressed as

$$\gamma_m(C) = -\int_S d\boldsymbol{S} \cdot \boldsymbol{V}_m(\boldsymbol{R}) \ , \tag{31.30}$$

where

$$\boldsymbol{V}_m(\boldsymbol{R}) \equiv \operatorname{Im} \sum_{n\neq m} \frac{\langle m, \boldsymbol{R}|\nabla_{\boldsymbol{R}} H|n, \boldsymbol{R}\rangle \times \langle n, \boldsymbol{R}|\nabla_{\boldsymbol{R}} H|m, \boldsymbol{R}\rangle}{(E_m(\boldsymbol{R}) - E_n(\boldsymbol{R}))^2} \ . \tag{31.31}$$

Starting from (31.19) we have used the fact in (31.31) that $\langle m, \boldsymbol{R}|\nabla_{\boldsymbol{R}}|n, \boldsymbol{R}\rangle$ is imaginary, i.e., $\langle m, \boldsymbol{R}|\nabla_{\boldsymbol{R}}|m, \boldsymbol{R}\rangle = \mathrm{i}\,\operatorname{Im}[\langle m, \boldsymbol{R}|\nabla_{\boldsymbol{R}}|m, \boldsymbol{R}\rangle]$.

We are now going to study a by now standard example in which we obtain a nonzero flux in (31.30). Note that if two energy eigenvalues cross in parameter space, e.g., $E_m(\boldsymbol{R}^*) = E_n(\boldsymbol{R}^*)$ for some $\boldsymbol{R}^*$, something interesting will certainly occur. It is this kind of spin degeneracy ($\boldsymbol{r} \equiv \boldsymbol{\sigma}$, $\sigma' = \pm 1$, $E_+(0) = E_-(0)$) that will appear in the sequel. The parameter space is given by the magnetic field $\boldsymbol{R}(t) \equiv \boldsymbol{B}(t)$ and a spin 1/2 particle is cyclically transported in this space. The Hamiltonian that describes the time development in $\boldsymbol{R}$-space is given by

$$H\big(\boldsymbol{R}(t)\big) = -\frac{\mu}{2}\boldsymbol{\sigma} \cdot \boldsymbol{R}(t) = -\frac{\mu}{2}\begin{pmatrix} Z(t) & X(t) - \mathrm{i}Y(t) \\ X(t) + \mathrm{i}Y(t) & -Z(t) \end{pmatrix} . \tag{31.32}$$

The energy eigenvalues follow from

$$\begin{vmatrix} Z - E & X - \mathrm{i}Y \\ X + \mathrm{i}Y & -Z - E \end{vmatrix} = 0$$

or

$$E_\pm(\boldsymbol{R}) = \pm\frac{\mu}{2}\sqrt{X^2 + Y^2 + Z^2} = \pm\frac{\mu}{2}R . \tag{31.33}$$

Hence, there exists a degeneracy where $\boldsymbol{R} = 0$.

Now, in (31.31) we need

$$\nabla_{\boldsymbol{R}} H(\boldsymbol{R}) = \nabla_{\boldsymbol{R}}\left(-\frac{\mu}{2}\boldsymbol{\sigma} \cdot \boldsymbol{R}\right) = -\frac{\mu}{2}\boldsymbol{\sigma} ,$$

and, assuming $\sigma'_z = -1$ as initial spin state,

$$\frac{\langle -|\nabla_{\boldsymbol{R}} H|+\rangle \times \langle +|\nabla_{\boldsymbol{R}} H|-\rangle}{(E_- - E_+)^2} ,$$

where we temporarily rotate the z-axis so as to point along $\boldsymbol{R}$. In the numerator we have in components

$$\begin{aligned}
&\frac{\mu^2}{4}\begin{pmatrix} \langle -|\sigma_x|+\rangle \\ \langle -|\sigma_y|+\rangle \\ \langle -|\sigma_z|+\rangle \end{pmatrix} \times \begin{pmatrix} \langle +|\sigma_x|-\rangle \\ \langle +|\sigma_y|-\rangle \\ \langle +|\sigma_z|-\rangle \end{pmatrix} \\
&\quad = \frac{\mu^2}{4}\begin{pmatrix} \langle -|\sigma_y|+\rangle\langle +|\sigma_z|-\rangle - \langle -|\sigma_z|+\rangle\langle +|\sigma_y|-\rangle \\ \langle -|\sigma_z|+\rangle\langle +|\sigma_x|-\rangle - \langle -|\sigma_x|+\rangle\langle +|\sigma_z|-\rangle \\ \langle -|\sigma_x|+\rangle\langle +|\sigma_y|-\rangle - \langle -|\sigma_y|+\rangle\langle +|\sigma_x|-\rangle \end{pmatrix}
\end{aligned}$$

$$\begin{aligned} \sigma_z|\pm\rangle &= \pm|\pm\rangle , \\ \sigma_x|\pm\rangle &= |\mp\rangle , \\ \sigma_y|\pm\rangle &= \pm\mathrm{i}|\mp\rangle ; \end{aligned} \quad = \frac{\mu^2}{4}\begin{pmatrix} -\mathrm{i}\langle -|-\rangle\langle +|-\rangle + \mathrm{i}\langle -|+\rangle\langle +|+\rangle \\ \langle -|+\rangle\langle +|+\rangle + \langle -|-\rangle\langle +|-\rangle \\ -\mathrm{i}\langle -|-\rangle\langle +|+\rangle - \mathrm{i}\langle -|-\rangle\langle +|+\rangle \end{pmatrix} = \frac{\mu^2}{4}\begin{pmatrix} 0 \\ 0 \\ -2\mathrm{i} \end{pmatrix} .$$

So we find, according to (31.31),

$$V_-(\boldsymbol{R}) = \frac{\mu^2}{4}\frac{1}{(\mu R)^2}\begin{pmatrix} 0 \\ 0 \\ -2 \end{pmatrix} = -\frac{1}{2R^2}\begin{pmatrix} 0 \\ 0 \\ 1 \end{pmatrix}$$

or, reverting to unrotated axes again:

$$V_-(\boldsymbol{R}) = -\frac{1}{2}\frac{\boldsymbol{R}}{R^3} . \tag{31.34}$$

Similarly for the spin state $\sigma'_z = +1$

$$V_+(\boldsymbol{R}) = +\frac{1}{2}\frac{\boldsymbol{R}}{R^3} . \tag{31.35}$$

Our findings correspond to the potential of a "magnetic monopole" of strength $\pm 1/2$ at the origin in parameter space; $\boldsymbol{R} = 0$ is the site of the degeneracy.

For the Berry phase we obtain

$$\gamma_\pm(C) = -\int_S d\boldsymbol{S} \cdot \boldsymbol{V}_\pm(\boldsymbol{R}) = \mp\tfrac{1}{2}\Omega(C) , \tag{31.36}$$

where $\Omega(C)$ is the solid angle subtended by the closed path as seen from the origin of parameter space, $\boldsymbol{R} = 0$. We can verify this result more explicitly by using the representation of our spin state

$$|\uparrow, \boldsymbol{R}\rangle = \begin{pmatrix} \cos(\theta/2) \\ \sin(\theta/2)\,\mathrm{e}^{\mathrm{i}\varphi(t)} \end{pmatrix} , \tag{31.37}$$

which is taken along the $\boldsymbol{R}$-field whose cyclic adiabatic motion is given by

$$\boldsymbol{R}(t) = R\big(\sin\theta\,\cos\varphi(t),\ \sin\theta\,\sin\varphi(t),\ \cos\theta\big) \tag{31.38}$$

with θ fixed and $\varphi(t) = \omega t$, so that $\boldsymbol{R}(t)$ moves on a cone of half angle θ with period $2\pi/\omega$. The spin state (31.37) is, of course, an eigenstate of the Hamiltonian (31.32), since

$$\begin{aligned} \boldsymbol{\sigma} \cdot \boldsymbol{R}(t)|\uparrow, \boldsymbol{R}\rangle &= R\begin{pmatrix} \cos\theta & \sin\theta\,\mathrm{e}^{-\mathrm{i}\varphi} \\ \sin\theta\,\mathrm{e}^{\mathrm{i}\varphi} & -\cos\theta \end{pmatrix}\begin{pmatrix} \cos(\theta/2) \\ \sin(\theta/2)\,\mathrm{e}^{\mathrm{i}\varphi} \end{pmatrix} \\ &= R\begin{pmatrix} \cos(\theta/2) \\ \sin(\theta/2)\,\mathrm{e}^{\mathrm{i}\varphi} \end{pmatrix} . \end{aligned} \tag{31.39}$$

Using (31.37), we finally obtain for the Berry phase, according to (31.14):

$$\begin{aligned} \gamma_\uparrow(t) &= \mathrm{i}\int_0^t dt'\,\big(\cos(\theta/2),\ \sin(\theta/2)\,\mathrm{e}^{-\mathrm{i}\varphi}\big)\begin{pmatrix} 0 \\ \mathrm{i}\dot{\varphi}(t')\,\sin(\theta/2)\,\mathrm{e}^{\mathrm{i}\varphi} \end{pmatrix} \\ &= \mathrm{i}\int_0^t dt'\,\sin^2(\theta/2)\,\mathrm{i}\frac{d\varphi(t')}{dt'} , \\ &= -\frac{1}{2}(1-\cos\theta)\int_0^t dt'\frac{d\varphi}{dt'} . \end{aligned}$$

With $\varphi = \omega t$, we find for a complete cylce $T = 2\pi/\omega$

$$\gamma_\uparrow(C) = -\frac{1}{2}(1-\cos\theta)2\pi = -\frac{1}{2}\Omega(C) . \tag{31.40}$$

This coincides with (31.36).

Another well-known phenomenon that finds its explanation in a Berry phase is the Aharonov–Bohm effect. For this reason, consider a thin magnetic flux tube. Furthermore, let a quantal system consist of a charged particle confined to a box which is not penetrated by the flux tube. The box is located at a distance $\boldsymbol{R}$ away from the flux tube. Now the box is transported around the flux tube on a closed classical path C. In this case, it is not even necessary that the cyclic round-trip be adiabatic.

The fast variable $\boldsymbol{r}$ is the location of the charged particle as measured from the flux tube. After the box has completed a full circle, the particle will have picked up a phase – the Berry phase – which is equal to the phase difference in comparing charged particles passing on opposite sites of the flux tube with subsequent recombination.

The amount of flux encircled by the box is given by

$$\int d\boldsymbol{a} \cdot \boldsymbol{B} = \int d\boldsymbol{a} \cdot (\nabla \times \boldsymbol{A}) = \oint_C d\boldsymbol{R} \cdot \boldsymbol{A}(\boldsymbol{R}) = \Phi \,. \tag{31.41}$$

If there is no flux line present ($\boldsymbol{A} \equiv 0$), the particle Hamilton operator depends only on the canonical position $\boldsymbol{r}$ and momentum $\boldsymbol{p}$: $H = H(\boldsymbol{p}, \boldsymbol{r} - \boldsymbol{R})$, and the wave function has the form $\psi_n(\boldsymbol{r} - \boldsymbol{R})$ with the energy E_n independent of $\boldsymbol{R}$.

The situation changes if we have a nonzero flux line inside C. Then the energy states $|n(\boldsymbol{R})\rangle$ satisfy the Schrödinger equation

$$H\left(\boldsymbol{p} - \frac{q}{c}\boldsymbol{A}(\boldsymbol{r}), \boldsymbol{r} - \boldsymbol{R}\right)|n, \boldsymbol{R}\rangle = E_n|n, \boldsymbol{R}\rangle \,. \tag{31.42}$$

As we know from Chap. 21, the solution of this problem can be obtained from the flux-free problem with the aid of a gauge factor. Since $\boldsymbol{A}(\boldsymbol{r})$ can be gauged to zero (at least locally) outside the flux tube, i.e., it can be represented as the gradient of a scalar function $\chi(\boldsymbol{r})$, we must multiply our free ($\boldsymbol{A} \equiv 0$) wave function by an appropriate phase factor $\exp\{\frac{iq}{\hbar c}\chi(\mathrm{r})\}$:

$$\boldsymbol{A}(\boldsymbol{r}) \to \boldsymbol{A}(\boldsymbol{r}) - \nabla\chi(\boldsymbol{r}) = 0 \,, \quad |m\rangle \to \mathrm{e}^{(\mathrm{i}q/\hbar c)\chi(\boldsymbol{r})}|m\rangle \,.$$

In our case we get for the wave function in presence of the flux tube

$$\begin{aligned} \langle \boldsymbol{r}|n, \boldsymbol{R}\rangle &= \exp\left\{\frac{\mathrm{i}q}{\hbar c}\int_{\boldsymbol{R}}^{\boldsymbol{r}} d\boldsymbol{r}' \cdot \boldsymbol{A}(\boldsymbol{r}')\right\}\psi_n(\boldsymbol{r} - \boldsymbol{R}) \\ &=: \mathrm{e}^{\mathrm{i}\Lambda}\psi_n(\boldsymbol{r} - \boldsymbol{R}) \,, \end{aligned} \tag{31.43}$$

where $\psi_n(\boldsymbol{r} - \boldsymbol{R})$ denotes the wave function if the flux line is absent. We repeat:

$$\chi(\boldsymbol{r}) = \int_{\boldsymbol{R}}^{\boldsymbol{r}} d\boldsymbol{r}' \cdot \boldsymbol{A}(\boldsymbol{r}') \Rightarrow \boldsymbol{A}(\boldsymbol{r}) = \nabla\chi(\boldsymbol{r}) \,.$$

The solution (31.43) is single-valued in $\boldsymbol{r} \in$ box and (locally) in R. The energies are not affected by the vector potential.

We are now going to transport the box along the closed classical orbit C around the flux line. As result we expect a geometrical phase which can be calculated with the aid of formula (31.19). For this we need

$$
\begin{aligned}
\langle n, \boldsymbol{R}|\nabla_{\boldsymbol{R}}|n, \boldsymbol{R}\rangle &= \int d^3\boldsymbol{r}\,\langle n, \boldsymbol{R}|\boldsymbol{r}\rangle\langle\boldsymbol{r}|\nabla_{\boldsymbol{R}}|n, \boldsymbol{R}\rangle \\
&= \int d^3\boldsymbol{r}\,\mathrm{e}^{-\mathrm{i}\wedge}\psi_n^*(\boldsymbol{r}-\boldsymbol{R})\nabla_{\boldsymbol{R}}\{\mathrm{e}^{\mathrm{i}\wedge}\psi_n(\boldsymbol{r}-\boldsymbol{R})\} \\
&= \int d^3\boldsymbol{r}\,\mathrm{e}^{-\mathrm{i}\wedge}\psi_n^*(\boldsymbol{r}-\boldsymbol{R})\Big[-\frac{\mathrm{i}q}{\hbar c}\boldsymbol{A}(\boldsymbol{R}) \\
&\quad\times \mathrm{e}^{\mathrm{i}\wedge}\psi_n(\boldsymbol{r}-\boldsymbol{R}) + \mathrm{e}^{\mathrm{i}\wedge}\nabla_{\boldsymbol{R}}\psi_n(\boldsymbol{r}-\boldsymbol{R})\Big] \\
&= -\frac{\mathrm{i}q}{\hbar c}\boldsymbol{A}(\boldsymbol{R}) + \int d^3\boldsymbol{r}\,\psi_n^*(\boldsymbol{r}-\boldsymbol{R})\nabla_{\boldsymbol{R}}\psi_n(\boldsymbol{r}-\boldsymbol{R})\,. \qquad (31.44)
\end{aligned}
$$

The second term in (31.44) vanishes. This follows from

$$
\begin{aligned}
\int d^3\boldsymbol{r}\,\psi_n^*(\boldsymbol{r}-\boldsymbol{R})\nabla_{\boldsymbol{R}}\psi_n(\boldsymbol{r}-\boldsymbol{R}) \underset{\boldsymbol{r}\to\boldsymbol{r}+\boldsymbol{R}}{=} &\int d^3\boldsymbol{r}\,\langle n|\boldsymbol{r}\rangle\nabla_{\boldsymbol{r}}\langle\boldsymbol{r}|n\rangle \\
&= \nabla_{\boldsymbol{r}}\langle n|n\rangle = 0\,.
\end{aligned}
$$

Therefore, (31.44) reduces to the desired expression for the Berry connection:

$$
\mathrm{i}\langle n, \boldsymbol{R}|\nabla_{\boldsymbol{R}}|n, \boldsymbol{R}\rangle = \frac{q}{\hbar c}\boldsymbol{A}(\boldsymbol{R})\,. \qquad (31.45)
$$

At last we obtain for the Berry phase

$$
\gamma_n(C) = \frac{q}{\hbar c}\oint d\boldsymbol{R}\cdot\boldsymbol{A}(\boldsymbol{R}) = \frac{q}{\hbar c}\Phi \qquad (31.46)
$$

which we interpret in the following way: if the particle (confined in the box) travels once around the flux line, it accumulates a phase $\frac{q}{\hbar c}\Phi$. If $\frac{q}{\hbar c}\Phi = 2\pi n$, $n = 0, \pm 1, \pm 2, \ldots$, there is no Aharonov-Bohm effect. Also note that the result as stated in (31.46) is independent of the particle's energy state in the box. Furthermore, $\gamma_n(C)$ is invariant under continuous deformations of the path C: the Aharonov-Bohm phase is not only "geometrical" (as is any Berry phase), but even "topological".

We began this chapter on the Berry phase by following the Schrödinger state vector $|\psi(t)\rangle$ on its cyclic evolution in parameter space. But this treatment is not in accordance with the representation which we have favored so far, namely, the development of time-dependent base state vectors which satisfy the "wrong sign" Schrödinger equation. So let us return to the Heisenberg picture and study once more the whole problem with emphasis on transition amplitudes.

In our convention, operators and base state vectors respond under unitary transformation as

$$
\bar{X} = U^{-1}XU\,, \quad \overline{\langle|} = \langle|U\,, \quad UU^+ = \mathbb{1} = U^+U\,. \qquad (31.47)
$$

Let us then consider a class of Hamiltonians $H(\boldsymbol{B})$ which describe the interaction of our quantum system with an external uniform magnetic field, e.g., $H(\boldsymbol{B}) = -\mu\boldsymbol{J}\cdot\boldsymbol{B}$. The Hamiltonian $H(\boldsymbol{B})$ is taken to be invariant upon simultaneous rotation of the quantum system ($\boldsymbol{J}$) and the external field $\boldsymbol{B}$:

$$H(\boldsymbol{B}) = \bar{H}(R\boldsymbol{B}) = U^{-1}H(R\boldsymbol{B})U \ , \tag{31.48}$$

$$U = \exp\left\{\frac{\mathrm{i}}{\hbar}\boldsymbol{J}\cdot\hat{\mathrm{e}}\gamma\right\} \ . \tag{31.49}$$

The initial field configuration is given by $\boldsymbol{B}_0 = \boldsymbol{B}(t_0) = \text{const.}$, so that

$$\boldsymbol{B}(t) = R(t)\boldsymbol{B}_0 \ .$$

Now let us generate a time-dependent Hamiltonian by operating with the time-dependent rotation $R(t) \equiv R(\hat{\mathrm{e}}(t), \gamma(t))$ on the initial field vector $\boldsymbol{B}_0$:

$$H(t) \equiv H(\boldsymbol{B}(t)) = H(R(t)\boldsymbol{B}_0) \tag{31.50}$$

or

$$H(t) \underset{(48)}{=} U(\boldsymbol{R}(t))H(\boldsymbol{B}_0)U^+(\boldsymbol{R}(t)) \ . \tag{31.51}$$

In order to solve the Schrödinger equation for time-dependent base state vectors,

$$\mathrm{i}\hbar\frac{\partial}{\partial t}\langle\,;t| = \langle\,;t|H(t) \ , \tag{31.52}$$

we switch over to "rotated frame states" via the unitary transformation

$$\overline{\langle\,;t|} = \langle;t|U(t) \tag{31.53}$$

$$\langle\,;t| = \overline{\langle\,;t|}U^+(t) \ . \tag{31.54}$$

Inserting (31.54) into (31.52) we obtain

$$\mathrm{i}\hbar\frac{\partial}{\partial t}(\overline{\langle\,;t|}U^+) = \mathrm{i}\hbar\left(\frac{\partial}{\partial t}\overline{\langle\,;t|}\right)U^+ + \mathrm{i}\hbar\overline{\langle\,;t|}\frac{\partial U^+}{\partial t} = \overline{\langle\,;t|}U^+H(t) \tag{31.55}$$

or, upon multiplying from the right by $U(t)$:

$$\mathrm{i}\hbar\frac{\partial}{\partial t}\overline{\langle\,;t|} + \mathrm{i}\hbar\overline{\langle\,;t|}\frac{\partial U^+}{\partial t}U = \overline{\langle\,;t|}U^+H(t)U \underset{(51)}{\equiv} \overline{\langle\,;t|}H(\boldsymbol{B}_0) \tag{31.56}$$

i.e.,

$$\mathrm{i}\hbar\frac{\partial}{\partial t}\overline{\langle\,;t|} = \overline{\langle\,;t|}\left(H(\boldsymbol{B}_0) - \mathrm{i}\hbar\frac{\partial U^+}{\partial t}U\right) \ . \tag{31.57}$$

We can also rewrite the last term in (31.57) by making use of the unitarity condition:

$$U^+U = \mathbb{1} \Rightarrow \frac{\partial U^+}{\partial t}U + U^+\frac{\partial U}{\partial t} = 0 \Rightarrow \frac{\partial U^+}{\partial t}U = -U^+\frac{\partial U}{\partial t} \ . \tag{31.58}$$

Hence we obtain

$$\mathrm{i}\hbar\frac{\partial}{\partial t}\overline{\langle\,;t|} = \overline{\langle\,;t|}\big(H(\boldsymbol{B}_0) + H_1(t)\big) \ , \tag{31.59}$$

where

$$H_1(t) = \mathrm{i}\hbar U^+ \frac{\partial U}{\partial t} \ . \tag{31.60}$$

$H_1(t)$ in Eq. (31.59) contains the explicit time dependence. The unitarity condition can again be applied to show that $H_1(t)$ is a Hermitean operator:

$$H_1^+(t) = \left(\mathrm{i}\hbar U^+ \frac{\partial U}{\partial t}\right)^+ = -\mathrm{i}\hbar \frac{\partial U^+}{\partial t} U \underset{(58)}{=} \mathrm{i}\hbar U^+ \frac{\partial U}{\partial t} = H_1(t) \ ,$$

Now we need to calculate $H_1(t)$. For this reason we choose $\boldsymbol{B}_0$) to point in the direction of $\hat{X}_1$, a unit vector that lies in the $x - z$ plane of an x, y, z-coordinate System:

$$\boldsymbol{B}_0 = B\hat{X}_1 \ , \quad |\boldsymbol{B}_0| = B \ .$$

The angles $\varphi(t)$ and $\vartheta(t)$ are used to describe the curve drawn by $\boldsymbol{B}(t)$ during its adiabatic round trip in the spherical coordinate system. So we have for the initial field configuration

$$\boldsymbol{B}_0 = \hat{X}_1 B = R(\hat{y}, \vartheta_0)\hat{z}B \tag{31.61}$$

and, furthermore, using $\hat{z}B = R^{-1}(\hat{y}, \vartheta_0)\boldsymbol{B}_0 = R(\hat{y}, \ -\vartheta_0)\boldsymbol{B}_0$, we find

$$\begin{aligned} \boldsymbol{B}(t) &= R(\hat{z}, \varphi)R(\hat{y}, \vartheta)\hat{z}B \ , \\ &= R(\hat{z}, \varphi)R(\hat{y}, \vartheta)R(\hat{y}, \ -\vartheta_0)\boldsymbol{B}_0 \\ &= R(\hat{z}, \varphi(t))R\big(\hat{y}, \vartheta(t) - \vartheta_0\big)\boldsymbol{B}_0 =: R\boldsymbol{B}_0 \ . \end{aligned} \tag{31.62}$$

For one period T we have

$$\boldsymbol{B}(T) = \boldsymbol{B}_0, \ \vartheta(T) = \vartheta_0, \ \varphi(T) = 2\pi \ .$$

At last we turn to the computation of $H_1(t)$:

$$\begin{aligned} H_1(t) = \mathrm{i}\hbar U^+ \frac{\partial U}{\partial t} &= \mathrm{i}\hbar \lim_{\Delta t \to 0} \frac{1}{\Delta t} U^{-1}(R(t))\big(U(R(t + \Delta t)) \quad U(R(t))\big) \\ &= \mathrm{i}\hbar \lim_{\Delta t \to 0} \frac{1}{\Delta t}\big(U(R^{-1}(t)R(t + \Delta t)) - 1\big) \ . \end{aligned} \tag{31.63}$$

Using the definition of R in (31.62), we can calculate

$$\begin{aligned} R^{-1}(t)R(t + \Delta t) = &R\big(\hat{y}, \ -\vartheta(t) + \vartheta_0\big)R(\hat{z}, \ -\varphi(t)) \\ &\times R(\hat{z}, \varphi(t + \Delta t))R\big(\hat{y}, \vartheta(t + \Delta t) - \vartheta_0\big) \ . \end{aligned}$$

Here we substitute

$$\begin{aligned} \varphi(t + \Delta t) &= \varphi(t) + \Delta\varphi \ , \\ \vartheta(t + \Delta t) &= \vartheta(t) + \Delta\vartheta \ , \end{aligned}$$

to obtain

$$\begin{aligned} R^{-1}(t)R(t+\Delta t) &= R\big(\hat{y}, -\vartheta(t)+\vartheta_0\big)R\big(\hat{z}, -\varphi(t)\big) \\ &\quad \times R\big(\hat{z}, \varphi(t)+\Delta\varphi\big)R\big(\hat{y}, \vartheta(t)+\Delta\vartheta-\vartheta_0\big) \\ &= R\big(\hat{y}, -\vartheta(t)+\vartheta_0\big)R(\hat{z}, \Delta\varphi)R\big(\hat{y}, \vartheta(t)+\Delta\vartheta-\vartheta_0\big) \end{aligned}$$

This result is needed in (31.63):

$$\begin{aligned} U\big(R^{-1}(t)R(t+\Delta t)\big) &= \exp\left\{\frac{\mathrm{i}}{\hbar}J_y(-\vartheta+\vartheta_0)\right\} \\ &\quad \times \exp\left\{\frac{\mathrm{i}}{\hbar}J_z\Delta\varphi\right\}\exp\left\{\frac{\mathrm{i}}{\hbar}J_y(\vartheta-\vartheta_0+\Delta\vartheta)\right\} \\ &= \exp\left\{-\frac{\mathrm{i}}{\hbar}J_y(\vartheta-\vartheta_0)\right\}\left(1+\frac{\mathrm{i}}{\hbar}J_z\Delta\varphi+\ldots\right) \\ &\quad \times\left(1+\frac{\mathrm{i}}{\hbar}J_y\Delta\vartheta+\ldots\right)\exp\left\{\frac{\mathrm{i}}{\hbar}J_y(\vartheta-\vartheta_0)\right\} \\ &= 1+\exp\left\{-\frac{\mathrm{i}}{\hbar}J_y(\vartheta-\vartheta_0)\right\}\frac{\mathrm{i}}{\hbar}J_z\Delta\varphi \\ &\quad \times\exp\left\{\frac{\mathrm{i}}{\hbar}J_y(\vartheta-\vartheta_0)\right\}+\exp\left\{-\frac{\mathrm{i}}{\hbar}J_y(\vartheta-\vartheta_0)\right\} \\ &\quad \times\frac{\mathrm{i}}{\hbar}J_y\Delta\vartheta\ \exp\left\{\frac{\mathrm{i}}{\hbar}J_y(\vartheta-\vartheta_0)\right\}+\ldots \end{aligned}$$

Multiplying the last line by $\frac{1}{\Delta t}$ and letting $\Delta t \to 0$, we note that the 1 is cancelled by the -1 in (31.63) so that we are left with

$$\frac{\mathrm{i}}{\hbar}\exp\left\{-\frac{\mathrm{i}}{\hbar}J_y(\vartheta-\vartheta_0)\right\}J_z\dot{\varphi}\ \exp\left\{\frac{\mathrm{i}}{\hbar}J_y(\vartheta-\vartheta_0)\right\}+\frac{\mathrm{i}}{\hbar}J_y\dot{\vartheta}\,.$$

So far we have arrived at the following expression for $H_1(t)$:

$$\begin{aligned} H_1(t) &= \mathrm{i}\hbar\left(\frac{\mathrm{i}}{\hbar}\dot{\varphi}\,\mathrm{e}^{-\frac{\mathrm{i}}{\hbar}J_y(\vartheta-\vartheta_0)}J_z\,\mathrm{e}^{\frac{\mathrm{i}}{\hbar}J_y(\vartheta-\vartheta_0)}+\frac{\mathrm{i}}{\hbar}J_y\dot{\vartheta}\right) \\ &= -\dot{\varphi}\,\mathrm{e}^{-\frac{\mathrm{i}}{\hbar}J_y(\vartheta-\vartheta_0)}J_z\,\mathrm{e}^{\frac{\mathrm{i}}{\hbar}J_y(\vartheta-\vartheta_0)}-\dot{\vartheta}J_y\,. \end{aligned} \tag{31.64}$$

It is convenient to go to the $\hat{X}_i$-frame defined by

$$\hat{X}_1=\frac{\boldsymbol{B}_0}{B}\,,\quad \hat{X}_2=\hat{y}\times\frac{\boldsymbol{B}_0}{B}\,,\quad \hat{X}_3=\hat{y}\,.$$

In particular we have

$$\boldsymbol{J}\cdot\hat{X}_1=J_{X_1}=\mathrm{e}^{\frac{\mathrm{i}}{\hbar}\vartheta_0 J_y}J_z\,\mathrm{e}^{-\frac{\mathrm{i}}{\hbar}\vartheta_0 J_y}$$

which can be used in (31.64) to give us

$$H_1(t)=-\dot{\varphi}\,\mathrm{e}^{-\frac{\mathrm{i}}{\hbar}J_y\vartheta}J_{X_1}\,\mathrm{e}^{\frac{\mathrm{i}}{\hbar}J_y\vartheta}-\dot{\vartheta}J_{X_3}$$

or

$$H_1(t) = -\dot{\varphi}\Big(\cos\vartheta(t) J_{X_1} - \sin\vartheta(t) J_{X_2}\Big) - \dot{\vartheta} J_{X_3} \,. \tag{31.65}$$

At this point we assume that the Hamiltonian is given by

$$H(\boldsymbol{B}_0) = -\mu \boldsymbol{J}\cdot\boldsymbol{B}_0 = -\mu J_{X_1} B \,. \tag{31.66}$$

The eigenstates satisfy

$$\begin{aligned} H(\boldsymbol{B}_0)|m\rangle &= E_m|m\rangle \,, \\ J_{X_1}|m\rangle &= \hbar m|m\rangle \,. \end{aligned} \tag{31.67}$$

Let the initial state be an eigenstate of $H(\boldsymbol{B}_0)$:

$$\langle m; t=0| = \overline{\langle m; t=0|} = \langle m| \,. \tag{31.68}$$

Then the adiabatic approximation is defined by keeping only the diagonal part in (31.65):

$$\big(H_1(t)\big)_{\text{ad.}} = -\dot{\varphi}\cos\vartheta(t) J_{X_1} \,. \tag{31.69}$$

Integrating (31.59) between $0 \le t \le T$, we obtain

$$\begin{aligned} \overline{\langle m; T|} &= \overline{\langle m; t=0|} \exp\left[-\frac{\mathrm{i}}{\hbar}\int_0^T dt\big(H(\boldsymbol{B}_0) + H_1(t)\big)\right] \\ &= \mathrm{e}^{-\frac{\mathrm{i}}{\hbar}E_m T} \exp\left[\mathrm{i}m\int_0^T dt\,\dot{\varphi}\cos\vartheta(t)\right]\langle m| \,. \end{aligned}$$

Here we recall (31.53), i.e.,

$$\langle m; T| = \overline{\langle m; T|} U^+(R(T)) \overset{(49)}{\underset{(67)}{=}} \overline{\langle m; T|}\, \mathrm{e}^{-\frac{\mathrm{i}}{\hbar}\hbar m 2\pi}$$

so that we get

$$\begin{aligned} \overline{\langle m; T|} &= \mathrm{e}^{-\frac{\mathrm{i}}{\hbar}E_m T} \underbrace{\mathrm{e}^{-\mathrm{i}2\pi m}}_{=\exp\left[-\mathrm{i}m\int_0^T dt\,\dot{\varphi}\right]} \exp\left[\mathrm{i}m\int_0^T dt\,\cos\vartheta(t)\dot{\varphi}(t)\right]\langle m, 0| \\ &= \mathrm{e}^{-\frac{\mathrm{i}}{\hbar}E_m T}\exp\left[-\mathrm{i}m\int_0^T dt(1-\cos\vartheta(t))\dot{\varphi}\right]\langle m, 0| \\ &= \mathrm{e}^{-\frac{\mathrm{i}}{\hbar}E_m T}\mathrm{e}^{\mathrm{i}\gamma_m(C)}\langle m, 0| \end{aligned}$$

with

$$\begin{aligned} \gamma_m(C) &= -m\int_0^T dt(1-\cos\vartheta(t))\dot{\varphi}(t) \\ &= -m\int_0^T dt\,\dot{\varphi}\int_0^{\vartheta(t)} d\,\sin\vartheta \\ &= -m\iint \sin\vartheta\, d\vartheta\, d\varphi = -m\Omega(C) \,, \end{aligned} \tag{31.70}$$

where $\Omega(C)$ is the solid angle enclosed by the curve C.

Here, then, is our final answer for the transition amplitude of an energy eigenstate being adiabatically transported in a closed loop:

$$\langle m, T|n, 0\rangle = \mathrm{e}^{-\frac{\mathrm{i}}{\hbar}E_m T}\,\mathrm{e}^{\mathrm{i}\gamma_m(C)}\delta_{mn}\,, \tag{31.71}$$

$$|\langle m, T|n, 0\rangle|^2 = \delta_{mn}\,. \tag{31.72}$$

So the system will be found, with certainty in the energy state of the same quantum number n, but in addition to the dynamical phase $-\frac{1}{\hbar}E_m T$ it will have accumulated a geometrical, the Berry phase, $\gamma_m(C)$, in going through a complete cycle from 0 to T.

Note that the Berry phase can be written as the circulation of a classical gauge field along the closed loop. It was Berry who first noted that this gauge field is that of a magnetic monopole of charge m. This follows from writing

$$m\int_0^T dt(1-\cos\vartheta)\frac{d\varphi}{dt} = m\oint_C d\varphi(1-cos\vartheta) = \oint \boldsymbol{A}_m\cdot \boldsymbol{dl} =: -\gamma_m(C)$$

so that

$$\mathrm{i}\gamma_m(C) = -\mathrm{i}\oint \boldsymbol{A}_m\cdot \boldsymbol{dl} \tag{31.73}$$

or

$$\langle m, T|n, 0\rangle = \mathrm{e}^{-\frac{\mathrm{i}}{\hbar}E_m T}\exp\left[-\mathrm{i}\oint \boldsymbol{A}_m\cdot \boldsymbol{dl}\right]\delta_{mn}\,. \tag{31.74}$$

32. Classical Analogues to Berry's Phase

In the last chapter we saw how a quantum system can give rise to a Berry phase, by studying the adiabatic round trip of its quantum state on a certain parameter space. Rather than considering what happens to states in Hilbert space, we now turn to classical mechanics, where we are interested instead in the evolution of the system in configuration space.

To be more specific, let us consider a point particle constrained to move in a two-dimensional plane. In this plane, the particle moves under the influence of a cylindrically symmetric potential $V = V(r)$. Finally, let the plane (plate) move on a different manifold, e.g., on a sphere. This latter motion may be caused by an external force. The equations for a particle moving on the plate, where the plate continually changes its orientation, are generally very complicated. Therefore it will be our main goal, when dealing with the motion of a particle, to introduce a local inertial frame; this greatly simplifies the equations of motion. (Later we shall change this instantaneous inertial system adiabatically, so that velocity-dependent forces cannot act from one to the next "infinitesimally slowly" reached neighboring frame. This does not mean, of course, that after a finite but long time duration, nothing will have changed. On the contrary, it is precisely this effect that is of interest to us.)

The orientation of the plane is defined by a unit vector $\boldsymbol{S}$ that is perpendicular to it. Let this vector change slowly when the external force acts. After an adiabatic round trip, $\boldsymbol{S}(t)$ returns to its origin. Meanwhile, it has continually changed its orientation, as did the other two vectors, $\boldsymbol{N}(t)$ and $\boldsymbol{B}(t)$, which, together with $\boldsymbol{S}(t)$, form an orthonormal triad (dreibein) accompanying the curve:

$$\boldsymbol{B} = \boldsymbol{S} \times \boldsymbol{N} , \quad |\boldsymbol{B}| = |\boldsymbol{S}| = |\boldsymbol{N}| = 1 .$$

The time dependence of the frame defined by $\boldsymbol{S}$, $\boldsymbol{N}$ and $\boldsymbol{B}$ is characterized by Frenet's formulae

$$\begin{aligned}
\frac{d\boldsymbol{S}}{dt} &= \chi \boldsymbol{N} , \\
\frac{d\boldsymbol{N}}{dt} &= -\chi \boldsymbol{S} + \tau \boldsymbol{B} , \\
\frac{d\boldsymbol{B}}{dt} &= -\tau \boldsymbol{N} .
\end{aligned} \tag{32.1}$$

The first equation in (32.1) says that the change of $\boldsymbol{S}$ is chosen in the direction of $\boldsymbol{N}$. The time-dependent parameters χ and τ denote curvature and torsion of the curve. An adiabatic change of the dreibein means that χ and τ are small.

If $\boldsymbol{S}$ were time independent, then we would have, as equation of motion for our mass point,

$$m\ddot{\boldsymbol{r}} = -\hat{\boldsymbol{r}}\frac{dV(r)}{dr} , \qquad |\hat{\boldsymbol{r}}| = 1 , \tag{32.2}$$

where $\boldsymbol{r}$ is a two-dimensional vector in the plane, and $\boldsymbol{r} \cdot \boldsymbol{S} = 0$. As soon as $\boldsymbol{S}(t)$ changes in time, however, we will certainly obtain equations of motion more complicated than (32.2). In order to find out what they look like, we need the following equations:

$$\begin{aligned}
\ddot{\boldsymbol{N}} &= -\dot{\chi}\boldsymbol{S} - \chi\dot{\boldsymbol{S}} + \dot{\tau}\boldsymbol{B} + \tau\dot{\boldsymbol{B}} \\
&= -\dot{\chi}\boldsymbol{S} - \chi(\chi\boldsymbol{N}) + \dot{\tau}\boldsymbol{B} + \tau(-\tau\boldsymbol{N}) \\
&= -\dot{\chi}\boldsymbol{S} - (\chi^2 + \tau^2)\boldsymbol{N} + \dot{\tau}\boldsymbol{B}
\end{aligned} \tag{32.3}$$

$$\begin{aligned}
\ddot{\boldsymbol{B}} &= -\dot{\tau}\boldsymbol{N} - \tau\dot{\boldsymbol{N}} = -\dot{\tau}\boldsymbol{N} - \tau(-\chi\boldsymbol{S} + \tau\boldsymbol{B}) \\
&= -\dot{\tau}\boldsymbol{N} + \tau\chi\boldsymbol{S} - \tau^2\boldsymbol{B} .
\end{aligned} \tag{32.4}$$

A vector $\boldsymbol{r}_0$ in the plane can be written with respect to the $\{\boldsymbol{N}, \boldsymbol{B}\}$ basis as

$$\boldsymbol{r}_0 = x_0\boldsymbol{N} + y_0\boldsymbol{B} . \tag{32.5}$$

The equations of motion for the components x_0 and y_0 are then obtained by taking the following steps:

$$\begin{aligned}
\dot{\boldsymbol{r}}_0 &= \dot{x}_0\boldsymbol{N} + x_0\dot{\boldsymbol{N}} + \dot{y}_0\boldsymbol{B} + y_0\dot{\boldsymbol{B}} \\
\Rightarrow \ddot{\boldsymbol{r}}_0 &= \ddot{x}_0\boldsymbol{N} + \dot{x}_0\dot{\boldsymbol{N}} + \dot{x}_0\dot{\boldsymbol{N}} + x_0\ddot{\boldsymbol{N}} + \ddot{y}_0\boldsymbol{B} + 2\dot{y}_0\dot{\boldsymbol{B}} + y_0\ddot{\boldsymbol{B}} \\
&= \ddot{x}_0\boldsymbol{N} + 2\dot{x}_0(-\chi\boldsymbol{S} + \tau\boldsymbol{B}) + x_0\big(-\dot{\chi}\boldsymbol{S} - (\chi^2 + \tau^2)\boldsymbol{N} + \dot{\tau}\boldsymbol{B}\big) \\
&\quad + \ddot{y}_0\boldsymbol{B} + 2\dot{y}_0(-\tau\boldsymbol{N}) + y_0\big(-\dot{\tau}\boldsymbol{N} + \tau\chi\boldsymbol{S} - \tau^2\boldsymbol{B}\big) \\
&= \big(\ddot{x}_0 - x_0(\chi^2 + \tau^2) - 2\dot{y}_0\tau - y_0\dot{\tau}\big)\,\boldsymbol{N} \\
&\quad + \big(2\dot{x}_0\tau + x_0\dot{\tau} + \ddot{y}_0 - y_0\tau^2\big)\,\boldsymbol{B} \\
&\quad + \big(-2\dot{x}_0\chi - x_0\dot{\chi} + y_0\tau\chi\big)\boldsymbol{S} .
\end{aligned}$$

We assume that the mass point is held on the plane by external constraints. The motion in the $\boldsymbol{S}$-direction is therefore not of interest. Then the motion in the plane is given by

$$m\ddot{\boldsymbol{r}}_0 = -\nabla V\big(r_0\big) , \quad r_0 = \sqrt{x_0^2 + y_0^2} ,$$

or, in components,

$$m\left(\ddot{x}_0 - x_0\left(\chi^2+\tau^2\right) - 2\dot{y}_0\tau - y_0\dot{\tau}\right) = -\frac{\partial V}{\partial x_0}\,,$$
$$m\left(\ddot{y}_0 - \tau^2 y_0 + 2\tau\dot{x}_0 + \dot{\tau}x_0\right) = -\frac{\partial V}{\partial y_0}\,,$$

or, upon isolating the acceleration components,

$$\begin{aligned} m\ddot{x}_0 &= -\frac{\partial V}{\partial x_0} + m\left(\left(\chi^2+\tau^2\right)x_0 + 2\tau\dot{y}_0 + \dot{\tau}y_0\right) \\ m\ddot{y}_0 &= -\frac{\partial V}{\partial y_0} + m\left(\tau^2 y_0 - 2\tau\dot{x}_0 - \dot{\tau}x_0\right)\,. \end{aligned} \tag{32.6}$$

The terms proportional to the velocities $\dot{x}_0$ and $\dot{y}_0$ are the Coriolis forces.

The equations of motion (32.6) are obviously somewhat complicated. Therefore we shall try to simplify them by transforming to a local inertial system; here the velocity-dependent forces should vanish; of course, this applies only locally, i.e., in the immediate proximity of a point. Here, the coordinates are as close as possible to Euclidean coordinates; the velocity-dependent forces can be transformed away locally. Local inertial frames are defined as frames whose basis vectors undergo parallel transport. Parallel transport applied to an orthonormal set of vectors $\boldsymbol{U}_1$, $\boldsymbol{U}_2$ means that the change in the vectors has no components along the direction of the original vectors,

$$\boldsymbol{U}_i \cdot \frac{d\boldsymbol{U}_j}{dt} = 0\,, \quad i,j = 1,2\,. \tag{32.7}$$

It should be noted that on the basis of this definition, $\boldsymbol{N}$ and $\boldsymbol{B}$ do not undergo parallel transport, since

$$\boldsymbol{B}\cdot\frac{d\boldsymbol{N}}{dt} = \tau\,, \quad \boldsymbol{N}\cdot\frac{d\boldsymbol{B}}{dt} = -\tau\,.$$

The basis vectors that undergo parallel transport differ from $\boldsymbol{N}$ and $\boldsymbol{B}$ by a rotation

$$\begin{pmatrix}\boldsymbol{U}_1\\ \boldsymbol{U}_2\end{pmatrix} = \begin{pmatrix}\cos\beta & -\sin\beta\\ \sin\beta & \cos\beta\end{pmatrix}\begin{pmatrix}\boldsymbol{N}\\ \boldsymbol{B}\end{pmatrix}\,. \tag{32.8}$$

Let us determine the angle $\beta(t)$.

$$\dot{\boldsymbol{U}}_1 = \dot{\beta}\sin\beta\boldsymbol{N} + \cos\beta\dot{\boldsymbol{N}} - \dot{\beta}\cos\beta\boldsymbol{B} - \sin\beta\dot{\boldsymbol{B}}\,.$$

Forming

$$\begin{aligned} \dot{\boldsymbol{U}}_1 \cdot \boldsymbol{U}_2 &= (-\dot{\beta}\sin\beta\boldsymbol{N} + \cos\beta\dot{\boldsymbol{N}} - \dot{\beta}\cos\beta\boldsymbol{B} - \sin\beta\dot{\boldsymbol{B}}) \\ &\quad\cdot(\sin\beta\boldsymbol{N} + \cos\beta\boldsymbol{B}) \\ &= -\dot{\beta}\sin^2\beta + \tau\sin^2\beta + \tau\cos^2\beta - \dot{\beta}\cos^2\beta \\ &= \tau - \dot{\beta} = 0\,, \end{aligned}$$

we obtain

$$\frac{d\beta}{dt} = \tau\,. \tag{32.9}$$

The same result is obtained by forming $\dot{\boldsymbol{U}}_2 \cdot \boldsymbol{U}_1$. Use has been made of

$$\boldsymbol{N}^2 = 1\,, \quad \boldsymbol{N}\cdot\frac{d\boldsymbol{N}}{dt} = 0\,, \quad \boldsymbol{B}^2 = 1\,, \quad \boldsymbol{B}\cdot\frac{d\boldsymbol{B}}{dt} = 0\,.$$

So, for the change in time of the basis vectors $\boldsymbol{U}_i$, we have

$$\begin{aligned}\frac{d\boldsymbol{U}_1}{dt} &= \tau\sin\beta\boldsymbol{N} + \cos\beta(-\chi\boldsymbol{S} + \tau\boldsymbol{B}) - \tau\cos\beta\boldsymbol{B} - \sin\beta(-\tau\boldsymbol{N}) \\ &= -\chi\cos\beta\boldsymbol{S}\,,\end{aligned} \tag{32.10}$$

$$\begin{aligned}\frac{d\boldsymbol{U}_2}{dt} &= \dot\beta\cos\beta\boldsymbol{N} + \sin\beta\dot{\boldsymbol{N}} - \dot\beta\sin\beta\boldsymbol{B} + \cos\beta\dot{\boldsymbol{B}} \\ &= \tau\cos\beta\boldsymbol{N} + \sin\beta(-\chi\boldsymbol{S} + \tau\boldsymbol{B}) - \tau\sin\beta\boldsymbol{B} + \cos\beta(-\tau\boldsymbol{N}) \\ &= -\chi\sin\beta\boldsymbol{S}\,,\end{aligned} \tag{32.11}$$

or

$$\begin{aligned}\dot{\boldsymbol{U}}_1 &= -\chi\cos\beta\boldsymbol{S}\,, \\ \dot{\boldsymbol{U}}_2 &= -\chi\sin\beta\boldsymbol{S}\,.\end{aligned} \tag{32.12}$$

From this we get, via another time derivative,

$$\begin{aligned}\ddot{\boldsymbol{U}}_1 &= -\dot\chi\cos\beta\boldsymbol{S} - \chi(-\dot\beta\sin\beta\boldsymbol{S} + \cos\beta\dot{\boldsymbol{S}}) \\ &= (\chi\dot\beta\sin\beta - \dot\chi\cos\beta)\boldsymbol{S} - \chi^2\cos\beta\boldsymbol{N}\,,\end{aligned} \tag{32.13}$$

$$\begin{aligned}\ddot{\boldsymbol{U}}_2 &= -\dot\chi\sin\beta\boldsymbol{S} - \chi(\dot\beta\cos\beta\boldsymbol{S} + \sin\beta\dot{\boldsymbol{S}}) \\ &= -(\chi\dot\beta\cos\beta + \dot\chi\sin\beta)\boldsymbol{S} - \chi^2\sin\beta\boldsymbol{N}\,.\end{aligned} \tag{32.14}$$

In general, a local inertial frame can only be defined in the direct proximity of a point. In the present example, however, the motion takes place in a plane and not on a general curved surface; this enables us to extend our local inertial frame to the entire plane.

With respect to our inertial frame, a vector $\boldsymbol{r}$ can then be expressed as

$$\boldsymbol{r} = x\boldsymbol{U}_1 + y\boldsymbol{U}_2\,. \tag{32.15}$$

We need

$$\dot{\boldsymbol{r}} = \dot x\boldsymbol{U}_1 + x\dot{\boldsymbol{U}}_1 + \dot y\boldsymbol{U}_2 + y\dot{\boldsymbol{U}}_2$$

$$\begin{aligned}
\Rightarrow \ddot{\boldsymbol{r}} &= \ddot{x}\boldsymbol{U}_1 + 2\dot{x}\dot{\boldsymbol{U}}_1 + x\ddot{\boldsymbol{U}}_1 + \ddot{y}\boldsymbol{U}_2 + 2\dot{y}\dot{\boldsymbol{U}}_2 + y\ddot{\boldsymbol{U}}_2 \\
&= \ddot{x}\boldsymbol{U}_1 - 2\dot{x}\chi \cos\beta\boldsymbol{S} + x(\chi\dot{\beta}\sin\beta - \dot{\chi}\cos\beta)\boldsymbol{S} - x\chi^2\cos\beta\boldsymbol{N} \\
&\quad + \ddot{y}\boldsymbol{U}_2 - 2\dot{y}\chi\sin\beta\boldsymbol{S} - y(\chi\dot{\beta}\cos\beta + \dot{\chi}\sin\beta)\boldsymbol{S} - y\chi^2\sin\beta\boldsymbol{N} \\
&= \ddot{x}\boldsymbol{U}_1 + (x\chi\dot{\beta}\sin\beta - x\dot{\chi}\cos\beta - 2\dot{x}\chi\cos\beta)\boldsymbol{S} - x\chi^2\cos\beta\boldsymbol{N} \quad (32.16) \\
&\quad + \ddot{y}\boldsymbol{U}_2 - (y\chi\dot{\beta}\cos\beta + y\dot{\chi}\sin\beta - 2\dot{y}\chi\sin\beta)\boldsymbol{S} - y\chi^2\sin\beta\boldsymbol{N}\,.
\end{aligned}$$

The motion is to be restricted, as before, to the plane. Thus, we shall again suppress the $\boldsymbol{S}$-component. But first, $\boldsymbol{N}$ should be expressed as function of the basis vectors $\boldsymbol{U}_i$. According to (32.8), it holds that

$$\begin{aligned}
\boldsymbol{U}_1 &= \cos\beta\boldsymbol{N} - \sin\beta\boldsymbol{B}\,, \\
\boldsymbol{U}_2 &= \sin\beta\boldsymbol{N} + \cos\beta\boldsymbol{B}\,.
\end{aligned}$$

Multiplying the first line by $\cos\beta$ and the second line by $\sin\beta$ and adding the result yields

$$\cos\beta\boldsymbol{U}_1 + \sin\beta\boldsymbol{U}_2 = \boldsymbol{N}\,. \tag{32.17}$$

If we substitute this expression for $\boldsymbol{N}$ in (32.16) and suppress the $\boldsymbol{S}$ component, we obtain for the equation of motion relative to the inertial frame ($r = \sqrt{x^2 + y^2}$):

$$\begin{aligned}
m\ddot{x} &= -\frac{\partial V(r)}{\partial x} + m\chi^2\left(x\cos^2\beta + y\sin\beta\cos\beta\right) \\
m\ddot{y} &= -\frac{\partial V(r)}{\partial y} + m\chi^2\left(y\sin^2\beta + x\sin\beta\cos\beta\right)\,.
\end{aligned} \tag{32.18}$$

So we have reached our goal: relative to the inertial frame, the Coriolis forces no longer appear; they are automatically included in the rotation of the basis vectors $\boldsymbol{U}_1$ and $\boldsymbol{U}_2$. The dependence of τ enters (32.19) via the angle β. In the $\boldsymbol{U}_i$-system, the terms which are proportional to χ^2 correspond to the centrifugal force which is caused by the rotation of the plane. The equations of motion (32.19) can be simplified further by introducing a modified potential $U = V + W$, where W is given by

$$W = -\frac{m}{2}\chi^2(x\cos\beta + y\sin\beta)^2 = -\frac{m}{2}\omega^2 r^2\,. \tag{32.19}$$

In this way, (32.19) can be written in the simple form

$$\begin{aligned}
m\ddot{x} &= -\frac{\partial U}{\partial x}\,, \\
m\ddot{y} &= -\frac{\partial U}{\partial y}\,.
\end{aligned} \tag{32.20}$$

Let us now assume that the plane is rotated adiabatically; i.e., χ is small so that χ^2 in (32.19) can be neglected. No restrictions are made with respect to τ. Then we obtain, in adiabatic approximation (relative to the $\boldsymbol{U}_i$-frame),

$$m\ddot{x} = -\frac{\partial V}{\partial x} ,$$
$$m\ddot{y} = -\frac{\partial V}{\partial y} . \tag{32.21}$$

These equations of motion are identical to those of a particle moving in a fixed plane. Coriolis as well as centrifugal forces have been eliminated. The only effect of the rotation that remains is hidden in the time dependence of the basis vectors $\boldsymbol{U}_i$. If the potential does not depend on the direction, then all that can be observed is that the parallel transport will cause $\boldsymbol{U}_1$, $\boldsymbol{U}_2$ to be rotated with respect to $\boldsymbol{N}$, $\boldsymbol{B}$ by an angle $\beta = \int dt\, \tau(t)$.

We shall now apply the above formulae to the special case of the Foucault pendulum. Here, the normal vector of the plane at one point on the Earth's surface, is represented as

$$\boldsymbol{S} = \big(\sin\theta\, \cos\phi(t),\ \sin\theta\, \sin\phi(t),\ \cos\theta\big) .$$

To satisfy $\boldsymbol{N} \cdot \boldsymbol{S} = 0$, we require

$$\boldsymbol{N} = \big(-\sin\phi(t),\ \cos\phi(t),\ 0\big) . \tag{32.22}$$

θ measures the angle from the North Pole. The Earth rotates beneath our plane with constant angular velocity, $\omega = d\phi/dt$.

With respect to the original dreibein, it holds that

$$\frac{d\boldsymbol{S}}{dt} = \chi \boldsymbol{N} , \qquad \frac{d\boldsymbol{N}}{dt} = -\chi \boldsymbol{S} + \tau \boldsymbol{B} , \qquad \frac{d\boldsymbol{B}}{dt} = -\tau \boldsymbol{N} .$$

With (32.22), it follows that for $\boldsymbol{B}(t) = \boldsymbol{S} \times \boldsymbol{N}$,

$$\begin{aligned}
B_1 &= S_2N_3 - S_3N_2 = \sin\theta\, \sin\phi \cdot 0 - \cos\theta\, \cos\phi(t) \\
B_2 &= S_3N_1 - S_1N_3 = -\cos\theta\, \sin\phi(t) - 0 \\
B_3 &= S_1N_2 - S_2N_1 = \sin\theta\, \cos^2\phi + \sin\theta\, \sin^2\phi = \sin\theta .
\end{aligned}$$

So we obtain

$$\boldsymbol{B} = \big(-\cos\theta\, \cos\phi(t),\ -\cos\theta\, \sin\phi(t),\ \sin\theta\big) ,$$

from which we get

$$\begin{aligned}
\frac{d\boldsymbol{B}}{dt} &= (\cos\theta\, \sin\phi\dot{\phi},\ -\cos\theta\, \cos\phi\dot{\phi},\ 0) \\
&= -\tau\boldsymbol{N} = -\tau(-\sin\phi,\ \cos\phi,\ 0) .
\end{aligned}$$

From the last two equations we then obtain the relation

$$\tau = \dot{\phi}\, \cos\theta = \omega\, \cos\theta , \tag{32.23}$$

which yields, with the aid of (32.9):

$$\beta(t) = \phi(t) \cos\theta . \tag{32.24}$$

Let $\phi(0) = 0$ and $t = T$; i.e., after exactly one rotation of the Earth, $\phi(T) = 2\pi$. Then (32.22) and (32.24) tell us that while $\boldsymbol{N}$ has rotated by -2π, the inertial system has rotated by the angle $\beta = 2\pi \cos\theta$ with respect to $\boldsymbol{N}$. Therefore the net effect of the rotation of the local inertial frame is

$$\Omega = 2\pi(1 - \cos\theta) . \tag{32.25}$$

This is exactly the solid angle of the cap bounded by the curve, which the tip of $\boldsymbol{S}$ traces on the unit sphere.

In our second example we are going to deal with the motion of a free rigid body. "How much a free rigid body rotates" can be answered by solving Euler's equations of motion. It is, however, possible to look at the whole question in a completely new light by stressing the geometrical rather than the dynamical side of the problem.

It is well known that the angular momentum $\boldsymbol{J}$ of a free rigid body is constant in an inertial (space-fixed) reference frame. Relative to a body-fixed frame, the motion of $\boldsymbol{J}$ is periodic in time. Let the period be T. In one such period, the body, as viewed from the inertial frame, rotates about $\boldsymbol{J}$. In the sequel we want to prove that the rotation angle is given by

$$\Delta\theta = 2\frac{E}{J}T - \Omega_b . \tag{32.26}$$

The first term on the right-hand side is the dynamical angle, while the second term is the geometric (Hannay) angle. E is the kinetic energy of the initial condition; J is the magnitude of the angular momentum vector; T is the period of motion relative to the body frame; and Ω_b is the solid angle swept out by $\boldsymbol{J}$ relative to the body frame. We can think of our rigid body as of an asymmetric top with moments of inertia, $I_1 < I_2 < I_3$. The three components J_i of $\boldsymbol{J}$ relative to the body frame span a three-dimensional parameter space. The body angular momentum undergoes a closed circuit. In one such period, the body, looked at from the inertial frame, rotates about its angular momentum vector by an amount $\Delta\theta$. We want to emphasize that the formula for $\Delta\theta$ is exact. There is no adiabatic approximation involved.

The motion of a rigid body is described by a time-dependent 3×3 rotation matrix $g = g(t)$ $(\equiv R(t))$. If $\boldsymbol{X}$ is the position of a point on the reference body, then

$$\boldsymbol{x}(t) = g(t)\boldsymbol{X} , \quad g(t) \in SO(3) , \quad \boldsymbol{X} \equiv \boldsymbol{X}_b = \text{const.} \tag{32.27}$$

is its position in the inertial frame. The angular momentum of a free rigid body is constant in time: $\dot{\boldsymbol{M}} = 0$, where

$$\boldsymbol{M} = \sum_a \boldsymbol{x}_a \times \boldsymbol{p}_a = \text{const.}$$

The sum runs over the body's particles. The angular momentum, as viewed from the body-fixed frame, is no longer constant; $\dot{\boldsymbol{M}}_b \neq 0$. The relation between $\boldsymbol{M}$ and $\boldsymbol{M}_b$ is given by

$$M = gM_b \,. \tag{32.28}$$

In particular,

$$\|M_b\|^2 = \|M\|^2 \tag{32.29}$$

since

$$gg^T = \mathbb{1} \,, \quad \det g = 1 \quad \text{(no reflection)} \,. \tag{32.30}$$

The set of rotations can be parametrized by, e.g., the Euler angles. Equation (32.29) says that M_b moves on the surface of a sphere. The kinetic energy of the motion is given by

$$H = \tfrac{1}{2} M_b^T I_b^{-1} M_b \,, \tag{32.31}$$

where I_b is the moment of inertia tensor. It is a symmetric positive definite matrix. Both M and H are constants of the motion. This means that M_b moves along a curve defined by intersecting the sphere defined by (32.29) with the ellipsoid defined by (32.31). Now let us assume that $M = J$ and $H = E$ are typical, so that these intersecting curves are in fact closed. Let T be the period of M_b's oscillation along the curve:

$$M_b(t_0 + T) = M_b(t_0) \,. \tag{32.32}$$

Then, with the aid of (32.28) we obtain

$$g(t_0 + T)^{-1} M = g(t_0)^{-1} M \,, \quad M = J \,,$$

so that

$$g(t_0 + T) g(t_0)^{-1} J \equiv RJ = J \,. \tag{32.33}$$

This means that $R = g(t_0 + T)g(t_0)^{-1}$ is a rotation about the J axis.

Note that

$$g(t_0 + T) = Rg(t_0) \,, \tag{32.34}$$

so that R describes the rotation of the body in the space-fixed frame (i.e., relative to the inertial frame) after each closed orbit (full cycle) of its angular momentum in the body-fixed frame. The angle of the rotation is $\Delta\theta$ and is given explicitly in (32.26), as we now shall prove.

Let us start at $t = t_0 = 0$. Then $g(0) = \mathbb{1}$, $M_b(0) = J$ and

$$z(0) = (\mathbb{1}, J) \tag{32.35}$$

are initial conditions for the motion of the rigid body in phase space. The phase space trajectory $z(t)$ through $z(0)$ consists of pairs:

$$\big(g(t), M_b(t)\big) \,. \tag{32.36}$$

Consider the following two curves in the phase space of the rigid body, both beginning at $z(0)$:

$$C_1(t) = z(t) , \quad 0 \leqslant t \leqslant T . \tag{32.37}$$

This part of the trajectory describes the dynamical evolution starting at $z(0)$.

$$C_2(\theta) , \quad 0 \leqslant \theta \leqslant \Delta\theta \tag{32.38}$$

denotes a counterclockwise spatial rotation of the body about the $\boldsymbol{J}$ axis by θ radians. These two curves intersect when $t = T$ and $\theta = \Delta\theta$. $C = C_1 - C_2$ is thus a closed curve, starting and ending at $z(0)$.

Now it is convenient to introduce the one-form pdq on phase space:

$$pdq = \sum_a \boldsymbol{p}_a \cdot d\boldsymbol{x}_a , \quad d\boldsymbol{x}_a = d\boldsymbol{\beta} \times \boldsymbol{x}_a , \quad d\boldsymbol{\beta} = \hat{\boldsymbol{\beta}} d\beta , \quad |\hat{\boldsymbol{\beta}}| = 1$$

so that

$$pdq = \left(\sum_a \boldsymbol{x}_a \times \boldsymbol{p}_a \right) \cdot d\boldsymbol{\beta} = \boldsymbol{M} \cdot d\boldsymbol{\beta} . \tag{32.39}$$

We shall now evaluate the line integral $\int pdq$ for the two special curves C_1 and C_2 that make up our curve C. The curve C_1 is parametrized by the time t:

$$d\boldsymbol{\beta} = \boldsymbol{\omega}_b dt \quad \text{along } C_1 .$$

Here, $\boldsymbol{\omega}_b$ is the angular velocity; $\omega_b = 2\pi/T$, where T is the period of motion of $\boldsymbol{J}$ relative to the body frame. Hence we obtain

$$pdq = \boldsymbol{M}_b \cdot \boldsymbol{\omega}_b \, dt = \boldsymbol{\omega}_b \cdot I_b \boldsymbol{\omega}_b \, dt = 2E \, dt$$

or

$$pdq = 2E \, dt \quad \text{along } C_1 . \tag{32.40}$$

The curve C_2 is parametrized in radians θ:

$$d\boldsymbol{\beta} = \hat{\boldsymbol{\beta}} d\theta .$$

Moreover, $\hat{\boldsymbol{\beta}} = \boldsymbol{J}/J$ and $\boldsymbol{M} = \boldsymbol{J}$ on C_2, so that

$$pdq = \boldsymbol{M} \cdot d\boldsymbol{\beta} = \boldsymbol{J} \cdot \frac{\boldsymbol{J}}{J} d\theta = J d\theta \quad \text{along } C_2 . \tag{32.41}$$

At this point we make use of Stokes' theorem,

$$\int_{C=\partial\Sigma} \alpha = \int_{\Sigma} d\alpha . \tag{32.42}$$

Here, α denotes any one-form, and d stands for the exterior derivative. According to (32.40) and (32.41), the left-hand side of (32.42) is given by

$$\int_C \overset{(1)}{\alpha} = \int_{C_1} pdq - \int_{C_2} pdq = 2ET - J\Delta\theta \,. \tag{32.43}$$

We still need the surface integral of the 2-form, $\int_\Sigma d\alpha$. Σ is contained in the three-dimensional submanifold defined by $\boldsymbol{M}(q, p) = \boldsymbol{J}$ of our six-dimensional phase space.

On our way to determining $d\alpha$ we are going to expand pdq in terms of Euler angles $(\varphi, \vartheta, \psi)$ for the rotation group. The complete rotation matrix is then written as

$$g = g(\varphi, \vartheta, \psi) = g_3(\varphi)g_2(\vartheta)g_3(\psi) \,. \tag{32.44}$$

$g_i(\theta) \equiv R(\theta_i, \hat{\boldsymbol{e}}_i)$ denotes counterclockwise rotation about the i-th coordinate axis by an angle of θ_i radians, for $i = 1, 2, 3$. Let

$$\boldsymbol{J} = J\hat{\boldsymbol{e}}_3 \,, \quad |\hat{\boldsymbol{e}}_3| = 1 \,. \tag{32.45}$$

Now in (32.27) we had

$$\boldsymbol{x}(t) = g(t)\boldsymbol{X} \,, \quad \boldsymbol{X} = \text{const.} \tag{32.46}$$

From here we obtain

$$d\boldsymbol{x} = (dg)\boldsymbol{X} \,, \quad d\boldsymbol{x} = d\boldsymbol{\beta} \times \boldsymbol{x} \tag{32.47}$$

so that

$$(dg)\boldsymbol{X} \underset{(46)}{=} (dg)g^{-1}\boldsymbol{x} \underset{(47)}{=} d\boldsymbol{\beta} \times \boldsymbol{x} \,. \tag{32.48}$$

Let $R(\beta, \hat{\boldsymbol{\beta}})$ denote the counterclockwise rotation about the $\hat{\boldsymbol{\beta}}$-axis by β rad, where $\hat{\boldsymbol{\beta}}$ is fixed. Then (32.48) can also be rewritten as

$$(dR)\boldsymbol{X} = d\beta\hat{\boldsymbol{\beta}} \times \boldsymbol{x} = d\beta\hat{\boldsymbol{\beta}} \times R\boldsymbol{X} \tag{32.49}$$

or

$$(dR)R^{-1}\boldsymbol{x} = d\beta\hat{\boldsymbol{\beta}} \times \boldsymbol{x} \,. \tag{32.50}$$

Likewise we write

$$g_3(\varphi) \equiv R(\varphi, \hat{\boldsymbol{e}}_3) \,, \quad g_2(\vartheta) \equiv R(\vartheta, \hat{\boldsymbol{e}}_2) \,. \tag{32.51}$$

Now let us differentiate

$$\begin{aligned} g &= g_3(\varphi)g_2(\vartheta)g_3(\psi) \,. \\ (dg) &= \left[dg_3(\varphi)\right]g_2(\vartheta)g_3(\psi) + g_3(\varphi)\left[dg_2(\vartheta)\right]g_3(\psi) \\ &\quad + g_3(\varphi)g_2(\vartheta)\left[dg_3(\psi)\right] \,. \end{aligned} \tag{32.52}$$

Multiply from the right by

$$
\begin{aligned}
g^{-1} &= g_3(\psi)^{-1} g_2(\vartheta)^{-1} g_3(\varphi)^{-1} : \\
[dg]g^{-1} &= \big[dg_3(\varphi)\big] g_3(\varphi)^{-1} + g_3(\varphi)\Big\{\big[dg_2(\vartheta)\big] \underbrace{g_3(\psi) g_3(\psi)^{-1}}_{=1} g_2(\vartheta)^{-1}\Big\} g_3(\varphi)^{-1} \\
&\quad + g_3(\varphi) g_2(\vartheta)\Big\{\big[dg_3(\psi)\big] g_3(\psi)^{-1}\Big\} g_2(\vartheta)^{-1} g_3(\varphi)^{-1} .
\end{aligned}
$$

The result is applied to $\boldsymbol{x}$:

$$
\begin{aligned}
(dg)g^{-1}\boldsymbol{x} &= \underbrace{\big(dg_3(\varphi) g_3(\varphi)^{-1}\boldsymbol{x}}_{d\varphi \hat{\boldsymbol{e}}_3 \times \boldsymbol{x}} + g_3(\varphi) \underbrace{\Big\{\big(dg_2(\vartheta)\big) g_2(\vartheta)^{-1}\Big\} g_3(\varphi)^{-1}\boldsymbol{x}}_{d\vartheta \hat{\boldsymbol{e}}_2 \times g_3(\varphi)^{-1}\boldsymbol{x}} \\
&\quad + \big(g_3(\varphi) g_2(\varphi)\big) \underbrace{\Big\{\big(dg_3(\psi) g_3(\psi)^{-1}\Big\} g_2(\vartheta)^{-1} g_3(\varphi)^{-1}\boldsymbol{x}}_{d\psi \hat{\boldsymbol{e}}_3 \times \big[g_3(\varphi) g_3(\vartheta)\big]^{-1}\boldsymbol{x}} . \qquad (32.53)
\end{aligned}
$$

Now it holds for any rotation g ($\det g = 1$)

$$
g(\boldsymbol{v} \times \boldsymbol{\omega}) = (g\boldsymbol{v} \times g\boldsymbol{\omega}) . \qquad (32.54)
$$

Therefore (32.53) goes into

$$
\begin{aligned}
(dg)g^{-1}\boldsymbol{x} &= d\varphi \hat{\boldsymbol{e}}_3 \times \boldsymbol{x} + d\vartheta g_3(\varphi)\hat{\boldsymbol{e}}_2 \times \boldsymbol{x} + d\psi g_3(\varphi) g_2(\vartheta)\hat{\boldsymbol{e}}_3 \times \boldsymbol{x} \\
&= \big[d\varphi \hat{\boldsymbol{e}}_3 + d\vartheta g_3(\varphi)\hat{\boldsymbol{e}}_2 + d\psi g_3(\varphi) g_2(\vartheta)\hat{\boldsymbol{e}}_3\big] \times \boldsymbol{x} \\
&\underset{(48)}{=} d\boldsymbol{\beta} \times \boldsymbol{x} , \quad \forall \boldsymbol{x} . \qquad (32.55)
\end{aligned}
$$

This identifies $d\boldsymbol{\beta}$:

$$
d\boldsymbol{\beta} = d\varphi \hat{\boldsymbol{e}}_3 + d\vartheta g_3(\varphi)\hat{\boldsymbol{e}}_2 + d\psi g_3(\varphi) g_2(\vartheta)\hat{\boldsymbol{e}}_3 \qquad (32.56)
$$

and, upon using

$$
\boldsymbol{M} = \boldsymbol{J} = J\hat{\boldsymbol{e}}_3 ,
$$

we find

$$
\begin{aligned}
pdq &\underset{(41)}{=} \boldsymbol{M} \cdot d\boldsymbol{\beta} = J\big\{d\varphi + d\psi \hat{\boldsymbol{e}}_3 \cdot \underbrace{g_2(\vartheta)\hat{\boldsymbol{e}}_3}_{\cos\vartheta \hat{\boldsymbol{e}}_3 + \sin\vartheta \hat{\boldsymbol{e}}_1}\big\} \\
&= J\{d\varphi + d\psi \cos\vartheta\} .
\end{aligned}
$$

So at last we arrive at

$$
\alpha = pdq = J\{d\varphi + d\psi \cos\vartheta\} \qquad (32.57)
$$

so that

$$d\alpha = J\{\underbrace{d^2\varphi}_{=0} + \underbrace{d^2\psi}_{=0} \cos\varphi + d\psi d(\cos\vartheta)\}$$

or

$$d\alpha = -J \sin\vartheta d\vartheta d\psi = -J d\Omega_{\text{inert.}} \,. \tag{32.58}$$

Finally we want to rewrite (32.58) in terms of the solid angle in the space of the body angular momentum. We begin with $\boldsymbol{M} = g\boldsymbol{M}_b$. Then we obtain explicitly

$$\begin{aligned}
\boldsymbol{M}_b = g^{-1}\boldsymbol{M} = g^{-1}\underbrace{\boldsymbol{J}}_{J\hat{\boldsymbol{e}}_3} &= Jg_3(\psi)^{-1}g_2(\vartheta)^{-1}\underbrace{g_3(\varphi)^{-1}\hat{\boldsymbol{e}}_3}_{=\hat{\boldsymbol{e}}_3} \\
&= Jg_3(\psi)^{-1}g_2(\vartheta)^{-1}\hat{\boldsymbol{e}}_3 = Jg_3(\psi)^{-1}\left[\cos\vartheta\hat{\boldsymbol{e}}_3 - \sin\vartheta\hat{\boldsymbol{e}}_1\right] \\
&= J\left[\cos\vartheta\hat{\boldsymbol{e}}_3 - \sin\vartheta g_3(\psi)^{-1}\hat{\boldsymbol{e}}_1\right] \\
&= J\{\cos\vartheta\hat{\boldsymbol{e}}_3 - \sin\vartheta\left[\cos\psi\hat{\boldsymbol{e}}_1 - \sin\psi\hat{\boldsymbol{e}}_2\right]\} \\
&= J\{-\sin\vartheta\cos\psi\hat{\boldsymbol{e}}_1 + \sin\vartheta\sin\psi\hat{\boldsymbol{e}}_2 + \cos\vartheta\hat{\boldsymbol{e}}_3\} \,.
\end{aligned}$$

This says that our ψ and ϑ are related to spherical coordinates on the body angular momentum space by $\psi = -\psi_b$, $\vartheta = -\vartheta_b$. This is an orientation-reversing coordinate transformation, so that

$$d\alpha = J d\Omega_b \underset{(58)}{=} -J d\Omega_{\text{inert.}} \,. \tag{32.59}$$

So finally we obtain the right-hand side of (32.42):

$$\int_\Sigma d\alpha = \int_\Sigma J d\Omega_b = J\Omega_b \tag{32.60}$$

and this completes the proof of our original claim (32.26):

$$\Delta\theta = 2\frac{E}{J}T - \Omega_b \,. \tag{32.61}$$

As our final example we consider the classical adiabatic motion of charged particles in strong magnetic fields. This so-called guiding center motion (cf. Chap. 10) also exhibits an anholonomic (nonintegrable) phase similar to Berry's (Hanny's).

The motion of the charged particle takes place in a nonuniform magnetic field $\boldsymbol{B} = \boldsymbol{B}(\boldsymbol{r})$. The equations of motion are nonlinear and in general nonintegrable; they show chaotic behavior. For strong magnetic fields, however, it is possible to perform a separation of time scales in the three degrees of freedom. The most rapid time scale arises from the gyration of the particle about a magnetic field line. It is like an oscillator being adiabatically transported by the slower degree of freedom. This can best be seen after a Hamiltonian formulation has been set up and a canonical perturbation analysis, otherwise known as "guiding center expansion", has been established – a highly nontrivial task.

In the case of a uniform field, $\boldsymbol{B} = B_0\hat{z}$, the particle is bound to follow a helical orbit around a field line. The frequency of motion in a plane orthogonal to $\boldsymbol{B}$ is given by the gyrofrequency $\Omega = eB_0/mc$. The guiding center position $\boldsymbol{X}$ moves along the field line (z-axis) with constant velocity. The vector running orthogonally from $\boldsymbol{X}$ on the z-axis to the particle position we call $\boldsymbol{r}$, the gyroradius vector. Finally, we define the "gyrophase" θ as the angle in the perpendicular plane between some reference direction $\hat{\boldsymbol{e}}$ and $\boldsymbol{r}$. Notet that for the case of a uniform field, $\hat{\boldsymbol{e}}$ may be chosen any constant perpendicular direction, such as $\boldsymbol{r}$.

Now we are going to look at nonuniform but time-independent magnetic fields. In this case the guiding center no longer follows a field line, but slowly drifts away from it: $\boldsymbol{X} = \boldsymbol{X}(t)$. This makes it necessary to consider arbitrary paths of transport, not just along field lines. Also, Ω becomes space dependent. Now, the dominant contribution in the time evolution of the gyrophase is given by the dynamical phase

$$\theta(t) = \int_0^t dt'\, \Omega(\boldsymbol{X}(t'))\ . \tag{32.62}$$

Note that in (32.62) we have evaluated the gyrofrequency at the guiding center position rather than at the particle position.

Of course, it is no longer possible to choose a constant reference direction to represent the origin of the gyrophase. To see this more clearly, let us elaborate for a while on the geometrical rather than the dynamical picture of the problem and introduce an orthonormal triad (dreibein) of unit vectors $(\hat{\boldsymbol{e}}_1, \hat{\boldsymbol{e}}_2, \hat{\boldsymbol{b}})$, $\hat{\boldsymbol{b}} = \boldsymbol{B}(\boldsymbol{r})/B$. All three vectors are functions of position $\boldsymbol{r}$. We are going to fill up space with such triads, so that we can talk about a field of orthonormal frames.

Now, when $\hat{\boldsymbol{b}}$ moves along the field line, the $\hat{\boldsymbol{e}}_i$ are constrained to stay perpendicular to $\hat{\boldsymbol{b}}$. But the $\hat{\boldsymbol{e}}_i$ are also free to rotate around $\hat{\boldsymbol{b}}$ by an arbitrary angle. Therefore the dynamical phase cannot be the only contribution to the gyrophase, because the definition of the gyrophase depends on the choice $(\hat{\boldsymbol{e}}_1, \hat{\boldsymbol{e}}_2)$, and we do not see any such dependence in Eq. (32.62). There is no best choice of $(\hat{\boldsymbol{e}}_1, \hat{\boldsymbol{e}}_2)$. One could pick

$$\hat{\boldsymbol{e}}_1 = \frac{\hat{\boldsymbol{b}} \cdot \nabla\hat{\boldsymbol{b}}}{|\hat{\boldsymbol{b}} \cdot \nabla\hat{\boldsymbol{b}}|}\ , \qquad \hat{\boldsymbol{e}}_2 = \hat{\boldsymbol{b}} \times \hat{\boldsymbol{e}}_1\ , \tag{32.63}$$

i.e., the principal normal and binormal vectors of the field line. But this is only a particular choice out of many. In fact we can call Eq. (32.63) a choice of gauge. Then every other choice of gauge is related to (32.63) by a rotation in the instantaneous perpendicular plane. So let us define a "gyrophase transformation":

$$\begin{aligned} \hat{\boldsymbol{e}}_1'(\boldsymbol{r}) &= \hat{\boldsymbol{e}}_1(\boldsymbol{r}) \cos\psi(\boldsymbol{r}) + \hat{\boldsymbol{e}}_2(\boldsymbol{r}) \sin\psi(\boldsymbol{r})\ , \\ \hat{\boldsymbol{e}}_2'(\boldsymbol{r}) &= -\hat{\boldsymbol{e}}_1(\boldsymbol{r}) \sin\psi(\boldsymbol{r}) + \hat{\boldsymbol{e}}_2(\boldsymbol{r}) \cos\psi(\boldsymbol{r})\ . \end{aligned} \tag{32.64}$$

Next, we want to know which quantitics are gyrogauge invariant, and how those which are not transform. Quantities which can be expressed purely in terms of $\hat{\boldsymbol{b}}$, B, etc. are gyrogauge invariant. The gyrophase itself is not gyrogauge invariant; it transforms according to

$$\theta' = \theta + \psi(\boldsymbol{r}) . \tag{32.65}$$

(We follow the gyrophase in clockwise direction, the same as the direction of rotation of a positively charged particle.) Returning to the dynamical consideration as stated in (32.62), we know that since $\dot{\theta}$ is not gyrogauge invariant, θ cannot be either, so there must be terms other than the dynamical phase reflecting this fact. Indeed, lengthy nontrivial calculations show that the guiding center expansion results in

$$\dot{\theta}(t) = \Omega(\boldsymbol{X}(t)) + \boldsymbol{R} \cdot \dot{\boldsymbol{X}}(t) + \quad \text{gyrogauge invariant terms} \tag{32.66}$$

with

$$\boldsymbol{R}(\boldsymbol{r}) = \nabla\hat{\boldsymbol{e}}_1(\boldsymbol{r}) \cdot \hat{\boldsymbol{e}}_2(\boldsymbol{r}) . \tag{32.67}$$

Equation (32.66) is actually an averaged equation; all fast oscillations have been averaged out.

The term of special interest to us is given by $\boldsymbol{R} \cdot \dot{\boldsymbol{X}}$, which is reminiscent of $\boldsymbol{A} \cdot \boldsymbol{v}$ in the Aharonov–Bohm effect. $\boldsymbol{R}$ is not gyrogauge invariant. In fact, we now want to demonstrate that $\boldsymbol{R}$ responds under the gyrogauge transformation (32.64) according to

$$\boldsymbol{R}' = \boldsymbol{R} + \nabla\psi . \tag{32.68}$$

Proof: first let us note that

$$\boldsymbol{R} = \left(\nabla\hat{\boldsymbol{e}}_1\right) \cdot \hat{\boldsymbol{e}}_2 \equiv \left(\partial_i e_{1j}\right) e_{2j} , \quad \hat{\boldsymbol{e}}_i = \hat{\boldsymbol{e}}_i(\boldsymbol{r}) .$$

Then let $\boldsymbol{a}$ be an arbitrary constant vector, so that it holds that

$$(\boldsymbol{a} \cdot \boldsymbol{R}) = \left[(\boldsymbol{a} \cdot \nabla)\hat{\boldsymbol{e}}_1\right] \cdot \hat{\boldsymbol{e}}_2 \stackrel{!}{=} -\left[(\boldsymbol{a} \cdot \nabla)\hat{\boldsymbol{e}}_2\right] \cdot \hat{\boldsymbol{e}}_1 , \quad \forall \boldsymbol{a} .$$

To see this, we start with

$$\begin{aligned}
&(\boldsymbol{a} \cdot \nabla)\left(\hat{\boldsymbol{e}}_1 \cdot \hat{\boldsymbol{e}}_2\right) = 0 \quad \text{or} \quad \nabla_{\boldsymbol{a}}\left(\hat{\boldsymbol{e}}_1 \cdot \hat{\boldsymbol{e}}_2\right) = 0 . \\
&\quad \Rightarrow \left(\nabla_{\boldsymbol{a}}\hat{\boldsymbol{e}}_1\right) \cdot \hat{\boldsymbol{e}}_2 + \left(\nabla\hat{\boldsymbol{e}}_2\right) \cdot \hat{\boldsymbol{e}}_1 = 0 \\
&\quad \Rightarrow \left[(\boldsymbol{a} \cdot \nabla)\hat{\boldsymbol{e}}_1\right] \cdot \hat{\boldsymbol{e}}_2 = -\left[(\boldsymbol{a} \cdot \nabla)\hat{\boldsymbol{e}}_2\right] \cdot \hat{\boldsymbol{e}}_1 , \quad \forall \boldsymbol{a} \\
&\quad \Rightarrow \left(\nabla\hat{\boldsymbol{e}}_1\right) \cdot \hat{\boldsymbol{e}}_2 = -\left(\nabla\hat{\boldsymbol{e}}_2\right) \cdot \hat{\boldsymbol{e}}_1 .
\end{aligned} \tag{32.69}$$

Of course, we have

$$\left(\nabla\hat{\boldsymbol{e}}_{\substack{1\\2}}\right) \cdot \hat{\boldsymbol{e}}_{\substack{1\\2}} = 0 . \tag{32.70}$$

Now we turn to $\boldsymbol{R}' = (\nabla\hat{\boldsymbol{e}}_1') \cdot \hat{\boldsymbol{e}}_2'$ and substitute (32.64):

$$
\begin{aligned}
R' &= \left[\nabla\left(\hat{e}_1 \cos\psi + \hat{e}_2 \sin\psi\right)\right] \cdot \left[-\hat{e}_1 \sin\psi + \hat{e}_2 \cos\psi\right] \\
&= \left[\cos\psi\left(\nabla\hat{e}_1\right) - \sin\psi(\nabla\psi)\hat{e}_1 + \sin\psi\left(\nabla\hat{e}_2\right)\right. \\
&\quad \left. + \cos\psi(\nabla\psi)\hat{e}_2\right] \cdot \left[-\hat{e}_1 \sin\psi + \hat{e}_2 \cos\psi\right] \\
&= -\sin\psi \cos\psi \underbrace{\left(\nabla\hat{e}_1\right) \cdot \hat{e}_1}_{=0} + \cos^2\psi\left(\nabla\hat{e}_1\right) \cdot \hat{e}_2 \\
&\quad + \sin^2\psi(\nabla\psi) - \sin\psi \cos\psi(\nabla\psi) \underbrace{\hat{e}_1 \cdot \hat{e}_2}_{=0} \\
&\quad \underbrace{- \sin^2\psi\left(\nabla\hat{e}_2\right) \cdot \hat{e}_1}_{= \sin^2\psi\left(\nabla\hat{e}_1\right)\cdot\hat{e}_2} + \sin\psi \cos\psi \underbrace{\left(\nabla\hat{e}_2\right) \cdot \hat{e}_2}_{=0} \\
&\quad - \sin\psi \cos\psi(\nabla\psi) \underbrace{\hat{e}_1 \cdot \hat{e}_2}_{=0} + \cos^2\psi(\nabla\psi) \\
&= \left(\sin^2\psi + \cos^2\psi\right)\Big[\underbrace{\left(\nabla\hat{e}_1\right) \cdot \hat{e}_2}_{=R} + \nabla\psi\Big] = R + \nabla\psi \,.
\end{aligned}
$$

Finally, using the result (32.68), we can find the transformation property of $\dot{\theta}$:

$$\dot{\theta}' = \Omega(X) + R' \cdot \dot{X} = (\Omega(X) + R \cdot \dot{X}) + \nabla\psi \cdot \dot{X}$$

or

$$\dot{\theta}' = \dot{\theta} + \nabla\psi \cdot \dot{X} \,, \tag{32.71}$$

which is completely consistent with (32.65):

$$\dot{\theta}' = \dot{\theta} + \dot{\psi}(r) = \dot{\theta} + \dot{r} \cdot \nabla\psi = \dot{\theta} + \dot{X} \cdot \nabla\psi \,, \tag{32.72}$$

after averaging, $\langle r \rangle = X$.

The time integral of the second term of (32.66) can be written as a line integral along the guiding center trajectory:

$$\Delta\theta = \int_{X_0}^{X_1} R \cdot dX \underset{\substack{\text{for closed}\\ \text{loops}}}{=} \int da \cdot (\nabla \times R) \,. \tag{32.73}$$

So long as $\nabla \times R \neq 0$, this angle will be anholonomic, i.e., dependent on the path $X(t)$ of the guiding center. We see that the (fast) gyration of the particle and X are coupled. X represents the slow change for the environment of the rapidly oscillating particle. The guiding center coordinate X is the analogue of the Berry (Hannay) connection A.

Note that the gyrophase is an angle in real physical space, which is endowed with a metric. The changes in $\hat{e}_i(X)$ as the guiding center X moves through space are comprised of two parts. First, the $\hat{e}_i$ are forced to stay perpendicular to $\hat{b}$, and since $\hat{b}$ is changing along the X-trajectory, the $\hat{e}_i$ must change. However, during the transport the $\hat{e}_i$ can also be rotated in their plane. Hence the increment in the vectors $\hat{e}_i(X)$ on moving X to $X+dX$ consists of a part parallel to $\hat{b}$ and a part perpendicular to $\hat{b}$, representing the angle of rotation $d\varphi$ within the moving plane. For instance, the increment $d\hat{e}_1$ of $\hat{e}_1$ can be decomposed in two components:

$$\begin{aligned} d\hat{e}_1 &= dX \cdot \nabla\hat{e}_1 = dX \cdot \nabla\hat{e}_1 \cdot \mathbb{1} = \big(dX \cdot \nabla\hat{e}_1\big) \cdot \big(\hat{e}_1\hat{e}_1 + \hat{e}_2\hat{e}_2 + \hat{b}\hat{b}\big) \\ &\underset{(70)}{=} \big(dX \cdot \underbrace{\nabla\hat{e}_1 \cdot \hat{e}_2}_{=R}\big)\hat{e}_2 + \big(dX \cdot \underbrace{\nabla\hat{e}_1 \cdot \hat{b}}_{=-(\nabla\hat{b})\cdot\hat{e}_1}\big)\hat{b} \\ &= (R \cdot dX)\hat{e}_2 - \big(dX \cdot \nabla\hat{b} \cdot \hat{e}_1\big)\hat{b}\,. \end{aligned} \tag{32.74}$$

Now we see that the angle $d\varphi$ can be written as $d\varphi = R \cdot dX$, which is identical to the increment in the gyrophase arising from the $R \cdot \dot{X}$-term in (32.66) when the guiding center X is displaced by dX. So we have shown that the analogue of Berry's (Hannay's) angle for guiding center motion is exactly the accumulation of the angle of rotation as the guiding center moves from some initial to some final point, as stated in (32.73).

33. Berry Phase and Parametric Harmonic Oscillator

Our concern in this section is once more with the time-dependent harmonic oscillator with Lagrangian

$$L = \tfrac{1}{2}\dot{x}^2 - \tfrac{1}{2}\omega^2(t)x^2 \ .$$

To present a coherent picture of the whole problem, let us briefly review some of the results of Chap. 20. There we found the propagation function

$$K\left(x_2, t_2; x_1, t_1\right) = \left[\frac{\sqrt{\dot{g}_1\dot{g}_2}}{2\pi i\hbar \sin\phi(t_2, t_1)}\right]^{1/2} e^{(i/\hbar)S_{cl}} \ , \tag{33.1}$$

where the classical action is given by

$$\begin{aligned} S_{cl} = \frac{1}{2}\Bigg[&\frac{\dot{\varrho}_2 x_2^2}{\varrho_2} - \frac{\dot{\varrho}_1 x_1^2}{\varrho_1} + \left(\dot{g}_2 x_2^2 + \dot{g}_1 x_1^2\right)\cot\phi(t_2, t_1) \\ &- 2\sqrt{\dot{g}_1\dot{g}_2}x_2x_1\frac{1}{\sin\phi(t_2, t_1)}\Bigg] \ . \end{aligned} \tag{33.2}$$

The various functions that enter (33.1) and (33.2) are defined by

$$\ddot{\varrho} + \omega^2(t)\varrho - \frac{1}{\varrho^3} = 0 \tag{33.3}$$

$$\dot{g} = \frac{1}{\varrho^2} \ , \quad \phi(t_2, t_1) = g(t_2) - g(t_1) \ . \tag{33.4}$$

In the limiting case of $\omega = \text{const.}$, we obtain

$$\omega = \text{const.} : \quad \varrho(t) = \omega^{-1/2} \ , \quad g(t) = \omega t \ , \quad \phi(t_2, t_1) = \omega(t_2 - t_1) \ . \tag{33.5}$$

Consider, then, the trace of the propagator (33.1):

$$\begin{aligned} G(t_2, t_1) :&= \int_{-\infty}^{+\infty} dx K\left(x, t_2; x, t_1\right) \\ &= \left[\frac{\sqrt{\dot{g}_1\dot{g}_2}}{2\pi i\hbar \sin\phi(t_2, t_1)}\right]^{1/2} \int_{-\infty}^{+\infty} dx \exp\Bigg\{\frac{i}{\hbar}\frac{x^2}{2}\Bigg[\left(\frac{\dot{\varrho}_2}{\varrho_2} - \frac{\dot{\varrho}_1}{\varrho_1}\right) \\ &\quad + \frac{1}{\sin\phi(t_2, t_1)}\left[\left(\dot{g}_2 + \dot{g}_1\right)\cos\phi(t_2, t_1) - 2\sqrt{\dot{g}_1\dot{g}_2}\right]\Bigg]\Bigg\} \ . \end{aligned}$$

Here we meet a Gauss-type integral,

$$\int_{-\infty}^{+\infty} dx \exp\left\{-\frac{x^2}{2i\hbar}[./.]\right\} = \left(\frac{\pi}{[./.]/2i\hbar}\right)^{1/2} = \left(\frac{[./.]}{2\pi i\hbar}\right)^{-1/2} .$$

This allows us to write

$$\begin{aligned} G(t_2, t_1) &= \left[\frac{(2\pi i\hbar)\sin\phi}{\sqrt{\dot{g}_1\dot{g}_2}}\right]^{-1/2} \left(\frac{1}{2\pi i\hbar}\right)^{-1/2} \left[\left(\frac{\dot{\varrho}_2}{\varrho_2} - \frac{\dot{\varrho}_1}{\varrho_1}\right)\right. \\ &\quad \left. + \frac{1}{\sin\phi}\left[(\dot{g}_2 + \dot{g}_1)\cos\phi - 2\sqrt{\dot{g}_1\dot{g}_2}\right]\right]^{-1/2} \\ &= \left[\frac{\sin\phi(t_2, t_1)}{\sqrt{\dot{g}_1\dot{g}_2}}\left(\frac{\dot{\varrho}_2}{\varrho_2} - \frac{\dot{\varrho}_1}{\varrho_1}\right) + \frac{\cos\phi(t_2, t_1)}{\sqrt{\dot{g}_1\dot{g}_2}}(\dot{g}_2 + \dot{g}_1) - 2\right]^{-1/2} . \end{aligned} \tag{33.6}$$

Using (33.4) we have

$$\frac{1}{\sqrt{\dot{g}}} = \varrho , \quad \frac{1}{\sqrt{\dot{g}_1\dot{g}_2}} = \varrho_1\varrho_2$$

so that

$$\frac{1}{\sqrt{\dot{g}_1\dot{g}_2}}\left(\frac{\dot{\varrho}_2}{\varrho_2} - \frac{\dot{\varrho}_1}{\varrho_1}\right) = \varrho_2\varrho_1\left(\frac{\dot{\varrho}_2}{\varrho_2} - \frac{\dot{\varrho}_1}{\varrho_1}\right) = \varrho_1\dot{\varrho}_2 - \varrho_2\dot{\varrho}_1 \tag{33.7}$$

and

$$\frac{1}{\sqrt{\dot{g}_1\dot{g}_2}}(\dot{g}_2 + \dot{g}_1) = \varrho_2\varrho_1\left(\frac{1}{\varrho_2^2} + \frac{1}{\varrho_1^2}\right) = \frac{\varrho_1}{\varrho_2} + \frac{\varrho_2}{\varrho_1} . \tag{33.8}$$

These expressions are substituted in (33.6) and so we obtain

$$\begin{aligned} G(t_2, t_1) = \Bigg[& (\varrho_1\dot{\varrho}_2 - \varrho_2\dot{\varrho}_1)\sin\phi(t_2, t_1) \\ & + \left(\frac{\varrho_1}{\varrho_2} + \frac{\varrho_2}{\varrho_1}\right)\cos\phi(t_2, t_1) - 2\Bigg]^{-1/2} . \end{aligned} \tag{33.9}$$

Here we should be able to extract some known results for the case of $\omega(t) \equiv \omega =$ const. According to (33.5) we have

$$\dot{\varrho}_{1,2} = 0 , \quad \varrho_{1,2} = \omega^{-1/2} , \quad \left(\frac{\varrho_1}{\varrho_2} + \frac{\varrho_2}{\varrho_1}\right) = 2 , \quad \phi = \omega(t_2 - t_1) \equiv \omega T$$

so that (33.9) simplifies to

$$G(T) = [2\{\cos \omega T - 1\}]^{-1/2} = \Bigg[-2 \cdot 2 \underbrace{\frac{1-\cos(\omega T)}{2}}_{=\sin^2(\omega T/2)}\Bigg]^{-1/2}$$

$$= \frac{1}{2\mathrm{i}\,\sin(\omega T/2)} = \frac{1}{\exp[\mathrm{i}(\omega T/2)] - \exp[-\mathrm{i}(\omega T/2)]}$$

$$= \frac{\exp[-\mathrm{i}(\omega T/2)]}{1-\mathrm{e}^{-\mathrm{i}\omega T}} = \exp[-\mathrm{i}(\omega T/2)] \sum_{n=0}^{\infty} \mathrm{e}^{-\mathrm{i}n\omega T}$$

$$= \sum_{n=0}^{\infty} \mathrm{e}^{-\mathrm{i}(n+1/2)\omega T} = \sum_{n=0}^{\infty} \exp\left[-\frac{\mathrm{i}}{\hbar}\hbar\omega\left(n+\frac{1}{2}\right)T\right],$$

which identifies the energy spectrum of the linear harmonic oscillator as $E_n = \hbar\omega(n+1/2)$.

To draw some nontrivial consequences of our result as stated in (33.9), we shall now consider the special case of a parametric oscillator for which ω is constant in the remote past and in the distant future, and we shall calculate the vacuum persistence amplitude , i.e., the amplitude for the ground state to remain in the ground state while $\omega(t)$ is acting for a finite time duration.

At $t_1 \to -\infty$ and $t_2 \to +\infty$ we set

$$\omega(t_1) \equiv \omega_1 \to \omega = \text{const.}\,, \qquad \omega_2 \to \omega = \text{const.} > 0\,. \tag{33.10}$$

In obtaining the vacuum transition amplitude we need the general solution of

$$\ddot{\varrho} + \omega^2(t)\varrho - \frac{1}{\varrho^3} = 0\,. \tag{33.11}$$

As time-independent solution for (33.11) at $t_1 \to -\infty$ we choose

$$\varrho_1 \equiv \varrho(-\infty) = \frac{1}{\sqrt{\omega}}\,, \quad \dot{\varrho}_1 = \left.\frac{d\varrho}{dt}\right|_{t_1=-\infty} = 0\,. \tag{33.12}$$

Then, however, for $t_2 \to +\infty$ we have to use the most general solution of (33.11) for constant ω:

$$\varrho(t_2) = \gamma_1 \frac{1}{\sqrt{\omega}}\left[\cosh\delta + \gamma_2 \sinh\delta\, \sin(2\omega t_2 + \varphi)\right]^{1/2}, \tag{33.13}$$

where $\gamma_{1,2} = \pm 1$ and φ is a real phase constant. The meaning of the real parameter δ becomes clear when we go back to (33.11) – with constant ω – which we multiply by $\dot{\varrho}$ to obtain

$$\frac{1}{2}\frac{d}{dt}\left(\dot{\varrho}^2\right) + \frac{\omega^2}{2}\frac{d}{dt}\varrho^2 + \frac{1}{2}\frac{d}{dt}\left(\frac{1}{\varrho^2}\right) = 0$$

or

$$\frac{d}{dt}\left\{\dot{\varrho}^2 + \omega^2\varrho^2 + \frac{1}{\varrho^2}\right\} = 0$$

which can be integrated with the result

$$\dot{\varrho}^2 + \omega^2\varrho^2 + \frac{1}{\varrho^2} = 2\omega \cosh\delta\,. \tag{33.14}$$

The integration constant on the right-hand side of (33.14) is chosen so that ϱ will be real for all values of the real parameter δ.

Let us check quickly that (33.13) is indeed a solution of (33.14). We need

$$\varrho^2 = \frac{1}{\omega}[\cosh\delta + \gamma_2 \sinh\delta \sin(2\omega t + \varphi)]\,, \quad \frac{1}{\varrho^2} = \omega[./.]^{-1} \tag{33.15}$$

$$\dot{\varrho} = \frac{\gamma_1}{\sqrt{\omega}}\frac{1}{2}[./.]^{-1/2}(\gamma_2 \sinh\delta \cos(2\omega t + \varphi))2\omega \tag{33.16}$$

$$\dot{\varrho}^2 = \omega[./.]^{-1} \sinh^2\delta \cos^2(2\omega t + \varphi)\,. \tag{33.17}$$

When substituted in (33.14) we find

$$\omega[./.]^{-1} \sinh^2\delta \cos^2(2\omega t + \varphi) + \omega[./.] + \omega[./.]^{-1} \stackrel{!}{=} 2\omega \cosh\delta$$

or

$$\sinh^2\delta \cos^2(2\omega t + \varphi) + [./.]^2 + 1 \stackrel{!}{=} [./.]2 \cosh\delta\,. \tag{33.18}$$

Since

$$[./.]^2 - 2[./.]\cosh\delta + \underbrace{1}_{=\cosh^2\delta - \sinh^2\delta} = ([./.] - \cosh\delta)^2 - \sinh^2\delta$$

we can continue to write for (33.18)

$$\begin{aligned}\sinh^2\delta \underbrace{[1 - \cos^2(2\omega t + \varphi)]}_{=\sin^2(2\omega t+\varphi)} &= ([./.] - \cosh\delta)^2 \\ &\equiv \left(\cosh\delta + \gamma_2 \sinh\delta \sin(2\omega t + \varphi) - \cosh\delta\right)^2 \\ &= \sinh^2\delta \sin^2(2\omega t + \varphi)\,.\end{aligned}$$

In (33.9) we also need

$$\phi(t_2, t_1) = g(t_2) - g(t_1) = \int_{t_1}^{t_2} dt\, \dot{g}(t) = \int_{t_1}^{t_2} dt \frac{1}{\varrho^2(t)}\,. \tag{33.19}$$

Here it is convenient to split up the last integral in (33.19) in three parts:

$$\phi(t_2, t_1) = \int_{t_1}^{t_2} dt \frac{1}{\varrho^2(t)} = \int_{t_1}^{t_i} \frac{dt}{\varrho^2(t)} + \int_{t_i}^{t_f} \frac{dt}{\varrho^2(t)} + \int_{t_f}^{t_2} \frac{dt}{\varrho^2(t)}\,. \tag{33.20}$$

The first and the third integral are to be integrated between times for which we encounter free evolution of our system ($\omega = \text{const.}$). $\dot{\omega} \neq 0$ is important for the time interval of the second integral on the right-hand side: $t \in [t_i, t_f] \subset [t_1 \to -\infty, t_2 \to \infty]$. Also note that this integral is constant with resepct to t_1 and t_2. Thus, for $t_1 \to -\infty$, $t_2 \to +\infty$, we have

$$\begin{aligned}
\phi(t_2, t_1) &= -\int^{t_1} dt\, \omega + \text{const.} + \omega \int^{t_2} dt \\
&\quad \times \frac{1}{\cosh\delta + \gamma_2 \sinh\delta \sin(2\omega t + \varphi)} \\
&= \frac{1}{2}\int^{x=2\omega t_2+\varphi} dx \frac{1}{\cosh\delta + \gamma_2 \sinh\delta \sin x} - \omega t_1 + \text{const.}
\end{aligned}$$

To compute the integral we recall the formula

$$\int \frac{dx}{a + b\sin x} = \frac{2}{\sqrt{a^2 - b^2}} \arctan\left(\frac{a\tan(x/2) + b}{\sqrt{a^2 - b^2}}\right) .$$

This brings us to

$$\begin{aligned}
\phi(t_2, t_1) &= \arctan\left[\cosh\delta \tan\left(\omega t_2 + \frac{\varphi}{2}\right)\right. \\
&\quad \left. + \gamma_2 \sinh\delta\right] - \omega t_1 + \text{const.}
\end{aligned} \tag{33.21}$$

For reasons which will become clear in a moment, we are interested in the limit $t_2 \to -\mathrm{i}\infty$. Therefore let us write $t_2 = -\mathrm{i}\tau_2$, so that

$$\begin{aligned}
\tan\left(\omega t_2 + \frac{\varphi}{2}\right) &= \tan\left(-\mathrm{i}\omega\tau_2 + \frac{\varphi}{2}\right) = -\tan\left(\mathrm{i}\omega\tau_2 - \frac{\varphi}{2}\right) \\
&= \mathrm{i}\frac{\mathrm{e}^{\mathrm{i}(\mathrm{i}\omega\tau_2 - \varphi/2)} - \mathrm{e}^{-\mathrm{i}(\mathrm{i}\omega\tau_2 - \varphi/2)}}{\mathrm{e}^{\mathrm{i}(\mathrm{i}\omega\tau_2 - \varphi/2)} + \mathrm{e}^{-\mathrm{i}(\mathrm{i}\omega\tau_2 - \varphi/2)}} \underset{\tau_2 \to \infty}{\longrightarrow} \mathrm{i}\frac{0 - \mathrm{e}^{\omega\tau_2}\,\mathrm{e}^{\mathrm{i}(\varphi/2)}}{0 + \mathrm{e}^{\omega\tau_2}\,\mathrm{e}^{\mathrm{i}(\varphi/2)}} = -\mathrm{i} .
\end{aligned}$$

Thus we obtain for $\tau_2 \to \infty$

$$\phi(t_2, t_1) \to \arctan[\gamma_2 \sinh\delta - \mathrm{i}\cosh\delta] + \mathrm{i}\omega\tau , \tag{33.22}$$

$$\equiv \alpha + \mathrm{i}\omega\tau_1 . \tag{33.23}$$

Writing

$$\alpha = \arctan\beta , \quad \beta \equiv \gamma_2 \sinh\delta - \mathrm{i}\cosh\delta \tag{33.24}$$

we have

$$\alpha = \frac{1}{2\mathrm{i}}\ln\frac{1 + \mathrm{i}\beta}{1 - \mathrm{i}\beta} = -\frac{1}{2\mathrm{i}}\ln\frac{1 - \mathrm{i}\beta}{1 + \mathrm{i}\beta} = -\frac{1}{\mathrm{i}}\ln\left(\frac{1 - \mathrm{i}\beta}{1 + \mathrm{i}\beta}\right)^{1/2}$$

so that

$$e^{-i\alpha} = \left(\frac{1-i\beta}{1+i\beta}\right)^{1/2} = \left[\frac{1-i\gamma_2 \sinh\delta + \cosh\delta}{1+i\gamma_2 \sinh\delta + \cosh\delta}\right]^{1/2} . \tag{33.25}$$

After these intermediate calculations we return to (33.9):

$$G(t_2, t_1) = \left[\omega^{-1/2}\frac{d\varrho(t_2)}{dt_2}\sin\phi + \left(\frac{\omega^{-1/2}}{\varrho(t_2)} + \frac{\varrho(t_2)}{\omega^{-1/2}}\right)\cos\phi - 2\right]^{-1/2} . \tag{33.26}$$

The next step involves taking the following limiting processes:

$$\varrho(t_2 = -i\tau_2) = \gamma_1\omega^{-1/2}\big[\cosh\delta + \gamma_2\sinh\delta\,\underbrace{\sin(-2i\omega\tau_2+\varphi)}_{\underset{\tau_2\to\infty}{\longrightarrow}\exp[2\omega\tau_2+i\varphi]/2i}\big]^{1/2}$$

$$\underset{\tau_2\to\infty}{\longrightarrow} \gamma_1\omega^{-1/2}\left(\frac{\gamma_2\sinh\delta}{2i}\right)^{1/2}\exp\left[\omega\tau_2 + i\frac{\varphi}{2}\right].$$

From here we obtain

$$\frac{\varrho(t_2)}{\omega^{-1/2}} \to \gamma_1\left(\frac{\gamma_2\sinh\delta}{2i}\right)^{1/2}\exp\left[\omega\tau_2 + i\frac{\varphi}{2}\right], \quad \frac{\omega^{-1/2}}{\varrho(t_2)} \to 0 .$$

We also need

$$\frac{d\varrho(t_2)}{dt_2} = \frac{d\varrho_2}{d(-i\tau_2)} = i\frac{d\varrho_2}{d\tau_2} = i\omega\varrho_2$$

so that

$$\omega^{-1/2}\frac{d\varrho(t_2)}{dt_2} = i\omega^{1/2}\varrho(t_2) = i\frac{\varrho_2}{\omega^{-1/2}} .$$

When we substitute these quantities in (33.26), we get

$$i\frac{\varrho_2}{\omega^{-1/2}}\sin\phi + \left(0 + \frac{\varrho_2}{\omega^{-1/2}}\right)\cos\phi = \frac{\varrho_2}{\omega^{-1/2}}[\cos\phi + i\sin\phi] = \frac{\varrho_2}{\omega^{-1/2}}e^{i\phi}$$

$$= \gamma_1\left(\frac{\gamma_2\sinh\delta}{2i}\right)^{1/2}\exp\left[\omega\tau_2 + i\frac{\varphi}{2}\right]e^{i\phi}$$

or introducing the Euclidean propagator

$$G_E(\tau_2, \tau_1) \to \left[\gamma_1\left(\frac{\gamma_2\sinh\delta}{2i}\right)^{1/2}\exp\left[\omega\tau_2 + i\frac{\varphi}{2}\right]e^{i\phi} - 2\right]^{-1/2}$$

where

$$e^{i\phi} \underset{(23)}{=} \exp\big[i(\alpha + i\omega\tau_1)\big] = \exp[i\alpha]\exp\big[-\omega\tau_1\big]$$

$$\underset{(25)}{=} \exp\big[-\omega\tau_1\big]\left[\frac{(1+\cosh\delta) + i\gamma_2\sinh\delta}{(1-\cosh\delta) - i\gamma_2\sinh\delta}\right]^{1/2} . \tag{33.27}$$

What emerges as our final expression is

$$G_E(\tau_2,\tau_1) \to \left[\gamma_1\left(\frac{\gamma_2 \sinh\delta}{2\mathrm{i}}\right)^{1/2} \mathrm{e}^{\mathrm{i}(\varphi/2)}\,\mathrm{e}^{\omega(\tau_2-\tau_1)} \times \left[\frac{(1+\cosh\delta)+\mathrm{i}\gamma_2\sinh\delta}{(1-\cosh\delta)-\mathrm{i}\gamma_2\sinh\delta}\right]^{1/2}\right]^{-1/2} . \tag{33.28}$$

Now let us search for a relation that connects (33.28) with the vacuum-to-vacuum amplitude. Generally speaking, if we have an external source J acting on the system between $t' \equiv t_i$ and $t'' \equiv t_f$, where $t_1 < t' < t'' < t_2$, we obtain for the transition amplitude

$$\langle x_2,t_2|x_1,t_1\rangle^J = \int dx'\,dx''\langle x_2,t_2|x'',t''\rangle_{\text{free}}\langle x'',t''|x',t'\rangle^J\langle x',t'|x_1,t_1\rangle_{\text{free}} .$$

Upon using

$$\begin{aligned}\langle x_2,t_2|x'',t''\rangle_{\text{free}} &= \langle x_2|\mathrm{e}^{-\mathrm{i}H_0(t_2-t'')}|x''\rangle\\ &= \sum_m \phi_m(x_2)\phi_m^*(x'')\,\mathrm{e}^{-\mathrm{i}E_m(t_2-t'')}\end{aligned}$$

and, likewise for $\langle x',t'|x_1,t_1\rangle$, where $H_0 = H(J=0)$, $H_0\phi_m = E_m\phi_m$, we get

$$\begin{aligned}\langle x_2,t_2|x_1,t_1\rangle^J &= \sum_{n,m}\mathrm{e}^{-\mathrm{i}E_m(t_2-t'')}\,\mathrm{e}^{-\mathrm{i}E_n(t'-t_1)}\phi_m(x_2)\phi_n^*(x_1)\\ &\quad\times\int dx'\,dx''\phi_m^*(x'')\langle x'',t''|x',t'\rangle^J\phi_n(x') .\end{aligned}$$

Taking the trace in x-space we find

$$\begin{aligned}G(t_2,t_1) &= \int dx\langle x,t_2|x,t_1\rangle\\ &= \sum_n \mathrm{e}^{-\mathrm{i}E_n(t_2-t''+t'-t_1)}\\ &\quad\times\int dx'\,dx''\phi_n^*(x'')\langle x'',t''|x',t'\rangle^J\phi_n(x')\\ &= \sum_n \mathrm{e}^{-\mathrm{i}E_n(t_2-t_1)}\,\mathrm{e}^{+\mathrm{i}E_n(t''-t')}\\ &\quad\times\int dx\,dx''\phi_n^*(x'')\langle x'',t''|x',t'\rangle^J\phi_n(x') .\end{aligned}$$

At this point we perform a Wick rotation $\tau_{1,2} = \mathrm{i}t_{1,2}$; we do not rotate t' and t''! This brings us to

$$\begin{aligned}G_E(\tau_2,\tau_1) &= \sum_n \mathrm{e}^{-E_n(\tau_2-\tau_1)}\,\mathrm{e}^{\mathrm{i}E_n(t''-t')}\\ &\quad\times\int dx'\,dx''\phi_n^*(x'')\langle x'',t''|x',t'\rangle^J\phi_n(x')\end{aligned}$$

$$\underset{\substack{\tau_2\to+\infty\\ \tau_1\to-\infty}}{\longrightarrow} \mathrm{e}^{-E_0(\tau_2-\tau_1)}\,\mathrm{e}^{\mathrm{i}E_0(t''-t')}$$

$$\times \underbrace{\int dx'\,dx''\phi_0^*(x'')\langle x'',t''|x',t'\rangle^J\phi_0(x')}_{\equiv\langle\phi_0,t''|\phi_0,t'\rangle^J} .$$

The above formulae for G and its "Euclidean" counterpart G_E are valid for $t_2 > t''$ and $t_1 < t'$, where $t' \equiv t_i$ and $t'' = t_f$ are fixed. Under these conditions, G_E depends only on the difference $\tau_2 - \tau_1$, despite the fact that the system is not translational invariant. Here, then, are the formulae we were looking for:

$$\lim_{\substack{\tau_2\to+\infty\\ \tau_1\to-\infty}} \frac{G_E(\tau_2,\tau_1)}{\mathrm{e}^{-E_0(\tau_2-\tau_1)}} = \mathrm{e}^{\mathrm{i}E_0(t_2-t_1)}\,\langle\phi_0,t_f|\phi_0,t_i\rangle^J$$

or, in terms of the "vacuum persistence amplitude" $\langle 0_+|0_-\rangle^J$,

$$\begin{aligned} P_{00} :&= |\langle 0_+|0_-\rangle^J|^2 \equiv |\langle\phi_0,t_f|\phi_0,t_i\rangle^J|^2 \\ &= \lim_{\substack{\tau_2\to+\infty\\ \tau_1\to-\infty}} \left|\frac{G_E(\tau_2,\tau_1)}{\mathrm{e}^{-E_0(\tau_2-\tau_1)}}\right|^2 . \end{aligned} \tag{33.29}$$

In terms of the effective action Γ and with the external source $J \equiv \omega(t)$, we obtain

$$|G_E|^2 = \exp\left\{-2\mathrm{Im}\Gamma^E_{\tau_2,\tau_1}[\omega]\right\} \tag{33.30}$$

$$P_{00} = \lim_{\substack{\tau_2\to+\infty\\ \tau_1\to-\infty}} \exp\Big[-2\Big\{\mathrm{Im}\Gamma^E_{\tau_2,\tau_1}[\omega] - E_0\big(\tau_2-\tau_1\big)\Big\}\Big] . \tag{33.31}$$

It is (33.29) together with (33.28) that enable us to compute the probability for the vacuum to remain the vacuum under the influence of the parametric perturbation $\omega(t)$. But first we have to calculate ($E_0 = \omega/2$)

$$\begin{aligned} |G_E\big(\tau_2,\tau_1\big)|^2 &\to \left[\frac{1}{2}\sinh\delta\,\mathrm{e}^{2\omega(\tau_2-\tau_1)}\left|\frac{(1+\cosh\delta)+\mathrm{i}\gamma_2\sinh\delta}{(1-\cosh\delta)-\mathrm{i}\gamma_2\sinh\delta}\right|\right]^{-1/2} \\ &= \sqrt{2}\,\mathrm{e}^{-\omega(\tau_2-\tau_1)}\left[\sinh\delta\left|\frac{(1+\cosh\delta)+\mathrm{i}\gamma_2\sinh\delta}{(1-\cosh\delta)-\mathrm{i}\gamma_2\sinh\delta}\right|\right]^{-1/2} \end{aligned}$$

and thus

$$P_{00} = \sqrt{2}\left[\sinh\delta\left|\frac{(1+\cosh\delta)+\mathrm{i}\gamma_2\sinh\delta}{(1-\cosh\delta)-\mathrm{i}\gamma_2\sinh\delta}\right|\right]^{-1/2} . \tag{33.32}$$

After a few more elementary steps, the outcome for the right-hand side of (33.32) is

$$P_{00} = \left(\frac{2}{1+\cosh\delta}\right)^{1/2} , \tag{33.33}$$

which is the result we wanted to derive. It expresses the quantum mechanical probability P_{00} in terms of the parameter δ, which can be found from the asymptotic form (33.13) of the classical equation (33.11).

Now we turn to the main topic of this chapter: computation of Berry's phase contribution to the vacuum decay amplitude for the generalized parametric harmonic oscillator.

Let us briefly review some of the elements necessary to set up our problem stated in the Hamiltonian of the generalized harmonic oscillator:

$$H(t) = \tfrac{1}{2}\left[X(t)x^2 + Y(t)(xp + px) + Z(t)p^2\right], \tag{33.34}$$

with slowly varying parameters (X, Y, Z) (t). The system characterized by the time-dependent Hamiltonian (33.34) allows for an Hermitean invariant $I(t)$, which is given by (33.34):

$$I(t) = \frac{1}{2}\left\{\frac{x^2}{\varrho^2} + \left[\varrho\left(p + \frac{Y}{Z}x\right) - \frac{x}{Z}\dot{\varrho}\right]^2\right\} \tag{33.35}$$

with

$$\frac{dI(t)}{dt} \equiv \mathrm{i}[H, I] + \frac{\partial I(t)}{\partial t} = 0$$

and $\varrho(t)$ a c-number solution of the auxiliary equation

$$\frac{1}{\varrho}\frac{d}{dt}\left(\frac{\dot{\varrho}}{Z}\right) - \left[\frac{d}{dt}\left(\frac{Y}{Z}\right) - \frac{XZ - Y^2}{Z} + \frac{Z}{\varrho^4}\right] = 0\,. \tag{33.36}$$

The instantaneous eigenstates of $I(t)$ are defined by

$$I(t)|\lambda_n, t\rangle = \lambda_n|\lambda_n, t\rangle\ , \tag{33.37}$$

where the eigenvalues λ_n are time independent, $(\partial/\partial t)\lambda_n = 0$. The system (33.34) develops according to the Schrödinger equation ($\hbar = 1$)

$$\mathrm{i}\frac{\partial}{\partial t}|\psi(t)\rangle = H(t)|\psi(t)\rangle$$

whose solution can be expressed in terms of the eigenstates $|\lambda_n, t\rangle$:

$$|\psi(t)\rangle = \sum_n C_n\, \mathrm{e}^{\mathrm{i}\alpha_n(t)}|\lambda_n, t\rangle\ . \tag{33.38}$$

The constant coefficients C_n have to be determined from the initial conditions. According to the general theory of Lewis and Riesenfeld, the phase angles $\alpha_n(t)$ can be obtained from the equation

$$\alpha_n(t) = \int_0^t dt'\langle\lambda_n, t'|\mathrm{i}\frac{\partial}{\partial t'} - H(t')|\lambda_n, t'\rangle\ . \tag{33.39}$$

In our particular case this can be evaluated to yield

$$\alpha_n(t) = -\left(n + \frac{1}{2}\right) \int_0^t dt' \frac{Z(t')}{\varrho^2(t')} . \tag{33.40}$$

The eigenvalue spectrum of I is given by $\lambda_n = n + 1/2$, $n = 0, 1, 2, \ldots$.

We are now going to introduce the effective action $\Gamma[X(t), Y(t), Z(t)]$ in the spirit of field theory. One must recognize that it is the effective action that properly addresses questions like vacuum persistence amplitude of a quantum system, a topic we are now going to concentrate on. In a certain sense we are dealing with a toy model simulating particle creation in relativistic field theory by a prescribed external field (QED), or cosmological particle creation by a time-dependent metric.

The effective action is defined by the path integral representation

$$e^{i\Gamma[X,Y,Z]} = \int \mathcal{D}p(t)\mathcal{D}x(t) \exp\left\{i \int_{t_1}^{t_2} dt[p\dot{x} - H(p, x; X, Y, Z)]\right\} , \tag{33.41}$$

where the integration is to be performed over all paths satisfying $x(T) = x(0)$ and $T \to \infty$ at the end, meaning an adiabatically closed cycle. The effective action Γ itself (or, for finite T, Γ_T) can be computed with the aid of the Feynman propagator $K(x_2, t_2|x_1, t_1)$ in presence of the "external field" $(X, Y, Z)(t)$ by a similar path integral with terminal conditions $x(t_1) = x_1$, $x(t_2) = x_2$. We are specifically interested in the "loop contribution", i.e., the trace of the diagonal part of K in x-space:

$$G(T) \equiv e^{i\Gamma_T[X,Y,Z]} = \int_{-\infty}^{+\infty} dx\, K(x, T|x, 0) . \tag{33.42}$$

At this point we recall that the imaginary part of Γ_∞ is related to the vacuum persistence amplitude. Instead of explicitly computing the path integral, we now make substantial use of the Lewis–Riesenfeld theory to determine K. We claim that the equation for the kernel,

$$\left(i\frac{\partial}{\partial t} - H_{x_2}(t)\right) K\big(x_2, t|x_1, 0\big) = 0 , \quad t \neq 0$$

with the boundary condition $K(x_2, 0|x_1, 0) = \delta(x_2 - x_1)$ is solved by

$$K\big(x_2, t|x_1, 0\big) = \sum_n e^{i\alpha_n(t)} \langle x_2|\lambda_n, t\rangle\langle\lambda_n, 0|x_1\rangle . \tag{33.43}$$

That this statement is true can be recognized from the fact that $K_{(x_1,0)}(x_2, t)$ is a wave function of the type (33.38) for a special choice of the C_n's. Let us quickly check our claim. Equation (33.43) obviously reduces to

$$K\big(x_2, 0|x_1, 0\big) = \sum \langle x_2|\lambda_n, 0\rangle\langle\lambda_n, 0|x_1\rangle = \langle x_2|x_1\rangle = \delta\big(x_2 - x_1\big)$$

since (33.40) implies $\alpha_n(0) = 0$ and the eigenstates of $I(t)$ form a complete set for all t. Furthermore,

$$\begin{aligned}\left(\mathrm{i}\frac{\partial}{\partial t} - H_{x_2}(t)\right) K\big(x_2, t|x_1, 0\big) \\ = \sum_n \langle x_2 \Big| \left(\mathrm{i}\frac{\partial}{\partial t} - H(t)\right) \mathrm{e}^{\mathrm{i}\alpha_n(t)} \Big| \lambda_n, t\rangle \langle \lambda_n, 0|x_1\rangle = 0\end{aligned}$$

following from the result by Lewis and Riesenfeld:

$$\left(\mathrm{i}\frac{\partial}{\partial t} - H(t)\right) \mathrm{e}^{\mathrm{i}\alpha_n(t)} |\lambda_n, t\rangle = 0 \,.$$

Thus we obtain

$$\begin{aligned}G(T) &\equiv \int_{-\infty}^{+\infty} dx\, K(x, T|x, 0) \\ &= \int_{-\infty}^{+\infty} dx \sum_n \mathrm{e}^{\mathrm{i}\alpha_n(t)} \langle x|\lambda_n, T\rangle \langle \lambda_n, 0|x\rangle \\ &= \sum_n \mathrm{e}^{\mathrm{i}\alpha_n(t)} \int_{-\infty}^{+\infty} dx \langle \lambda_n, 0|x\rangle \langle x|\lambda_n, T\rangle \\ &= \sum_n \mathrm{e}^{\mathrm{i}\alpha_n(t)} \langle \lambda_n, 0|\lambda_n, T\rangle = \mathrm{e}^{\mathrm{i}\Gamma_T} \,. \end{aligned} \tag{33.44}$$

Next we turn to the adiabatic limit of our so far exact treatment. Let us assume that the external parameters (X, Y, Z) perform an adiabatic excursion during the time T in the parameter space so that $(X, Y, Z)(0) = (X, Y, Z)(T)$. In the adiabatic limit, the $\dot{\varrho}$-term in the auxiliary equation (33.36) may be ignored; then we obtain

$$\frac{Z}{\varrho^2} = \omega_D \left[1 - \frac{Z}{\omega_D^2}\frac{d}{dt}\left(\frac{Y}{Z}\right)\right]^{1/2} \,.$$

The frequency ω_D can be obtained by rewriting the Hamiltonian (33.34) in terms of action-angle variables. The result is a linear relation $H = \omega_D J$, with

$$\omega_D = \frac{\partial H}{\partial J} = \sqrt{XZ - Y^2} \,, \quad XZ > Y^2 \,.$$

Furthermore, expanding with respect to

$$\frac{Z}{\omega_D^2}\frac{d}{dt}\left(\frac{Y}{Z}\right) \ll 1 \,,$$

we obtain

$$\frac{Z}{\varrho^2} = \omega_D \left[1 - \frac{Z}{2\omega_D^2}\frac{d}{dt}\left(\frac{Y}{Z}\right)\right] = \omega_D - \frac{Z}{2\omega_D}\frac{d}{dt}\left(\frac{Y}{Z}\right) \,. \tag{33.45}$$

When this adiabatic expression is substituted into (33.40), the Lewis–Riesenfeld phase goes over to the Berry phase:

$$\alpha_n(T) = -\left(n + \frac{1}{2}\right)\int_0^T dt\, f(t) \tag{33.46}$$

where

$$f(t) \equiv \omega_D(t) - \frac{Z}{2\omega_D}\frac{d}{dt}\left(\frac{Y}{Z}\right).$$

Because the external parameters return to their starting point at $t = T$, so does the adiabatic solution (33.45) as well as the operator $I(t)$ and its eigenstates. Hence it holds that $\langle\lambda_n, 0|\lambda_n, T\rangle = \langle\lambda_n, 0|\lambda_n, 0\rangle = 1$. In this way we obtain for the adiabatic approximation of the effective action

$$\begin{aligned}\exp\{\mathrm{i}\Gamma_T[X, Y, Z]\} &= \sum_{n=0}^{\infty} \mathrm{e}^{-\mathrm{i}(n+1/2)\phi(T)} \\ &= 2^{-1/2}[\cos\phi(T) - 1]^{-1/2}\,,\end{aligned} \tag{33.47}$$

where the total phase collected during one cycle of adiabatic excursion is given by

$$\begin{aligned}\phi(T) &= \int_0^T dt\,\omega_D(t) - \int_0^T dt \frac{Z}{2\omega_D}\frac{d}{dt}\left(\frac{Y}{Z}\right) \\ &= \int_0^T dt\,\omega_D(t) - \oint_C d\boldsymbol{R}\cdot\frac{Z}{2\omega_D}\nabla_{\boldsymbol{R}}\left(\frac{Y}{Z}\right), \quad \boldsymbol{R} = (X, Y, Z)\,,\end{aligned} \tag{33.48}$$

where the first term is the dynamical phase and the second is the geometrical Berry phase, i.e., only dependent on the path in parameter space. By the way, we can easily rediscover the standard result for the time-independent harmonic oscillator by recognizing that the phase function is then given by $\phi(T) = \omega T$. As can be seen from (33.47), the effective action is augmented by an "anomalous" geometric phase contribution,

$$\Gamma[C] = -\left(n + \frac{1}{2}\right)\oint_C d\boldsymbol{R}\cdot\left[-\frac{Z}{2\omega_D}\nabla_{\boldsymbol{R}}\left(\frac{Y}{Z}\right)\right], \tag{33.49}$$

not unlike the appearance of anomalies in gauge field theories.

Now let us assume that the oscillator is in its ground state ("vacuum") in the remote past, $t \to -\infty$. What, then, is the probability $|\langle 0_+|0_-\rangle^{\boldsymbol{R}}|^2$ for the oscillator to be still in the ground state in the distant future, $t \to +\infty$? Quite generally, given the traced Feynman kernel

$$G(t_2, t_1) = \int_{-\infty}^{+\infty} dx\, K(x, t_2|x, t_1)\,,$$

the vacuum persistence amplitude can be calculated as given in (33.39):

$$P_{00} \equiv |\langle 0_+|0_-\rangle|^2 = \lim_{\substack{\tau_2\to+\infty \\ \tau_1\to-\infty}} \left|\frac{G_E(\tau_2,\tau_1)}{\mathrm{e}^{-E_0(\tau_2-\tau_1)}}\right|^2 , \tag{33.50}$$

where E_0 is the ground state energy of the unperturbed system. Thus, initially and finally, the oscillator is a simple harmonic oscillator in its ground state $E_0 = \omega/2$. (The above formula still holds if we put $\tau_1 = 0$, as was done previously.) P_{00} is related to the imaginary part of the effective action as stated in (33.31)

$$P_{00} = \lim_{\substack{\tau_2\to+\infty \\ \tau_1\to-\infty}} \exp\left[-2\left\{\mathrm{Im}\Gamma_{\tau_2,\tau_1}[X,Y,Z] - E_0\left(\tau_2-\tau_1\right)\right\}\right] .$$

Let us consider $\langle 0_+|0_-\rangle^{\boldsymbol{R}}$ of the parametrically excited oscillator for a periodic path (period $T \to \infty$) in the space of the external parameters $\boldsymbol{R} = (X,Y,Z)(t)$. If the time evolution is truly adiabatic, no excitation ("particle creation") will occur, and $P_{00} = 1$. Knowing $\Gamma_T[\boldsymbol{R}]$, we can compute the deviation from $P_{00} = 1$ for very slow, but nonadiabatic changes of the parameters. The result is

$$P_{00} = \lim_{T\to\infty} \mathrm{e}^{2E_0T}\left|\sum_{n=0}^{\infty} \exp\left[-\left(n+\frac{1}{2}\right)\int_0^T d\tau\, f(-\mathrm{i}\tau)\right]\right|^2 . \tag{33.51}$$

One can justify that the integral in the exponential of (33.51) has a positive real part, so that only the $n = 0$ term contributes for $T \to \infty$. Here, then, is our final result for the probability of the ground state to remain in the ground state:

$$P_{00} = \exp\left[-\mathrm{Re}\int_0^{\infty} d\tau\left\{\omega_D(-\mathrm{i}\tau) - 2E_0\right\} + \mathrm{Re}\int_0^{\infty} d\tau \left\{\frac{Z}{2\omega_D}\frac{d}{dt}\left(\frac{Y}{Z}\right)\right\}(t=-\mathrm{i}\tau)\right] , \tag{33.52}$$

which exhibits explicitly the contributions arising from the dynamical and geometrical (Berry) amplitude. The transitions occur by almost adiabatic motion and are contained in a dynamical and geometrical (Berry) part, where the latter is the analytic continuation (in time) of the Berry phase.

34. Topological Phases in Planar Electrodynamics

section is meant to be an extension of Chap. 31 on the quantal Berry phases. In particular, we are interested in studying the electromagnetic interaction of particles with a nonzero magnetic moment in $D = 2 + 1$ dimensions and of translational invariant configurations of $(D = 3 + 1)$-dimensional charged strings with a nonzero magnetic moment per unit length. The whole discussion is based on our article in Physical Review **D44**, 1132 (1991).

We begin by recalling that the Lagrangian density of electrodynamics is given by

$$\mathcal{L} = \frac{1}{2}(\boldsymbol{E}^2 - \boldsymbol{B}^2) - \varrho\phi + \boldsymbol{j} \cdot \boldsymbol{A} \,, \tag{34.1}$$

where the particle-field interaction is contained in

$$L' = \int d^2\boldsymbol{x}\mathcal{L}' = \int d^2x\{-\varrho\phi + \boldsymbol{j} \cdot \boldsymbol{A}\} \,. \tag{34.2}$$

As pointed out above, we consider $D = 3 + 1$ with translation invariance along the x^3-axis and $D = 2 + 1$ in parallel. In the former case all quantities $(L, \varrho, \dots)$ are understood to be "per unit length". Hence we write D-vectors as $x^\mu = (x^0, x^i)$ where i $= 1, 2$ for $D = 2 + 1$ (particles) and i $= 1, 2, 3$ for $D = 3 + 1$ (strings oriented parallel to the x^3-axis). Let us consider a classical model for the magnetic moment. Then in its rest frame the current density $j_0^\mu = (\varrho_0, j_0^i)$ of a particle located at $\boldsymbol{x} = \boldsymbol{x}_p$ has the following form:

$$j_0^\mu(\boldsymbol{x}) = \begin{pmatrix} \varrho_0 \\ j_0^i \end{pmatrix} = \begin{pmatrix} e\delta^2(\boldsymbol{x} - \boldsymbol{x}_p) \\ \mu\varepsilon^{ij}\partial_j\delta^2(\boldsymbol{x} - \boldsymbol{x}_p) \end{pmatrix} \;; \quad i, j = 1, 2 \,. \tag{34.3}$$

The 2 component vectors $\boldsymbol{x}, \boldsymbol{x}_p$ lie in the x^1-x^2 plane.

Let us quickly check that the point source (34.3) gives rise to the magnetic moment μ:

$$\begin{aligned}
&\frac{1}{2}\int d^2\boldsymbol{x}\, \boldsymbol{x} \times \boldsymbol{j}(\boldsymbol{x}) = \frac{1}{2}\int d^2\boldsymbol{x}\; \varepsilon^{ik}x_i j_k(\boldsymbol{x}) \\
&= \frac{1}{2}\underbrace{\varepsilon^{ik}\varepsilon_{kl}}_{=-\delta^i_l}\,\mu\,\underbrace{\int d^2\boldsymbol{x}\; x_i\partial_l\delta^2(\boldsymbol{x} - \boldsymbol{x}_p)}_{=-\delta_{il}} = \frac{1}{2}\mu\;\delta_{ll} = \mu \,.
\end{aligned}$$

Now we assume that the particle or the string moves with the velocity $\boldsymbol{v}_p = \dot{\boldsymbol{x}}_p$ relative to the laboratory frame. (It is understood that $j_0^3 = 0$ if $D = 3 + 1$.) The resulting current distribution is obtained by boosting j_0^μ from the particle's rest frame. Since eventually we are mainly interested in the adiabatic limit, it is sufficient to keep only the terms linear in the velocity ("Galileo boost"). Hence one has in the laboratory frame

$$\begin{aligned} \varrho &= \varrho_0 + \boldsymbol{v}_p \cdot \boldsymbol{j}_0 + O\left(\boldsymbol{v}_p^2\right) , \\ \boldsymbol{j} &= \boldsymbol{j}_0 + \boldsymbol{v}_p \varrho_0 + O\left(\boldsymbol{v}_p^2\right) , \end{aligned} \tag{34.4}$$

or

$$\begin{aligned} \varrho(\boldsymbol{x}) &= e\delta^2(\boldsymbol{x} - \boldsymbol{x}_p) + \mu\varepsilon^{ij} v_{pi} \partial_j \delta^2(\boldsymbol{x} - \boldsymbol{x}_p) + O\left(\boldsymbol{v}_p^2\right) , \\ j^i(\boldsymbol{x}) &+ \mu\varepsilon^{ij} \partial_j \delta^2(\boldsymbol{x} - \boldsymbol{x}_p) + e v_p^i \delta^2(\boldsymbol{x} - \boldsymbol{x}_p) + O\left(\boldsymbol{v}_p^2\right) . \end{aligned} \tag{34.5}$$

Therefore the interaction with an external field $A^\mu = (\phi, \boldsymbol{A})$, $A_3 = 0$, is given by

$$\begin{aligned} L' &= \int d^2x \{-\varrho\phi + \boldsymbol{j} \cdot \boldsymbol{A}\} \\ &= \int d^2x \left[-\varrho_0\phi - \boldsymbol{v}_p \cdot \boldsymbol{j}_0\phi + \boldsymbol{j}_0 \cdot \boldsymbol{A} + \varrho_0 \boldsymbol{v}_p \cdot \boldsymbol{A}\right] \\ &= \int d^2x \Big[-e\phi(\boldsymbol{x})\delta^2(\boldsymbol{x} - \boldsymbol{x}_p) - \underbrace{\mu v_{p_i} \varepsilon^{ij} \partial_j \delta^2(\boldsymbol{x} - \boldsymbol{x}_p) \phi(\boldsymbol{x})}_{= +\mu v_{pi} \varepsilon^{ij} \partial_j \phi(\boldsymbol{x}) \delta^2\left(\boldsymbol{x} - \boldsymbol{x}_p\right) = \mu \boldsymbol{v}_p \times \nabla\phi(\boldsymbol{x}) \delta^2\left(\boldsymbol{x} - \boldsymbol{x}_p\right)} \\ &\quad + \underbrace{\mu A_i \varepsilon^{ij} \partial_j \delta^2(\boldsymbol{x} - \boldsymbol{x}_p)}_{= \mu\varepsilon^{ji} \partial_j A_i(\boldsymbol{x}) \delta^2\left(\boldsymbol{x} - \boldsymbol{x}_p\right)} + e\delta^2(\boldsymbol{x} - \boldsymbol{x}_p) \boldsymbol{v}_p \cdot \boldsymbol{A}(\boldsymbol{x}) \Big] \end{aligned}$$

or

$$L' = -e\phi(\boldsymbol{x}_p) + e\boldsymbol{v}_p \cdot \boldsymbol{A}(\boldsymbol{x}_p) + \mu B(\boldsymbol{x}_p) + \mu \boldsymbol{E}(\boldsymbol{x}_p) \times \boldsymbol{v}_p \tag{34.6}$$

with the electric field $\boldsymbol{E} = -\nabla\phi$ and the magnetic field $B = \nabla \times \boldsymbol{A} = \varepsilon^{ij} \partial_i A_j$. Assuming that the field A^μ is generated by another particle, either of the four terms on the r.h.s. of (34.6) can give rise to a topological phase. Hence, let us first calculate the ϕ- and $\boldsymbol{A}$-fields generated by the other particle. To do so, we distinguish particles (or strings) with $e \neq 0$ and $\mu = 0$ and refer to them as "charges", and particles with $e = 0$ and $\mu \neq 0$, which we call "magnetic moments" for short. Then we perform the following four experiments:

(1) A magnetic moment is transported adiabatically around a charge which is at rest in the origin. The effect on the wave function of the magnetic moment is considered.

(2) As in (1), but now the effect on the wave function of the charge at rest is considered.

(3) A charge moves adiabatically around a magnetic moment which is at rest in the origin. The effect on the wave function of the magnetic moment is considered.

(4) As in (3), but the effect on the wave function of the charge is considered.

By "considering the effect on the wave function" we have in mind the following gedanken experiment due to Berry. In the first experiment, (1), for instance, we assume that (by means of some additional interaction) the wave function of the magnetic moment is confined to a small box centered around the position $\boldsymbol{x} = \boldsymbol{x}_p(t)$ of the particle. Then, invoking the general philosophy of Berry phases, the contents of the box are considered the proper "system" or the "rapid degrees of freedom", whereas the field generated by the charge in the origin is considered a set of external parameters or "slow degrees of freedom". The Berry phase obtains as a response of the wave function inside the box to an adiabatic excursion in the space of external parameters. In the case at hand, this is tantamount to a motion of the box around the second particle. Similarly, in all the gedanken experiments listed above, one of the two particles, namely the one whose wave function is considered, defines the "system" living within the "box", whereas the other serves as a source of time-dependent external fields. The respective topological phases are easily computed.

Experiment (1). This experiment coincides with the standard Aharonov–Casher (AC) setup in which a neutron moves around a charge. The interaction term of interest is the last term in (34.6): $L_1' = \mu \boldsymbol{E} \times \boldsymbol{v}$. To find the electric field by the charge at the origin, we recall from elementary electrostatics (in 2 dimensions):

$$\nabla \cdot \boldsymbol{E} = -\nabla^2 \phi = \varrho \, .$$

In terms of the Green's function $G(\boldsymbol{x}, \boldsymbol{x}')$ of ∇^2 the solution reads

$$\begin{aligned}\phi(\boldsymbol{x}) &= -\int d^2x' \; G(\boldsymbol{x}, \boldsymbol{x}') \; \varrho(\boldsymbol{x}') \\ &= -\frac{1}{2\pi} \int d^2x' \; \ln|\boldsymbol{x} - \boldsymbol{x}'| \; \varrho(\boldsymbol{x}') \, .\end{aligned}$$

Substituting $\varrho(\boldsymbol{x}') = e\delta^2(\boldsymbol{x}' - \boldsymbol{x}_p)$ we obtain

$$\phi(\boldsymbol{x}) = -\frac{e}{2\pi} \ln|\boldsymbol{x} - \boldsymbol{x}_p|$$

so that

$$-E_i = \partial_i \phi(\boldsymbol{x}) = -\frac{e}{2\pi} \frac{1}{2} \underbrace{\partial_i \ln\big(\boldsymbol{x} - \boldsymbol{x}_p\big)^2}_{= \frac{2(\boldsymbol{x} - \boldsymbol{x}_p)_i}{2(\boldsymbol{x} - \boldsymbol{x}_p)^2}}$$

or

$$\nabla \phi(\boldsymbol{x}) = -\boldsymbol{E}(\boldsymbol{x}) = -\frac{e}{2\pi} \frac{(\boldsymbol{x} - \boldsymbol{x}_p)}{|\boldsymbol{x} - \boldsymbol{x}_p|^2} \, .$$

Hence the electric field due to the charge at the origin ($x_p = 0$) is given by

$$E(x) = \frac{e}{2\pi}\frac{|x|}{|x|^2} . \tag{34.7}$$

Consequently,

$$L_1' = \frac{e}{2\pi}\mu\frac{x \times v}{|x|^2} = \frac{e\mu}{2\pi}\frac{x^i\varepsilon_{ij}\dot{x}^j}{|x|^2} . \tag{34.8}$$

Then, during one round trip along the path C, we accumulate the following Berry phase:

$$\theta_1 \equiv \int_0^T dt\, L_1' = e\mu \oint_C \frac{x \times dx}{2\pi|x|^2} = e\mu . \tag{34.9}$$

Here the value of the integral in (34.9) is one, as can be seen as follows:

$$I = \oint_C \frac{x \times dx}{2\pi|x|^2} \equiv \oint_C F_i dx_i$$

where

$$F_i = \frac{1}{2\pi|x|^2}x_k\varepsilon_{ki} = \frac{1}{2\pi}\varepsilon_{ki}\frac{1}{2}\partial_k \ln|x|^2$$

and therefore

$$\begin{aligned}\nabla \times F = \varepsilon_{li}\partial_l F_i &= \underbrace{\varepsilon_{ki}\varepsilon_{li}}_{=2\delta_{kl}}\frac{1}{4\pi}\partial_k\partial_l \ln|x|^2 \\ &= \frac{1}{4\pi}\nabla^2 \ln|x|^2 \sim \delta^2(x) .\end{aligned}$$

We conclude that the value of I does not depend on the local details but solely on the homotopy class of C. For one revolution around the origin we may therefore choose a circle of radius $|x| = R$:

$$I = \frac{1}{\pi R^2}\underbrace{\frac{1}{2}\oint_{C_R} x \times dx}_{=\pi R^2} = 1 .$$

Experiment (2). Here, the relevant part of (34.6) is $L_2' = -e\phi(0)$, where ϕ is generated by the motion of the magnetic moment:

$$\phi(\boldsymbol{x}) = -\frac{1}{2\pi}\int d^2\boldsymbol{x}'\ \ln|\boldsymbol{x}-\boldsymbol{x}'|\ \underbrace{\varrho(x')}_{=\underbrace{\varrho_0(\boldsymbol{x}')}_{=0}+\underbrace{\boldsymbol{v}_p\cdot\boldsymbol{j}_0(\boldsymbol{x}')}_{=\mu v_{pi}\varepsilon^{ij}\partial_j'\delta^2(\boldsymbol{x}'-\boldsymbol{x}_p)}}$$

$$= -\frac{1}{2\pi}\mu v_{pi}\varepsilon^{ij}\underbrace{\int d^2\boldsymbol{x}'\ \ln|\boldsymbol{x}-\boldsymbol{x}'|\ \partial_j'\delta^2(\boldsymbol{x}'-\boldsymbol{x}_p)}_{=-\int d^2\boldsymbol{x}'\delta^2(\boldsymbol{x}'-\boldsymbol{x}_p)\underbrace{\frac{1}{2}\partial_j'\ln(\boldsymbol{x}-\boldsymbol{x}')^2}_{=-\frac{2(\boldsymbol{x}-\boldsymbol{x}')_j}{2(\boldsymbol{x}-\boldsymbol{x}')^2}}}$$

$$= -\frac{\mu}{2\pi}\frac{v_{pi}\varepsilon^{ij}(x-x_p)_j}{|\boldsymbol{x}-\boldsymbol{x}_p|^2} = \frac{\mu}{2\pi}\frac{(\boldsymbol{x}-\boldsymbol{x}_p)\times\boldsymbol{v}_p}{|\boldsymbol{x}-\boldsymbol{x}_p|^2}\ .$$

We need

$$\phi(0) = -\frac{\mu}{2\pi}\frac{\boldsymbol{x}_p\times\boldsymbol{v}_p}{|\boldsymbol{x}_p|^2}$$

and so we end up again for the phase for one revolution

$$\theta_2 \equiv \int_0^T dt\ L_2' = \frac{e\mu}{2\pi}\int_0^T dt\frac{\dot{\boldsymbol{x}}\times\dot{\boldsymbol{x}}}{|\boldsymbol{x}|^2} = e\mu\ . \tag{34.10}$$

Experiment (3). The relevant interaction term is now $L_3' = \mu B(0)$, where $B(0)$ is the magnetic field generated by the orbital motion of the charge acting upon the magnetic moment at the origin.

Starting from $\nabla^2\boldsymbol{A} = -\boldsymbol{j}$, we obtain first the vector potential (recall $\boldsymbol{j}_0 = 0$ for $\mu = 0$)

$$\boldsymbol{A}(\boldsymbol{x}) = -\frac{1}{2\pi}\int d^2\boldsymbol{x}'\ \ln|\boldsymbol{x}-\boldsymbol{x}'|\ \underbrace{\boldsymbol{j}(\boldsymbol{x}')}_{=\boldsymbol{j}_0+\boldsymbol{v}_p\varrho_0 = e\delta^2(\boldsymbol{x}'-\boldsymbol{x}_p)\boldsymbol{v}_p}$$

$$= -\frac{e\boldsymbol{v}_p}{2\pi}\ln|\boldsymbol{x}-\boldsymbol{x}_p|\ .$$

From here we obtain

$$B(\boldsymbol{x}) = \varepsilon^{ij}\partial_i A_j = -\frac{e}{2\pi}v_{pj}\underbrace{\varepsilon^{ij}}_{=-\varepsilon^{ji}}\frac{1}{2}\partial_i\ln\left(\boldsymbol{x}-\boldsymbol{x}_p\right)^2$$

so that

$$B(x) = \frac{e}{2\pi} \frac{\dot{x}_p \times (x - x_p)}{|x - x_p|^2}$$

and

$$B(0) = \frac{e}{2\pi} \frac{x_p \times \dot{x}_p}{|x_p|^2} .$$

Hence the Berry phase is

$$\theta_3 \equiv \int_0^T dt\, L_3' = e\mu \int_0^T dt\, \frac{x_p \times \dot{x}_p}{2\pi |x_p|^2} = e\mu . \tag{34.11}$$

Experiment (4). This is the situation of the Aharonov–Bohm effect discussed in Chap. 31. The interaction is $L_4' = e v_p \cdot A(x_p)$, where A is the vector potential generated by the magnetic moment at the origin. To obtain this potential we compute

$$A^i(x) = -\frac{1}{2\pi} \int d^2x' \ln|x - x'| \underbrace{j_0^i(x')}_{= \mu \varepsilon^{ij} \partial_j' \delta^2 (x' - x_p)}$$

$$= \frac{1}{2\pi} \mu \varepsilon^{ij} \int d^2x'\, \delta^2 (x' - x_p)\, \partial_j' \frac{1}{2} \ln (x - x')^2$$

$$= -\frac{\mu}{2\pi} \frac{\varepsilon^{ij} (x - x_p)_j}{|x - x_p|^2} .$$

For $x_p = 0$, we find

$$A_i(\mathrm{x}) = -\frac{\mu}{2\pi} \frac{\varepsilon_{ij} x^j}{|x|^2} \tag{34.12}$$

so that again the phase for one circuit is

$$\theta_4 \equiv \int_0^T dt\, L_4' = e\mu \int_0^T dt\, \frac{x \times \dot{x}}{2\pi |x|^2} = e\mu , \tag{34.13}$$

where in the $(3+1)$-dimensional interpretation, μ coincides with the flux throught the solenoid.

As we see, all four phases coincide numerically. In all the above experiments, we encounter an effective interaction $L' = \mathcal{A} \cdot v$ with a vortex potential $\mathcal{A}(\mathcal{A}_i \sim \varepsilon_{ij} x_j / |x|^2)$, where $\mathcal{A}$ does not necessarily stand for the magnetic vector potential.

Finally, let us consider a set of interacting nonrelativistic particles with charges e_p and magnetic moments μ_p, $p = 1, 2, \ldots N$. In addition to the Coulomb and Lorentz forces acting between them, there will be a Berry phase part of the $\mathcal{A} \cdot v$-type interaction which receives contributions from all four effects discussed above. The relevant part of the Lagrangian is obtained by starting from (34.6) for one particular particle, and inserting the expression for the fields generated by the other particles:

$$L' = \mu \boldsymbol{E}^{(1)} \times \boldsymbol{v} - e\phi^{(2)} + \mu B^{(3)} + e\boldsymbol{v} \cdot \boldsymbol{A}^{(4)} ,$$

where we have to insert (cf. the four experiments listed above)

$$(1) \quad \boldsymbol{E}^{(1)}(\boldsymbol{x}) = \frac{e}{2\pi} \frac{\boldsymbol{x} - \boldsymbol{x}_p}{|\boldsymbol{x} - \boldsymbol{x}_p|^2} .$$

$$(2) \quad \phi^{(2)}(\boldsymbol{x}) = \frac{\mu}{2\pi} \frac{(\boldsymbol{x} - \boldsymbol{x}_p) \times \boldsymbol{v}_p}{|\boldsymbol{x} - \boldsymbol{x}_p|^2} .$$

$$(3) \quad B^{(3)}(\boldsymbol{x}) = \frac{e}{2\pi} \frac{\boldsymbol{v}_p \times (\boldsymbol{x} - \boldsymbol{x}_p)}{|\boldsymbol{x} - \boldsymbol{x}_p|^2} .$$

$$(4) \quad A_i^{(4)}(\boldsymbol{x}) = -\frac{\mu}{2\pi} \frac{\varepsilon_{ij}(\boldsymbol{x} - \boldsymbol{x}_p)_j}{|\boldsymbol{x} - \boldsymbol{x}_p|^2} .$$

In this way we obtain for a two-particle system

$$L_{2-\text{body}} = L_V + L_S ,$$

with the $\boldsymbol{A}$-type and ϕ-type interactions contained in L_V and L_S, respectively. (V and S stands for "vector" and "scalar", respectively.) More explicitly:

$$2L_V = e_1\boldsymbol{v}_1 \cdot \underbrace{\boldsymbol{A}(\boldsymbol{x}_1)}_{\sim\mu_2} + \mu_1 \underbrace{B(\boldsymbol{x}_1)}_{\sim e_2 v_2} + e_2\boldsymbol{v}_2 \cdot \underbrace{\boldsymbol{A}(\boldsymbol{x}_2)}_{\sim\mu_1} + \mu_2 \underbrace{B(\boldsymbol{x}_2)}_{\sim e_1 v_1}$$

$$2L_S = -e_1 \underbrace{\phi(\boldsymbol{x}_1)}_{\sim\mu_2 v_2} + \mu_1 \underbrace{\boldsymbol{E}(\boldsymbol{x}_1)}_{\sim e_2} \times \boldsymbol{v}_1 - e_2 \underbrace{\phi(\boldsymbol{x}_2)}_{\sim\mu_1 v_1} + \mu_2 \underbrace{\boldsymbol{E}(\boldsymbol{x}_2)}_{\sim e_1} \times \boldsymbol{v}_2 .$$

If we now insert the expressions (1), ... , (4), we obtain

$$\begin{aligned} 2L_V &= -\frac{e_1\mu_2}{2\pi} \frac{v_1^i \varepsilon_{ij}(\boldsymbol{x}_1 - \boldsymbol{x}_2)_j}{|\boldsymbol{x}_1 - \boldsymbol{x}_2|^2} + \frac{\mu_1 e_2}{2\pi} \frac{\boldsymbol{v}_2 \times (\boldsymbol{x}_1 - \boldsymbol{x}_2)}{|\boldsymbol{x}_1 - \boldsymbol{x}_2|^2} \\ &\quad - \frac{e_2\mu_1}{2\pi} \frac{v_2^i \varepsilon_{ij}(\boldsymbol{x}_2 - \boldsymbol{x}_1)_j}{|\boldsymbol{x}_1 - \boldsymbol{x}_2|^2} + \frac{\mu_2 e_1}{2\pi} \frac{\boldsymbol{v}_1 \times (\boldsymbol{x}_2 - \boldsymbol{x}_1)}{|\boldsymbol{x}_1 - \boldsymbol{x}_2|^2} \\ &= -\frac{e_1\mu_2}{\pi} \frac{\boldsymbol{v}_1 \times (\boldsymbol{x}_1 - \boldsymbol{x}_2)}{|\boldsymbol{x}_1 - \boldsymbol{x}_2|^2} + \frac{e_2\mu_1}{\pi} \frac{\boldsymbol{v}_2 \times (\boldsymbol{x}_1 - \boldsymbol{x}_2)}{|\boldsymbol{x}_1 - \boldsymbol{x}_2|^2} . \end{aligned}$$

Hence the A-type interaction yields

$$L_V = -\frac{1}{2\pi} \left\{ e_1\mu_2\boldsymbol{v}_1 - e_2\mu_1\boldsymbol{v}_2 \right\} \times \frac{(\boldsymbol{x}_1 - \boldsymbol{x}_2)}{|\boldsymbol{x}_1 - \boldsymbol{x}_2|^2} . \tag{34.14}$$

Similarly, we compute

$$\begin{aligned} 2L_S &= -\frac{e_1\mu_2}{2\pi} \frac{(\boldsymbol{x}_1 - \boldsymbol{x}_2) \times \boldsymbol{v}_2}{|\boldsymbol{x}_1 - \boldsymbol{x}_2|^2} + \frac{\mu_1 e_2}{2\pi} \frac{(\boldsymbol{x}_1 - \boldsymbol{x}_2) \times \boldsymbol{v}_1}{|\boldsymbol{x}_1 - \boldsymbol{x}_2|^2} \\ &\quad - \frac{e_2\mu_1}{2\pi} \frac{(\boldsymbol{x}_2 - \boldsymbol{x}_1) \times \boldsymbol{v}_1}{|\boldsymbol{x}_1 - \boldsymbol{x}_2|^2} + \frac{\mu_2 e_1}{2\pi} \frac{(\boldsymbol{x}_2 - \boldsymbol{x}_1) \times \boldsymbol{v}_2}{|\boldsymbol{x}_1 - \boldsymbol{x}_2|^2} \\ &= -\frac{\mu_1 e_2}{\pi} \frac{\boldsymbol{v}_1 \times (\boldsymbol{x}_1 - \boldsymbol{x}_2)}{|\boldsymbol{x}_1 - \boldsymbol{x}_2|^2} + \frac{\mu_2 e_1}{\pi} \frac{\boldsymbol{v}_2 \times (\boldsymbol{x}_1 - \boldsymbol{x}_2)}{|\boldsymbol{x}_1 - \boldsymbol{x}_2|^2} \end{aligned}$$

so that the ϕ-type contribution gives

$$L_S = -\frac{1}{2\pi}\{\mu_1 e_2 \boldsymbol{v}_1 - \mu_2 e_1 \boldsymbol{v}_2\} \times \frac{(\boldsymbol{v}_1 - \boldsymbol{x}_2)}{|\boldsymbol{x}_1 - \boldsymbol{x}_2|^2} . \tag{34.15}$$

For equal charges and magnetic moments, we obtain ($e_1 = e_2 = e,\ \mu_1 = \mu_2 = \mu$):

$$L_V = L_S = -\frac{e\mu}{2\pi} \frac{(\boldsymbol{v}_1 - \boldsymbol{v}_2) \times (\boldsymbol{x}_1 - \boldsymbol{x}_2)}{|\boldsymbol{x}_1 - \boldsymbol{x}_2|^2} . \tag{34.16}$$

Writing for the relative position $\boldsymbol{x} = \boldsymbol{x}_1 - \boldsymbol{x}_2$, we finally arrive at

$$L_{2\text{-body}} = \frac{\theta}{\pi} \frac{\boldsymbol{x} \times \dot{\boldsymbol{x}}}{|\boldsymbol{x}|^2} , \qquad \theta = e\mu . \tag{34.17}$$

The appearance of the interaction term (34.17) is remarkable for the following reason. One of the simplest field theory models showing the phenomenon of fractional ("anyonic") statistics consists of a commuting or anticommuting matter field coupled to a U(1) Chern–Simons gauge field. These systems provide an interesting laboratory for the investigation of fractional spin and statistics which, in $2 + 1$ dimensions, are possible due to the fact that the rotation group SO(2) is Abelian and that the first homotopy group of the many-particle configuration space is a braid group. Moreover, anyons of this type also made their appearance in the theory of the fractional quantum Hall effect and of high-T_c superconductivity. To capture the essence of the "anyonization" via Chern–Simons gauge fields it is not really necessary to describe the matter sector by a (relativistic) field theory; for many considerations it is sufficient to consider nonrelativistic point particles (of mass m and charge e) whose dynamics is governed by the action

$$S = \int dt \sum_{p=1}^{N} \left(\frac{m}{2} \dot{\boldsymbol{x}}_p^2 + e\dot{\boldsymbol{x}}_p \cdot \boldsymbol{A}\big(t, \boldsymbol{x}_p(t)\big) - eA_0\big(t, \boldsymbol{x}_p(t)\big) \right) + \Gamma_{CS} \tag{34.18}$$

with the Chern–Simons term

$$\Gamma_{CS} = \frac{1}{2}\kappa \int d^3x\ \varepsilon^{\mu\nu\varrho} A_\mu(x) \partial_\nu A_\varrho(x) . \tag{34.19}$$

Since no Maxwell term is included in the gauge field action its only effect is to change the statistics of the originally bosonic particles. It can be shown that, because of the Chern–Simons term, each particle of charge e also carries a magnetic flux $\Phi = -(e/\kappa)$. We can visualize these $(2 + 1)$-dimensional flux-carrying particles as $(3 + 1)$-dimensional flux tubes ("solenoids") cut by a plane perpendicular to the magnetic field. When the world lines of two particles wind around each other, due to the Aharonov–Bohm effect, their wave function will pick up a phase factor $\exp\{ie \oint \boldsymbol{A} \cdot d\boldsymbol{x}\} = \exp\{ie\Phi\} = \exp\{-ie^2/\kappa\}$. Since the exchange of two particles corresponds to one-half of a revolution of the particle around the other (followed by a translation) the phase factor associated to it is $\exp(i\theta)$ with the "statistics angle"

$\theta = -(e^2/2\kappa)$. The origin of this phase is most easily understood if one eliminates the gauge field from (34.18) by means of its equation of motion. One obtains the following effective Lagrangian:

$$L_{\text{eff}} = \sum_P \frac{m}{2}\dot{\boldsymbol{x}}_p^2 - \frac{e^2}{2\pi\kappa}\sum_{p<q}\frac{(\boldsymbol{x}_p - \boldsymbol{x}_q) \times (\dot{\boldsymbol{x}}_p - \dot{\boldsymbol{x}}_q)}{|\boldsymbol{x}_p - \boldsymbol{x}_q|^2} . \tag{34.20}$$

More generally, whenever in some two-particle system, say, the interaction Lagrangian contains a piece which has the form of the second term on the right-hand-side of (34.20),

$$L_\theta = \frac{\theta}{\pi}\frac{\boldsymbol{x} \times \dot{\boldsymbol{x}}}{|\boldsymbol{x}|^2} , \tag{34.21}$$

where $\boldsymbol{x} \equiv \boldsymbol{x}_1 - \boldsymbol{x}_2$ is the relative separation of the two-particles, an Aharonov–Bohm-type phase will appear if one particle is moved around the other. Equation (34.21) yields, for a full circuit,

$$\int dt\, L_\theta = 2\theta \oint \frac{\boldsymbol{x} \times d\boldsymbol{x}}{2\pi|\boldsymbol{x}|^2} = 2\theta , \tag{34.22}$$

so that θ is indeed the angle related to the exchange of the two particles. In deriving (34.17) we have shown that, at low energies, particles with non-zero charge and magnetic moment undergo a kind of "self-anyonization". Without explicitly introducing a Chern–Simons term, their effective Lagrangian contains the term (34.17) which is of the form (34.21) with the "statistics angle" given by $\theta = e\mu$.

References

Aharonov, Y., Casher. A.: Phys. Rev. Lett. **53**, 319 (1984)

Arnol'd, V.I.: *Mathematical Methods of Classical Mechanics* (Springer, New York 1989)

Berry, M.V.: Proc. R. Soc. **A392**, 45 (1984); J. Phys. **A18**, 15 (1985)

Berry, M.V.: "Regular and Irregular Motion," in *American Institute of Physics Conference Proceedings*, Vol. 46, ed. by S. Jorna (1978); Proc. R. Soc. **A392**, 45 (1984)

Born, M.: *Vorlesungen über Atommechanik* (Springer, Berlin Heidelberg New York 1925, Reprint 1976)

Bouchiat, C.: J. Physique **48**, 1627 (1987)

Casati, G., Ford, J. (eds.): "Stochastic Behavior in Classical and Quantum Hamiltonian Systems," in *Springer Lecture Notes in Physics 93* (1979)

Chirikov, B.: Physics Reports **52**, 263 (1979)

Doughty, N.A.: *Lagrangian Interaction* (Addison-Wesley, 1990)

Dunne, G.V., Jackiw, R., Trugenberger, C.A.: Phys. Rev. **D41**, 661 (1990)

Felsager, B.: *Geometry, Particles and Fields* (Odense University Press, 1981)

Feynman, R.P., Hibbs, A.R.: *Quantum Mechanics and Path Integrals* (McGraw-Hill, 1965); *Statistical Mechanics* (Benjamin, 1972)

Flaschka, H., Chirikov, B. (eds.): "Nonlinear Phenomena," Physica **D33**, in *Progress in Chaotic Dynamics* (North-Holland, 1988)

J. Ford: "A Picture Book of Stochasticity", in *American Institute of Physics Conference Proceedings*, Vol. 46, ed. by S. Jorna (1978)

Gibbons, G.W.: Phys. Lett. **60A**, 385 (1977)

Goldstein, H.: *Classical Mechanics* (Addison-Wesley, 1980) Gozzi, E., Thacker, W.D.: Phys. Rev. **D35**, 2398 (1987)

Gozzi, E., Thacker, W.D.: Phys. Rev. **D35**, 2388 and 2398 (1987)

Gutzwiller, M.C.: *Chaos in Classical and Quantum Mechanics* (Springer, New York 1992)

Hannay, J.H.: J. Phys. **A18**, 221 (1985)

Hansson, T.H., Sporre, M., Leinaas, J.M.: Mod. Phys. Lett. **A6**, 45 (1991)

Holstein, B.R.: *Topics in Advanced Quantum Mechanics* (Addison-Wesley, 1992)

Holstein, B.R., Swift, A.R.: Am. J. Phys. **50**, 829 (1982); Am. J. Phys. **57**, 1079 (1989)

Khandekar, D.C., Lawande, S.V.: J. Math. Phys. **16**, 384 (1975); Phys. Repts. **137**, 115–229 (1986)

Kleinert, H.: *Path Integrals in Quantum Mechanics, Statistics and Polymer Physics* (World Scientific, Singapore 1990)

Kleppner, D., Kolenkow, R.J.: *An Introduction to Mechanics* (McGraw-Hill, 1973)

Kugler, M.: Am. J. Phys. **57**, 247 (1989)

Kuypers, F.,: *Klassische Mechanik* (Physik-Verlag, Weinheim 1982)

Leinaas, J.M., Myrheim, J.: Nuovo Cim. **37B**, 1 (1977)

Lewis, H.R., Riesenfeld, W.B.: J. Math. Phys. **10**, 1458 (1968)

Lichtenberg, A.J., Lieberman, M.A.: *Regular and Stochastic Motion* (Springer, New York 1983); 2nd Ed.: *Regular and Chaotic Dynamics* (Springer, New York 1992)

Littlejohn, R.G., Robbins, J.M.: Phys. Rev. **A36**, 2953 (1987)

Littlejohn, R.G.: Phys. Rev. **A38**, 6034 (1988)
Marsden, J., Montgomery, R., Ratiu, T.: *Reduction, symmetry and Berry's phase in mechanics*, Mem. Am. Math. Soc. **88** (1990)
Montgomery, R.: Am. J. Phys. **59**, 394 (1991)
Moore, E.N.: *Theoretical Mechanics* (Wiley, 1983)
Ozorio de Almeida, A.M.: *Hamiltonian Systems* (Cambridge University Press, 1988)
Park, D.: *Classical Dynamics and its Quantum Analogues* (Springer, Berlin, Heidelberg 1990)
Pauli, W.: *Lectures on Physics*, Vol. 6 (MIT Press 1983); German original: Feldquantisierung (1950–51)
Percival, I., Richards, D.: *Introduction to Dynamics* (Cambridge University Press, 1987)
Ramond, P.: *Field Theory: A Modern Primer* (Addison-Wesley, 1989)
Reuter, M.: Phys. Rev. **D42**, 1763 (1990)
Reuter, M.: Phys. Rev. **D44**, 1132 (1991)
Rivers, R.J.: *Path Integral Methods in Quantum Field Theory* (Cambridge University Press, 1988)
Sagdeev, R.Z. (ed.): *Nonlinear Phenomena in Plasma Physics and Hydrodynamics* (Physics Series, MIR Publishers, Moscow 1986)
Saletan, E.J., Cromer, A.H.: *Theoretical Mechanics* (Wiley, 1971)
Scheck, F.: *Mechanics* (Springer, Berlin, Heidelberg 1990)
Scheck, F.: *Mechanik* (Springer, Berlin, Heidelberg 1994)
Schulman, L.S.: *Techniques and Applications of Path Integration* (Wiley, 1981)
Schuster, H.G.: *Deterministic Chaos* (VCH, Weinheim 1987)
Schwabl, F.: *Quantum Mechanics* (Springer, Berlin, Heidelberg 1992)
Schwabl, F.: *Quantenmechanik* (Springer, Berlin, Heidelberg 1994)
Schwinger, J.: *Quantum Kinematics and Dynamics* (Addison-Wesley, Reading, MA 1991)
Smith, G.R., Kaufman, A.N.: Phys. Rev. Lett. **34**, 1613 (1975); Phys. Fluids **21**, 2230 (1978)
Tabor, M.: *Chaos and Integrability in Nonlinear Dynamics: An Introduction* (Wiley, 1989)
Verhulst, F.: *Nonlinear Differential Equations and Dynamical Systems* (Springer, Berlin, Heidelberg 1990)
Whittaker, E.T.: *A Treatise on the Analytical Dynamics of Particles and Rigid Bodies* (Cambridge University Press, 1989)
Wiegel, F.W.: *Introduction to Path-Integral Methods in Physics and Polymer Science* (World Scientific, Singapore 1986)
Witten, E.: Phys. Lett. **117B**, 324 (1982)

Index